谨以此书献给清华大学建校 110 周年

杨金龙 黄勇 著
Jinlong Yang
Yong Huang

陶瓷新型胶态
成型工艺（修订版）

Novel
Colloidal Forming
of Ceramics (2nd Ed.)

清华大学出版社
北京

内容简介

本书讨论高性能、复杂形状陶瓷零部件的低成本、近净尺寸的先进制备技术,反映了作者多年来对陶瓷胶态注射成型工艺研究的成果,主要以陶瓷胶态注射成型工艺为主线,介绍了陶瓷胶态注射成型的理论基础、研究重点、材料的可靠性、Si_3N_4 和 SiC 的胶态注射成型、低毒体系、胶态成型工艺的应用以及相关研究进展,还讨论了可控释放高价反离子原位凝固陶瓷浓悬浮体、陶瓷分散剂失效原位凝固陶瓷浓悬浮体等新技术。

本书可供高校和科研院所非金属材料等专业的师生以及材料制造企业的技术人员阅读。

For sale and distribution in the mainland of the People's Republic of China exclusively.
此版本仅限于中国大陆地区销售。
本书海外版由清华大学出版社授权 Springer 在中国大陆以外地区出版发行。
ISBN 978-981-15-1871-3

本书封面贴有清华大学出版社防伪标签,无标签者不得销售。
版权所有,侵权必究。举报:010-62782989,beiqinquan@tup.tsinghua.edu.cn。

图书在版编目(CIP)数据

陶瓷新型胶态成型工艺 = Novel Colloidal Forming of Ceramics: 2nd Ed / 杨金龙,黄勇著. —修订版. —北京:清华大学出版社,2021.3
ISBN 978-7-302-57591-7

Ⅰ. ①陶… Ⅱ. ①杨… ②黄… Ⅲ. ①陶瓷-成型 Ⅳ. ①TQ174.6

中国版本图书馆 CIP 数据核字(2021)第 030123 号

责任编辑:黎 强
封面设计:何凤霞
责任印制:杨 艳

出版发行:清华大学出版社
    网 址:http://www.tup.com.cn, http://www.wqbook.com
    地 址:北京清华大学学研大厦 A 座 邮 编:100084
    社 总 机:010-62770175 邮 购:010-62786544
    投稿与读者服务:010-62776969,c-service@tup.tsinghua.edu.cn
    质量反馈:010-62772015,zhiliang@tup.tsinghua.edu.cn
印 装 者:小森印刷(北京)有限公司
经 销:全国新华书店
开 本:155mm×235mm 印 张:36.75
版 次:2010 年 7 月第 1 版 2021 年 3 月第 2 版 印 次:2021 年 3 月第 1 次印刷
定 价:198.00 元

产品编号:080966-01

About the Authors

Jinlong Yang born in 1966, graduated from Beijing Institute of Technology in 1987, graduated and got the master degree from North University of China (NUC) in 1990, got the doctor degree from Tsinghua University in 1996. He had the post-doctoral experience at Department of Materials in Swiss Federal Institute of Technology (ETH-Zurich) from 1999 to 2000. He became a full professor in Tsinghua University in 2004, and had the position of special-engaged professor and director of laboratory of advanced ceramics in NUC since 2006.

Research fields include: structured ceramics, ceramic matrix composites, colloidal forming technology of ceramics and laser green machining of ceramics. He had won some top honors, including the second class prize of National Technology Invention for a research program by State Council of China, and "Science and Technology Advancement Prize" for other three programs awarded by the Ministry of Education and the government of Beijing respectively. He has delivered more than 100 papers, and got 30 state patents for invention.

Yong Huang born in 1937, graduated from Tsinghua University in 1962. He has been engaging in inorganic nonmetallic material teaching and research for many years. He was the visiting scholar at the Department of Materials Science and Engineering in the University of Michigan in USA, the research scientist at the material center of Massachusetts Institute of Technology in America, the visiting professor at the Department of Materials in Swiss Federal Institute of Technology (ETH Zurich), and also carried out a cooperative research with the University of Monash in Australia. His research interests include the composing, structure and performance of the advanced ceramic, as well as the preparation of high performance ceramic and the strengthening mechanism. He has won many top honors, including the second class National Technology Invention Award for a research program by State Council of China and "Science and Technology Advancement Prize" for other nine programs awarded by the Ministry of Education and the government of Beijing, Fujian Province respectively. He has delivered more than 400 papers, got 25 state patents for invention, published 14 teaching materials, text book, handbook etc. He was rewarded a national excellent text book and the first class award of excellent text book for building materials. He also was rewarded for the excellent teacher of Beijing, advanced person of the State Plan for Development of Basic Research and the winner of special government allowance.

Introduction

In ancient times, ceramic vessels or crafts were usually manufactured using clay-based natural raw materials. It is well known that the mud mixed clay with water has good plasticity and can be easily processed into products of various shapes. However, the forming techniques used earlier were mainly manual procedures. Therefore, to a large extent, the forming of ceramic wares was just a kind of skill or workmanship. By the 1960s, new ceramic materials had already developed into an independent scientific system. At the same time, raw materials that were used to manufacture ceramics began to transition from the clay-based system to one with accurate chemistry composition. In particular, for preparing high-performance ceramics, synthetic chemical raw materials, such as Al_2O_3, ZrO_2, Si_3N_4, and SiC, were mainly used. These ceramics had excellent properties because of their structural characteristics of covalent bond and ionic bond, and were widely considered as candidate materials in many fields requiring high-temperature resistant, wear-resistant, and corrosion resistant substances. It had been predicted that such materials would be developed rapidly, and various new types of materials with excellent properties would be explored.

By the late 1970s, the emergence of the worldwide oil crisis caused many developed countries led by America and Japan to draft national development plans for high temperature structural ceramic materials used in the field of internal combustion engines, especially automobile engines. With excellent properties such as resistance to high temperature, wear, and corrosion, high-performance ceramics were considered the optional material for non-water cooling and adiabatic ceramic engine parts. The forming technique of ceramics was also among the top-ranking research topics.

During the period of the '7th Five-Year Plan' and '8th Five-Year Plan' in China, around the key components of ceramic insulation engines, an in-depth research on ceramic injection molding, extrusion molding, slip casting, and pressure filtration was done. Moreover, a few engine parts samples with high performance were prepared. However, because of the high cost, poor performance repeatability, and low yield, the process of industrialization of high-performance ceramics was greatly restricted. After years of research and exploration, there was a growing recognition as the key technology for forming high-performance ceramic materials and parts. The forming technique is not only the precondition for materials design and formula, but also the important factor in reducing manufacturing costs and improving the yield and the performance repeatability of products. Simultaneously, several research booms were set off around the new forming technique of ceramics at home and abroad. During the period of the '9th Five-Year Plan' in China, in order to achieve high-tech ceramic industrialization at the earliest, the ceramic forming technique with high performance and low cost was granted special funds from the 863 Program.

Along with the research upsurge in the field of ceramic engines, there was also considerable interest in the field of injection molding of ceramics. On the basis of the theory of plastic injection molding, thermoplastic, thermosetting, and water-soluble organic compounds were used as binders, and then mixed with ceramic powder to prepare a suspension with high volume loading. In addition, ceramic parts with high size precision and complex shape can be prepared by injection molding. It is suitable for automotive and large-scale production. After several decades, owing to in-depth studies on ceramic injection molding, it developed into an integrated science and technology involving rheology, the dynamic molding process of injection suspension and thermal degradation of organic compounds as well as other interdisciplinary technology. However, some problems caused by organic enrichment or particle rearrangement were exposed during the time and energy consuming process of debindering, such as poor uniformity and easy cracking. Therefore, debindering was considered the key issue to be solved, and the solution to lower the organic content gradually became an important research topic. In order to simplify the process of debindering, low-pressure injection molding with some small molecular organics was paid more attention.

After the 1990s, quickset injection molding was invented by B. E. Novich of the U.S. The pore fluid was used as a carrier in the process, the volume of which did not change with temperature. After the suspension was injected into the container, the carrier was sublimated, and then the green body was solidified by controlling the temperature and pressure. Because of avoiding the polymer organic carrier with large molecule, the problem of organic debindering was solved ingeniously. Due to significant advantages such as high automation and good size precision, injection molding continues to be highly used and is considered a highly competitive forming process.

In the mid-1980s, in order to avoid the difficulty of debindering in injection molding, which was caused by the large number of organic binders, traditional slip casting was again paid more attention, as it involved less organics and low cost. Moreover, the operation and control were easy in this method. However, because of the green body with low green density and poor strength, it was not suitable for the preparation of high-performance ceramics. On the basis of traditional slip casting, the pressure filtration and centrifugal casting techniques were developed thereafter. The green body's density and strength were improved by applied pressure and centrifugal force, and at the same time the complicated debindering process was avoided. However, such processes were also unable to meet the green bodies with high reliability and high performance due to poor uniformity of the green bodies.

After the 1990s, in order to improve the uniformity and reliability of ceramic bodies, forming in situ techniques such as gelcasting, temperature-induced flocculation, colloidal vibration casting, and direct coagulation casting were developed. The in situ solidification process is highly regarded because it is the precondition to ensure uniformity and is an important way to improve the reliability of ceramics.

Gelcasting, a novel colloidal forming technology of ceramics, was first invented by Oak Ridge National Laboratory (ORNL), USA, in 1990. In the gelcasting process, about 2–4 wt.% acrylamide monomer is added into the ceramic suspension, and then it polymerizes in situ by the interaction of catalysts and initiators. Furthermore, the drying process should take place at room temperature and high humidity for a long time; otherwise, the green bodies would easily crack. In addition, the degree of automation and industrialization of the gelcasting process is poor when compared to injection molding. The ceramic bodies prepared by gelcasting have the obvious advantages of high strength and excellent machinability. Thus, some ceramic parts with complex shape or that are difficult to be demolded, just like inside thread, can be passed through green body machining after drying to achieve the required shape and precision. As a kind of brittle and difficult-to-machine material, it is very important, even necessary for the bodies to be machined partly, which also provides people a very good idea.

Temperature-induced flocculation was developed by L. Bergstrom, a professor at the Institute of Surface Chemistry of Stockholm University, Sweden, in 1993. A special amphoteric polyester surfactant or dispersant changing with temperature is introduced into the concentrated suspension to make particles disperse. One end of the dispersant is adsorbed on the particle surface, and the other end goes inside the solvent. As the temperature is reduced, the solubility of the dispersant declines, the function of dispersion fails, and then the suspension is flocculated in situ. The outstanding advantage of this method is the recycling of the unqualified green bodies, but it is restricted to use such dispersant for the different ceramic system.

Colloidal vibration casting molding was first developed by F. F. Lange, a professor at the University of California, Santa Barbara, USA, in 1993. In this method, the prepared dilute suspension (20–30% by volume) with high ionic concentration is pressure-filtered or centrifuged to obtain a green body with high solid loading, which is in the solid state under static conditions but is in the fluid state if certain external forces (such as vibration) are applied. Then, the suspension will solidify in situ under static conditions, after be poured. The outstanding advantage is the use of the thixotropic property when the concentrated suspension has a high ionic strength. Moreover, the concentrated suspension need not be prepared with high solids loading. However, the green bodies have poor strength and are prone to cracking and deformation.

Direct coagulation casting was invented by the research team under Prof. Gauckler from the Swiss Federal Institute of Technology in Zurich, Switzerland in 1994. In this method, first, a biological enzyme and a substrate are introduced into the ceramic concentrated suspensions at low temperature. At this time, the enzyme is in the inactivated state and does not almost react with the substrate. Then, the temperature of the suspension rises to 20–40 °C, and the enzyme is activated and reacts with the substrate. By adjusting the pH value to the isoelectric point or by increasing the internal ionic strength, the suspension is coagulated in situ. This method results in wet bodies with enough strength to be demolded. In the process, it is necessary that the solid loading of the suspension is more than 50% by volume, but the strength of the green body is low. However, the body is extremely uniform and does not contain any organic substance. Thus, it is suitable for preparing high-reliability ceramics, for which the Weibull modulus can reach 40.

Thus, after the 1990s, research on the in situ forming technique had become a new hot spot in the field of high-performance ceramics. People gradually realized that the forming technique is very important in the whole study of ceramic materials, and that the industrialization process of high-performance ceramics would be promoted greatly through in-depth study on the forming technique and its basic theory. In the mid-1980s, studies were conducted on injection molding and pressure filtration in the Tsinghua University in China. After the 1990s, the study of gelcasting and direct coagulation casting had progressed significantly, especially with the invention of the novel injection molding technique for water-based nonplastic slurry of ceramics, also called colloidal injection molding. A colloidal injection molding machine was also developed. The development of the gel tape casting process and gelcasting with low toxicity system was studied in detail in the nearly five years. Recently, techniques such as freeze-gelcasting and colloidal forming for ultralight and high-strength porous ceramics have been invented, which further enrich the theory and technique of ceramic forming. The industrialization process of high-performance ceramics is also being promoted greatly. In China, other universities and institutes such as Shanghai Institute of Ceramics, Chinese Academy of Sciences, and Tianjin University are conducting similar research. The key technology and development trends regarding the colloidal forming of ceramics include the following aspects.

(1) The preparation of a concentrated suspension with low viscosity and high solids loading is the precondition to guarantee the density, uniformity, and strength of the green bodies. With high density, the shrinkage of the green bodies can be decreased in drying, and deformation and cracking can be avoided in the sintering process. Therefore, the foundation of colloidal forming is to prepare the suspension with high solids loading.

(2) The in situ consolidation technique: In a sense, a new consolidation technique of suspension means a new forming process. It is very important for colloidal forming to look for a new solidification method of suspension. For forming in situ, the relative position of particles does not change during the solidification of suspension. It is a necessary condition to guarantee body uniformity, which is also a key factor in improving the reliability of ceramic materials. Currently, the development of an in situ consolidation technique requires the suspension to contain a few or no any organic substances. By varying the charged and dispersant characteristic between colloidal particles, the viscosity

of the suspension is increased and this leads to in situ solidification of the green bodies. At the same time, with enough strength to be demolded, the ceramic green bodies can be easily transported during large-scale production.
(3) It is very important to avoid residual stress in green bodies by using colloidal forming. Normally, there is no shrinkage in the green body when the suspension is formed, and no residual stress also. Thus, we believe that colloidal forming without shrinkage will become an important trend in the next 10 or 20 years.
(4) Near net-size forming: Ceramic substances are hard-machined materials with high hardness and brittleness. The sintered bodies should be close to the actual size of the final parts in order to decrease machining of ceramics.

The preparation process of high-performance ceramic materials and parts is the sticking point of development and application of ceramics. It involves the preparation of high-performance ceramic powders, and the forming and sintering processes. The quality of the green and sintered body is directly impacted by the quality of the powder, because the rheological behaviors and solids loading of concentrated suspension are directly determined by the properties of the powder during colloidal forming. Moreover, the quality of green bodies guarantees the quality of the sintered body. The preparation of powders, and the forming and sintering of ceramics are mutual constraints and complementary. However, according to the current development of ceramic forming, the forming process, which plays an important role in the preparation process of ceramic materials and is related with the industrialization and scale of production, is the critical step to guarantee performance reliability and repeatability as well as yield of ceramics. Hence, research on forming techniques for ceramics involving low cost and having high reliability is of great significance. This will promote the industrialization of high-performance ceramics in China and abroad. This is not only the requirement of governments around the world and of industries, but also the urgent need of ceramic scientists.

To sum up, the green bodies prepared by an ideal forming process should have a good uniformity, a high green density, enough strength to be demolded, and no residual internal stress. It can obtain ceramic material with high performance, guarantee the homogeneous of shrinkage, and avoid deformation during the sintering process. In order to meet the above requirements, the forming process from a technical perspective should have the following characteristics:

(1) Adding no or as few as possible polymers (0.1–4 wt.%).
(2) Solids loading of suspension more than 55 vol.%.
(3) The solidification of suspension is a rapid, in situ technology.
(4) The suspension has no shrinkage and internal stress during the transformation from liquid to solid state.
(5) A kind of near-net-size forming process.

The first point is to avoid the problem of debindering. The second point is to improve the density and the strength of the green bodies. The third point is to guarantee the uniformity and the dimensional accuracy of the green bodies. The fourth point is to avoid deformation and cracking of the green bodies during burn-out and sintering, in order to improve the reliability. The fifth point is to guarantee the dimensional accuracy of the sintered products, so that there is no or less machining.

There are eight chapters in this book. The main contents involve our research achievements and developments in more than a decade. Recent international research results in the field of ceramic forming are also introduced. To summarize, the basic problems and the future development trends of the forming process are proposed and discussed in detail.

Contents

1 Aqueous Colloidal Injection Molding of Ceramics (CIMC) Based on Gelation 1
   1.1 What is Colloidal Injection Molding? 3
      1.1.1 Colloidal Injection Molding of Ceramics (CIMC) 3
      1.1.2 The Flowchart of CIMC 4
      1.1.3 The Machine of Colloidal Injection Molding of Ceramics 5
   1.2 Pressure-Induced Forming 6
      1.2.1 Effect of Hydrostatic Pressure on Solidification 6
      1.2.2 Homogeneity of the Green Bodies 7
      1.2.3 Controlling the Inner Stress in the Green Body 7
   1.3 Storage Stability of Ceramic Slurries 10
      1.3.1 The Importance of Storage Stability of Slurry 10
      1.3.2 Chemical Stability 10
      1.3.3 Inhibitor for Slurry Storage 12
   1.4 To Prepare High Reliability Ceramic Parts with Complex Shapes: Aqueous Colloidal Injection Molding 13
   1.5 Summary 14
   References 16

2 Gel Tape Casting of Ceramic Substrates 17
   2.1 Fundamental Principle and Processing of Aqueous Gel Tape Casting 19
      2.1.1 Tape Casting Types and the Raw Materials Used 19
      2.1.2 Polymerization of the Monomer 23
      2.1.3 Influence Factors on Polymerization of the Monomer ... 31
      2.1.4 Processing of the Gel Tape Casting 37

	2.2	Characteristics of Slurries Used for Aqueous Gel Tape Casting	40
		2.2.1 Properties of the Aqueous Ceramic Slurries with Binder	40
		2.2.2 Influence of Dispersants on Stability and Rheology of Aqueous Ceramic Slurries with Organic Monomer	45
		2.2.3 Influence of Plasticizer on Properties of Aqueous Ceramic Slurry with Organic Monomer	49
		2.2.4 Influence of PH Value on Properties of Slurry with Organic Monomer	50
		2.2.5 Effects of Surfactant on Wetting and Green Tape Releasing (Separating)	51
		2.2.6 Foam and Pore Elimination	53
		2.2.7 Sintering of Green Tape Prepared by Slurry	54
	2.3	Aqueous Gel Tape Casting with Styrene-Acrylic Latex Binder	55
		2.3.1 The Importance of Binders in Gel Tape Casting Process	55
		2.3.2 The Forming Film Mechanism of Latex Binder	58
		2.3.3 Rheological Properties of the Alumina Slurries with Binder	60
		2.3.4 The Physical Properties and Microstructure of Green Tapes	61
	2.4	A Gel Tape Casting Process Based on Gelation of Sodium Alginate	63
		2.4.1 Why Study on Tape Casting of Sodium Alginate	63
		2.4.2 The Preparation of Aqueous Alumina Suspensions with Sodium Alginate and Calcium Phospher Tribasic	65
		2.4.3 Control of the Gelation of Sodium Alginate	66
		2.4.4 Characterization of Green Tapes	68
	2.5	Spray Trigger Fast-Curing for Gel Tape Casting Process	70
		2.5.1 The Idea of Spray Trigger Fast-Curing	70
		2.5.2 Outline of the New Process	70
	2.6	Summary	71
	References		75
3	**Gelation Forming Process for Toxicity-Free or Low-Toxicity System**		**79**
	3.1	Gelation Forming of Ceramic Suspension with Agarose	80
		3.1.1 Characteristics of Agarose	80
		3.1.2 The Effect of Agarose Contents on the Rheology of Aqueous Ceramic Suspensions	82
		3.1.3 The Forming Courses of the Aqueous Ceramic Suspensions with Agarose	84

3.2	Alumina Casting Based on Gelation of Gelatin		88
	3.2.1	Characteristics of Gelatin	88
	3.2.2	The Gelation Process of the Ceramic Slurry with Gelatin Solution	91
	3.2.3	Preparation of Green Body Using Slurry with Gelatin Solution	93
3.3	A Casting Forming for Ceramics by Gelatin and Enzyme Catalysis		95
	3.3.1	Research Background	95
	3.3.2	Gelation Mechanism of Gelatin Solution with Urea Under Enzyme Catalysis	97
	3.3.3	Rheology and Zeta Potential of Alumina Suspension Containing Gelatin and Urea	99
	3.3.4	Coagulation Forming and Microstructure of Green Body	100
3.4	Alumina Forming Based on Gelation of Sodium Alginate		102
	3.4.1	Research Background	102
	3.4.2	Gelation Principle of Sodium Alginate	104
	3.4.3	Preparation Process of Alumina Green Bodies and Samples by Sodium Alginate	107
3.5	Gelcasting of Silicon Carbide Based on Gelation of Sodium Alginate		110
	3.5.1	Research Background	110
	3.5.2	Effect of Dispersant on the Colloidal Behavior of the SiC Suspension	112
	3.5.3	Rheological Property of SiC Suspension	113
	3.5.4	Sedimentation Behavior of the SiC Suspension	115
	3.5.5	Gelation Principle and Process of the Alginate Solution	116
	3.5.6	Gelation of the SiC Suspension with Alginate	116
3.6	Alumina Gelcasting with a Low-Toxicity System of HEMA		119
	3.6.1	Academic Idea and Research Program	119
	3.6.2	Colloidal Chemistry and Rheological Property	120
	3.6.3	Binder Burnout and Application of the New System	122
3.7	A Synergistic Low-Toxicity Gelcasting System by Using HEMA and PVP		124
	3.7.1	Academic Idea and Research Program	124
	3.7.2	ζ Potentials and Rheological Properties	125
	3.7.3	Activation Energies and Solidification	128

		3.7.4	Green Strengths and Microstructures	129

- 3.7.4 Green Strengths and Microstructures ... 129
- 3.7.5 Exfoliation Elimination Effect and Analysis of the Interaction Between PVP and HEMA Molecules ... 131
- References ... 134

4 Generation, Development, Inheritance, and Control of the Defects in the Transformation from Suspension to Solid ... 139

- 4.1 Rheological Behaviors of Aqueous Ceramic Suspensions ... 141
  - 4.1.1 Rheological Behaviors of Aqueous Alumina Suspensions ... 142
  - 4.1.2 Effect of Rheological Properties of Suspension on Mechanical Strength of Ceramics ... 145
  - 4.1.3 Effects of Solid Volume Fraction on Colloidal Forming ... 153
- 4.2 Generation and Development of Defects ... 158
  - 4.2.1 Generation Mechanisms of Agglomerations in Ceramic Suspensions ... 158
  - 4.2.2 Influences of Idle Time on Microstructures and Mechanical Properties of Green Bodies by Direct Coagulation Casting ... 165
- 4.3 Effect of Ionic Conductance on Preparation of Highly Concentrated Suspension ... 174
  - 4.3.1 Academic Idea and Research Program ... 174
  - 4.3.2 The Relationship Between Ion Conductivity Constants and Solids Volume Loading ... 176
- 4.4 Control of Inner Stress in Green Body ... 181
  - 4.4.1 Origin, Transformation and Control of Inner Stress in Green Body ... 181
  - 4.4.2 Release and Control of Inner Stresses in Ceramic Green Body ... 187
- 4.5 Suppression of Surface Exfoliation with the Addition of Organic Agents ... 195
  - 4.5.1 Suppression of Surface Exfoliation by Introducing Polyacrylamide (PAM) into a Monomer System in Suspension ... 195
  - 4.5.2 Suppression of Surface Exfoliation by Introducing Polyethylene Glycol (PEG) into Monomer System in Suspension ... 203
- References ... 221

5 Gelcasting of Non-oxide Ceramics 225
5.1 Effects of Powder Surface Modification on Concentrated
Suspension Properties of Silicon Nitride 226
5.1.1 Contributing Factor and Elimination of Macropores
in Silicon Nitride Green Bodies 226
5.1.2 Effect of Foreign Ions on Concentrated Suspension
of Silicon Nitride................................. 232
5.1.3 Effects of Acid Cleaning and Calcinations
on the Suspension Properties of Silicon Nitride 238
5.1.4 Effects of Liquid Medium and Surface Group
on Dispersibility of Silicon Nitride Powder 249
5.2 Gelcasting of Silicon Nitride Ceramics 255
5.2.1 Preparation of Silicon Nitride Ceramics
with Surface-Coated Silicon Nitride Powder 255
5.2.2 Preparation of Silicon Nitride Ceramics
with Surface-Oxidized Silicon Nitride Powder........... 267
5.2.3 Preparation of Silicon Nitride Ceramics Using
Combination Processing............................ 276
5.3 Gelcasting of Silicon Carbide Ceramic and Silicon
Nitride-Bonded Silicon Carbide Ceramic 287
5.3.1 Gelcasting of Concentrated Aqueous Silicon Carbide
Ceramic 287
5.3.2 Gelcasting of Aqueous Slurry with Silicon
Nitride-Bonded Silicon Carbide 294
References 306

6 Applications of New Colloidal-Forming Processes................ 311
6.1 Ceramic Microbeads.................................... 311
6.1.1 The Forming Principle of Ceramic Microbeads
Based on Gelcasting 311
6.1.2 The Process of Preparing Microbeads 314
6.1.3 The Properties of Ceramic Microbeads 314
6.1.4 Summary 328
6.2 Improving the Breakdown Strength of the Rutile Capacitor...... 329
6.2.1 The Influence of Sintering Additives on the Flow
Behavior....................................... 331
6.2.2 Calcining of the Rutile Mixture 333
6.2.3 The Rheological Behavior of the Calcined Rutile
Mixture.. 334
6.2.4 Gelcasting of the Calcined Rutile Mixture.............. 335
6.2.5 Summary 336

	6.3	Thin-Wall Rutile Tube for Ozone Generator with High Dielectric Constant..............................	338
		6.3.1 Results and Discussions......................	339
		6.3.2 Summary................................	341
	6.4	Refractory Nozzle of Zirconia..........................	342
		6.4.1 Rheological Behaviors of Zirconia Suspensions with Different Dispersants......................	343
		6.4.2 Sediment Stability of Zirconia Suspension with Different Dispersants......................	345
		6.4.3 Preparation of Zirconia Refractory Nozzles............	346
		6.4.4 Summary................................	346
	6.5	Water-Based Gelcasting of Lead Zirconate Titanate...........	348
		6.5.1 Colloidal Chemistry and Rheological Behavior.........	350
		6.5.2 Microstructure and Properties....................	354
		6.5.3 Summary................................	357
	References..		357
7	**New Methods and Techniques Based on Gelation**...............		359
	7.1	Development Overview and Application of Solid Freeform Fabrication.......................................	361
		7.1.1 Development Overview of Solid Freeform Fabrication....	361
		7.1.2 Application of Solid Freeform Fabrication.............	362
	7.2	Development Overview and Application of Freeze-Gelcasting....	371
		7.2.1 The Combination of Gelcasting and Freeze-Casting Technique...................................	371
		7.2.2 Fabrication of Ceramics with Special Porous Structures...................................	372
		7.2.3 Microstructure and Properties of Porous Alumina Ceramics...................................	378
		7.2.4 Mechanical Properties and Applications of Alumina Ceramics with Ultra-Low Density...................	382
	7.3	Solidification of Concentrated Silicon Nitride Suspensions for Gelcasting by Ultrasonic Effects.......................	387
		7.3.1 The Forming Method of Gelcasting Using Ultrasonic Effects......................................	387
		7.3.2 Preparation of Concentrated Silicon Nitride Suspensions..................................	388
		7.3.3 Ultrasonic Accelerated Solidification.................	389
		7.3.4 Comparison of Thermal- and Ultrasonic-Activated Solidifications................................	392

	7.4	Novel Laser Machining Technology for Alumina Green Ceramic	394
		7.4.1 Laser Machining Technology	394
		7.4.2 Practical Application of Laser Machining Technology	395
		References	402
8	**Novel In-situ Coagulation Casting of Ceramic Suspensions**	405
	8.1	Direct Coagulation Casting of Ceramic Suspension by High Valence Counter-Ions	407
		8.1.1 Direct Coagulation Casting by Using Calcium Iodate as Coagulating Agent	407
		8.1.2 Direct Coagulation Casting by Using Calcium Phosphate as Coagulating Agent	423
		8.1.3 Direct Coagulation Casting by Using Thermo-Sensitive Liposomes as Coagulating Agent	432
		8.1.4 Direct Coagulation Casting from Citrate Assisted by pH Shift	442
		8.1.5 Direct Coagulation Casting via High Valence Counter-Ions from Chelation Reaction	474
	8.2	Dispersion Removal Coagulation Casting	501
		8.2.1 Dispersant Reaction Method	501
	8.3	Dispersant Hydrolysis Method	521
	8.4	Dispersant Separation Method	532
		References	541

Appendix A: The Testing, Analyzing and Sintering Methods Used in Authors' Research 549

Appendix B: The Raw Materials Used in Authors' Research 551

Index of Scholars 553

Index of Terms 561

Postscript 567

Chapter 1
Aqueous Colloidal Injection Molding of Ceramics (CIMC) Based on Gelation

Abstract In this chapter, aqueous colloidal injection molding (CIMC), which is based on gelation of monomer polymerization, like a reactive injection molding of double slurry of polyester, is discussed systematically. The solidification mechanism is based on gelation of monomer polymerization. It was found that pressure can induce the polymerization of the monomer. The gelation time can be effectively controlled by both temperature and pressure. The testing equipment of gel point was developed and used to test the gel reaction of the monomer in the suspension under different pressures. A machine of for ceramic colloidal injection molding was developed. It was observed that copper accelerated the polymerization of the monomer. Therefore, copper parts are forbidden for use in a CIMC machine.

Keywords Gelcasting · Injection molding · Colloidal injection molding · Pressure-induced forming · Fast mixing with double suspensions · Ceramic slurry

Nowadays, high-performance ceramics are becoming increasingly more important in several fields, such as spaceflight, energy, mechanics, and bio-techniques. However, the high cost and low reliability have prevented high-performance ceramics from being utilized on a large scale until now. To solve this problem, more attention is being paid to the forming process of ceramics. In the industrialization of high-performance ceramics, forming has become a bottleneck not to overstep. Colloidal forming is an important forming technique. It includes slip casting, tape casting, direct coagulation casting (DCC), injection molding, and gelcasting. Among these techniques, gelcasting and injection molding are considered the two possible solutions to address the industrialization of high-performance ceramics. Though both have many advantages, there are still several problems to be solved in the industrialization process.

Gelcasting is a new ceramic forming technique that is generating worldwide attention. The process is based on the casting of slurry containing powder, water, and water-soluble organic monomers. After casting, the mixture is then polymerized to form gelled parts. Drying, burning out, and sintering complete the manufacturing process. The process is generic and can be used for a wide range of ceramic and metallic powders. It is a suitable technique for the fabrication of near net shape prototypes or small series using inexpensive molds. In contrast with slip casting, the gelled parts are more homogeneous and have a much higher green strength. Gelcast parts contain only a low percent of organic components, thereby making binder removal much less critical compared to injection molding. The advantages of gelcasting can be summarized as follows:

(1) Capability of producing complex parts like injection molding.
(2) Ease of implementation, owing to its similarity with other well-established processes like slip casting.
(3) Low capital equipment cost.
(4) Possibility to use low-cost mold materials.
(5) Capability of implementation for mass production.
(6) High green strength.
(7) Excellent green machinability that allows machining much finer details than in CIPed parts.
(8) Highly homogeneous material properties.
(9) Low organic content that translates into easy binder removal.
(10) Generic method applicable for both ceramic and metal powders.

Meanwhile, there are still some disadvantages that make it difficult to realize the industrialization by gelcasting. Most important of all, low automation prevents gelcasting from being used in the industrialization of high-performance ceramics (Omatete et al. 1991; Gilissen et al. 2000).

Injection molding, however, has been used in the ceramic industry for several years for its high automation. Ceramic injection molding is a well-established shaping technique, which involves the mixing of ceramic powder with a large concentration of a melt polymer (up to 50–60 vol%). The carrier polymer provides very high viscosity, and so very high pressures (10 ± 150 MPa) and temperatures (120 ± 200 °C) are needed for injection. In addition to the high cost owing to the use of organics, the major problem arises from the debindering step that can easily lead to defects and failure of the sintered body. In the preparation of products with complex shape and big cross section, the problem is more obvious. These disadvantages prevent injection molding from being used in the industrialization of high-performance ceramics (Novak et al. 2000; Wen-Cheng et al. 2000; Liu 1999; Krug et al. 2000).

The aim of this study is to develop a new forming technique to meet the needs of industrialization. The new technique is called Colloidal Injection Molding of Ceramics (CIMC).

1.1 What is Colloidal Injection Molding?

1.1.1 Colloidal Injection Molding of Ceramics (CIMC)

Traditional injection molding can be used to form complex ceramic parts with high dimensional precision, and can be used in the automation of large-scale production of ceramic products in industries. The problem with this technique is that due to the high organic content, debindering becomes very difficult. Colloidal forming processes, such as gelcasting and DCC can avoid this disadvantage, and improve the uniformity of green bodies and the reliability of the ceramics products. These approaches, however, have shortcomings. For example, they need manual operation and thus are difficult to be used in ceramic industries, where automation is a necessity. An optimal approach would be a combination of both injection molding and colloidal forming processes, bringing together their advantages. The problem is that injection molding is a plastic forming process, and the colloidal forming process starts from slurry. Since the slurry is of no plasticity in colloidal forming, it is normally impossible to apply injection molding.

We have developed a new technique, namely CIMC (Fig. 1.1) (Wen-Cheng et al. 2000), combining both injection molding and the colloidal forming process. This technique can be used to produce complex ceramic parts of high reliability, by using a colloidal injection molding machine invented by the authors. The new approach (with our machine) provides high automation of injection molding, high homogeneity of products, and low organic content in the colloidal forming process. It is possible to produce high-performance ceramic parts of complex shapes with high reliability by this newly developed technique, and thus it can be applied in ceramics industries.

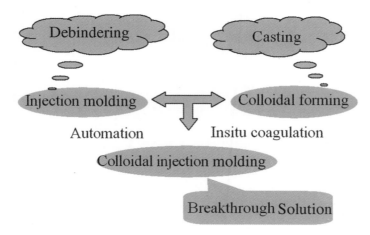

Fig. 1.1 The idea for colloidal injection molding of ceramics

1.1.2 The Flowchart of CIMC

The fast, uniform mixing and controllable colloidal forming process of ceramics was to divide the ceramic suspension into two components A and B. The monomer was added into A, and the initiator was added into B. In this case, there was no reaction in either A or B because of segregation of the monomer and the initiator. The suspensions maintained good fluidity until they were mixed quickly and uniformly. Once A and B were mixed together, the encounter of the monomer and initiator resulted in gel reaction in the suspension. The initiator quantity and the applied pressure were the factors dominating the solidification reaction. Figure 1.2 shows the schematic graph of this process. The catalyst in suspension A could not generate a reaction with the monomer. Its main function was to accelerate the reaction when the monomer and initiator encountered.

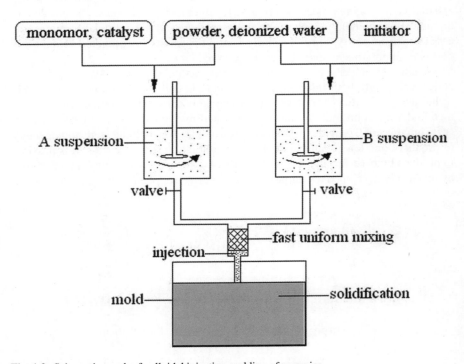

Fig. 1.2 Schematic graph of colloidal injection molding of ceramics

1.1.3 The Machine of Colloidal Injection Molding of Ceramics

Special equipment was manufactured for the process.

Figure 1.3 shows the flow chart for the working of the machine. The prepared suspensions were transferred into the A and B containers. In the containers, the suspensions were vacuumized through the vacuum pumps to completely remove the gas solved in them. To avoid the deposition of the suspensions, a stirring paddle was used in each of the containers to continuously intermix the slurries. Furthermore, with the use of bellows pumps, the slurries could circulate in two routes: inner circle and outer circle. This circular transportation in addition to the stirring action on the slurries completely prevented the occurrence of deposition. The suspensions' state was monitored through the testing valves, which were positioned on the outer circle line.

Before injection molding, the metering pumps quantified two slurries of the same volume. The two slurries were simultaneously and quickly injected into the static admixer with a high pressure of 5–8 MPa. The core status of the static admixer can be seen in Fig. 1.3. It is similar to a bridge connecting the machine and the mold. Its main function was to realize fast uniform mixing of A and B suspensions. The inner structure of the device can be seen in Fig. 1.4. From this chart, we can observe that the mixing process was realized by a series of mixing units with different shapes in the hollow tube. These units made the penetrating slurries rapidly levo-rotate and dextral-rotate alternately. These frequent changes of the slurries' flowing directions could help in obtaining a good mixing effect, which is called fast, uniform mixing.

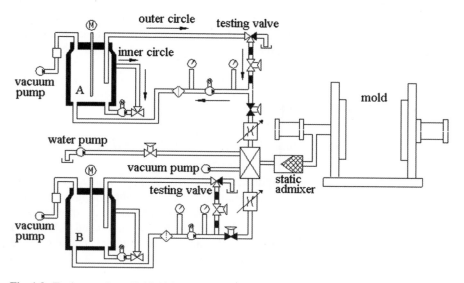

Fig. 1.3 Equipment for colloidal injection molding of ceramics

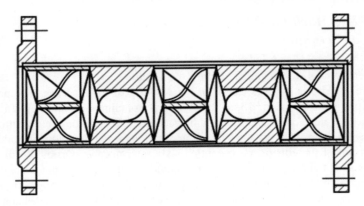

Fig. 1.4 Inner structure of the static admixer

These mixed suspensions were injected into the mold under a pressure of 3–8 MPa. The antisticking agent is usually pre-coated onto the wall of the stainless steel mold. The initiator volume and injection pressure are the dominative factors to control the solidification process. The separation of the monomer and initiator guaranteed no reaction in the separated suspensions, which had gained enough operating time for the process. By adjusting the initiator volume and injection pressure in the process, the solidification speed of the suspension in the mold could be effectively controlled, which explains the meaning of 'controllable colloidal forming'. In the process, the crucial impact of temperature on gel reaction was completely avoided.

It is well known that the pressure distribution is homogenous in flowing suspensions, and the fast, uniform mixing results in physical uniformity in the suspensions. The combination of these two points enabled us to ensure simultaneous solidification in different parts of the suspensions. The synchronous solidification consumedly decreased the inner stress in the green bodies.

1.2 Pressure-Induced Forming

1.2.1 Effect of Hydrostatic Pressure on Solidification

A pressure induction unit was then added to apply an external pressure after injection. The gelation curves of Al_2O_3 slurry under different pressures with a mold temperature of 25 and 36 °C are shown in Fig. 1.5. Under increased pressures, the onset time of gelation becomes shorter and the gelation speed increases significantly. At higher temperature, the influence of the pressure becomes more significant and the gelation finishes almost immediately. Pressure-induced solidification has distinct advantages over temperature-induced solidification.

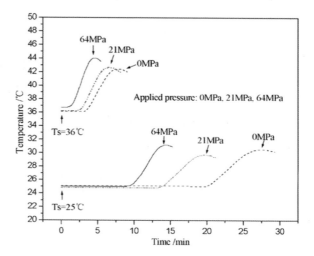

Fig. 1.5 The effect of external pressure on the gelation of Al$_2$O$_3$ slurry with a solid loading of 50 vol%; T$_s$ is the starting temperature

1.2.2 Homogeneity of the Green Bodies

An alumina (Al$_2$O$_3$) green body of turbine shape with a diameter of 105 mm was prepared by this process. The spatial distribution of density is compared with that obtained by the conventional process by using the same injection molding system shown in Fig. 1.6. The value in each part represents the density in g/cm^3. The relative standard deviation of density for colloidal injection molding is 0.2% compared to 0.7% for conventional injection molding. The results show that the colloidal injection molding process significantly improves the homogeneity in the density of the green body. Nevertheless, the solidification in this process is associated with the surface temperature of the mold. The gradient of the temperature will inevitably cause nonuniform solidification and inner stress in the green body. A new strategy has been developed to minimize the inner Fig. 1.5. The effect of external pressure on the gelation of Al$_2$O$_3$ slurry with a solid loading of 50 vol%; T$_s$ is the starting temperature stress in green bodies prepared in this process.

1.2.3 Controlling the Inner Stress in the Green Body

The inner stress in green bodies is often responsible for the initiation of microcracks during the subsequent drying and debindering processes. During colloidal injection molding or other gelcasting processes, monomers in the ceramic suspension polymerize and then form gel networks to hold the ceramic particles. The solidification speed increases with increasing temperature for a given composition. Nonuniform solidification occurs due to the temperature gradient in the ceramic suspension and results in the development of inner stress in the green body. Theoretical analysis

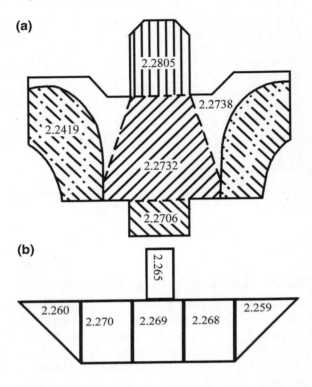

Fig. 1.6 Comparison of density distribution in Al_2O_3 green bodies of turbine shape prepared by **a** conventional injection molding and **b** colloidal injection molding in the same system

has indicated that the magnitude of the inner stress increases with the stiffness of the green body. Based on this finding, a fraction of the monomers is replaced by a moderator, hydroxylethylacrylate, in the suspension during gelcasting or colloidal injection molding to control the stiffness of the gel network. Test bars with dimensions of 6 × 6 × 42 mm^3 were prepared with 50 vol% solid loading and various amounts of the moderator from 0 to 100 wt% in the total monomers. The samples were dehydrated at about 70 °C in an electrical furnace. The flexural strength and elastic modulus of the green body are evaluated using three-point bending. The results are shown in Fig. 1.7 against the amount of the moderator. Both the flexural strength and elastic modulus of green body decrease simultaneously with the amount of the moderator. This reveals that the strength of the polymer network is reduced when the harder polymer chain is relaxed by the incorporation of the shorter chain molecules of the moderator. Five alumina disks, a, b, c, d, and e, were produced under the same conditions. The diameter of the samples is about 50 mm and the thickness is about 4 mm. In Fig. 1.7, the identifiers of the disk samples are marked above the corresponding amount of the moderator. Figure 1.8 shows the surface patterns of the alumina green bodies after drying and debindering to remove all the water and organic binders. Radial cracks were found on samples c, d, and e, while samples a and b with higher amount of moderator were free of visible cracks. It is likely that the inner stress was initiated at the forming stage and magnified during the drying process when the

1.2 Pressure-Induced Forming

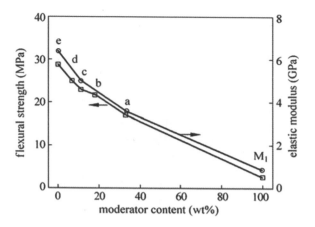

Fig. 1.7 Strength and elastic modulus of alumina green bodies versus the amount of moderator

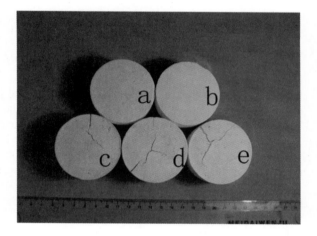

Fig. 1.8 Surface patterns of alumina green bodies after drying and debindering

preform becomes harder. The intrinsic strength of the green body after drying and debindering is mainly determined by the nature of the ceramic powder and the solid loading. Observations of the surface patterns confirm that both the inner stress and the elastic modulus of the green body decrease with the amount of the moderator. Choosing a proper amount of the moderator can effectively reduce the occurrence of cracks during the subsequent processes.

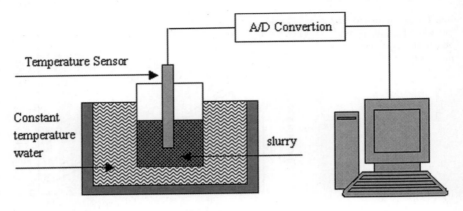

Fig. 1.9 Metering system for polymerization process

1.3 Storage Stability of Ceramic Slurries

1.3.1 The Importance of Storage Stability of Slurry

In this article, idle time of polymerization was used to judge the store stability of the slurry. Since acrylamide polymerization is an exothermic reaction, gelation can be determined by the change of temperature. The metering system (Liguo et al. 2002) is as shown in Fig. 1.9. At first, ceramic slurries with high solid loading were prepared by long-term. Then, the slurry was poured into the mold, and the mold was placed in constant-temperature water until its temperature was equal to the scheduled temperature. The initiator and catalyst were added to the slurry later. The period from then to the beginning of polymerization is defined as idle time. When the monomer was polymerized, the temperature of the slurry rose because of exothermal polymerization. During this process, the temperature of slurry was detected by a temperature detector and recorded by a computer. Based on this temperature, the polymerization process could be analyzed.

1.3.2 Chemical Stability

During the research on gelcasting, it was found that after long-term storage, even though the slurry had not been polymerized, the solidification properties of the ceramic slurry had been modified greatly. Sometimes, the idle time was prolonged, and sometimes it was shortened. This affected the repeatability and the stability of the forming process. Hence, it was an obstacle for industrialization of gelcasting.

It is known that acrylamide in the slurry is prone to hydrolysis and thereby acrylic acid is generated. If the slurry were stored under strong light or there were some

1.3 Storage Stability of Ceramic Slurries

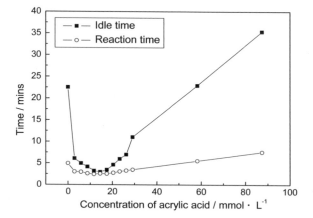

Fig. 1.10 Influence of acrylic acid on acrylamide polymerization

impurities in the slurry, the hydrolysis would be accelerated. In our research, it was found that acrylic acid has a great effect on the solidification of ceramic slurry. The influence of minute amounts of acrylic acid on the idle time and reaction time of acrylamide polymerization is shown in Fig. 1.10.

From Fig. 1.10, it can be seen that the solidification process is accelerated under very low concentration of acrylic acid. However, the reaction is decelerated when the acrylic acid concentration is more than 15 mmol/l. Both its idle time and reaction time were prolonged. The mechanism may be explained as follows. Acrylic acid is weak acid. Even if there is only a minute amount of acrylic acid in the slurry, the acidity of the slurry still increases. When the acrylic acid concentration is 3 mmol/l, the pH of slurry changes to about 3.5. Increase in the acrylic acid concentration leads its pH change to 3.1 slowly. It has been found that the pH of slurry has a significant effect on polymerization initiated with ammonium APS and TEMED. On the one hand, hydrogen ion acted as a catalyst for the decomposition of APS; on the other hand, hydrogen ion could react with TEMED and generate stable ammonium, after which the decomposition speed rate of APS decreases. Thus, when the concentration of acrylic acid is greater than a certain value, 15 mmol/l, the rate of polymerization decreases and the idle time becomes longer. Therefore, the ceramic slurry for gelcasting should be protected from light. Before the slurry is used after long-term storage, its solidification properties have to be tested.

In this research, it was found that the materials of the storage container also have an influence on the storage time of the slurry. If the storage container is made of copper or copper alloy, even at very low temperature, the slurry is polymerized. The mechanism can be explained as follows. Copper in air is prone to oxidation. Therefore, copper oxide and cuprous oxide are generated on the surface of the storage container. Copper oxide has no effect on the stability of slurry. Nevertheless, the reducibility of cuprous oxide is strong. It can react with the impurities present in the ceramic slurry, thereby generating free radicals. When all inhibiting agents are consumed, polymerization reaction occurs. Thus, the storage container for gelcasting should not be made of copper or copper alloy.

1.3.3 Inhibitor for Slurry Storage

The activity of acrylamide is so high that it can self-polymerize under certain conditions. If the season were summer or large milling balls were used, the temperature of the ceramic slurry could reach more than 50 °C. The slurries are often polymerized during or storage. Even if the initiator and catalyst were not added into the slurry, some free radicals would still be generated in the slurry. These free radicals react with oxygen in the ceramic slurry prior. The free peroxide radical is so stable that it cannot initiate polymerization. However, there is only about 20 wt% water in the high solid loading ceramic slurry, and therefore, there is only little oxygen dissolved in the slurry. As the environment temperature rises, the consumption rate of oxygen also increases. When all oxygen dissolved in the ceramic slurry is exhausted, the slurry solidifies unexpectedly.

In order to avoid this problem, inhibitors for gelcasting were studied. Many kinds of inhibitors, such as ferric chloride, phenol, and diphenylamine, have been investigated. A compound inhibitor of phenothiazine and catechol has the best effect among these inhibitors at room temperature. The best mixture ratio for phenothiazine and catechol was found to be about 1:5. Table 1.1 shows the inhibiting effect of this compound inhibitor on acrylamide polymerization.

It has been observed that the idle time was extended greatly after the compound inhibitor was added into the slurry. Now, the ceramic slurry can store for a long time even at higher temperatures. Since both phenothiazine and catechol are organic substances, no impurity was left in the ceramic components after sintering. The influence of phenothiazine and catechol on the rheological property of ceramic slurry was also discussed. The viscosity of the slurry decreased with the addition of phenothiazine and catechol. Therefore, this compound inhibitor does not influence the viscosity of ceramic slurry and is suitable for gelcasting.

Table 1.1 Inhibiting effect of phenothiazine-catechol on acrylamide polymerization [APS] = 2.76 mmol/l, [TEMED] = 40 m mol/l

Serial number	Temperature (°C)	Phenothiazine (m mol l^{-1})	Catechol/m mol l^{-1}	Idle time τ/min
1	30	0	0	2.6
2	30	0.025	0.125	5.8
3	30	0.05	0.25	7.2
4	30	0.075	0.375	12

1.4 To Prepare High Reliability Ceramic Parts with Complex Shapes: Aqueous Colloidal Injection Molding

The repeatability of the mechanical properties of the final products has been found to be greatly improved. A typical zirconia-toughened alumina (ZTA-50) and several grades of silicon nitride-based ceramics (Si_3N_4) were formed by colloidal injection molding. The ZTA was sintered at 1,580 °C in air for 2 h. The Si_3N_4 green bodies were further cold isostatic pressed to improve the green density prior to a two-stage gas pressure sintering process. The first stage was performed at 1,750 °C or 1.5 h under a nitrogen pressure of 0.3 MPa and the second stage at 1,900 °C under a pressure of 6 MPa in N_2 for various holding times from 1 to 2.5 h. The flexural strength was measured by a standard three-point bending method for ceramics. The Weibull modulus was derived from 20 such tests on each material. As shown in Fig. 1.11, the Weibull modulus of ZTA-50 materials is as high as 24, and the optimum of Si_3N_4-based materials is 34 for the optimum holding time (2 h). As a reference, the typical value of these materials prepared by conventional forming processes is about 15. Therefore, the repeatability of the mechanical properties has been improved greatly. The microstructures of the sintered ZTA-50 and Si_3N_4 bodies were observed by using a SEM as shown in Fig. 1.12. The micrograph of ZTA-50 confirms that the ZrO_2 particles are uniformly dispersed. Similarly, in the sintered Si_3N_4, the elongated β-Si_3N_4 grains are also well distributed. The improved uniformity of the microstructure is responsible for the increased repeatability of the fracture strength.

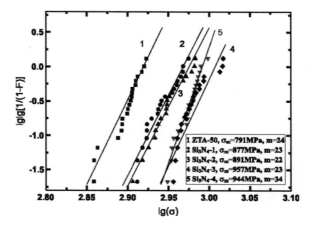

Fig. 1.11 Strength and Weibull modulus of ceramics; σ is the flexural strength, F is the probability of rupture, $σ_m$ is the mean flexural strength, m is the Weibull modulus

Fig. 1.12 Microstructure of the sintered bodies by SEM **a** ZTA-50 and **b** Si_3N_4

1.5 Summary

As a new technique of ceramic forming, CIMC combines the advantages of gelcasting and injection molding, which helps to realize the injection molding of a ceramic aqueous suspension. It not only solves the problem of low automation in gelcasting, but also reduces the difficulty of debindering in injection molding. However, CIMC is not a simple combination of gelcasting and injection molding. Pressure-induced forming is used in CIMC in order to eliminate the inhomogeneous polymerization due to the temperature gradient. Pressure-induced solidification occurs simultaneously within the slurry, by which the homogeneity of the green body can be improved and the inner stress be reduced. The hydrostatic pressure helps to suppress structural defects, such as delamination or cracks. The solidification process can be achieved at room temperature, so that the process can be simplified and easily automated for continuous production of a large batch. The pressure can be controlled precisely to optimize the speed of solidification and minimize the structural defects.

The inner stress in green bodies caused by nonuniform solidification in gelcasting processes can be significantly reduced by the addition of a moderator in the ceramic suspension. Combining this strategy with the colloidal injection molding process is a promising solution to form inner-stress-free ceramic parts of complex shapes. These processing strategies have led to the improvement of the reliability of the products.

The process of fast, uniform mixing and controllable colloidal forming of ceramics was invented to completely avoid the crucial effect of temperature on the colloidal forming of ceramics. The process is to divide the ceramic suspension into two components A and B. The monomer is added into A, and the initiator is added into B. Under this condition, there is no reaction in either A or B because of segregation of the monomer and the initiator. The suspensions maintained good fluidity until they were mixed quickly and uniformly. The mixture rapidly solidifies under the influence of the dominative factors of initiator volume and injection pressure. The new process has enough operating time and can be controlled expediently. Highly automatic equipment were also developed for the process. Al_2O_3 grinding ballings with perfect physical properties were prepared by the corresponding equipment.

Storage of ceramic slurry has been one of the key problems in the industrialization of gelcasting. Except for the sedimentation of the ceramic particles, solidification properties of slurries were also affected after long-term storage. The idle time was prolonged or shortened because the acrylamide in the slurry hydrolyzed into acrylic acid. Even minute concentrations of acrylic acid have a considerable influence on the polymerization of acrylamide. The material of the storage container also influences the storage time of the slurry. The storage container should not be made of copper or copper alloy. In order to store the ceramic slurry in a much safer manner, an inhibitor should be added to the slurry. A compound inhibitor of phenothiazine and catechol is the best option.

The advantages of CIMC can be summarized as follows.

(a) Easy debindering,
(b) Homogeneity,
(c) Complex-shape parts with all kinds of cross section,
(d) Near net-size forming, and
(e) High automation.

With these advantages, CIMC can meet better the needs of industrialization. High-performance ceramics parts of complex shapes with low cost can be prepared by CIMC.

References

Gilissen, R., Erauw, J.P., Smolders, A., Vanswijgenhoven, E., & Luyten, J. (2000). Gelcasting: A near net-shape technique. *Materials and Design, 21*, 251–257.

Krug, S., Evans, J. R. G., & ter Maat, J. H. H. (2000). Residual stresses and cracking in large ceramic injection moldings subjected to different solidification schedules. *Journal of the European Ceramic Society, 20*, 2535–2541.

Liu, D.-M. (1999). Control of yield stress in low-pressure ceramic injection moldings. *Ceramics International, 25*, 587–592.

Liguo, Ma., Yong, Huang, Jinlong, Yang, & Liang, Su. (2002). *Journal of Chengdu University, 21*, 5–10.

Novak, S., Dakskobler, A., & Ribitsch, V. (2000). The effect of water on the behavior of alumina-paraffin suspensions for low-pressure injection molding (LPIM). *Journal of the European Ceramic Society, 20*, 2175–2181.

Omatete, O. O., Janney, M. A., & Strehlow, P. A. (1991). Gelcasting: A new ceramic forming process. *American Ceramic Society Bulletin, 70*(10), 1641–1649.

Wen-Cheng, J.W., Rongyuan, W., & Sah-Jai, H. (2000) Effects of pressure parameters on alumina made by powder injection molding. *Journal of the European Ceramic Society, 20*, 1301–1310.

Chapter 2
Gel Tape Casting of Ceramic Substrates

Abstract In this chapter, aqueous gel tape casting (in short form, novel gel-tape-casting) processing is introduced systematically. The solidification mechanism and factors influencing the gel-tape-casting are analyzed. Alumina green tapes have been prepared through gel-tape-casting successfully. Experimental results show that the efficiency of styrene-acrylic binder is obvious for water-based tape casting. Sodium alginate has been used successfully to consolidate alumina tapes from aqueous suspensions. Based on gel-tape-casting, another novel process, spray inducing instant solidification of gel tape casting, was developed to overcome the limitation of operating time. This process is ideal for forming ceramic sheets with thickness less than 0.3 mm. Research studies have demonstrated that the ceramic sheet products fabricated by the new gel-tape-casting process have good quality, reduce pollution, lower costs, and yield significant social and economic benefits.

Keywords Gel-tape-casting · Slurry · Ceramic sheet · Substrate · Styrene-acrylic latex · Sodium alginate

Ceramics substrates (ceramic sheets) are important electrical component materials that possess many outstanding characteristics such as high temperature resistance, wear resistance, corrosion resistance, anti-oxidation, lightweight property, and insulation. Therefore, they are widely applied in integrated circuits, ceramic capacitors, piezoelectric ceramic devices, layered composite material, and many other areas. Generally, ceramic green sheets can be produced through various molding methods (Zhengqi et al. 1996) as follows: (1) extrusion forming; (2) dry pressing; (3) rolling film forming; (4) screen printing forming; (5) tape casting, etc. However, all the methods used presently have their limitations for mass production of ceramic sheets even though some new molding methods and technologies will be further developed and perfected. Tape casting is considered the main manufacturing method for large-scale ceramic sheet production.

Tape casting technology first appeared in the 1940s during the Second World War when there was a serious shortage of supplies for producing mica capacitors. To solve this problem, people began using tape casting to produce ceramic sheets for ceramic capacitors.

Tape-casting (cast extension molding) is also known as doctor blading (or knife-coating). During the period 1943–1945, G. N. Howatt and a few others from the Massachusetts Institute of Technology at Fort Monmouth Signal Laboratory pioneered tape-casting and built the world's first cast extension molding machine. This technology of ceramics forming was reported publicly in 1945 (Howatt et al. 1945).

Tape casting was formally applied to industrial production of ceramic capacitors in 1947. G. N. Howatt received the franchise in 1952. This was the first ceramic patent (Howatt 1952) for tape casting for production.

In 1967, an Al_2O_3 film was successfully prepared by using tape-casting by Stetson and Gyurk (1967). In the same year, IBM Corporation announced that layer packaging materials for use of computer had been made by using the tape-casting technology (Schwartz and Wilcox 1967). In the 1970s, ultra-fine powder tape-casting study began to appear. As the technology advanced, many new products were successfully developed, and the number of tape-casting applications grew. In 1996, capacitors with 5 μm films were fabricated successfully by tape casting in Japan. By 1997, tape-casting machinery, which could form films of 5 μm thickness, began to emerge in the Japanese and U.S. markets. In 1998, researchers announced that a film with thickness of 3 μm had been obtained through tape-casting technology.

Tape-casting, as a processing method, is a multidisciplinary technique involving ceramic technology, physical chemistry of powder surfaces, gel chemistry, and organic and polymer chemistry. As the tape-forming process had been put forward for several decades, and extensive studies of the process were subjected to the attention of the scientific community, it has undergone continuous improvement and enhancement.

Although the traditional tape-casting technology for ceramic sheets has many advantages including suitability for large-scale continuous molding of ceramic sheets with simple and easy-to-operate equipment, several organic compounds are necessary, such as solvents, binder agent, plasticizers, and other additives. Most of these organic compounds are to some degree toxic, especially the organic solvents. They are volatilized during the forming process and cause environmental pollution. Since the 1970s, increasingly stringent environmental requirements have been established by countries across the world. Therefore, increasingly more researchers are engaged in the investigation of water-based tape casting systems.

Although water-based tape-casting is a less costly, non-toxic, and nonflammable technology, there are also a number of issues to be addressed. These issues include low rate of water evaporation, and drying, high concentration of the binder in slurry, slurry sensitivity to process parameters, difficulty in forming a smooth, dense surface on the ceramic film, and propensity to form cracks upon drying. In addition, the ceramic powder in the water reunite flocculation easily and therefore it is difficult to form a stable slurry of good dispersion.

A water-based system is not available at present for industrial-scale forming due to the several drawbacks mentioned above, which need to be overcome. The slurry curing principle should be studied thoroughly in order to make water-based tape casting mass production a reality. For this purpose, gel-casting is introduced into the tape-casting process and then a fast gel-curing, water-based tape casting or aqueous gel-tape-casting process was proposed by the structural ceramic research team at the Tsinghua University.

In this chapter, the novel gel tape casting technology is introduced comprehensively covering the curing mechanism and kinetics, slurry preparation and characteristics, and the slurry rheological behavior and other research aspects.

2.1 Fundamental Principle and Processing of Aqueous Gel Tape Casting

2.1.1 Tape Casting Types and the Raw Materials Used

(1) **Tape casting types**

(1) Two traditional tape casting technologies
Tape casting is a well-established technique used for large-scale fabrication of ceramic-based substrates and multilayered structures, such as multilayered ceramic capacitors and multichip ceramic modules (Mister 1998). The applications of this technique also have been tested in many other fields, including solid fuel cells, photovoltaic cells, and laminated object molding (Mister 1998).

Tape casting is the basic process for the fabrication of high-quality ceramic sheets, such as Al_2O_3 or AlN substrates for circuitry, and $BaTiO_3$ for multilayer capacitors (Mistler 1998; Fiori and Portu 1986; Hyatt 1986). The tape casting process can be divided into non-aqueous and aqueous systems according to the types of solvents used (Hotza et al. 1995; Moreno 1990; Nahass et al. 1990). Non-aqueous solvents typically have lower boiling points and can prevent hydration of the ceramic powder, but require special precautions concerning toxicity and inflammability. Aqueous solvents, on the other hand, have the advantages of incombustibility, non-toxicity, and low cost, and are therefore good substitutes for non-aqueous solvents.

However, an aqueous system is subject to high sensitivity to minor process variables changes such as drying conditions, casting composition, and film thickness (Nahass et al. 1992; Mistler et al. 1978). Therefore, cracks, bubbles, and other defects usually appear in the products. Due to these issues, aqueous systems are seldom used for industrial applications up to today. To overcome the drawbacks of tape casting mentioned above, some researchers have studied alternative methods to upgrade the process (Carisey et al. 1995; Landham et al. 1987; Grader 1993; Yoshikawa et al. 1988; Lee et al. 1993; Smith et al. 1990; Waack et al. 1988). A possibility to form tapes by gel-casting is discussed later in this chapter.

(2) New Aqueous Gel-Tape-Casting
Based on the advantages and shortcomings of the traditional tape casting technology, a structural ceramic research team from Tsinghua University (Jun-hui 2001a, b; Yong 2000a, b), have combined the techniques of gel-casting and tape-casting and proposed

a new aqueous gel tape casting process. Gel tape-casting with a water-based ceramic slurry was achieved through water-soluble organic monomer polymerization and curing under certain conditions. In the water-based gel tape casting process, ceramic powder and some additives are added into the premixed solution that is prepared with organic monomer and cross-linker dissolved in deionized water. The ceramic slurry with a solid content of greater than 55 vol% or 60 vol% containing the initiator and catalyst takes on good liquidity and moderate viscosity. After tape casting, the casted slurry is heated at appropriate temperature, and the organic monomer is polymerized due to the existing initiator and catalyst. Therefore, the slurry is cured and formed to green tape.

This technology can be applied for different ceramic powders. The green tapes have less organic content that can be removed through special degreasing processes and then directly sintered sequentially. The shaped green tapes have smooth surfaces, high strength, and good flexibility. However, oxygen is to impede free radical polymerization of the organic monomer. As a result, the gel tape casting process adopts either inert gas protection or other oxygen isolation devices.

(2) **Effects of solvent systems on the tape casting (cast extension molding) process**

Owing to its lower boiling point, the organic solvent system is more favorable to dry green tapes, during which the organic solvents will not subject to hydrolysis reaction with ceramic particles so as to ensure the stability of slurry. The disadvantages of this process include the following. (1) Most solvents used, such as toluene, xylene, and trichloroethylene are toxic to a certain extent and thus lead to deterioration in production conditions, cause environmental pollution, and increase production costs. (2) It is more difficult to prepare the high solid content of the slurry, which requires high doses of binder, plasticizers, and other organic compounds to ensure a high liquidity demands of the slurry. (3) Green tapes are liable to crack and deform due to more organic binders and then lower density. The consequent poor quality of products causes low production efficiency and high energy consumption (Medowski and Sutch 1976; Schuetz et al. 1987; Kita et al. 1982; Burnfield and Peterson 1992; Nagata 1993; Hotza et al. 1995; Hyatt 1995).

A water-based system using water as solvent that takes on the lower dosage of organic matters is noncombustible, non-toxic, and low-cost. However, because water is less volatile, a water-based system is not conducive to the drying of green tapes, and has low production efficiency. During the drying process, the evaporation of water is often accompanied by the segregation of organic compounds and fine particle, leading to nonuniform microstructure of the green tapes. Drying shrinkage with anisotropy in the clear will easily lead to cracking of products. In addition, hydration reaction will take place on certain ceramic powders such as AlN, affecting the rheological behavior of the slurry and the final green body composition (Egashira et al. 1991; Greil et al. 1994; Hotza 1995). Owing to these problems, water-based systems have been studied for many years, but are not yet widely applied.

The slurry formulas commonly used in the organic solvent and water-based slurry systems are listed in Tables 2.1 and 2.2, (Zhengqi et al. 1996). It can be seen that the

2.1 Fundamental Principle and Processing of Aqueous Gel Tape …

Table 2.1 The slurry formula used commonly in organic solvent, for example (Zhengqi et al. 1996)

Raw material	Function	Dosage (g)	Process
Alumina	Ceramic powder	194.40	First: Mixing by ball-milling for 24 h
Magnesia	Grain growth inhibitor	0.49	
Menhaden fish oil	Suspending agent	3.56	
Trichloroethylene	Solvent agent	75.81	
Ethanol (alcohol)	Solvent agent	29.16	
Polyethylene butyral	Binder	7.78	Second: Blending for short time after joining additives
Polyethylene glycol (PEG)	Plasticizer	8.24	
Octyl phthalic acid	Plasticizer	7.00	

Table 2.2 The slurry formula used commonly in a water-based system, for example (Zhengqi et al. 1996)

Raw material	Functions	Dosage (g)	Process
Distilled water	Solvent agent	31.62	Premixing
Magnesia	Grain growth inhibitor	0.25	
Polyethylene glycol	Plasticizer	7.78	
Butyl benzyl phthalates	Plasticizer	57.02	
Non-ionic octyl phenoxy ethanol	Wetting agent	0.32	
Propylene-based sulfoacid	suspending agent	4.54	
Alumina	ceramic powder	123.12	Milling for 24 h after joining the premixed solution
Acrylic resin emulsion system	Binder	12.96	Milling for 0.5 h after joining the above-mentioned premixture
Paraffin emulsion	Defoamer	0.13	Milling for 3 min after joining the above-mentioned premixture

organic compound concentration is 40.3 wt% and that of ceramic powder content is 59.7 wt% in organic solvents; whereas these concentrations are respectively 34.8 wt% and 51.9 wt% in addition to 13.3 wt% moisture of in a water-based system. The amount of both organic matter and ceramic powder are found to be low in a water-based system.

In general, the strength of a water-soluble adhesive is significantly lower than that of the binder used in an organic solvent system (Morris 1986). To ensure that the required green body strength is achieved and to avoid cracking, a high binder dosage is necessary in a water-based system, which reduces the content of ceramic powder accordingly.

On the other hand, organic solvents have lower surface tension and better wettability on the surface of ceramic particles. Therefore, a ceramic slurry with a stable, good dispersion is easier to form. However, the water surface tension is much higher than that of common organic solvents, and poor wettability on the surface of ceramic particles leads to decrease in solid loading in suspensions. By adding a wetting agent in the slurry system, the wettability between water and the ceramic particles can be improved and the solid loading should be increased in slurry also.

To explore the molding quality of green tapes to different solvent systems, contrast research for tape casting of Al_2O_3 slurries, which were prepared by a water-based system and an organic solvents system, were carried out by Nahass et al. (1990). The results showed that the slurry with organic solvents had lower sensitivity on the change of the molding process parameters, and it was relatively easy to receive high intensity, uniform structure of the green tape. While the slurry with water-based system was sensitive to the dry conditions, the composition of slurry and the film thickness for minor changes of technical parameters. Green tapes with uniform, crack-free zones can be fabricated only when the conditions are accurate and the control during the forming process is stable (Nahass et al. 1990; Hotza et al. 1995).

Drying of ceramic green tapes is the most difficult procedure in the aqueous tape casting process (Scherer 1990; Briscoe et al. 1997; Ford 1986). Water-based slurry in the drying process is easier to have cracking, curling and to tend segregation of organism and small particles, leading to peeling and other defects. It is obvious that the green bodies dried faster. Therefore, process conditions such as temperature, relative humidity, and air fluid speed should be controlled precisely according to the thickness of the green tape and the slurry compositions so that the water evaporation rate slows down. The pores in the full tape must be eliminated as much as possible to avoid uneven regional contraction caused by deformation or cracking and green tape curly (Briscoe et al. 1997; Ford 1986; Descamps et al. 1995). The drying rate of the thick tape should be controlled especially in the drying late.

For a water-based slurry system, the green body becomes brittle after water evaporation. The moisture content of the green body after drying should generally be maintained at around 2–5 wt% (Descamps et al. 1995).

In recent years, tape casting has generated considerable attention due to environmental and health aspects. Organic solvents have lower boiling points, but need special precautions concerning flammability and toxicity. Therefore, slurries using water as the solvent have appeared (Chertier 1993; Kristoffersson and Carlstrom 1997a, b). Aqueous-based systems have the advantages of nonflammability, non-toxicity, and lower cost (Hotza et al. 1995).

However, some problems occur in water-based systems. For example, it is difficult to obtain stable slurries with appropriate rheological properties and high solids loading, which would be suitable to give uniform tapes with high green density. Furthermore, the drying procedures of the slurries are not easily controllable to obtain a favorable range of thickness (Chartier and Bruneau 1993) and to decrease green tape defects that result from the drying process (Hotza et al. 1995).

A novel process, gel-tape-casting, has been developed to overcome the problems described. This process is a combination of tape-casting and gel-casting, based on the mechanism of polymerization of soluble monomers. The slurry prepared using this process contains a solids loading >50 vol%, and it can solidify in a few tens of minutes under a nitrogen atmosphere, at a specific temperature.

In general, aqueous slurry has less tolerance to minor changes in type and content of additives, drying conditions, casting composition, and film thickness (Hotza et al. 1995). Therefore, more attention must be paid to how additives and other conditions influence the properties of the slurry and green tape.

(3) **Starting materials usually used**

The ceramic powder usually used is oxide materials such as α-Al_2O_3, MgO, TiO_2, ZrO_2, $MgAl_2O_4$, and Mullite, and nonoxide materials such as Si_3N_4, and SiC. Sintering assistants used are Kaolin powder, $CaCO_3$, and SiO_2.

Acrylamide (AM) is used as organic monomer. The cross-linker is N, N′-methylene bisacrylamide (MBAM). Ammonium persulfate is used as the initiator of polymerization. The catalyst used is N, N, N′, N′ -tetramethyl ethylenediamine.

The premixed solution was prepared with monomer and cross-linker dissolved in deionized water. Ammonium persulfate and N, N, N′, N′, -tetramethyl ethylenediamine were used as polymerization initiator and catalyst, respectively.

The dispersants usually used are tetramethylammonium hydroxide and ammonium citrate. The plasticizer usually used is glycerol and polyethylene glycol (PEG). Surfactants usually used are dodecylbenzene sulfonic acid, sodium salt, Tween 80, and tributyl phosphate.

(4) **Organic compounds commonly used in slurry for the tape-casting process**

The organisms commonly used in slurry are shown in Table 2.3.

2.1.2 *Polymerization of the Monomer*

(1) Gel-Tape-Casting Mechanism
For organic tape casting, a green tape is prepared by the evaporation of an organic solution. However, preparation of a green tape by the gel-tape-casting is based on the mechanism of polymerization of aqueous monomers. During the casting, the polymerization of monomers is initiated at elevated temperature, and a polymer network is formed (Wayne 1969).

When the ambient temperature increases, the initiator in the slurry is decompounded into initiating radicals:

$$(NH_4)_2S_2O_8 \rightarrow 2NH_4^+ + 2SO_4^- \tag{2.1}$$

Table 2.3 Organic matter commonly used in slurry for tape casting (Zhengqi et al. 1996)

		Solvents	Binders	Plasticizer	Suspending agent	Wetting agent
Organic solvent		Acetone Butanol Benzene Bromochloromethane Di-acetone ethanol Propanol Ding B ethyl enone Toluene Trichloroethylene Xylene;dimethybenzene	Cellulose acetate butene; Cellulose ether; Petroleum resin; Polyethylene; **Polyacrylic ester;** Poly-propylene; **Polyvinyl alcohol (PVA);** Polyethylene butyral; **Ethylene chloride;** Polymethacrylate; **Ethyl cellulose;** Rosin resin acid;	butylbenzyl phthalic acid; Dibutyl phthalate acid; Dimethyl phthalate acid; Phthalates mixture; Polyethylene glycol-dielectric; **Tritolyl phosphate;**	**fatty acid;** Natural fish oil; Synthetic surfactants; **benzene sulfonic acid;** Fish Oil; Oleic acid; Methanol; Octane;	Poly-propylene-based ethanol; Polyethylene glycol ethyl ether; Ethyl benzene glycol; Polyoxyethylene(PEO); Single-oleic acid glycerol; Three oleic acid glycerol; Ethanol category
Water-based system		Paraffin wax systems **Organic silicon systems** Ethanol non-ionic surfactants	Acrylic polymer systems Acrylic polymer emulsion systems Ethylene oxide polymer Hydroxyethyl cellulose Methyl cellulose **Polyvinyl alcohol (PVA)** Isocyanate Paraffin wax lubricant Urethane (water-soluble) Methyl acrylate copolymer salt Paraffin wax emulsion Ethylene—vinyl acetate copolymer emulsion	**Butyl benzyl phthalate** **Dibutyl phthalate** **Ethyl toluene sulfonamide glycerin** Alkyl-glycerine **Triethylene glycol** Three-N-butyl phosphate Gasoline **Polylol**	Phosphate Phosphate complex salt Allyl sulfoacid Natural sodium salt Acrylic copolymer	Non-ionic octyl phenoxy ethanol Ethanol-type nonionic surfactants

The initiating radical reacts with the monomer, and a monomer-free radical is produced:

$$SO_4^- + CH_2=CH \rightarrow SH_4CH_2CH- \atop | \atop CONH_2 \quad (2.2)$$

That can act with other monomers and result in a propagating radical:

$$SO_4CH_2\underset{CONH_2}{CH-} + nCH_4{=}\underset{CONH_2}{CH} \rightarrow SO_4CH_2\underset{CONH_2}{CH} + CH_2\underset{CONH_2}{CH} + nCH_2\underset{CONH_2}{CH-} \quad (2.3)$$

As the reaction proceeds, monomers polymerize into polymer chains. In most cases, the C=C double bond of the cross-linker, MBAM, breaks down and adheres to the heads of two polymer chains, so that these polymer chains join. Finally, cross-linked polyacrylamide (PAM), which fixes ceramic powders in its framework structure, is formed.

The slurry cannot be solidified to green tape when exposed to air because of the antipolymerizing effect of oxygen. Oxygen has a greater tendency to react with a radical formed in the process than AM and hence it produces an inert radical. Thus, the polymerization stops. Therefore, a cover filled with nitrogen is required to keep out the oxygen.

However, the green tape formed from monomer polymerization is too brittle to be rolled. Plasticizer G containing hydroxyl is introduced into this system. The hydroxyl in the plasticizer G tends to form a hydrogen bonding with oxygen/nitrogen atoms in the polymer chains, thus substituting for a part of the cross-linker molecules to build a soft organic bridge between the polymer chains. Thus, the flexibility of the green tape is improved (Moreno 1992a, b, c).

In the gel-tape-casting process, an AM monomer solution is used as the solvent, and the slurry is prepared by adding ceramic powders and additives to the solvent. During casting, the initiator and catalyst are added to the slurry and polymerization proceeds. The slurry solidifies and a green tape is obtained. The structural formula of AM is

$$CH_2 = CH - \underset{NH_2}{C} = O \quad (2.4)$$

Polymerization of AM may occur as follows. First, the initiator is decomposes into primary free radicals, as shown by

$$(NH_4)_2S_2O_8 \rightarrow 2NH_4^+ + 2SO_4^- \quad (2.5)$$

The primary free radicals can be combined with the monomer molecule, which produces monomer free radicals, as in the following reaction

$$SO_4^- + CH_2=CH\underset{|}{} \longrightarrow SO_4CH_2-CH-\underset{|}{} \quad (2.6)$$
$$CONH_2 CONH_2$$

These free radicals can combine with monomers and result in chain-free radicals.

$$SO_4CH_2-\underset{CONH_2}{\underset{|}{C}H}- + nCH_2=\underset{CONH_2}{\underset{|}{C}H}$$
$$\longrightarrow SO_4CH_2CH\text{-}[CH_2CH]_n CH_2CH- \quad (2.7)$$
$$CONH_2 \;\; CONH_2 \;\; CONH_2$$

This reaction proceeds continuously and the AM molecules polymerize into polymer chains. A network is formed by polymer chains by two ways, crosslinking (2.8) of polymer chains and bridging (2.10) between the polymer chains by cross-linker molecules.

$$\begin{array}{c}
+\!\!\!\!\text{[}CH_2-\underset{CONH_2}{\underset{|}{C}H}\text{]}_n CH_2-\underset{CONH_2}{\underset{|}{C}H}- \\
+ \\
+\!\!\!\!\text{[}CH_2-\underset{CONH_2}{\underset{|}{C}H}\text{]}_n CH_2-\underset{CONH_2}{\underset{|}{C}H}- \\
\\
+\!\!\!\!\text{[}CH_2-\underset{CONH_2}{\underset{|}{C}H}\text{]}_n CH_2-\underset{CO}{\underset{|}{C}H}- \\
|\\
\longrightarrow NH \quad + \; NH_3 \\
|\\
CO\\
+\!\!\!\!\text{[}CH_2-\underset{CONH_2}{\underset{|}{C}H}\text{]}_n CH_2-CH-
\end{array} \quad (2.8)$$

In order to make the monomers polymerize completely, MABM [molecular formula shown in (2.9)] is added as cross-linker into the monomer solution.

$$\underset{\underset{O}{\overset{||}{C}}-NH-CH_2-NH-\underset{O}{\overset{||}{C}}}{\overset{CH_2=CHCH=CH_2}{||}} \quad (2.9)$$

MABM has two C=C double bonds and can be connected between two polymer chains by a bridge effect, which is beneficial for network formation (2.10).

2.1 Fundamental Principle and Processing of Aqueous Gel Tape … 27

$$2\left\{ \begin{array}{c} -CH_2CH + CH_2CH \frac{}{n} CH_2CH - \\ | \quad\quad | \quad\quad | \\ CONH_2 \quad CONH_2 \quad CONH_2 \end{array} \right\}$$

$$+ \quad \begin{array}{c} CH_2=CH \quad\quad\quad CH=CH_2 \\ | \quad\quad\quad\quad\quad\quad | \\ C-NH-CH_2-NH-C \\ \| \quad\quad\quad\quad\quad\quad \| \\ O \quad\quad\quad\quad\quad\quad O \end{array}$$

$$\longrightarrow \quad \begin{array}{c} -CH_2CH + CH_2CH \frac{}{n} CH_2CH - CH_2CH - \\ | \quad\quad | \quad\quad | \quad\quad | \\ CONH_2 \quad CONH_2 \quad CONH_2 \quad CO \\ \quad\quad\quad\quad\quad\quad\quad\quad\quad\quad | \\ \quad\quad\quad\quad\quad\quad\quad\quad\quad\quad NH \\ \quad\quad\quad\quad\quad\quad\quad\quad\quad\quad | \\ \quad\quad\quad\quad\quad\quad\quad\quad\quad\quad CH_2 \\ \quad\quad\quad\quad\quad\quad\quad\quad\quad\quad | \\ \quad\quad\quad\quad\quad\quad\quad\quad\quad\quad NH \\ CONH_2 \quad CONH_2 \quad CONH_2 \quad CO \\ | \quad\quad | \quad\quad | \quad\quad | \\ -CH_2CH + CH_2CH \frac{}{n} CH_2CH - CH_2CH - \end{array} \quad (2.10)$$

The ceramic slurry solidifies by polymerization of AM, and thus a ceramic green tape is obtained. The network of PAM acts as a frame that fixes ceramic powders by adsorption. The details of the combination between the ceramic powder and PAM are shown in Fig. 2.1.

It should be mentioned that the slurry exposed to air could not be solidified to a green tape because of the anti-polymerizing effect of oxygen, therefore, during the casting process the slurry should be protected by an inert atmosphere of nitrogen or argon.

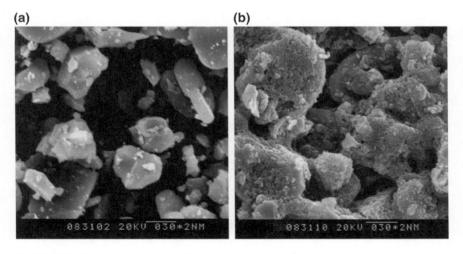

Fig. 2.1 SEM image of ceramic particles **a** before polymerization, **b** after polymerization

(2) Study of polymerization kinetics

The polymerization rate of ceramic slurry depends on the properties of the slurry, such as the concentration and pH value of the monomer solution, and the reaction conditions, such as polymerization temperature and amount of initiator and catalyst. To obtain a green tape with high strength and suitable flexibility, slurry preparation and polymerization conditions should be controlled precisely.

Experimental results show that pores are always present in the green tape if the slurry solidifies too fast, because the gas released during polymerization cannot be eliminated from the green tape completely. Pores in the green tape would decrease the strength of the green tape and of the sintered sample. Therefore, it is necessary to keep the polymerization rate within a suitable range. In other words, the slurry would polymerized to a greater extent within a few minutes than a few seconds.

(1) Concentration of the monomer solution

The relationship between the polymerization rate and the concentration of the monomer solution is shown in Fig. 2.2. The higher the concentration is, the shorter the polymerization time, i.e., the higher the polymerization rate.

The polymerization time decreases sharply with the increase of the monomer solution concentration up to 40 wt% after which it remains practically constant. Thus, the polymerization rate can be adjusted more easily by controlling the monomer solution concentration if the concentration is less than 40 wt%. For the Al_2O_3 slurry, in order to obtain a suitable polymerization rate in the gel-tape-casting, the concentration of the monomer solution in the premixed solution should be kept in the range of 30–40 wt%.

(2) pH value

The pH value of the slurry affects the polymerization rate significantly because the initiator $(NH_4)_2S_2O_8$ decomposes more quickly in an acidic condition than in an alkaline condition. As a result, the polymerization rate decreases with the increase of the pH value and the polymerization time increases almost linearly with increase in pH (Fig. 2.3). Besides influencing the polymerization rate, the pH value also influences the microstructure of the green tape. At low pH, polymerization of AM tends to release NH_3, which forms pores in the green tape and leads to an inhomogeneous

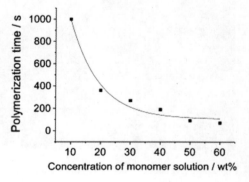

Fig. 2.2 Relationship between concentration of monomer solution and polymerization time

2.1 Fundamental Principle and Processing of Aqueous Gel Tape ...

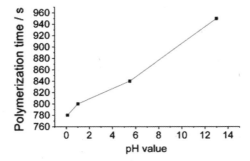

Fig. 2.3 Relationship between the pH of the ceramic slurry and polymerization time

microstructure (Fig. 2.4a). Experimental results show that a suitable pH value of Al_2O_3 slurry lies in the range 7–9, a condition beneficial for both inhibiting pore formation in the green tape and improving the dispersibility of the ceramic powder in the slurry. As a result, a green tape with a homogeneous microstructure can be obtained (Fig. 2.4b).

(3) Polymerization temperature

Temperature is one of the most important factors influencing the polymerization rate. Although a high temperature is beneficial for improving polymerization rate, at an excessively high temperature, the water in the slurry will volatilize so quickly that pores will be formed in the green tape.

Figure 2.5 shows the relationship between temperature and polymerization time. With the increase of temperature, the polymerization time is reduced dramatically. A suitable temperature for polymerization of the system lies in the 40–80 °C range.

Fig. 2.4 SEM image of the broken section of green tape made from slurries with different pH values

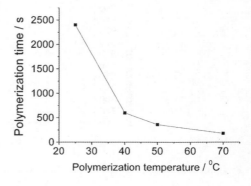

Fig. 2.5 Relationship between temperature and polymerization time

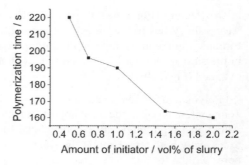

Fig. 2.6 Relationship between amount of initiator and polymerization time

During the forming process, it is expected that the slurry can be kept stable for a long time before casting, and it would polymerize very quickly after casting. According to the sensitivity of the polymerization rate on temperature, the onset of polymerization can be controlled accurately by keeping the ceramic slurry at low temperature (such as 25 °C) before casting, and heating it to 40–80 °C to complete polymerization after casting.

(4) Initiator and catalyst

It has been proved that the higher the polymerization rate, the more difficult it is to control the casting process. If the polymerization proceeds too quickly, the slurry may polymerize in the reservoir before passing the blade. The role of the initiator is to create primary free radicals to initiate the polymerization of the monomers. A large amount of initiator will increase the initiating rate and lead to a very fast polymerization. Figure 2.6 shows the relationship between the amount of initiator and the polymerization time of the Al_2O_3 slurry. A suitable amount of initiator lies in the range 0.5–1 vol% of slurry.

The catalyst can significantly lower the activation energy of the reaction and improve the polymerization rate. Experimental results show that the slurry cannot polymerize without a catalyst even if a certain amount of initiator is added. Nevertheless, a very high amount of catalyst makes the polymerization rate uncomfortable. For the studied Al_2O_3 slurry, an appropriate amount of catalyst is about 0.5 vol%

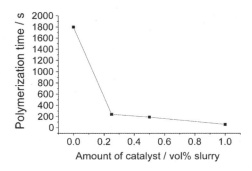

Fig. 2.7 Relationship between amount of catalyst and polymerization time

of the slurry. The relationship between the polymerization time and the amount of catalyst is shown in Fig. 2.7.

It can be seen from that the polymerization rate of AM depends on the properties of the slurry and on the reaction conditions. The concentration of the monomer solution should be kept within 30–40 wt% to obtain a suitable polymerization rate. A suitable range of pH values of the Al_2O_3 slurry and temperature are 7–9 and 40–80 °C, respectively. The sensitivity of AM to temperature can be used to control the onset of polymerization accurately. For the studied Al_2O_3 slurry, the suitable amounts of initiator and of catalyst are 0.5–1 vol% and ~0.5 vol% of the Al_2O_3 slurry, respectively.

2.1.3 Influence Factors on Polymerization of the Monomer

(1) Effects of dispersant

Ammonium citrate is used as a dispersant in this system to improve the dispersion of ceramic powders and to achieve a solids loading >50 vol%. Molecules of the dispersant are absorbed onto the surface of the ceramic powder when they are dissolved into the slurry; this changes the charge distribution on the particle surface. Generally, ceramic powders with high Zeta potential (absolute value) usually exhibit high stability in the slurry (Moreno 1992a, b, c). Addition of the dispersant can increase the absolute value of the zeta potential of the slurry at pH values of 8–10 (as shown in Fig. 2.8). Therefore, the dispersion of the ceramic powders in the slurry is improved.

An optimal amount of the dispersant is to be used in the slurry to achieve good fluidity. The apparent viscosity of slurries (55 vol%) containing various contents of the dispersant has been tested. The viscosity of the slurry decreases with an increase in dispersant at relatively lower dispersant concentration; however, it increases when the dispersant content exceeds a fixed value (1 wt% in this system). Graule and Grauckler (1993) reported the same phenomenon in an aqueous slurry by using Tiron as dispersant. For the dispersant content higher than a fixed value (as shown Fig. 2.9), the increase in viscosity can be attributed to an increase in ionic strength. The diffuse electrical double layer around each ceramic particle is compressed by the ions, and the repulsive potential is decreased.

Fig. 2.8 Absolute value of the zeta potential of a slurry with ammonium citrate dispersant (-▲-) is higher than that of a slurry without dispersant (-■-) in a pH range of 8–10

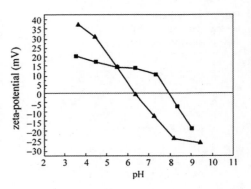

Fig. 2.9 Viscosity at 50/s of a slurry with a solids loading of 50 vol% versus the dispersant content; the viscosity of the slurry reaches the minimum value when the content of dispersant is 1 wt%

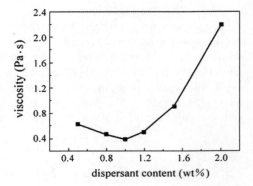

(2) Effects of plasticizer

The solidified green tape made from the slurry without plasticizer is too brittle to be rolled. A plasticizer is introduced into the slurry to make the green tape soft. The addition of the plasticizer affects the rheological behavior of the slurry. Three curves of viscosity have been compared that represent the slurries (55 vol%) with 10 wt% plasticizer G, 10 wt% PEG, and without plasticizer (as shown in Fig. 2.10). The addition of the plasticizer increases the slurry viscosity to a certain extent.

The relationship between the content of plasticizer G and the slurry viscosity (50 vol%) has also been studied. These viscosity curves indicate that the viscosity of the slurry increases with an increase in plasticizer G (as shown Fig. 2.11).

The flexibility of the green tape is judged by the number of times that the tape can sustain bending without breaking. The experiment shows that the flexibility improves (50 or more bends) gradually when the amount of plasticizer G increases. However, after the plasticizer G amount reaches ~5 wt%, the flexibility does not seem to change. Meanwhile, the fracture strength of the green tape decreases at high amounts of plasticizer G. Therefore, 5 wt% of plasticizer is considered the optimal amount in this system to balance the needs of flexibility and fracture strength.

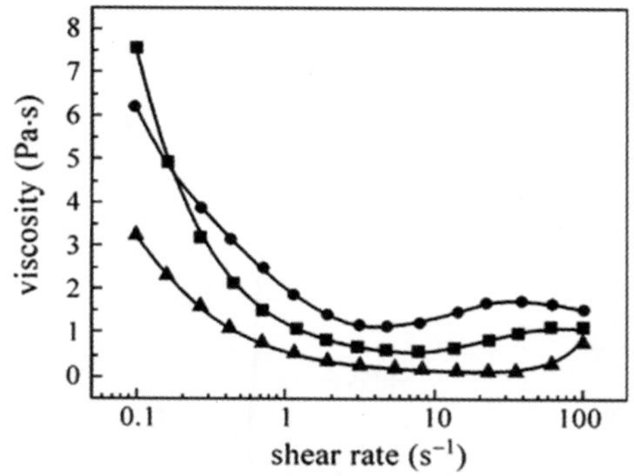

Fig. 2.10 Effect of various plasticizers on viscosity of the slurry (55 vol%) ((■) 10 wt% G, (●) 10 wt% PEG, and (▲) with no plasticizer); viscosity of slurry increases after addition of plasticizer

Fig. 2.11 Relationship between the amounts of plasticizer G and viscosity of the slurry ((■) 3 wt%, (●) 5 wt%, and (▲) 10 wt%); viscosity increases with increasing amounts of G at relatively low shear rate. However, the difference disappears at shear rates $>10\ s^{-1}$

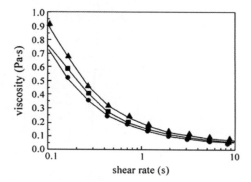

(3) Surfactant effects

During gel-tape-casting, it is difficult to wet the carrier film with the aqueous slurry, and, thus, a flat and homogeneous green tape cannot be formed. Therefore, a surfactant is needed in this system. From the wetting angles of slurries with the surfactant, it can be concluded that a small amount of surfactant in the slurry decreases the surface tension significantly. This results in decrease of the wetting angle of the slurry. For example, the wetting angle shifts from 109.8° to 72.2° after using surfactant D (as shown in Table 2.4).

The releasing property of the green tape made from organic slurry is perfect because of its strong adhesion. On the contrary, the green tape from aqueous gel-tape-casting is difficult or impossible to release from the carrier film. However, the green tape with surfactant D added easily releases from the carrier film and has smooth surfaces. The good releasing property is attributed to the structure of surfactant D.

In general, surfactant molecules gather on the surface of the solution during processing, with hydrophilic groups inserting in solution and hydrophobic group facing out. Common surfactants have an organic hydrophobic group, which has a strong attractive force to the plastic surface of the carried film. However, the hydrophobic group of surfactant D shows great electronegativity, and its radius is large enough to

Table 2.4 Wetting angles and surface tensions of slurries with various surfactants (Amount of surfactant in slurry is 1%. Addition of surfactant can decrease the wetting angle of the slurry and, thus, improve its wetting property)

Solution	Wetting angle (deg)	Surface tension (mN/m)
Deionized water	109.8	70.86
Slurry without surfactant	103.3	46.76
Slurry with surfactant A	69.5	36.05
Slurry with surfactant C	75.0	
Slurry with surfactant B	73.5	24.24
Slurry with surfactant D	72.2	26.96

cover the entire carbon chain of the surfactant D molecule (Tadros 1984). Thus, the active force between surfactant D and the carrier film is so small that the green tape can be released easily.

(4) Defoaming and deairating

Foams are introduced when a surfactant is added to a slurry. The amount of foam increases as the content of the surfactant increases. If foams are not removed from slurry before casting, pinholes and other defects appear in the green tape. Some defoamers are used to eliminate the foam from the slurry. Small amounts (<1 wt%) of defoamer decreases the amount of foam.

There are three main mechanisms of foam removal: decreasing the solubility of surfactant; spreading over the interface and displacing the stabilizing surfactant; and causing molecules in the liquid membrane to drain quickly (Myers 1992). Defoamers used in this experiment all are oleaginous and follow the second mechanism.

Some foam remains in the slurry even after a defoamer is used. To eliminate the residual foam thoroughly, the slurry is deairated for −30 min in a rotating flask under high vacuum. To avoid the polymerization of monomers, the de-airing process is conducted in an ice-water bath. The green-tape surface contains no evident foam after using the defoamer and de-airing under vacuum.

(5) Debinding and sintering

The green tape is dried at 120 °C for 24 h after it is released. TGA/DTA of the green tape is performed under an air atmosphere, as shown in Fig. 2.12. These analyses have been conducted at a heating rate of 5 °C/min to a maximum temperature of 800 °C. A large weight loss occurred between 285 and 579 °C, which is associated with the burnout of most of the organic components. The decomposition of some reagents and organic components with large molecular weight results in another weight loss between 660 and 750 °C.

An exothermic peak in the DTA curve starts at 660 and ends at 730 °C. This peak illustrates that reactions occur in this temperature range. The debinding temperature has to be increased to 800 °C to remove organic components completely before sintering.

2.1 Fundamental Principle and Processing of Aqueous Gel Tape …

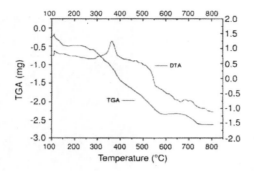

Fig. 2.12 TGA and DTA of green tape

Table 2.5 Properties of samples sintered under various systems

Sintering system	Line shrinkage (%)	Fracture strength (MPa)	Density (g/cm^3)	Appearance evaluation
1	10.24	266	3.63	Fluctuant
2	10.87	257	3.64	Fluctuant
3	11.56	281	3.72	
4	12.31	284	3.75	
5	14.68	292	3.80	Flat
6	15.13	296	3.83	Flat
7	15.95	315	3.84	Flat

Since the content of organic components is relatively lower than in organic tape casting, there is no need for a separate debindering process. Sintering systems with various debindering temperatures and holding times have been conducted, and the properties of the sintered samples by these sintering systems are listed.

Comparison of sintering results shows that a slow heating rate and a high sintering temperature are favorable to acquire sintered samples with high density and flat surfaces.

Debindering and sintering were performed and alumina substrates with high density (3.84 g/cm^3), high fracture strength (315 MPa), and flat surfaces were obtained.

Most organic components are burned out between 280 and 580 °C.

Weight loss that occurs between 660 and 750 °C is associated with an exothermic peak in the DTA curve. Therefore, it is necessary to increase the debinder temperature to 800 °C (Table 2.5).

The SEM photographs of the green tapes on the surface and underside are shown in Figs. 2.13 and 2.14.

Fig. 2.13 SEM photograph showing defects on underside of green tape due to ceramic particles adhering to carrier film

Fig. 2.14 SEM photograph of the green tapes on the surface **a** Green tape surface without bubbles after defoaming and de-airing; **b** Surface with pinholes caused by foam

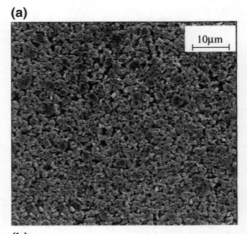

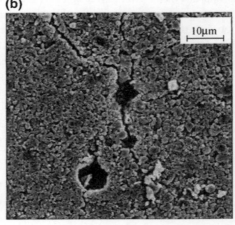

2.1.4 Processing of the Gel Tape Casting

(1) Technological characteristics and device for gel tape casting

Ammonium citrate was selected as dispersant. An organic substance containing hydroxyl was used as plasticizer (labeled as G). Several types of surfactants, including anionic A, cationic B, nonionic C, and special-type D, were chosen. Defoamers were olefin, tributyl phosphate, dipentylamine, and 2-ethylhexanol. The content of all reagents were based on the premixed solution.

The ceramic powder, organic binder, plasticizer, decentralized agents, and other additives added to solvents were mixed with solvents through ball-milling to form a slurry with uniformity, stability, and certain liquidity. According to the different solvents used, the cast extension molding system is divided into organic solvents and water-based system. The organic matter and their amounts in different slurry systems also varied. The organic compounds used commonly in the two kinds of systems are listed in Table 2.3 (Zhengqi et al. 1996).

Some additives were mixed with ceramic powders in the premixed solution to obtain a slurry with a solids loading of >50 vol%. The slurry was then ball-milled for 10–24 h to reach homogeneity. After the ball-mill process, the initiator and catalyst were mixed into the slurry, and the slurry was deaired in an ice-water bath for 30 min. The slurry was then ready for casting.

The gel tape-casting device was especially for this study is shown in Fig. 2.15a, b. It has the atmosphere of the protection device and temperature control system, and is significantly different from the ordinary tape casting machine. It is found that heaters beneath the support and a protecting cover above the support. During casting, the slurry was poured into a reservoir and then drawn, passing the blade on the carrier film, which moved into the cover filled with nitrogen atmosphere. The casting speed was constant at 1.2 cm/s, and the gap between the blade and the support was adjusted at several hundred microns. The polymerization of the monomers was then initiated,

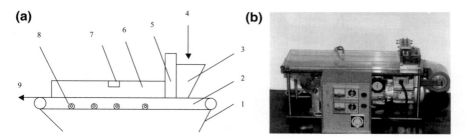

Fig. 2.15 **a** Schematic of gel tape casting device; **b** gel tape casting machine designed and fabricated by authors 1–baseband; 2–baseplate; 3–the material trough; 4–slurry; 5–doctor brade; 6–the protective shield of atmosphere; 7–venthole; 8–heating wire; 9–green tape

and the slurry was solidified to a green tape in <20 min under a nitrogen atmosphere and a temperature of ~40–70 °C. A green tape with adequate strength and flexibility could be prepared with a thickness in the range 0.1–1 mm.

The zeta potential of the slurry was characterized (Model ZetaPlus, Brookhaven Instruments Corporation, Holtsville, N.Y., USA) by using an electrophoretic light-scattering method at 25 °C. Rheological behaviors of the slurry were tested using a rheometer (Model MCR300, Paar Physica USA, Glen Allen, PA, USA.). The surface tension and wetting angle of the slurry were determined (Mode Processor Tensiometer K12, Kruss USA, Charlotte, N.C., USA). A SEM (Model S-450, Hitachi Corp., Tokyo) was used to characterize the microstructures of the green tape. TGA/DTA of the green tape to a maximum temperature of 800 °C was determined (Model Thermogravimetric, Setaram Inc., Mt. Laurel, N. J., USA).

The Processes for Preparing Ceramic Sheets are Discussed Below.

After milling and gas removal, the slurry can be used to cast extension molding. The slurry flow from the material trough 3 to the lower base-band 1, which can move to forward, tape film (tape band) thickness was controlled by the doctor brade 5. Blank film together with the base-band access to go around hot air drying room, the tape film after drying retain a certain solvents, together with the base-band together reel standby application. The solvent in the process of storage distributed evenly and eliminated moisture gradient. Finally, the dried green tape film was made into various products by cutting, drilling, or stamping according to the required shape.

(2) Flowchart for the gel-tape-casting process

The flowchart for the gel-tape-casting is presented in Fig. 2.16. A solution of monomer and cross-linker is prepared. Then, alumina powder, dispersant, plasticizer, and sintering assistants (Kaolin, $CaCO_3$, and SiO_2) are added to the solution to prepare slurry with solids loading more than 50 vol%. After 10–24 h of ball-milling, the slurry is deaired under vacuum (0.02 atm). Then, the initiator and catalyst are added and the slurry is ready for tape casting.

The prepared slurry is poured into the reservoir and drawn past the blade by the moving surface, and brought into the cover. The slurry polymerizes at about 50 °C under inert atmosphere. The cast tape has enough strength and flexibility to sustain shaping processes such as incision and punching. The cast tape is sintered by the following stages: it is heated up to 1,150 °C at a rate of 2 °C min^{-1}, then for 0.5 °C min^{-1} up to 1,550 °C with a plateau for 2 h.

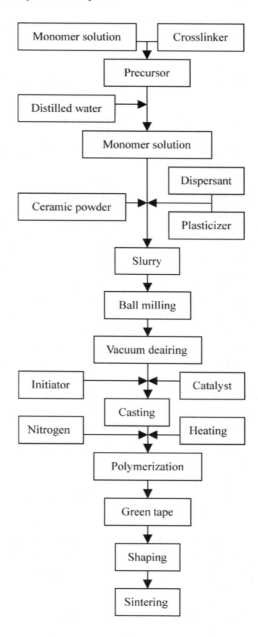

Fig. 2.16 Flowchart of the gel-tape-casting process

2.2 Characteristics of Slurries Used for Aqueous Gel Tape Casting

2.2.1 Properties of the Aqueous Ceramic Slurries with Binder

(1) The importance of additives in tape casting

Tape casting is one of the main fabrication methods of thin ceramic substrates (Mistler 1998; Hotza et al. 1995). Traditional tape casting requires large amounts of organic solvent in the slurry, which may cause defects such as (i) cracks and skins resulting from anisotropic drying shrinkage; (ii) holes inside the green tape during binder burnout; and (iii) inhomogeneous microstructures along the thickness of the tape. Moreover, the organic components used in tape casting are highly volatile, toxic, and expensive (Chartier and Bruneau 1993; Doreau et al. 1998; Briscoe et al. 1997).

More and more emphasis is focused on developing aqueous tape casting, and new molding mechanisms are being used to update traditional tape casting (Smith et al. 1990; Chartier et al. 1997). We developed a new process, aqueous gel-tape-casting, during this study. This process has the advantages of high solids loading (up to 60 vol%) and low toxicity. The aqueous slurry can be rapidly solidified by the polymerization of aqueous organic monomers.

However, it was observed that the aqueous slurry hardly wetted the surface of the organic carrier film, and that it was also hard for the green tape to be released from the carrier film. To overcome these problems, additives, such as surfactants, are used. As a result of adding surfactant into the slurry, foams are also introduced, which may destroy the surface and form pores inside the green tape. To eliminate foams, defoamers are employed. The influence of these additives on the gel-tape-casting process was investigated.

(2) Features of aqueous ceramic slurry with binder

Tape casting is a low-cost and useful technique for preparing thin ceramic sheets, and it has been used mainly to produce substrates, solid electrolytes, packages, and multilayer capacitors for the electronics industry (Hotza et al. 1995; Zeng et al. 2000; Cui et al. 2003; Kristoffersson and Carlstrom 1997a, b). In general, tape-casting slurry is a typical organic system. Since the volatility and toxicity of organic solvents are harmful to the environment and pose health hazards, water-based tape casting is already given much attention to the ceramics industry (Pagnoux et al. 1998; Doreau et al. 1998, 1999; Gutierrez and Moreno 2001; Briscoe et al. 1998).

An aqueous slurry must be adjusted in order to yield tapes that satisfy some quality criteria (Hotza et al. 1995), such as: (1) high solid loading in water; (2) no defects during drying; (3) high strength to allow the manipulation of dried sheets; (4) microstructural homogeneity; (5) good thermocompressibility (the object to laminate multilayer materials); (6) easy pyrolysis (burnout); and (7) high mechanical strength of green tapes. These require careful selection of the slurry additives together with accurate control of many processing parameters.

2.2 Characteristics of Slurries Used for Aqueous Gel Tape Casting

The water-based binder is very important because it provides strength to green tapes after evaporating water through the organic bridges between the ceramic particles. Some water-soluble binders can be used in aqueous tape casting, e.g., cellulose ethers (such as hydroxyethyl cellulose), polyvinyl alcohols, and latexes. All three types of binders have been successfully used in aqueous tape casting (Kristoffersson and Carlstrom 1997a, b; Pagnoux et al. 1998; Doreau et al. 1998, 1999; Gutierrez and Moreno 2001).

This paper focuses on the rheological behavior of the suspensions and drying process of aqueous slurries with a styrene-acrylic latex binder system.

(3) The properties of alumina powder and binder

The alumina powder used in this experiment had a median size of 3.84 μm as shown in Fig. 2.17a. It was produced by Henan Xinyuan Corp. The binder used in the aqueous tape casting process was a commercial styrene-acrylic latex that had low foam and viscosity to produce flexible green tapes with good mechanical properties and lamination behaviors. As shown in Fig. 2.17b, the styrene-acrylic latex was a stable nonionic latex at pH 7, which the isoelectric point (IEP) must be low pH 3.0 when the zeta potential arrived at zero. The latex was a dispersion of fine latex particles of d50 ≈ 300 nm and the solid loading was about 50 wt%. The glass transition temperature (Tg) of the latex was about 273 K, which allowed it to form a film at room temperature.

The fabricated slurry was degassed in vacuum to remove gas bubbles and then it was cast on the polymer film substrate and dried naturally.

(4) The effect of pH value and dispersant on characteristics of slurries with binder

A high-performance aqueous ceramic slurry can be acquired by adjusting the pH value and selecting the appropriate dispersant (Hotza et al. 1995; Zeng et al. 2000). The Zeta potential is a very important physical parameter that can decide the stability of the aqueous ceramic slurry. The higher the zeta potential, the higher the repulsive energy and the more stable the slurries. Figure 2.18 shows the relationship between the pH value and the zeta potential of alumina in water. The curve is the zeta potential of Al2O3 without any additive at different pH values; the IEP at which the zeta

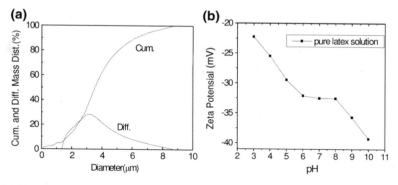

Fig. 2.17 a Distribution of alumina powder diameter and b Zeta potential of emulsion binder

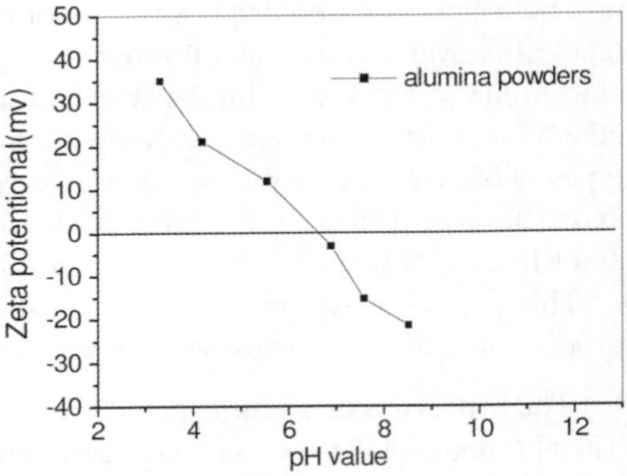

Fig. 2.18 Zeta potential of alumina in water

Fig. 2.19 Relation between the concentration of dispersant and the shear viscosity

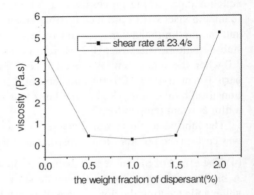

potential was zero was about pH 6.8. The IEP of the latex was low, 3.0; therefore, the pH value of the alumina slurry must be far from the range of pH 3-6.8. Some research results show well-dispersed ceramic slurries at pH 10 for aqueous slurry of alumina. In this experiment, the pH values of latex and alumina solution were adjusted to 10, and then mixed together. It was reported that the polyelectrolyte (PAA–NH4) could not dissociate at pH below 3.5, but could completely dissociate around pH 10. The result shows that the pH value greatly affects the zeta potential and the dispersive effects of the dispersant. For the pH value of alumina slurry was 10, the PAA–NH4 polyelectrolytes could get completely ionization and produced stable slurry. Therefore, the PAA–NH4 was an appropriate dispersant for latex slurry.

Figure 2.19 shows the relation between the concentration of the dispersant and the shear viscosity of slurries at shear rate of 20 s^{-1}, and the function curve appeared to be U-shaped. The reason is that PAA–NH4 is a long link molecule, which can absorb on the surface of particles to prevent particles from contacting each other and thus stabilizing the slurries. At some point, the long-chain molecule was bestrewing the surface of the alumina particles, which dispersant had the best effects in slurry. In this experiment, when the concentration of dispersant was around 1.0 wt% of alumina powders, the viscosity of the slurry was the lowest at the same shear rate. When the

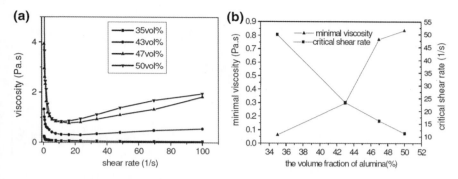

Fig. 2.20 **a** Rheological behaviors of the slurries with different solid fractions and **b** Relationship of the lowest viscosity and corresponding shear rate

concentration of dispersant was smaller than 1.0 wt%, the distribution of dispersant on the surface of alumina particles would be insufficient, and the dispersive effect would be not good. If the concentration of dispersant were more than 1.0 wt%, the long link molecules may entwist or flocculate each other, on the contrary, it would bring rising on the viscosity of slurry. In a word, the main experimental factors that influenced the stability of the slurry were the properties of dispersant and pH value.

(5) The rheological characteristic of slurries with binder

The rheological behaviors of the slurries prepared with latex binders always show a typical shear-thinning behavior with different solid volume fraction at first, and then exhibit shear-thickening from a critical point, which has the lowest viscosity (shown in Fig. 2.20a). This behavior agreed well with the requirements for the tape casting process when the shear rate was relatively low. At a shear rate of $21.7\ s^{-1}$, the viscosity of the suspension remained approximately 0.2–1.0 Pa s with fixed binder volume fraction of 30%. Hence, all the four slurries were suitable for tape casting (Pagnoux et al. 1998; Doreau et al. 1998, 1999; Gutierrez and Moreno 2001).

Figure 2.20b shows that the lowest viscosity of the suspensions and the corresponding shear rate increased with alumina solid content by holding constant the binder concentration (fixed solid fraction 30 vol%). The variation curves show the shear rate value at the critical point, which decreased with the increase of the solids fraction. The result might be from the increasing of flowing resistance with increment of solid fraction in suspensions. Therefore, the optimal alumina solid fraction of slurry is less than 50 vol%.

(6) The drying curves of alumina slurry with binder

Figure 2.21 shows the drying curves of the suspension, which included different drying temperature on the same slurry (solid fraction was 40 vol%, latex binder fraction was 30 vol%). As seen from the figure, the slope of the drying curve increased with the rise of the drying temperature before the transition point, which the water loss was hold invariability after this critical point. The slope was equal to the drying rate which was solvent water losing at one time unit. Hence, the larger the slope

Fig. 2.21 Drying curve with different temperatures

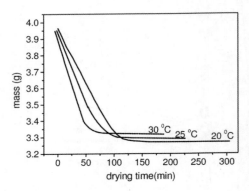

of the drying curve, the faster the drying rate. However, the transition point was ahead of time that the water loss began to hold invariability with the increment of temperature from 20 to 30 °C. In general, the drying stage before the transition point was defined the first drying process. All weight-loss curves in Fig. 2.21 followed a linear relationship with time during the first initial drying process. Therefore, it might be concluded that the evaporation process from the slurry interface should control the first drying process of the suspensions before around 60 min in this slurry. However, the drying rate of the surface of slurries was very high after the suspensions were cast on the support film. After some time, the drying rate became very low. The drying process might be controlled by the transportation of the water from inner to the surface of suspensions after the surface of suspensions dried. Therefore, the resistance of diffusion of the solvent water was greatly decided by the slurry coating thickness, which was larger than the evaporation resistance of the initial stage. The drying rate was slower than the initial drying stage, and Fig. 2.21 shows the results.

(7) The microstructure of green tape prepared by slurry with binder

During the drying process, the compressive force that was induced by evaporation would compress the latex particles distortion, and result in cross-linking structures in contact latex particles of the solidified tapes. Figure 2.22 shows the cross-linking structures in a SEM photograph of green tapes. Therefore, the drying process was one of the most important steps, which enables forming a solid tape from slurry. The film forming mechanisms usually include two necessary conditions: (1) the compressive force for latex particles results from solvent evaporation proceeds, and (2) the glass transition temperature of latex is lower than the work temperature (Xie et al. 2001; Mengkwang 1996a, b).

This study showed the efficiency of styrene-acrylic binder for water-based tape casting. The rheological behaviors of the slurries prepared with latex binders always showed a typical shear-thinning behavior with different solid volume fraction at first, and then exhibited shear-thickening from a critical point. The PAA–NH_4 was found to be an effective dispersant for the slurry of latex binder. The main experimental factors that influenced the stability of the slurry were the properties of the dispersant and the pH value. The research results show that the glass transition temperature of

2.2 Characteristics of Slurries Used for Aqueous Gel Tape Casting

Fig. 2.22 Cross-linking structures in SEM photograph of green tapes fracture

latex (Tg) is also important to enable film formation. The drying speed of slurries was mostly decided by the drying temperature and slurry coating thickness.

2.2.2 Influence of Dispersants on Stability and Rheology of Aqueous Ceramic Slurries with Organic Monomer

(1) Effect of dispersant on powder dispersion and slurry viscosity

In order to get a slurry with a higher solids loading, dispersants are used. A dispersant can change the charge distribution of the powder surface. Figure 2.23 shows the Zeta potential of the slurries with and without ammonium citrate. It is clear that the dispersant (ammonium citrate) can raise the absolute value of Zeta potential of the slurry when the pH value is kept between 8 and 10, and with a shift in the IEP from 8 to 6.5.

The effects of ammonium citrate on the Zeta potential can be considered as follows. In the range of pH 5.3, alumina reacts with OH^-, leaving AlO_2^- with negative Zeta potential:

$$Al_2O_3 + 2OH^- \rightarrow 2AlO_2^- + H_2O \tag{2.11}$$

When ammonium citrate dissolves in slurry, the concentration of protolyzable anions (citrate) is enhanced. Citrate anions tend to chemically bond to the surface of alumina particles. Since the absorption of citrate anions thickens the double layer around the alumina particle, the Zeta potential increases. On the other hand, to acquire an equal number of positively and negatively charged sites, a higher concentration of cations, mainly H^+ in aqueous solution, are needed. Thus, the pH_{IEP} shifts down (Moreno 1992a, b, c).

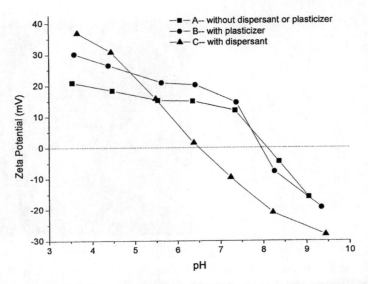

Fig. 2.23 Zeta potential of slurry: A–without dispersant or plasticizer; B–with plasticizer G; and C–with dispersant (ammonium citrate)

Rheological behaviors of slurries with different amounts of dispersant were tested at a shear rate of 50 s^{-1}, as shown in Fig. 2.24. The viscosity of the slurries decreases with increasing dispersant content at a relatively lower dispersant concentration. In this case, increasing the dispersant content gradually augments the concentration of citrate anions around the alumina particle. Thus, it can enhance the Zeta potential and reduce the viscosity of the slurry. However, when the content of dispersant is greater than a certain value of approximately 1 wt% based on the alumina powder,

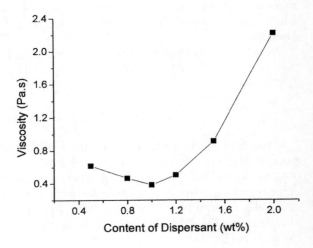

Fig. 2.24 Viscosity curve of slurries with varying content of dispersant at a shear rate of 50 s^{-1}

the concentration of citrate anions around the alumina particle reaches its maximum value. The redundant citrate anions just reduce the thickness of the double layer and the Zeta potential. As the dispersion of alumina particle deteriorates, the viscosity increases rapidly. A slurry with >55 vol% solids loading can maintain good fluidity with the help of 1 wt% dispersant.

(2) **The effects of dispersants on rheological property of slurry with monomer**

Slurry preparation is one of the most important stages in the gel-tape-casting process. The expected slurry has high solids loading with good fluidity and stability. Al_2O_3 powder used in this study has a wide distribution of particle size, which is helpful to get high solids loading. The solids loading of the slurry prepared by this Al_2O_3 powder can be more than 55 vol%. The fluidity and stability of the slurry are influenced by factors such as additives and pH value in the slurry. The effects of these factors on the rheological property of slurry are discussed respectively.

Dispersants are added to slurries to improve the dispersibility of ceramic powders. For ceramic slurries, there is a relationship between stability and Zeta potential of ceramic powders. Powders with high Zeta potential (absolute value) usually exhibit higher stability in slurries than powders with low Zeta potential do. Accordingly, the stability of slurries can be determined by testing the Zeta potential of powders in slurries.

The testing results of the Zeta potential of Al_2O_3 powders according to pH value in slurries are shown in Fig. 2.25. Slurries (solids loading is 55 vol%) added ammonium citrate (SA), tetramethylammonium hydroxide (SN), respectively, as dispersant are compared with the slurry (solids loading is 45 vol%) without dispersant. The absolute value of the Zeta potential of powders in slurries added dispersant is higher than that of powders in the slurry without dispersant. The curve of viscosity variation of these three slurries is shown in Fig. 2.26. It can be seen that the viscosity decreases considerably after the dispersants have been added. This result implies that SA and SN can effectively improve the dispersibility and stability of Al_2O_3 powders in slurries by

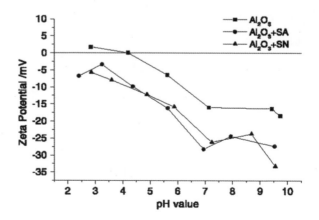

Fig. 2.25 Zeta potential of Al_2O_3 with different dispersants

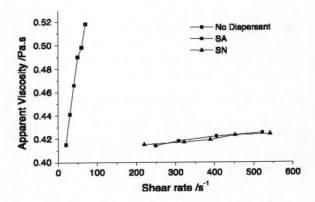

Fig. 2.26 Viscosity of Al$_2$O$_3$ slurry with different dispersants

enhancing the Zeta potential, and will lead to a low viscosity and increased stability of slurries.

It is also found that there is a relationship between the concentration of dispersant and the viscosity of slurries. In order to determine a suitable concentration, a study was carried out on the interaction of concentration and the viscosity. The amounts of dispersant (SA) added in Al$_2$O$_3$ slurries vary from 0.5 to 2 wt%. The relationship between the viscosity and the concentration is shown in Fig. 2.27. The viscosity of slurries decreases with increase in the amount of dispersant, and the minimum viscosity is obtained when the amount of dispersant is about 1 wt%. As the amount of dispersant increases beyond 1 wt%, the viscosity increases again. This phenomenon is due to the following reasons. SA is helpful for improving the Zeta potential of Al$_2$O$_3$ particles. With the increase of SA, Al$_2$O$_3$ particles exhibit an enhanced suspensibility and fluidity, and the slurry viscosity is decreased. When the amount of SA is more than 1 wt%, the concentration of ions of the slurry is increased and leads to the increase of slurry viscosity.

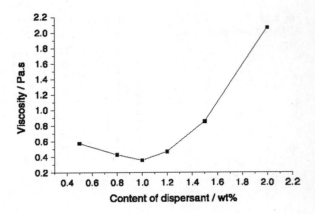

Fig. 2.27 Curve of viscosity of slurry versus content of dispersant (SA)

2.2.3 Influence of Plasticizer on Properties of Aqueous Ceramic Slurry with Organic Monomer

In general, the polymerized green body is too brittle to be rolled and shaped. A plasticizer is used to make the green body soft enough to sustain machining and shaping processes. Glycerol (GY) is selected as the plasticizer in this study. The relationship between the amount of GY added and the viscosity of slurries is shown in Table 2.6. The testing results indicate that with the increase in amount of plasticizer, the viscosity increases, and the green tape is likely to be broken while being peeled from the carrier.

A study was carried out to determine the optimal concentration of plasticizers that can yield a cast tape that is both soft and strong. The flexibility is determined by the number of times the tape can sustain being bent without breaking. With the amount of plasticizer increased from 0 to 7 wt%, the bending time is increased from 1 to more than 50 (Fig. 2.28).

The flexibility of the cast tape is improved gradually with the amount of plasticizer increasing and attains the optimal value when the amount is about 5 wt%. When the amount is more than 5 wt%, the flexibility is kept nearly normal, but the viscosity of slurry increases, which results in a low fluidity of slurry. For this reason, it is optimal to keep the amount of plasticizer less than 5 wt% in slurry to achieve a balance between the flexibility and the strength of cast tape.

The plasticizer has no evident influence on the Zeta potential of the alumina particle, as shown in Fig. 2.23. However, it influences the viscosity of the slurry. The apparent viscosities of slurries with 6, 10, and 20% plasticizer (based on premix solution) have been compared, and the results are shown in Fig. 2.29. The figure illustrates that the viscosity of slurry increases with plasticizer content, although it does not obviously affect the solids loading and maneuverability of the slurry.

Table 2.6 Viscosities of slurries contained organic monomer with different content of GY plasticizer (solid loading 30 vol%)

Shear rate (s^{-1})	Viscosity of slurries with different content of plasticizer (Pa s)		
	6 wt%	10 wt%	20 wt%
0.0999			
0.215			
0.464	0.00201		0.0032
1	0.0016	0.000209	0.0027
2.15	0.00116	0.00286	0.00176
4.64	0.00149	0.00157	0.00231
10	0.00147	0.00184	0.00228
21.5	0.00157	0.00199	0.00236
46.4	0.00162	0.00195	0.0024
100	0.00174	0.00202	0.00254

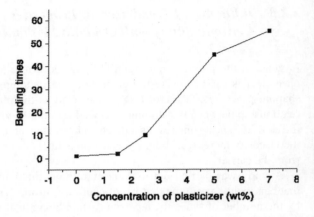

Fig. 2.28 Relationship of bending time and concentration of GY plasticizer

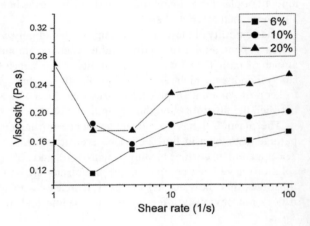

Fig. 2.29 Viscosity of the slurries with 6, 10, and 20% of plasticizer (G) based on the premixed solution. G—plasticizer containing hydroxyl

The increasing viscosity may be attributed to aggregation of alumina particles at higher concentrations of plasticizer.

Addition of a plasticizer is accompanied by a decrease in tape strength. To balance the requirement of strength and flexibility, 2 wt% plasticizer was chosen for this study. In this case, the green tape can be bent to 90° more than 50 times without cracking.

2.2.4 Influence of PH Value on Properties of Slurry with Organic Monomer

The influence of the pH value on the rheological property of slurries is very important and a suitable pH value would enhance the solids loading of slurry. The pH value is adjusted to review the variation of viscosity of the slurry. Table 2.7 shows the testing results of viscosity of Al_2O_3 slurries at different pH values. It can be seen that the

2.2 Characteristics of Slurries Used for Aqueous Gel Tape Casting

Table 2.7 Testing result of viscosity of slurries with monomer at different pH value (solid loading 55 vol%)

Shear rate (s^{-1})	Viscosity (Pa s)				
	pH 2.14	pH 3.9	pH 6.05	pH 7.8	pH 10.11
0.1	29	13	6.61	3.26	1.71
0.164	17	8.26	4.07	1.95	0.916
0.268	12	5.69	2.39	1.22	0.554
0.439	9.02	4.01	1.83	0.783	0.354
0.72	6.13	2.86	1.14	0.506	0.236
1.18	4.31	1.96	0.813	0.342	0.168
1.93	2.98	1.33	0.541	0.231	0.125
3.16	2.14	0.904	0.409	0.165	0.089
5.18	1.47	0.629	0.278	0.118	0.065
8.48	0.965	0.443	0.196	0.0864	0.0482
13.9	0.652	0.319	0.133	0.0668	0.0377
22.8	0.483	0.235	0.0921	0.0553	0.0317
37.3	0.391	0.174	0.0792	0.0488	0.0289
61.1	0.312	0.131	0.0710	0.0462	0.0291
100	0.254	0.101	0.0658	0.0449	0.0314

viscosity drops dramatically with the pH value increasing from 2.14 to 10.11. With the increase in pH value, the absolute value of Zeta potential of Al_2O_3 powders increases significantly (Fig. 2.25), resulting in an obvious improvement in dispersibility of the powders. Therefore, slurries exhibit lower viscosity at higher pH values.

It is obvious that low viscosity is beneficial for making the forming process easily operable. It also enhances the solids loading of slurry. According to the experimental results, the suitable range of the pH value for the Al_2O_3 slurry is about 7–9, and the solids loading of the slurry can reach more than 55 vol% with good fluidity and stability.

2.2.5 Effects of Surfactant on Wetting and Green Tape Releasing (Separating)

Aqueous slurry is hard to overspread on the organic carrier film. Thus, a surfactant is required to improve the wetting property of the aqueous slurry. Introduction of even small amounts of surfactant into slurry can obviously reduce surface tension, which leads to reduction of the contact angle at the interface of the solution and carrier film. Table 2.8 shows the contact angles of slurries on the carrier film when various surfactants are used. It can be concluded from these contact angles that the slurry without surfactant is a hard-to-wet carrier film because its contact angle is larger than

Table 2.8 Contact angle and surface tension of slurries with different surfactants on carrier film

Solution	Contact angle	Surface tension (mN/m)
Deionized water	109.8°	70.86
Slurry without surfactant	103.3°	46.76
Slurry with surfactant A	69.5°	36.05
Slurry with surfactant C	75.0°	–
Slurry with surfactant B	73.5°	24.24
Slurry with surfactant D	72.2°	26.96

Note Concentration of surfactant in solution is 1%, but that of surfactant D is 1%

90°. All these surfactants have an obvious effect on decreasing the contact angle to below 90°.

It is also found that the underside of the green tape often sticks to the carrier film due to the high interface tension, as shown in Fig. 2.30. Various surfactants show different effects on this releasing property. Surfactants A, B, and C did not solve this problem, but made the interface tension stronger. While using surfactant D, the green tapes could be easily released from carrier film and held very smooth surfaces. The difference in the releasing property results from the surfactant structures. In general, surfactant molecules are gathered on the surface of solution during the process – with the hydrophilic group in aqueous solution and the hydrophobic group on the surface – while most common surfactants have an organic hydrophobic group, which has a strong attractive force to the organic carrier film. However, the hydrophobic group of surfactant D demonstrated great electronegativity, and the chemical bonds in this surfactant were like an electrovalent bond to some extent. Therefore, the attractive force between surfactant D and the plastic film is so weak that the green tape can be stripped easily from the carrier film.

Fig. 2.30 Fracture surface of green tape that sticks to carrier film (×200)

2.2.6 Foam and Pore Elimination

It is observed that adding a surfactant into slurry may introduce foams. In particular, for surfactant D, there were many pinholes and cracked bubbles on the surface of the green tape when the foams were not removed from the slurry before casting. Figure 2.31a shows a tape surface with pinholes related to surfactant D.

Regarding the foams, when a surfactant is added into the aqueous solution, the molecules of the surfactant tend to aggregate at the liquid/air interface. As shown in Fig. 2.32a, a layer of liquid film encloses a pocket of air at the surface of the slurry during mixing. The surfactant anions along both sides of the liquid film repel each other to maintain a constant thickness of the liquid film. Thus, a stable foam is formed.

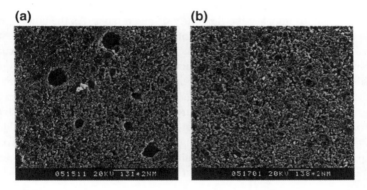

Fig. 2.31 Alumina green tape surface **a** with pinholes and **b** without any pores (×1,000)

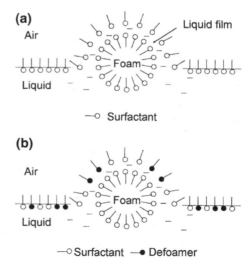

Fig. 2.32 Sketch illustration of **a** foam forming and **b** foam eliminating

Some methods are adopted to eliminate the foams in slurry. Various defoamers are mixed in slurry during ball-milling. When the defoamer molecules dissolve in slurry, they replace part of the surfactant molecules at the surface of the liquid film as shown in Fig. 2.32b. Since the electrostatic repulsion in this local area decreases greatly, the slurry in the liquid film descends by gravity until the film drains off. Then, the foam cracks and a small amount (<1 wt%) of defoamer shows a strong ability to reduce the number of foams.

Before tape casting, a vacuum de-airing process is required. Vacuum de-airing under 0.02 MPa for 25 min can remove almost all the residuary foams in slurry. The green tape after foam elimination has a highly homogeneous microstructure and no pinholes can be seen in the SEM image, as shown in Fig. 2.31b.

Alumina green tapes have been prepared through gel-tape-casting, which combines the merits of gelcasting and tape casting. A slurry with adequate fluidity can sustain a solids loading of more than 55 vol% when 1 wt% dispersant is used. A plasticizer is used to improve the flexibility of green tapes. As the content of plasticizer in slurry increases, the flexibility of green tape is enhanced, but its strength decreases to a certain extent. A surfactant can decrease the surface tension of aqueous slurry, obviously due to its molecular structure, so that the slurry can easily wet the carrier film. Surfactant D can reduce the interface tension between the green tape and organic carrier film dramatically, thus making it possible for the green tape to be released from the carrier film. Foams are produced during mixing, which disrupt the microstructure of the green tape. Foam removal processes, including addition of defoamers in slurry and vacuum de-airing, are adopted to eliminate the foams. A green tape with a high solids loading, high strength, flexibility, and good releasing property is thereby successfully fabricated.

2.2.7 Sintering of Green Tape Prepared by Slurry

To examine the performance of gel-tape-casting process, the cast tape is sintered at 1,550 °C for 2 h. The little amount of organic compounds in cast tape makes it possible for the sample to be sintered directly without a special burnout process. No crack is found in the sintered sample.

The relative density of the sintered sample is 97.7%. The variation of solids loading of slurry (53–55 vol%) shows no influence on the density of the sintered sample. The SEM photograph of the section of the sintered sample is shown in Fig. 2.33. A homogeneous microstructure without exaggerated grain growth is observed.

Ceramic thin sheets of 0.1–1 mm are successfully formed by the gel-tape-casting process. In this process, polymerization of organic monomers is used to form the network in which a ceramic green body is solidified. Special drying and burnout procedures, which are critical stages in conventional tape casting, are omitted, and the production efficiency is highly improved.

Gel-tape-casting was performed with the protection of an inert atmosphere in order to overcome the effect of oxygen anti-polymerization. Al_2O_3 slurry polymerizes

Fig. 2.33 SEM of section of sintered sample

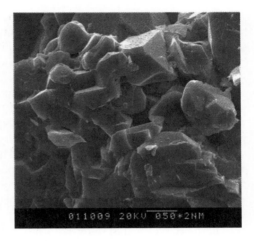

successfully in a few minutes, the temperature being around 50 °C with the protection of nitrogen gas. The relative density of the Al_2O_3 tape is about 61%.

The Al_2O_3 sample is sintered at 1,550 °C for 2 h and relative density of 97.7% is given. A homogeneous microstructure without exaggerated grain growth is observed. The variation of solids loading of slurry (53–55 vol%) shows no influence on the density of sintered sample.

2.3 Aqueous Gel Tape Casting with Styrene-Acrylic Latex Binder

2.3.1 The Importance of Binders in Gel Tape Casting Process

(1) The effects of binder on green tapes prepared by gel tape casting process

Tape casting is a low-cost and useful technique for preparing thin ceramic sheets, and it has been used mainly for producing substrates, solid electrolytes, packages, and multiplayer capacitors for the electronics industry (Hotza et al. 1995; Cui et al. 2003; Kristoffersson and Carlstrom 1997a, b). In general, tape-casting slurry is a typical organic system. Since the volatility and toxicity of organic solvents are harmful to the environment and human health, water-based tape casting is widely used in the ceramics technology (Pagnoux et al. 1998; Doreau et al. 1998, 1999; Gutierrez and Moreno 2001; Briscoe et al. 1998).

An aqueous slurry must be adjusted in order to yield tapes that satisfy some quality criteria, such as high solid loading in water, no defects during drying, cohesion to allow the manipulation of dried sheets, microstructural homogeneity, good thermocompressibility (the object to laminate multilayer materials), easy pyrolysis (burnout), and high mechanical strength of green tapes. This requires careful selection of the slurry additives together with accurate control of many processing parameters.

The binder is very important because it provides strength to green tapes after evaporating the solvent water through organic bridges among the ceramic particles. Usually, there are some water-soluble binders that can be used in aqueous tape casting, e. g, cellulose ethers (such as hydroxyethyl cellulose), polyvinyl alcohols and latexes. All three types of binders have successfully been used in aqueous tape casting. Kristoffersson and Carlstrom (1997a, b) and his colleagues studied the solid volume fraction, green density, and sintering behavior of alumina with the aforementioned binders system. In their experiments, they were able to achieve good surface quality, high green densities, and good laminated properties of green tapes with a pure acrylic emulsion binder.

The objective of this study is to produce good-quality green tapes by water-based tape casting with styrene-acrylic latex binder, and to study the forming film mechanism of latex binder.

(2) **The materials used and tape casting process**

(1) Raw materials

The alumina powder was produced by Xinyuan Company in Henan Province (China). The mean particle size was $D_{50} = 3.84$ μm. The binder was styrene-acrylic latex (Shandong Daming Water-proof Materials Company) that had low foam, viscosity, and good forming-film properties to produce green tapes with good mechanical and lamination behaviors, and decreased the additional organic additives such as plasticizer or defoamer. The glass transition temperature (T_g) is an important parameter for latex, and the T_g of this latex was below room temperature, about 273 K.

The dispersant used for stabilizing the alumina suspensions was PAA–NH$_4$, which is an ammonium polyacrylate salt solution with low molecular weight. The optimal amount of PAA–NH$_4$ was 1.0 wt% with respect to the alumina powder.

Three different concentrations of latex based on the gross volume (alumina powders + binder + water + additives) were 10 vol%, 20 vol%, and 30 vol%, respectively. The Al$_2$O$_3$ volume fraction was maintained at a constant level (43 vol%) and adjusted by deionized water.

(2) Preparation of the slurries and tape casting

The processing steps of the slurry preparation are shown in Fig. 2.34. The slurries were characterized by means of rheological measurements and tested at 25 °C with an MCR 300, GER, using a concentric-coaxial cylinder. The rheological behaviors of the suspensions were characterized by a rheometer in the shear rate range of 0.1–100 s^{-1}.

Tape casting experiments were carried out on a laboratory tape caster (manufactured by Tsinghua University). This device enabled the production of tapes of 120 mm in width and 500 mm in length by fixing the container with two blades. Micrometer screws allowed setting the gap of blades from 0.05 mm up to 2 mm.

2.3 Aqueous Gel Tape Casting with Styrene-Acrylic Latex Binder

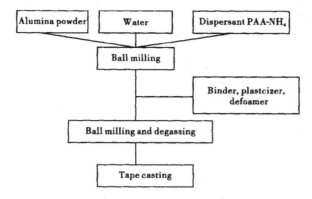

Fig. 2.34 Flowchart of tape casting process

In the process, the support film was of prime importance in order to be stripped easily and get continuous coiling tapes. Thus, we adopted the polyester film by silicon treatment whose thickness was 0.1 mm.

After some previous tests, a gap of 500 μm under the blade and a casting speed of 10 mm s^{-1} was selected for all experiments. According to the empirical Eq. (2.1), the shear rate generated under the blade was estimated to be 20 s^{-1}.

$$\gamma = \upsilon/h \tag{2.12}$$

where γ is the shear rate (1/s), υ the tape casting speed (m/s), and h the blade gap height (m).

The drying process was tested in a container on an electronic balance at room temperature. As shown in Fig. 2.35, the container was fabricated by plastic sheets that were bonded on the support film, whose dimensions were: length = 40 mm,

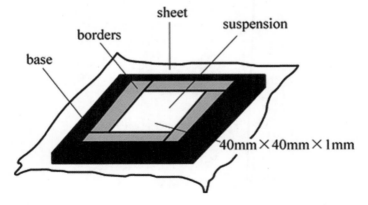

Fig. 2.35 Mold used in the drying experiment

width = 40 mm, and height = 1 mm. After casting the suspension in the container, a function relationship between the weight loss of the samples and time was recorded every two minutes.

The bulk density of the green tapes was calculated by measuring the thickness, diameter, and weight of the disk-shaped specimens. The average density of the unlaminated green sheets was estimated on the base of 10 sample disks that were punched along the dried coiling tapes.

2.3.2 The Forming Film Mechanism of Latex Binder

Latex can be defined as a stable colloidal dispersion of a polymeric substance in an aqueous solvent. The polymer particles are usually spherical in the size range of 30–500 nm. The volume fraction of polymer is generally in the range of 0.4–0.7. The styrene-acrylic latex acts as a binder that is absorbed on the surface of alumina particles in slurries. After the slurries are cast on the support film, the water continuously evaporates from the slurries all along. Finally, the latex particles coalesce and form a polymeric film on the alumina particles when the water is lost. The alumina particles would accumulate by bonding action of the polymer film, and then form the alumina tape. The latex binder forms a film on the surface of alumina particles as can be seen in Fig. 2.36.

The drying process is one of the most important steps in the tape casting process. It enables forming a solid tape from a liquid slip. The film forming mechanisms include two necessary conditions: the compressive force for latex particles results from the solvent evaporation proceed and the glass transition temperature of latex is lower than work temperature (Hong and Feng 2001 (in Chinese); Mengkwang 1996a, b). During

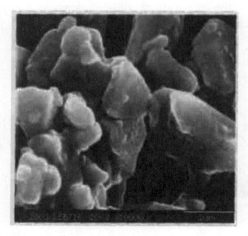

Fig. 2.36 SEM image of latex film on the surface of alumina particles

2.3 Aqueous Gel Tape Casting with Styrene-Acrylic Latex Binder

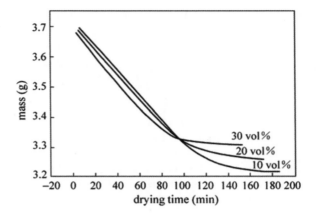

Fig. 2.37 Drying curve of suspensions containing different fractions of latex binder

this drying stage, the drying drive forces compress the latex particles to distort, and form the cross-linking structures in contacting latex particles of the solidified tapes.

Figure 2.37 shows a drying curve of the suspension, which contains different fractions of latex binder under the same conditions. As shown in Fig. 2.37, the gross drying time decreases with the increase in binder fraction in suspensions. The suspension with 30 vol% binder dries quickly and the water loss remains invariable after 160 min. However, the suspension with 10 vol% binder dries slowly and the water loss is held invariably after 200 min. All weight-loss curves follow a linear relationship with time during the initial drying process before 80 min. Therefore, it may be concluded that the latex binder cannot influence the first drying process of the suspensions because it is controlled by the evaporation process.

The drying curves deviate from the straight line after about 80 min. The drying curve of the suspension with 30 vol% binder is the earliest deviation curve, and the water loss rate is the slowest at the second stage among the three types of slurries. This is because the second drying process is controlled by the transportation of the water from the inner to the surface of suspensions after the surface of suspensions dried. The high fraction of latex binder should increase the resistance to diffusion at the second stage. Since the Al_2O_3 volume fraction is maintained at a constant level (43 vol%) and adjusted by deionized water, the suspension with 30 vol% would contain much less water than other suspensions, and therefore, the gross drying time is the shortest.

In summary, the solvent water removal from suspensions involves a two-stage process. The first stage is controlled by the evaporation of the water, while the second stage is controlled by the diffusion process of the water through the solidified part of the film with a decreasing drying rate because of the diffusion resistance. The surface of the tapes dries when the first stage has been finished fast, and therefore the second stage is slower than the first.

During the first drying stage, the solvent water starts to evaporate from the top surface of the tape, allowing particles to approach each other. As a consequence, the curvature of the liquid/vapor interface in the pore channels among particles progressively increases. For a liquid that completely wets the solid surface, the difference in pressure across these curved surfaces, ΔP is given by:

$$\Delta P = -\gamma_{LV}(1/r_1 + 1/r_2) \tag{2.13}$$

where, γ_{LV} is the liquid/vapor interfacial energy, r_1 and r_2 are the radius of curvature of the inside meniscus surface and outside meniscus surface, respectively. When the center of the curvature is in the vapor phase, the radius of curvature is negative and the liquid is in tension ($\Delta P > 0$). This tension of the liquid corresponds to a compressive force between the ceramic particles, which is responsible for shrinkage of the casting slurries. Hence, the ΔP is a driving force for suspension forming film.

During the drying process, the compressive force induced by evaporating will press the latex particles and lead to their distortion, resulting in cross-linking structures in the contacted latex particles of the solidified tapes. However, the ΔP is only one of the driving forces for film formation. If the latex particles are rigid balls (the glass transition temperature is very high), the compressive force cannot press the latex particles until they are distorted, even if the compressive force is very strong. Therefore, the glass transition temperature of latex (T_g) is also important for film formation and should be below room temperature by at least 25 K (Hong and Feng 2001 (in Chinese)).

The glass transition temperature T_g of styrene-acrylic latex is 5 °C, while, the working temperature T is about 25 °C. The difference in temperature ($T - T_g$) is enough to form a film by drying. Since the binder ensures the high strength of the green tapes, and the plasticizer ensures a good toughness of these tapes, they can coil continuously by a roller. The role of the plasticizer is to control the rate of evaporation of water in the drying process and to retain some water in the green tapes after drying.

Hence, the driving force ΔP and T_g are the two key factors for forming a film in the tape casting process.

2.3.3 Rheological Properties of the Alumina Slurries with Binder

The rheological behaviors of the slurries prepared with latex binders always show a typical shear-thinning behavior with different volume fractions of binders (Fig. 2.38a). This behavior agrees well with the requirements for the tape casting process. At a shear rate of 21.7 s^{-1}, the viscosity of the suspension remains approximately 0.2–0.4 Pa s with binder volume fraction of 20 and 30%. Hence, the two slurries are suitable for tape casting. Figure 2.38b shows that the viscosity of the suspensions increases with the styrene-acrylic binder content by keeping constant the solid concentration (constant solid fraction 43 vol% and constant shear rate).

2.3 Aqueous Gel Tape Casting with Styrene-Acrylic Latex Binder

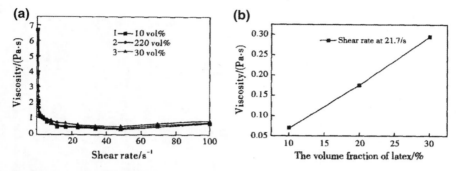

Fig. 2.38 **a** The relationship between viscosity and shear rate with different latex fractions; **b** The relationship between viscosity and different latex fractions at constant shear rate

Fig. 2.39 Density~latex fraction

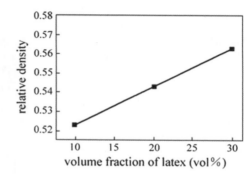

The increase of viscosity results from the increment of latex, thereby increasing the flowing resistance of suspensions.

2.3.4 The Physical Properties and Microstructure of Green Tapes

The bulk density of the alumina green tape increases with the increase in latex volume fraction from 10 to 30 vol% of the entire system. As shown in Fig. 2.39, it results from the increase in gross solid volume (alumina + solid particles of latex), and the compressive force for the latex particles become stronger because the drying driving force increases with the increase in latex volume.

During the drying stage, the compressive force acts on the latex particles and lead to their distortion, resulting in cross-linking structure in the contact latex particles of the solidified tapes. Figure 2.40 shows the homogeneous cross-linking structure in a fracture SEM image of the green tapes.

Fig. 2.40 SEM

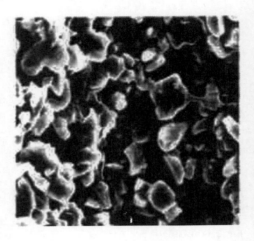

Fig. 2.41 Coiling alumina green tapes

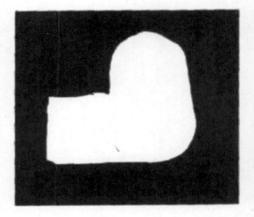

In this experiment, a smooth-surface and high-strength green tape was fabricated by aqueous tape casting with a latex binder system, as shown in Fig. 2.41.

During the experiment, well-dispersed aqueous slurries suitable for tape casting were prepared. The slurries prepared with latex binders exhibited an obvious shear-thinning behavior.

The water-based tape casting has been used successfully to fabricate a smooth-surface and high-strength green tape with styrene-acrylic latex binder. The film drying mechanism involving the following two necessary conditions was studied: the compressive force for latex particles results from solvent evaporation, and the glass transition temperature of latex is lower than the work temperature. The driving force ΔP and T_g are the two key factors for forming a film in the tape casting process.

2.4 A Gel Tape Casting Process Based on Gelation of Sodium Alginate

2.4.1 Why Study on Tape Casting of Sodium Alginate

(1) **Research background**

The tape casting process can be divided into non-aqueous and aqueous systems according to the solvents used. Non-aqueous solvents typically have lower boiling points and can prevent the hydration of some ceramic powders, but require special precaution concerning toxicity and inflammability. In recent years, the environmental and health aspects of the tape casting process have drawn more attention. Therefore, aqueous solvents, which have the advantages of incombustibility, non-toxicity, and low cost, are considered suitable substitutes for non-aqueous solvents. However, an aqueous suspension has a lower tolerance to minor changes in drying conditions. Moreover, the drying rate of the aqueous suspension is invariably much slower than that of the non-aqueous systems (Briscoe et al. 1998). Therefore, non-homogeneous microstructure and cracks usually occur in green tapes. To overcome these drawbacks, Xiang developed a gel tape casting process as an alternative to the process, in which a monomer solution was used as a substitute for organic solvents and binders to coagulate ceramic suspensions (Xiang 2002). However, the monomer used is acrylamide with neural toxicity. In this paper, a natural innoxious polymer, sodium alginate, was used to consolidate ceramic suspensions.

Sodium alginate is a type of gelling polysaccharide most commonly isolated from brown kelp. Sodium alginate can be dissolved in water at room temperature. it undergoes chemical gelation to form a three-dimensional (3D) network in the presence of multivalent cations, particularly calcium (Jin and Lin 1993). Alginates are usually used as processing aids in many traditional ceramic-forming processes, including dry molding, extrusion, and slip casting, where they impart plasticity, workability, suspension stability, and suitable wet and dry strength. In fact, the chemical gelation of sodium alginate has also been used in gel-casting and solid free form fabrication techniques to produce ceramics (Suresh et al. 1998; Xie et al. 2001). In this paper, sodium alginate was used to fabricate green tapes by using a sequestrant to control gelation behavior.

(2) **Experimental procedure**

(A) Starting materials

The starting ceramic powder used was α-Al_2O_3 (>99.7 wt% purity, Henan Xinyuan Aluminum Corporation, China) with a mean particle size of 2.9 μm, and specific surface area 0.434 m^2/g. All additives of the suspensions used in this work are listed in Table 2.9.

Table 2.9 Additives of gel tape casting suspensions

Binders	Sodium alginate
	Styrene-acrylic latex[a]
Dispersant	Ammonium citrate tribasic, $C_6H_{17}N_3O_7$
Calcium salt	Calcium phosphate tribasic, $Ca_3(PO_4)_2$
Sequestrant	Sodium hexameta phosphate, $(NaPO_3)_6$
Acid	Adipic acid, $C_6H_{10}O_4$
Plasticizer	Glycerol

[a]50 wt% polymer, glass transition temperature (T_g) 5 °C

(B) Suspension preparation and tape casting

The process flowchart is presented in Fig. 2.42. Sodium alginate was dissolved in the deionized water to form a solution, and then the styrene-acrylic latex was added into the solution and stirred for 30 min. After alumina powders, dispersant, plasticizer,

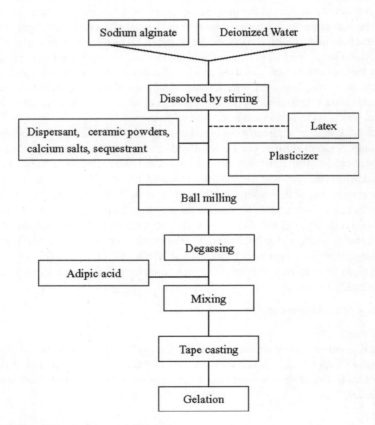

Fig. 2.42 Flowchart of gel tape casting process

2.4 A Gel Tape Casting Process Based on Gelation of Sodium Alginate

sequestrant, and calcium salts were added into the mixture of binders, the suspension was ball-milled for 24 h. Before casting, the resulting suspension was deaired under vacuum (0.02 atm) for 10–15 min and then adipic acid was added into it. The prepared suspension was cast at room temperature, and was subsequently consolidated to form a green tape.

(C) Characterization

The rheological properties of the suspensions were measured using a rheometer MCR300 (Physica Corp., Germany). The microstructure of the green tape was observed by using a scanning electron microscope (SEM, CSM 950). Mercury porosimetry was used to characterize the density of green tapes.

2.4.2 The Preparation of Aqueous Alumina Suspensions with Sodium Alginate and Calcium Phosphere Tribasic

Suspensions with 52 vol% solids loading (50 vol% alumina and 2 vol% calcium phosphate tribasic) and 1 wt% sodium alginate (based on water) were prepared. The effect of the amount of dispersant (based on ceramic powders) on the viscosity of the suspensions is shown in Fig. 2.43. It is obvious that the viscosity of the suspensions decreases with the increase in the dispersant amount. The minimum of viscosity, corresponding to the best dispersion of alumina particles in the suspensions, is observed when ~0.3 wt% dispersant is added. When the amount of dispersant increases beyond 0.3 wt%, the viscosity increases again, which implies that there exists an optimum concentration at which just enough dispersant is present to provide maximized coverage of the ceramic particles and any excess dispersant may be harmful in decreasing viscosity. This optimal dispersant content was used for the preparation of all suspensions in the following experiments.

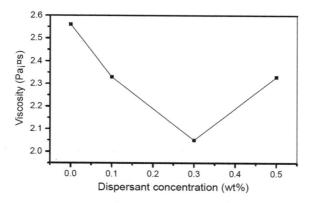

Fig. 2.43 Effects of dispersant content on the viscosity of Al_2O_3 suspensions (shear rate: 50 s^{-1})

Fig. 2.44 Viscosity curves of Al$_2$O$_3$ suspensions with different latex concentrations

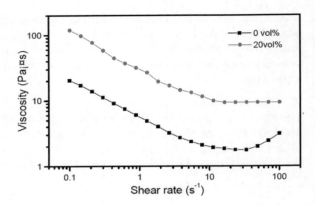

In the range of shear rate tested, the suspension with 52 vol% solids loading and 1 wt% sodium alginate showed an initial shear-thinning behavior at low shear rates followed by a shear-thickening behavior (shown in Fig. 2.44). Such behavior is unfavorable for tape casting because of the high shear stress generated under the blade (~50 s^{-1} in our conditions). The shear-thickening behavior usually occurs in a highly concentrated suspension. The structure of the suspension is determined by the interaction forces between particles (Bender and Wagner 1996; Granja et al. 1997). At low shear rates, the agitation action can gradually break down the structure in the suspension by overcoming the attractive energy and the viscosity of the suspension decrease. At high shear rates, the hydrodynamic forces between the particles become dominant. This will lead to the formation of clusters in suspensions and the viscosity of the suspension would increase.

The green tapes obtained using the above-mentioned suspensions were easily removable from the carrier and no obvious cracks were observed but the tapes were not flexible enough for handling. The plasticizer (glycerol) could not effectively improve the flexibility of green tapes. In our previous study, alumina suspensions with 20 vol% styrene-acrylic latex and 1 vol% glycerol were found to be suitable for tape casting (Cui et al. 2002). Therefore, 20 vol% styrene-acrylic latex (based on sodium alginate solution) was simultaneously added into the suspensions to improve the flexibility of the tapes here. As shown in Fig. 2.44, the addition of latex has a significant effect on the viscosity of the suspensions and the type of their rheological behavior. The shear-thickening behavior observed earlier disappears, and the suspensions containing latex exhibit a shear-thinning behavior more suitable for tape casting.

2.4.3 Control of the Gelation of Sodium Alginate

In order to complete casting processing under control, a controllable reaction of calcium salts with sodium alginate must be considered. Generally, it is difficult to control the gelation rate between calcium salts (such as CaC$_{l2}$ and CaC$_2$O4) and alginate,

2.4 A Gel Tape Casting Process Based on Gelation of Sodium Alginate

which makes it impossible to complete the casting processing in a suitable period of time. In this study, the reaction rate was adjusted by a sequestrant (($NaPO_3$)$_6$) and adipic acid ($C_6H_{10}O_4$). Generally, on adding $Ca_3(PO_4)_2$ and $(NaPO_3)_6$ into the ceramic suspensions simultaneously, the gelation reaction between the calcium salts and sodium alginate will be prevented because a stable complex is formed from the reaction between $(NaPO_3)_6$ and $Ca_3(PO_4)_2$. After $C_6H_{10}O_4$ is added into the ceramic suspensions, the complex decompose and calcium ions are released slowly, thus leading to gelation.

The gelation rate of suspension with 52 vol% solids loading was investigated. The onset of gelation in ceramic suspensions was determined by the changes in suspension temperature, since gelation is an exothermic reaction. The process was monitored in terms of idle time, t_{idle}, the time interval between the addition of the acid and the onset of gelation, which is equivalent to the time available for casting the suspensions during processing. Figures 2.45 and 2.46 show the variation of idle time with the amount of acid and with sequestrant, respectively. These data clearly show that the idle time is reduced by an increase in acid content and is increased

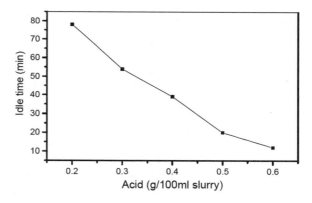

Fig. 2.45 Relationship between amount of acid and idle time

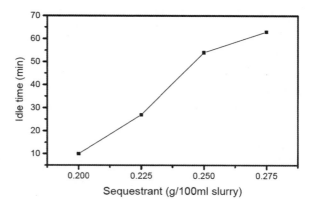

Fig. 2.46 Relationship between amount of sequestrant and idle time

by the increase in sequestrant content. The idle time, i.e., the release rate of calcium ions, can be easily adjusted by the concentration of the reagents, and thus the process can be easily completed under control.

2.4.4 Characterization of Green Tapes

The green densities and the flexibility of green tapes using three formulations of binders are summarized in Table 2.10. Although the tapes with only sodium alginate showed poor flexibility, the tapes with alginate and latex showed good flexibility, similar to the tapes with only latex. The relative density of the green tapes with latex is higher than with only alginate. However, the addition of alginate to the suspensions with latex showed no obvious influence on the density of the green tapes.

In the conventional aqueous tape casting process, the drying procedure is a critical stage. Usually, the drying of green tapes is inhomogenous. During drying, both the anisotropic migration of the solvent and the differential shrinkage will generate stresses in the green tapes and usually lead to cracking. An obvious difference in the microstructures of the upper and bottom layers is usually observed (Zeng et al. 2000; Cui et al. 2002; Bernd and Jurge 2002). In this process, the ceramic powders and organic additives are solidified in the 3D network due to the in situ gelation of sodium alginate simultaneously in the whole tape. Therefore, there is no need for a special drying procedure in which the temperature and humidity are to be controlled carefully in order to avoid crack and inhomogeneity due to migration of solvent carrying binders, plasticizers and fine powders. Figure 2.47 shows the microstructures of the upper and bottom surfaces of a green tape containing 20 vol% latex and 1 wt% sodium alginate. No obvious difference was observed between these two surfaces. This shows that a more homogenous microstructure can be obtained by the addition of sodium alginate.

A novel gel tape casting has been described. In this method, sodium alginate was used to consolidate alumina tapes from aqueous suspensions. However, because of low flexibility of the resulting tapes, styrene-acrylic latex was also added to enhance the flexibility. A special drying procedure, which is critical stages in conventional tape casting, is avoided.

The use of a dispersant can effectively improve the fluidity of the suspension. The optimal amount of dispersant is ~0.3 wt%. The styrene-acrylic latex strongly affects the rheological behavior of the suspensions. Suspensions containing styrene-acrylic latex exhibit a typical shear-thinning behavior suitable for this process.

Table 2.10 Green densities and flexibility of green tapes

Binders	Green density (% TD)	Observation
Alginate	~56	Poor flexibility
Latex	~58	Good flexibility
Alginate + latex	~58	Good flexibility

2.4 A Gel Tape Casting Process Based on Gelation of Sodium Alginate

Fig. 2.47 SEM micrograph of Al_2O_3 green tapes (**a**: upper surface; **b**: bottom surface)

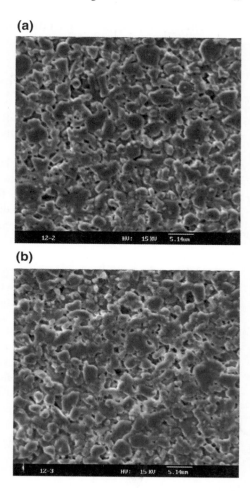

The time available for casting (idle time) can be controlled by varying the concentration of adipic acid and sequestrant. The idle time is reduced by an increase of acid content. It is increased by an increase of sequestrant content. The processing can be easily controlled by changing the quantities of adipic acid and sequestrant.

Green tapes with homogeneous microstructure were fabricated by this novel forming process.

2.5 Spray Trigger Fast-Curing for Gel Tape Casting Process

2.5.1 The Idea of Spray Trigger Fast-Curing

The gel tape casting process includes the following steps: joining the initiator and catalyst into the prepared slurry → placing the slurry into the trough of materials → flow into the tape → heat curing.

Acrylamide polymerization rate is related to temperature. Polymerization is slow at room temperature, and quickly accelerates when the temperature exceeds 40–50 °C. In the molding process, the use of this feature, the curing rate of the slurry can be controlled by adjusting the temperature, the slurry maintain a lower temperature (room temperature or even lower) before the gel casting, avoid curing in the trough of materials; the slurry will be fast solidified into the tape or sheet after being heated to 60–70 °C.

Since the slurry can still be polymerized slowly at low temperatures, the molding operation of the slurry must be completed in a certain period of time and have a certain limitations. Otherwise, the slurry solidification will not take place simultaneously and would lead to cracking or inner stress of the green bodies and in further breakage in the ensuing process. If the slurry comes into contact with the initiator in the course of flow extension, the operation of time constraints can be eliminated. In response to this request and improve and control the slurry polymerization, the spray trigger fast-curing cast process was proposed (Jun-hui 2001a, b; Yong 2000a, b).

This process is still based on gel tape casting. The slurry without initiator flow extension on the baseband first and then initiator is sprayed on the flow extension tape and trigger slurry polymerization, completed curing. In this manner, the onset of slurry solidification can be controlled and the operating hours can be an unlimited extension in theory.

According to the technological features of trigger fast-curing spray-forming extension, an assist device for the spray initiator is installed on the gel tape casting machine (as shown in Fig. 2.48), so that it can meet the needs of the spray trigger fast-curing process.

2.5.2 Outline of the New Process

Figure 2.49 shows the process flow of the Al_2O_3 slurry fast-curing molding by the usage of a spray initiator trigger. In comparison with gel tape casting, this process no longer exists in the operation time limit and thus is more flexible. Table 2.11 shows a comparison between the two processes.

In the spray trigger fast-curing cast extension molding process, the initiator is sprayed on the moving baseband in the course of slurry flow. As only the surface layer of slurry is in contact with the initiator, the thickness of the forming tape must

2.5 Spray Trigger Fast-Curing for Gel Tape Casting Process

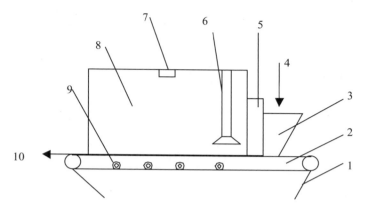

Fig. 2.48 The schematic diagram of the tape casting molding machine for spray trigger fast-curing 1–baseband; 2–baseplate; 3–the material trough; 4–slurry; 5–doctor blade; 6–nozzle; 7–gas inlet; 8–the protective atmosphere; 9–heating wire; 10–green tape

be controlled to ensure the slurry is completely cured. The results show that when the molding thickness is less than 0.3 mm, a smooth paste can be cured completely to form a green tape.

Green tape molding for larger thicknesses can be carried out repeatedly using many times the way.

It must be pointed out that the concentration of the initiator for spray trigger curing should be diluted relatively in order to avoid local polymerization, in fact equivalent to reduce the solid volume fraction in the slurry; therefore the density of the green tape is lower relatively than one of the gel tape casting.

In the spray trigger fast-curing cast molding process, the spray nozzle capable for shaping the quality of green tape is critical. If it is less atomized, larger droplets will exist in the initiator solution, resulting in a partial dilution phenomenon of the slurry, leading to formation of an uneven microstructure of the green tape.

This process is characterized by the following: usage of less organic matter, low level of environmental pollution, fast film-forming of the slurry (to achieve continuous production), shorter drying cycle, and lower production costs. At the same time to avoid limit time in the process of forming, more convenient operation.

2.6 Summary

In this chapter, the traditional tape casting and aqueous gel tape casting (in short form, novel gel-tape-casting) processes for ceramic green sheet production have been introduced systematically. Ceramic sheet products with good quality and high productivity and significantly reducing pollution, lowering costs, and having significant social and economic benefits, have been fabricated by this new gel-tape-casting

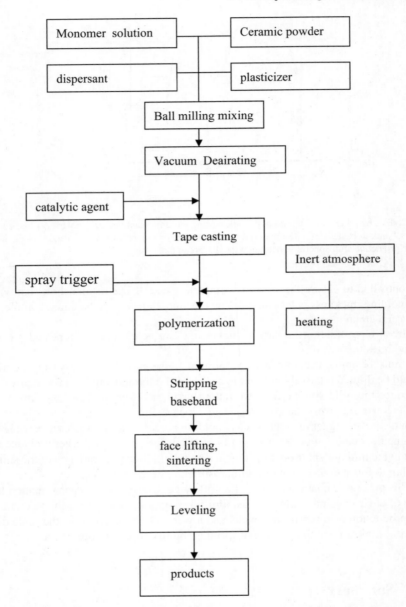

Fig. 2.49 Process flow for the fast-curing molding process by using a spray initiator trigger

2.6 Summary

Table 2.11 Comparison between spray trigger cast gel-forming and gel tape casting

Parameter	Spray trigger fast-curing cast extension molding	Gel tape casting
Solid volume fraction of slurry (vol%)	55	55
Initiator (vol% of slurry)	>1	0.5–1
Thickness of green tape (mm)	<0.3	0.1–1

process. The research studies have demonstrated that this new aqueous tape casting technology is universally applicable for a variety of ceramic powders.

The main features of the contents in this chapter are as follows:

(1) The solidification mechanism of the gel-tape-casting is analyzed. The polymerization rate of AM depends on the properties of the slurry and on reaction conditions. The concentration of the monomer solution should be kept in the range 30–40 wt% in the premixed solution to obtain a suitable polymerization rate. A suitable range of pH values of the Al_2O_3 slurry and temperature is 7–9 and 40–80 °C, respectively. The AM sensitivity to temperature can be used to control the onset of polymerization accurately. For the studied Al_2O_3 slurry, the suitable amounts of initiator and of catalyst were 0.5–1 vol% and ~0.5 vol% of the Al_2O_3 slurry, respectively.

(2) Ceramic thin sheets of 0.1–1 mm have been successfully formed by the gel-tape-casting process. In this process, polymerization of organic monomers is used to form the network in which a ceramic green body is solidified. Special drying and burnout procedures, which are critical stages in conventional tape casting, are omitted, and the production efficiency is highly improved.

SA is used as a dispersant in the Al_2O_3 slurry. It can efficiently improve the dispensability of the Al_2O_3 powder and the fluidity of the slurry. Accordingly, high solids loading slurry with 55 vol% can be obtained. The recommended amount of SA is 1 wt%. GY is used as a plasticizer to ensure that the cast tape is soft enough; however, it increases the viscosity of the slurry. The amount of GY added in the slurry should be kept less than 5 wt% in order to achieve a balance between the flexibility and the strength of the cast tape.

Gel-tape-casting was performed under the protection of inert atmosphere in order to overcome the effect of oxygen anti-polymerization. Al_2O_3 slurry is polymerized successfully in a few minutes, the temperature being around 50 °C with the protection of nitrogen gas. The relative density of the Al_2O_3 tape is about 61%.

The Al_2O_3 sample is sintered at 1,550 °C for 2 h and relative density of 97.7% is given. A homogeneous microstructure without exaggerated grain growth is observed. The variation of solids loading of slurry (53–55 vol%) shows no influence on the density of sintered sample.

(3) Alumina green tapes have been prepared through gel-tape-casting, which combines the merits of gelcasting and tape casting. A slurry with adequate fluidity

can sustain a solids loading of more than 55 vol% when 1 wt% dispersant is used. A plasticizer is used to improve the flexibility of green tapes. As the content of plasticizer in slurry increases, the flexibility of green tape is enhanced, but its strength decreases to a certain extent. A surfactant can decrease the surface tension of aqueous slurry, obviously due to its molecular structure, so that the slurry can easily wet the carrier film. Surfactant D can reduce the interface tension between the green tape and organic carrier film dramatically, thus making it possible for the green tape to be released from the carrier film. Foams are produced during mixing, which disrupt the microstructure of the green tape. Foam removal processes, including addition of defoamers in slurry and vacuum deairating, are adopted to eliminate the foams. A green tape with a high solids loading, high strength, flexibility, and good releasing property is thereby successfully fabricated.

(4) The efficiency of styrene-acrylic binder is obvious for water-based tape casting. The rheological behaviors of the slurries prepared with latex binders always showed a typical shear-thinning behavior with different solid volume fraction at first, and then exhibited shear-thickening from a critical point onward. PAA-NH4 was found to be an effective dispersant for a slurry of latex binder, and the main experimental factors that influenced the stability of the slurry were the properties of the dispersant and the pH value. The research results showed that the glass transition temperature of latex (Tg) was also important for film formation. The drying speed of slurries was largely decided by the drying temperature and the slurry coating thickness.

(5) The well-dispersed aqueous slurries suitable for tape casting are prepared. The slurries prepared with latex binders exhibit an obvious shear-thinning behavior.

Water-based tape casting has been used successfully to fabricate a smooth-surface and high-strength green tape with styrene-acrylic latex binder. The film drying mechanism is studied which involves two necessary conditions: the compressive force for latex particles results from solvent evaporation, and the glass transition temperature of latex is lower than the work temperature. The driving force ΔP and T_g are the two key factors for forming a filming in a tape casting processing.

(6) A novel gel tape casting has been described. Sodium alginate has been used to consolidate alumina tapes from aqueous suspensions. However, because of low flexibility of resulting tapes, styrene-acrylic latex was also added to enhance the flexibility. Special drying procedure, which is critical stages in conventional tape casting, is avoided.

The use of dispersant can effectively improve fluidity of the suspension. The optimal amount of dispersant is ~0.3 wt%. The styrene-acrylic latex strongly affects the rheological behavior of the suspensions. Suspensions containing styrene-acrylic latex exhibit a typical shear thinning behavior suitable for this process.

The time available for casting (idle time) can be controlled by varying the concentration of adipic acid and sequestrant. The idle time is reduced by an increase of acid content or increased by an increase of sequestrant content. The

2.6 Summary

processing can be easily controlled by changing the addition amounts of adipic acid and sequestrant.

Green tapes with homogeneous microstructure were fabricated by this novel forming process.

(7) On the base of gel-tape-casting, another novel process, spray inducing instant solidification of gel tape casting, was developed to take off the limitation of the operating time. This process is fit for the formation of ceramic sheets which thickness less than 0.3 mm.

References

Bender, J., & Wagner, N. J. (1996). Reversible shear thickening in monodisperse and bidisperse colloidal dispersions. *Journal of Rheology, 40,* 899–916.

Bernd, B., & Jurge, G. H. (2002). Aqueous tape casting of silicon nitride. *Journal of the European Ceramic Society, 22,* 2427–2434.

Briscoe, B. J., Lo Biundo, G., & Ozkan, N. (1997). Drying of aqueous ceramic suspensions. *Key Engineering Materials, 132–136,* 354–357.

Briscoe, B. J., Lo Biundo, G., & Ozkan, N. (1998). Drying kinetics of water-based ceramic suspensions for tape casting. *Ceramics International, 24,* 347–357.

Burnfield K. E., & Peterson B. C. (1992). Cellulose ethers in tape casting formulation. In M. J. Cima (Ed.), *Ceramic transactions*. Forming and technology for ceramics (Vol. 26, pp. 191–196). Westerville OH: American Ceramic Society.

Carisey, T., Laugier-Wert, A. H., & Brandon, D. G. (1995). *Journal of the European Ceramic Society, 15,* 1–8.

Chartier, T., & Bruneau, A. (1993). Aqueous tape casting of alumina substrates. *Journal of the European Ceramic Society, 12,* 243–247.

Chartier, T., Penarroya, R., Pagnoux, C., & Baumard, J. F. (1997). Tape casting using UV curable binders. *Journal of the European Ceramic Society, 17,* 765–771.

Cui, X. M., Ouyang, S. X., Yu, Z. Y., et al. (2002). The study of water-based tape casting with latex binder system. In *Proceedings of the 12th China National Conference on High Performance Ceramics*, Fuoshan, China (pp. 133–136).

Cui, X. M., Ouyang, S. X., et al. (2003). A study on LOM with water-based tape casting. *Materials Letters, 57,* 1300–1304.

Descamps, M., Mascart, M., Thierry, B., & Leger, D. (1995). How to control cracking of tape-cast sheets. *American Ceramic Society Bulletin, 74*(3), 89–92.

Doreau, F., Tari, G., et al. (1998). Processing of aqueous tape-casting of alumina with acrylic emulsion binders. *Journal of the European Ceramic Society, 18,* 311–321.

Doreau, F., Tari, G., et al. (1999). Mechanical and lamination properties of alumina green tapes obtained by aqueous tape casting. *Journal of the European Ceramic Society, 19,* 2867–2873.

Egashira, M., Shimizu, Y., & Takatsuki, S. (1991). Chemical surface treatments of aluminum nitride powder suppressing its reactivity with water. *Journal of Material Science Letters, 10,* 994–996.

Fiori, C., Portu, G. (1986). Tape casting: A technique for preparing and studying new materials. In R. W. Davidge (Ed.), *Novel ceramic processes and applications, British ceramic proceedings* (Vol. 38, pp. 213–225). Stoke-on-Trent, UK: The Institute of Ceramics.

Ford, R. W. (1986). *Ceramic drying*. Oxford: Pergamon Press.

Grader, G. (1993). *Journal of the American Ceramic Society, 76,* 1809–1814.

Granja, M. F. L., Doreau, F., & Ferreira, J. M. F. (1997). Aqueous tape-casting of silicon carbide. *Key Engineering Materials, 132–136,* 362–365.

Graule, T., & Grauckler, L. J. (1993). Electrostatic stabilization of aqueous alumina suspension by substituted phenols. In P. Duranand & J. F. Fernandez (Eds.), *Processing of Ceramics, Third Euro-Ceramics* (Vol. 1.1, pp. 491–500). Faenza Editrice Iberica.

Greil, P., Kulig, M., Hotza, D., Lange, H., & Tischtau, R. (1994). Aluminum nitride ceramics with high thermal conductivity from gas-phase synthesized powders. *Journal of the European Ceramic Society, 13*(3), 229–237.

Gutierrez, C. A., & Moreno, R. (2001). Influence of slip preparation and casting conditions on aqueous tape casting of Al_2O_3. *Materials Research Bulletin, 36,* 2059–2072.

Hong, X., & Feng, H. (2001). Doping chemistry. SPCC, 9–20 (in Chinese).

Hotza, D., & Greil, P. (1995). Review: Aqueous tape casting of ceramic powders. *Materials Science and Engineering A, 202,* 206–217.

Hotza, D., Sahling, O., & Greil, P. (1995). Hydrophobing of aluminum nitride powders. *Journal of Materials Science, 30,* 127–132.

Howatt, G. N. (1952). *Method of producing high-dielectric high-insulation ceramic plates*. U. S. Pat., 2,582,993.

Howatt, G. N., Breckenridge, R. G., & Brownlow, J. M. (October, 1945). Report No. 540, Div. 14, N. D. R. C. at Massachusetts Institute of Technology.

Hyatt, E. P. (1986). *American Ceramic Society Bulletin, 65,* 637–638.

Hyatt, T. P. (1995). Tape casting and roll compaction. *Ceramic Engineering Science Processing, 16*(3), 71–75.

Jin, J., & Lin, M. J. (1993). *The application and processing of marine alga*. Beijing: Science Press.

Jun-hui, X. (2001a). Study of a novel process-instant solidification of aqueous gel-tape-casting for ceramic sheet formation, in Chinese, Dissertation of Tsinghua University, Beijing (pp. 40–57).

Jun-hui, X. (2001b). Study of a novel process-instant solidification of aqueous gel-tape-casting for ceramic sheet formation, in Chinese, Dissertation of Tsinghua University, Beijing (pp. 104–106).

Kita, K., Fukuda, J., Ohmura, H., Sakai, T. (1982). *Green ceramic tapes and method of producing them*. U. S. Pat. No. 4,353,958.

Kristoffersson, A., & Carlstom, E. (1997b). Tape casting of alumina in water with an acrylic latex binder. *Journal of the European Ceramic Society, 17,* 289–297.

Kristoffersson, A., & Carlstrom, E. (1997a). Tape casting of alumina in water with an acrylic latex binder. *Journal of the European Ceramic Society, 17,* 56–59.

Landham, R. R., Nahass, P., Leung, D. K., Ungureit, M., Rhine, W. E., Bowen, H. K., et al. (1987). *American Ceramic Society Bulletin, 66,* 1513–1516.

Lee, H. D., Pober, R. L., Calvert, P. D., & Bowen, H. K. (1993). *Journal of Material Science Letters, 5,* 81–83.

Medowski, G. O., & Sutch, R. D., cited in Williams, J. C. (1976). Doctor-blade process. In F. F. Y. Wang (Ed.), *Ceramic fabrication processes. Treatise on materials science and technology*. (Vol. 9, pp. 173–198). Academic Press. New York.

Mengkwang, T. (1996a). *Dope industry* (Vol. 4, p. 32).

Mengkwang, T. (1996b). The role of coalescent of filmforming mechanism in latex pain. *Doping Industry, 4,* 32–35. (in Chinese).

Mister, R. E. (1990). Tape casting: The basic process for meeting the needs of the electronics industry. *American Ceramic Society Bulletin, 69*(6), 1022–1026.

Mister, R. E. (1998). Tape casting: Past, present, potential. *American Ceramic Society Bulletin, 77*(10), 82–86.

Mistler, R. E. (1998). *American Ceramic Society Bulletin, 77,* 82–86.

Mistler, R. E., Shanefield, D. J., & Runk, R. B. (1978). Tape casting of ceramics. In G. Onoda Jr. & L. L. Hench (Eds.), *Ceramic processing before firing* (pp. 411–448). New York: Wiley.

Moreno, R. (1990). *American Ceramic Society Bulletin, 69,* 1022–1026.

Moreno, R. (1992a). The role of slip additives in tape casting technology: Part I—solvents and dispersants. *American Ceramic Society Bulletin, 71*(10), 1521–1531.

Moreno, R. (1992b). The role of slip additives in tape-casting technology: Part II, solvents and plasticizers. *American Ceramic Society Bulletin, 71*(11), 1647–1657.

Moreno, R. (1992c). The role of slip additives in tape-casting technology: Part 1, solvents and plasticizers. American Ceramic Society Bulletin, *71*(11), 1521–1531.

Morris, J. Jr. (1986). *Organic component interactions in tape-casting slips of barium titanate*, Ph. D. thesis, The State University of New Jersey.

Myers, D. (1992). *Surfactant science and technology* (2nd ed., pp. 255–272). New York: VCH.

Nagata, K. (1993). Rheological behavior of suspension and properties of green sheet-effect of compatibility between dispersant and binder. *Journal of the Ceramic Society of Japan, 101*(10), 1271–1275.

Nahass, P., Rhine, W. E., Pober, R. L., Bowen, H. K., & Robbins, W. L. (1990). A comparison of aqueous and non-aqueous slurries for tape-casting. In K. M. Nair, R. Pohanka, & R. C. Buchanan (Eds.), *Ceramic transactions*. Materials and processes in microelectronic systems (Vol. 15, pp. 355–364). Westerville, OH: American Ceramic Society.

Nahass, P., Pober, R. L., Rhine, W. E., Robbins, W. L., & Bowen, H. K. (1992). *Journal of the American Ceramic Society, 75*, 2373–2378.

Pagnoux, C., Doreau, F., et al. (1998). Aqueous suspensions for tape-casting based on acrylic binders. *Journal of the European Ceramic Society, 18*, 241–247.

Scherer, G. W. (1990). Theory of drying. *Journal of the American Ceramic Society, 73*(1), 3–14.

Schuetz, J. E., Khoury, I. A., & DiChiara, R. A. (1987). Water-based binder for tape casting. *Ceramic Industry, 66*(10), 42–44.

Schwartz, B., & Wilcox, D. L. (1967). Laminated ceramics. In *Proceedings of Electronic and Computer Conference* (pp. 17–26).

Smith, D. J., Newnham, R. E., & Yoshikawa, S. (1990). UV curable system for ceramic tape casting. In *IEEE seventh international symposium on applications of ferroelectrics*. IL, USA: Urbana-Champaign.

Stetson, H. N., & Gyurk, W. J. (1967). Development of two microinch (CLA) as-fired alumina substrates. *The American Ceramic Society Bulletin, 46*(4), 387–389.

Suresh, B., Manupin, G. D., & Graff, G. L. (1998). Freeform fabrication of ceramics. *American Ceramic Society Bulletin, 7*, 53–58.

Tadros, T. F. (1984). *Surfactants* (pp. 287–322). New York: Academic Press.

Waack, R., Venkataswamy, K., Novich, B. E., Halloran, J. W., Egozy, A. R., Hodge, J. D., et al. (1988). *Polymerizable binder solutions for low viscosity, highly loaded particulate suspensions, and methods for making green articles there from by tape casting or injection molding*. U.S. Pat. PCT/US88/01232.

Wayne, N. J. (1969). *Chemistry of acrylamide*. Process Chemicals Department. Parsippany, N. J.: American Cyanamid Co.

Xiang, J. H. (2002). Processing of Al_2O_3 sheets by the gel tape casting process. *Ceramic International, 28*, 17–22.

Xie, Z. P., Huang, Y., Chen, Y. L., & Jia, Y. (2001). A new gel casting of ceramics by reaction of sodium alginate and calcium iodate at increased temperatures. *Journal of Material Science Letters, 20*, 1255–1257.

Yong, H., Junhui, X., Zhi-peng, X., Jin-long, Y. (2000a). *The thin ceramic green device body of rapid curing and forming in situ through spray initiator*. China Invent Patent ZL00107495.4.

Yong, H., Junhui, X., Zhi-peng, X., Jin-long, Y., Shaorong, C. (2000b). China Invent Patent ZL00102922.3.

Yoshikawa, S., Heartling, C., Smith, D. J., & Newnham, R. E. (1988). Patterned ceramic green films using UV curable binders. In G. L. Messing (Ed.), *Ceramic powder science III* (pp. 533–560). Westerville, OH: American Ceramic Society.

Zeng, Y. P., Jiang, D. L., & Greil, P. (2000). *Journal of the European Ceramic Society, 20*, 1691.

Zhengqi, Q., Xingnan, Q., Panfa, H. (1996). *New type of ceramic material manual*. Jiangsu Science and Technology Publishing House, 10.

Chapter 3
Gelation Forming Process for Toxicity-Free or Low-Toxicity System

Abstract In this chapter, the forming process of ceramics by gelling of agarose, gelatin, and sodium algaecide is presented, and another gelcasting system with low toxicity by using HEMA is also described. This study shows that gelled green bodies by the addition of agarose of 0.7–1.05 wt% and a solid loading of 53–55 vol% had a homogeneous density distribution and a green strength >3 MPa. Some simple and complex parts, such as a turbine rotor, were formed by this method. Gelatin can be dissolved in water above 40 °C, and the solutions gel at 15–20 °C. Wet green compacts can be formed from the alumina slurry containing 4–5 wt% gelatin (based on water) by gelation. A novel casting forming method by gelatin and enzyme catalysis is also described. As urease is added into the suspension, the gelatin molecules attract each other and change to a three-dimensional (3D) network structure through hydrogen bonding due to urea decomposition by urease. Then, the slurry containing ceramic powder and gelatin will be consolidated in situ by the gelation under ambient conditions. For another gel-casting process by alginate, in which solidified agent and chelator are added into sodium alginate solution simultaneously; thus, the gelation between calcium ions and sodium alginate is avoided in this stage before casting. Free calcium ions are released when hexanedioic acid is added into the suspension. They react with sodium alginate and form a 3D network. Therefore, the ceramic particles are fixed in this 3D network. The slurry is consolidated and near-net-shape green parts are formed. In addition, a new innoxious gelcasting system has been successfully developed by using HEMA or HEMA-PVP, by which dense and homogeneous complex-shaped alumina and silicon carbide parts have been successfully fabricated.

Keywords Ceramic slurry · Gelation · Agarose · Gelatin · Sodium algaecide · HEMA

Gelcasting has been widely studied during the last decade (Alford et al. 1987; Lange 1989; Reed 1994; Zhang et al. 1989; Xie et al. 1995). In this process, the slurry made from ceramic powder and a water-based monomer solution is poured into a mold, and then polymerized in situ. The gelled part is removed from the mold while still wet, and then it is dried and fired. The dried green body is strong enough to be machined. However, the process is not perfect in that the polymerization of monomers is difficult to control in the ceramic suspension. Reductive agents (acrylamide is the commonly used agent) usually restrain the free-radical polymerization. In addition, acrylamide is highly toxic. Therefore, a new gelcasting process that is toxicity-free or has reduced toxicity has been investigated in this study.

Many polymer solutions, such as agar, agarose, gelatin, carrageenan, and sodium algaecide, can gelate under suitable conditions (Whistler 1973; Amott et al. 1974). Some of them have been employed in the food industry (Amott et al. 1974). When the polymer is dissolved in water, the molecular chains attract each other and form a 3D network by hydrogen bonds or van der Walls forces. The gelling property of agar has been used in water-based injection molding (Hoffman 1972).

Millan et al. (2002a part I, b part II) studied the influence of aqueous solutions of agar, agarose, and carrageenan on the rheological properties of alumina slurries, and the mechanical and microstructural characterization of green alumina bodies. Gelcasting performance is a function of slurry viscosity, gelling behavior and time, and body deformation during drying. In general, carrageenan leads to higher shape distortion in the body. Better shape retention is achieved with precursor solutions with ≥ 3 wt% additive and total content of gel ≥ 0.5 wt%. Values of bend strength up to 4 MPa are obtained, which are significantly higher than those corresponding to slip-cast alumina without gelling additives, and they increase with the final concentration of the polysasccharide, while Young's modulus values are mainly influenced by the concentration of the additive in the precursor solution.

Santacruz et al. (2002) studied the rheological characterization of synergistic mixtures of carrageenan and locust bean gum for the aqueous gelcasting of alumina. The rheological behavior of 2 wt% solutions of these polysaccharides and their mixtures were measured under mixing conditions (60 °C) and by recording the viscosity and elastic modulus on cooling. The effect of the addition of these solutions to 50 vol% alumina slurries up to a concentration of 0.5 wt% has been studied. Although the gelling time increased, the resulting gels were stronger than for carrageenan alone. Gelcast alumina bodies with green and sintered densities of 57 and 97.6% of theoretical values have been obtained.

In this study, the forming process of ceramics by the gelling of agarose, gelatin, and sodium algaecide is presented, and another gelcasting system with low toxicity by using HEMA is also described.

3.1 Gelation Forming of Ceramic Suspension with Agarose

3.1.1 Characteristics of Agarose

Agarose is soluble in hot water and forms a gel on cooling. The characteristic strengths of the gels produced by these gelling agents vary greatly, and agarose is the strongest. Fanelli et al. (1989) examined the aqueous injection molding of mixes of agar, agarose, and alumina powder (33–39 vol%) with an apparent viscosity of

3.1 Gelation Forming of Ceramic Suspension with Agarose

Table 3.1 Chemical composition of alumina powder

Composition	Al_2O_3	$\alpha\text{-}Al_2O_3$	SiO_2	Fe_2O_3	Na_2O	B_2O_3	Weight loss
Content, wt%	99.7	≥95	0.05	0.03	0.05	0.04	0.10

300 Pa s or more. The purpose of the present work was to investigate the gelation casting of ceramic suspensions of low apparent viscosity with agarose as a binder for producing homogeneous complex-shaped green bodies.

The alumina powder used was a commercial grade An-05 (from Henan Xinyuan Alumina Industry Co. Ltd, China), prepared by calcining $\gamma\text{-}Al_2O_3$ in a high-temperature revolving furnace. The chemical composition is shown in Table 3.1. The average particle size was ~1.8 μm; PMAA-NH$_4$ of molecular weight ~10,000 was used as a dispersant.

Agarose, a purified derivative from agar, was used as a gelation agent. Analysis of the agarose revealed 0.2 wt% ionic impurities in the form of SO_4^{-2} and an ash content <1 wt%. In general, agarose molecules are stable and do not decompose over the pH range 4–10. Agarose absorbs water and swells in water at ambient temperature, and dissolves in hot water at ~80 °C. For the agarose used, the gelation strength of a solution with 1.5 wt% agarose reached more than 600 g cm^{-2}.

The zeta potentials of alumina powder with and without dispersant were calculated from the measured electrophoretic mobility by using electrophoretic light scattering (ELS) with a 5 mW laser source. A suspension of 0.05 vol% concentration was chosen.

The rheological properties of the agarose solution and ceramic suspensions, prepared at pH 7, were measured with a rotary rheometer (NSX-11, ChengDu). The gel strength of the agarose solution was measured using a technique commonly employed in the industrial gum industry, the details of which are described in the reference (Whistler 1973). The green body strength was measured by the three-point bending tests on sample bars of 6 × 5 × 30 mm^3. For comparison, green samples without agarose were prepared by centrifugal casting and had a density of 67.5% of the theoretical value. Thermogravimetric analysis was carried out by using a thermogravimetric analyzer (TGA92-18, SETARAM Corp., France).

Agarose is composed of alternate units of two forms of the sugar molecule. The polymer chains form rigid bundles of twisted helical chains, and by hydrogen bonding, the hydroxyl groups along the polymer backbone interact strongly with water molecules in the gel network (Amott et al. 1974). The apparent viscosity of an aqueous agarose solution is shown in Fig. 3.1. As can be seen, the viscosity increased with a decrease in temperature, and reached >10 000 MPa s near the gel point. At the gel point, gelation resulted in a tough mass. The gel strength increased linearly with the agarose concentration. For example, the gel strengths with 1.5 and 3 wt% agarose were ~650 and 1,300 g cm^{-2} respectively. Figure 3.2 illustrates the rheological behavior of a 3 wt% agarose solution at temperatures above the gelation point. Shear-thickening and slight increase in viscosity with the shear rate were observed. This is similar to the results on colloidal stable PVC suspensions reported earlier

Fig. 3.1 Apparent viscosity of a 3 wt% agarose solution

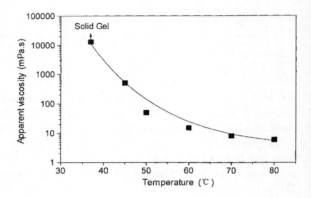

Fig. 3.2 Rheological behavior of a 3 wt% agarose solution

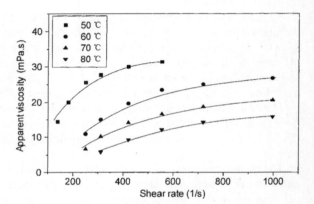

by Homann (1972). In general, shear-thickening often occurs in concentrated suspensions and monodisperse systems. Several researchers (Boersma et al. 1991) have suggested that shear-thickening can be explained in terms of the balance between the viscous forces and the interparticle repulsive forces.

3.1.2 The Effect of Agarose Contents on the Rheology of Aqueous Ceramic Suspensions

The variations of the zeta potential with the pH for as-received alumina and the suspension containing PMAA–NH_4 as dispersant are shown in Fig. 3.3. The isoelectric point (IEP) of the alumina powder was ~pH 5.3. After adding PMAA–NH_4, the IEP of the suspension changed to pH 3.1 due to adsorption of the dispersant on the alumina particles. In addition, the zeta potential increased exponentially in the range pH $\leq$ 5.3. This is very helpful for making slurries with low viscosity in deionized water, whose pH value is nearly 7.

3.1 Gelation Forming of Ceramic Suspension with Agarose

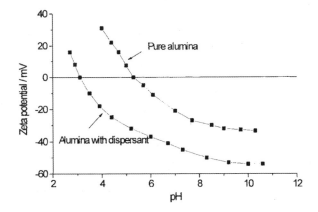

Fig. 3.3 Variation of zeta potential with the pH for as-received alumina and alumina with dispersant

The rheology of aqueous ceramic suspensions generally depends on solids content, particle characteristics, and binder type. Figure 3.4 shows that the viscosity varied with the shear rate for a solid loading of 55 vol% and agarose content of 0.7 wt% (Al_2O_3 basis) at a pH of 7. For the agarose solution, the suspension showed shear-thickening. This is due to a combination of the high alumina solid loading and the agarose solution. Previous studies (Yang et al. 1998) have also shown that alumina suspensions exhibit shear-thickening at high concentrations >50 vol%, and shear-thinning at low concentrations, respectively.

It is important to have low viscosity and high solid loading for the suspension used in gelation forming and these are influenced by the type and amount of binder. Figure 3.5 illustrates the effect of agarose additions on the viscosity of a suspension containing 53 vol% solid content. The viscosity was <1 Pa s above the gelation temperature for agarose additions from 0 to 0.7 wt% (Al_2O_3 basis). At 1.05 wt% addition, the viscosity increased considerably and casting became difficult. The agarose addition for solid loading >50 vol% should therefore be <1.05 wt%. The influence of

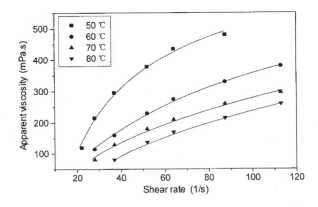

Fig. 3.4 Viscosity versus shear rate for a suspension with 55 vol% solid loading and 0.7 wt% agarose (Al_2O_3 basis)

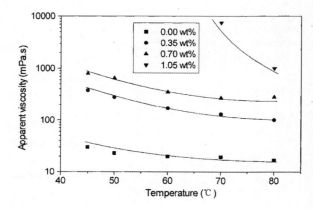

Fig. 3.5 Effect of agarose content on viscosity of suspension with 53 vol% solids loading

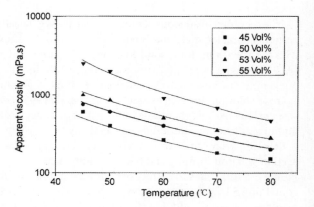

Fig. 3.6 Influence of alumina solid content on viscosity of suspension with 0.7 wt% agarose

solid content on viscosity is shown in Fig. 3.6. Suspensions containing <53 vol% and 0.7 wt% agarose exhibited good fluidity above the gel point.

3.1.3 The Forming Courses of the Aqueous Ceramic Suspensions with Agarose

(1) Forming

The flowchart of the forming process is shown in Fig. 3.7. Agarose and dispersant (PMAA–NH$_4$) were added to deionized water, and then mixed with alumina powder. The resulting slurry was then ball-milled for 24 h. After degassing in a rotary evaporator under vacuum, the slurry was heated to ~80 °C in a water bath to dissolve the agarose completely, to prevent water from evaporating. This was done in a closed chamber. The warm slurry was then cast into a nonporous mold, which could be

3.1 Gelation Forming of Ceramic Suspension with Agarose

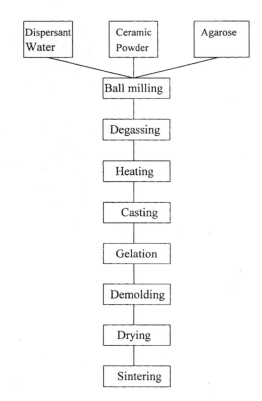

Fig. 3.7 Flowchart of the forming process for alumina slurries with agarose

made of metal, glass, or plastic. When cooling to the agarose gel point, the slurry gelled to the shape required, and then the wet green body was demolded.

A system containing 53 vol% solids content and 0.7 wt% agarose (Al_2O_3 basis) is suitable for gelation casting processing. Some simple and complex parts, such as turbine rotors (Fig. 3.8), were formed by this system. For forming of these parts, it should be noted that the mold needs to reach a temperature above the gel point in order to prevent the slurry fluidity decreasing significantly during the mold filling due to gelation. In addition, the wet green parts should be demolded after a tough gel is formed. In general, several hours (4–8) are needed to obtain a gel strength high enough for complex parts, such as turbine rotors and blades, to be removed from the mold. After demolding, the wet green parts have accurate dimensions and a very smooth surface, similar to an injection-molded body made with thermoplastic vehicles.

(2) **Drying and compact characteristics**

Linear shrinkage measurements were performed on gelled cake 40 mm in diameter and 20 mm thick. Figure 3.9 shows the average linear shrinkage in diameter and the resulting relative green density versus the solid loading during drying. The results illustrate that the green bodies with more than 50 vol% solid loading had shrinkage

Fig. 3.8 Turbine rotor (110 mm dia.) shaped by gelation forming

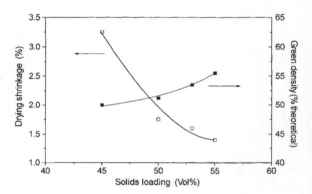

Fig. 3.9 Drying shrinkage and relative density versus the solid loading

lower than 1.7%, which increased to 3.2% for 45 vol% solid content. After drying, the green bodies gelled from the suspensions with 50, 53, and 55 vol% solid content, respectively, displayed 51, 54, and 56% theoretical densities. However, the green density for a solid loading of 45 vol% reached 50% theoretical density because of greater shrinkage upon its drying. In general, a larger shrinkage of the green body tends to cause deformation or cracking especially in complex-shaped parts. This implies that solid loading above 50 vol% is needed for successful processing.

The demolded wet green parts can be dried at ambient temperature or in a drying chamber without cracking or damage. Control over the starting alumina particle size and the resultant homogeneous pore size distribution in the green part is responsible for the rapid drying. Analysis by the Autoscan Mercury Porosimeter (Quantachrome Corp., USA) showed that a narrow, monodispersed pore size distribution was obtained in the green body, whereas the same alumina powder compacted by isostatic pressing exhibited a wide pore size distribution with double peaks (Yang et al. 1998).

A uniform density distribution was also observed for the complex-shaped green compacts. From density measurements of the turbine rotor shown in Fig. 3.10, for

3.1 Gelation Forming of Ceramic Suspension with Agarose

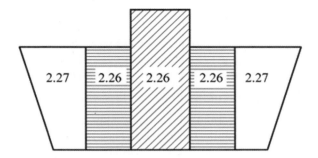

Fig. 3.10 Density distribution in green body of turbine rotor

example, the density of ~2.26 g cm^{-3} was the same in different parts of the rotor. This implies that gelation forming from the slurry to the green compact was completed homogeneously.

The green strength increased with agarose content as illustrated in Fig. 3.11. For example, the green strength, for a green body gelled from a concentrated suspension containing 0.7 wt% agarose (Al$_2$O$_3$ basis), reached 3.2 MPa, and compared with 1.5 MPa for a green sample formed by centrifugal casting without agarose.

(3) Sintering

Binder burnout is normally carried out before sintering for some forming methods such as injection molding, tape casting, and gel casting where the binder content ranges from 4–18 wt%. In the case of the present forming method, the gelled green body could be sintered directly because the organic binder addition was <1.05 wt%, Thermogravimetric analysis for a green body containing 53 vol% solids loading and 0.7 wt% agarose showed that weight loss started at 250 °C and was completed at ~500 °C. For the green turbine rotor, sintering was carried out at a heating rate of 60 K h^{-1} with a hold of 2 h at 1,600 °C. The sintered sample had a 97.8% theoretical density and a 16% linear shrinkage in diameter, which is comparable with that for samples made by die pressing. According to many reports, macro defects, such as

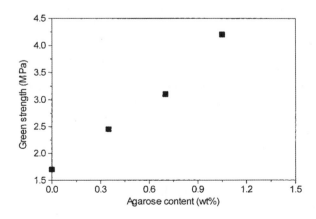

Fig. 3.11 Green strength versus agarose content (Al$_2$O$_3$ basis)

cracks, often occur in the center of turbine rotors prepared by injection molding by using organic-based vehicles due to nonuniformity in both density distribution and shrinkage. However, no cracks were observed in sections of the sintered rotor shaped by the gelation forming method. This is attributed to its uniform distribution of composition and density in the green state.

(4) **The basis of the above can be seen**

The gelation forming of ceramic compacts by agarose gelation has very good near net shape capability for complex-shaped parts requiring precise dimensions and very smooth surface finish. For the casting of alumina, an agarose addition of 0.7–1.05 wt% and a solid loading of 53–55 vol% were found to be successful. The gelled green bodies had a homogeneous density distribution, and a green strength >3 MPa. Green parts, such as turbine rotors, can be sintered directly without special binder burnout procedures, and without deformation or cracking in the finished component.

3.2 Alumina Casting Based on Gelation of Gelatin

3.2.1 Characteristics of Gelatin

(1) **Why study gelatin**

The gelation property of agarose has been used in water-based injection molding (Fanelli et al. 1989; Zhang et al. 1994) and gelcasting (Xie et al. 1999) of alumina. In these cases, a well-dispersed suspension of alumina and agarose powder is heated at 85–90 °C to dissolve the agarose completely and then this is cast into a nonporous mold. At about 35 °C, the agarose gels, enclosing the ceramic powder and water in its gel network to form a green body with the desired shape.

The drawback in the case of agarose is that the solving temperature is fairly high (~85 °C). During the process of heating, the water vaporizes heavily, resulting in agglomerates and flocculation, which increases the viscosity of the slurry.

Compared with agarose, gelatin (protein polymer composed of various amino acids) has some excellent properties (Ward and Courts 1977). The most attractive is that it can be dissolved quickly at a rather low temperature (about 40 °C). Our study investigated the gelling property of gelatin and the influence of gelatin on an alumina ceramic suspension. Rigid alumina green bodies were also prepared by in situ coagulation of slurries containing gelatin.

(2) **Properties of gelatin**

The gelatin powder with molecular weights ranging from 15 000 to 90 000 was provided by Shanghai Chemical Reagent Works. The analysis of the gelatin is shown in Table 3.2. A 1 wt% gelatin solution changes into a strong and stable gel over a wide pH range of 4–10 when cooled.

3.2 Alumina Casting Based on Gelation of Gelatin

Table 3.2 Impurities in gelatin powder

Composition	Chloride (Cl$^-$)	Heavy metal ash (Pb)	Ash	Sulfur dioxide (SiO$_2$)	Arsenide (As)	Iron (Fe)
Content (<wt%)	0.2	0.008	2	0.015	0.0003	0.007

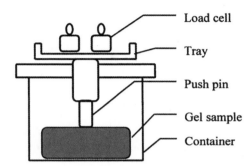

Fig. 3.12 Device for measuring gel strength

The composition of the alumina powder used here has been described in Sect. 3.1.1 of this chapter.

The rheological properties of the gelatin solution and mixed ceramic slurry were measured by a rotary rheometer (NXS-11, Chengdu, China). The gel strength was examined by a simple device usually used in the gum industry (Whistler 1973) (shown in Fig. 3.12). Supposing the total weight recorded on the load cell was W when the push pin broke the surface of the gel sample, and the cross-sectional area of the push pin was A, the gel strength could be calculated using the formula $G = W/A$, where the unit of G is gram per square centimeter (g cm^{-2}).

A dried green body was cut into sample bars of $6 \times 5 \times 36$ mm^3. A sintered plate was cut into samples of $3 \times 4 \times 36$ mm^3 and then polished. Their mechanical strengths at room temperature was determined by the three-point flexure test with space of 30 mm (crosshead speed of 0.5 mm min^{-1}). The microscopic morphology was observed by a scanning electron microscope (SEM) (OPTON CSM950).

(3) The effect of temperature change on viscosity of gelatin solution

The change in the apparent viscosity of a 6 wt% gelatin solution with temperature is shown in Fig. 3.13. Above 25 °C, the apparent viscosity changes only slightly, ranging from 2 to 8 MPa s. Above this temperature, the shape of the polymer molecule changed from random clew ball to single-helix chain. When the temperature dropped below 25 °C, the single helixes entangled with each other, resulting in an obvious rise in the apparent viscosity. With the apparent viscosity rising sharply to more than 20 Pa s, a 3D network was formed by the attraction between the molecular chains. At the joints of the gel network, the molecular chains were ordered into microcrystalline zones (Zhou et al. 1996; Jin and Lin 1993; Ward and Courts 1977).

Fig. 3.13 Change in apparent viscosity for a gelatin solution (6 wt%) with temperature

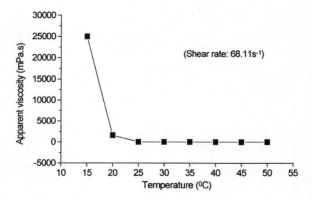

(4) The effect of the gelatin content on gel strength

The gel strength increased with the increase in the gelatin content (Fig. 3.14). As more microcrystalline zones were formed by the gelatin polymer chains, the structure of the gel network became tighter and the gel strength increased. Compared with agarose, the strength of the gelatin gel is relatively low (Fanelli et al. 1989; Zhang et al. 1994). With the content of 3 wt%, the strength of agarose gel is greater than 1,000 g cm^{-2}, while the strength of the gelatin gel is only 100 g cm^{-2}. Since the strength of the gel, along with the solids loading, ultimately determines the strength of the wet and dry green body, adding more gelatin into the slurry to increase the strength of the green body is more effective. However, in order to realize the forming process of gelcasting, the slurry should have an appropriate viscosity and fluidity. Therefore, the content of gelatin must also be determined from the rheology of the slurry besides the strength of green body.

Fig. 3.14 Variation in the gel strength with the gelatin concentration

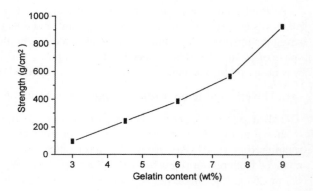

3.2.2 The Gelation Process of the Ceramic Slurry with Gelatin Solution

(1) The forming process of the alumina slurry with gelatin solution

The forming process is shown in Fig. 3.15. First, the alumina powder and dispersant were added to deionized water, ball-milled for more than 12 h, and then mixed with the gelatin solution prepared by dissolving gelatin powder in deionized water at 40–50 °C. After degassing by vibrating or stirring under a vacuum (10 mm Hg), the warm slurry was cast into a nonporous mold that could be made of steel, glass, or polyester plastics. Cooling down to the gel point of the gelatin, the slurry coagulated to form a green body. After demolding and drying under 80 °C for 24 h, the green parts could be sintered without special binder removal processing.

(2) Rheology of the alumina slurry with gelatin solution

Figure 3.16 shows the flow curves of the Al_2O_3 suspensions with different gelatin concentrations at 30 °C. The pure alumina slurry with a volume fraction of 53 vol%

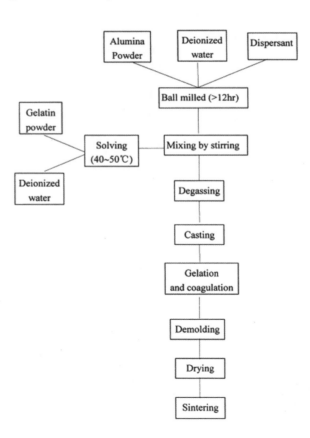

Fig. 3.15 Flowchart of the forming process for alumina slurries with gelatin

Fig. 3.16 Effect of the gelatin concentration on the viscosity of the suspension containing 53 vol% solids loading

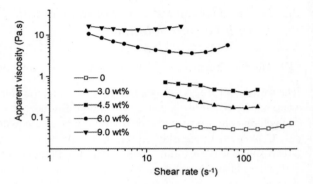

had an average viscosity of 50 mPa s. On adding gelatin, the viscosity increased greatly. When the content of gelatin was less than 4.5 wt% (based on water), the apparent viscosity was lower than 1 Pa s, which allowed the casting process to take place. Slurries with gelatin content greater than 6 wt% are thick and cannot flow into the mold easily, as their apparent viscosity is higher than 3 Pa s. Under such conditions, agglomerates or flocculation may occur and the homogeneous microstructure of the ceramic body will be destroyed.

The rheological properties of a slurry with 53 vol% Al_2O_3 and 3 wt% gelatin at temperatures are shown in Fig. 3.17. Before gelation, the slurry is a pseudoplastic fluid, because the gelatin polymer chains could turn and transform to decrease the shearing resistance. Since the slurry degassing and casting are usually performed under dynamic conditions rather than static, pseudoplasticity is helpful.

Fig. 3.17 Rheological behavior of the slurry with 53 vol% alumina and 3 wt% gelatin

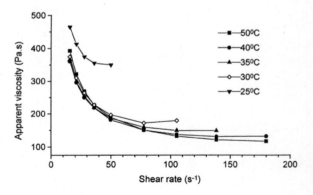

3.2.3 Preparation of Green Body Using Slurry with Gelatin Solution

(1) Forming of a wet green body

On casting the warm slurry of alumina and gelatin into a nonporous mold, it coagulated at the gel point of gelatin. The change in the apparent viscosity of the slurry (Fig. 3.18) is similar to that of a pure gelatin solution (Fig. 3.13). Above the gel point, the slurry had a relatively low viscosity (less than 200 MPa). At a temperature between 15 and 20 °C, the viscosity rose abruptly to more than 30 Pa s.

It has been reported that the network of gelatin gel contains many microcrystalline zones, which act as the joints of gel. After gelation, the polymer chains continued to rearrange in the network to form more microcrystalline zones, and thus increased the rigidity of the gel.

The wet green body was very weak and had low strength when just formed. It could not be demolded until the structure of the gelatin gel developed through the arrangement of polymer chains. A simple ceramic part, such as a disk 150 mm in diameter and 20 mm in thickness, could be demolded 6–8 h after slurry casting. For a complex compact, such as a turbine blade, nearly 24 h was required in order to avoid deforming or cracking of the wet body when demolding. In general, the demolded bodies had precise size and very smooth surface if the surface of the molds was smooth.

(2) Properties of the dry bodies and sintered bodies

The green body has a shrinkage of 2–3% (linear) during drying. Figure 3.19 shows an SEM photocopy of a dry body. It can be seen that the powders are connected by slender polymer chains, which are responsible for the strength of the green body. The influence of the gelatin content on the strength of the dry body is shown in Fig. 3.20. The strength is greater than 8 MPa when the gelatin content is more than 4.5 wt%. Although the body is stronger with more gelatin (with gelatin content higher than 4.5 wt%, the strength is greater than 8 MPa), the error bar is also enlarged, indicating

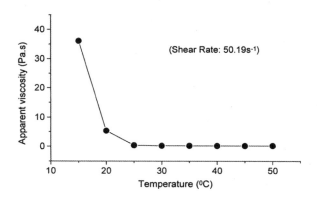

Fig. 3.18 Change in apparent viscosity for the slurry with 53 vol% alumina and 3 wt% gelatin

Fig. 3.19 SEM micrograph of a dry green body

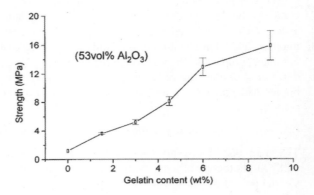

Fig. 3.20 Influence of the gelatin content on the strength of a dry green body

that homogeneity is reduced. This may be attributed to the incomplete degassing, inhomogeneous slurry mixing, or the presence of agglomerates resulting from the higher viscosity.

Regarding the fluidity of the slurry and the strength of the green body, 4–5 wt% gelatin (based on water) is suitable for the suspension containing 50–55 vol% alumina powder used in this research. Various ceramic green bodies were prepared by this forming process (shown in Fig. 3.21). Green parts can be sintered directly without special binder removal procedures because the organic binder content is 1 wt% based on the powder weight. Ceramics consisting of 95 wt% alumina and 5 wt% clay made by this method have linear shrinkage of 15–16% during firing. The theoretical density of the final product is 96.5% and the bending strength reaches 304.6 MPa.

Summary

(1) Gelatin can be dissolved in water above 40 °C, and the solution gels at 15–20 °C. Higher content of gelatin in the solution resulted in greater gel rigidity and green body strength. On the other hand, higher viscosity and poor uniformity of the

3.2 Alumina Casting Based on Gelation of Gelatin

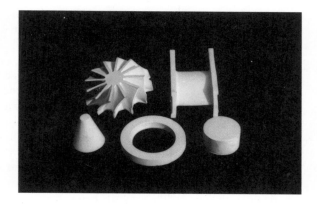

Fig. 3.21 Examples of ceramic green parts made by the forming method

slurry and green body were observed at a gelatin concentration higher than 6 wt% (based on water).

(2) Suitable gelatin content for suspensions of 50–53 vol% alumina powder is 4–5 wt% (based on water). Such slurries have apparent viscosities lower than 1 Pa and are pseudoplastic in nature, which facilitates casting.

(3) Wet green compacts can be formed from the alumina slurry by gelatin gelation. The strength of a coagulated body increases with the development of the gelatin gel structure. Therefore, a gelation time of several hours or more is needed from the start of the process until demolding is completed.

(4) The resultant green bodies have precise size and relative smooth surface with linear drying shrinkages of 2–3%. The strength of the dry body is higher than 8 MPa with about 1 wt% gelatin (based on the powder weight). The sintered samples have shrinkage of 15–16% and a bending strength of 304.5 MPa.

3.3 A Casting Forming for Ceramics by Gelatin and Enzyme Catalysis

3.3.1 Research Background

(1) **Academic idea**

We have reported that a ceramic suspension was gelled to a green body using agarose and gelatin in our previous study (Xie et al. 1998; Chen et al. 1999). In these cases, a well-dispersed ceramic suspension containing agarose or gelatin was ball-milled and heated at 80 or 40 °C to dissolve them completely, and then cast into a non-porous mold. During cooling down to room temperature, the suspension gelled and formed a wet green body with the desired shape. Some ceramic parts with complex shape, such as a turbine rotor, were produced by this method. However, the disadvantage of this process is that the suspension has to be heated to a certain temperature before casting, which might increase the viscosity of the slurry due to water vaporization during the heating process.

To overcome this drawback of the forming process, a new gelation process using gelatin and enzyme catalysis for forming a ceramic green body has been developed in this study. The forming method is based on the process in which gelatin, urea, and urease are used. Urea can prohibit a hydrogen bond from forming between the gelatin molecules when the hot gelatin solution cools down. As urease is added into the suspension, the gelatin molecules attract each other and change to a 3D network structure through hydrogen bonding due to urea decomposition by urease. Then, the suspension containing the ceramic powder and gelatin will be consolidated in situ by the gelation under ambient conditions. This process can prevent water vaporization and increase the viscosity before casting. Moreover, the process is more easily controlled to obtain a homogeneous microstructure in the green body.

(2) **Materials used**

The gelatin polymer with molecular weights ranging from 15,000 to 90,000 was provided by Shanghai Chemical Reagent Works. The analysis of the gelatin has been provided in one of our previous studies (Chen et al. 1999). The gelatin powder can be dissolved in water in less than 20 min at ~40 °C, and shows a pH value of 5.5–7 for 1 wt% solution. Urea with a purity of 99% is a commercial grand (Beijing Chemical Company). The alumina powder used here was a commercial grade An-05, and the details about the powder are described in the reference (Chen et al. 1999).

(3) **Process**

The forming process is described in Fig. 3.22. First, the urea and gelatin were dissolved in deionized water at ~40 °C. As the solution cooled down to room temperature, alumina powder and dispersant were added to the solution. Then, the mixture was ball-milled for more than 12 h to obtain a well-dispersed suspension with good fluidity. After degassing, the urease solution was added and mixed at a low temperature of 2–5 °C. Then, the slurry was cast into the mold under ambient conditions and coagulated to a green body through gelation due to the enzyme-catalyzed decomposition of urea.

(4) **Property measurement**

The rheological properties of the gelatin solution and the mixed ceramic slurry were measured by a rotary rheometer (NXS-11, Chengdu, China). The Zeta potential was calculated from the measured electrophoretic mobility by using ELS with a 5 mW laser source. A suspension of 0.05 vol% concentration was chosen. The pore size distribution in the green body was measured by using a mercury porosimeter (Autoscan Mercury Porosimeter, Quantachrome Company, USA). The microscopic morphology of the dried green body was observed by an SEM (OPTON CSM950).

3.3 A Casting Forming for Ceramics by Gelatin and Enzyme Catalysis

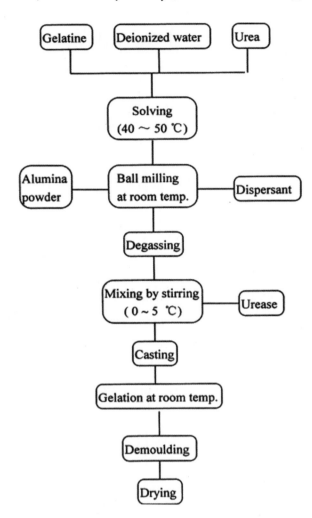

Fig. 3.22 Flowchart of the forming process. For alumina slurries with gelatin and urea

3.3.2 Gelation Mechanism of Gelatin Solution with Urea Under Enzyme Catalysis

(1) Prevention of urea on attraction of gelatin molecular

Gelatin, a kind of protein polymer, is mainly composed of various amino acids. Urea can act with amino acids with hydroxide radical and restrain the formation of hydrogen bonds between the polymer molecules or the polymer and water. Figure 3.23 shows that the viscosity of 2 wt% gelatin solution changes with the urea content under the cooling stage from 40 to 20 °C. The ratio of urea to gelatin is expressed as α. When $\alpha < 3$, the solution viscosity was high due to gelation of the gelatin. When α

Fig. 3.23 Effect of urea content on the viscosity of gelatin solution (2 wt%)

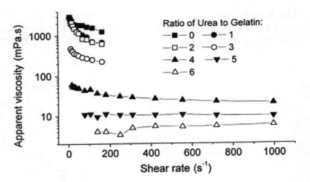

= 4, the solution exhibited some fluidity and still appeared in a viscous state, in which flocculation was observed. However, the viscosity decreased to 10 MPa s or lower when α > 5. In this case, sufficient urea prevented gelatin from gelling completely and a clear gelatin solution with perfect fluidity could be prepared. For a system with 3 wt% gelatin solution, which was also studied, the results almost similar to those explained above were observed.

(2) **Gelling of gelatin solution under urease catalysis**

When urease, as a catalyst, was added into the gelatin solution, the urea could be decomposed by urease catalysis. The reaction is as follows:

$$NH_2 - CO - NH_2 + H_2O + Urease \rightarrow NH_4^+ + HCO_3^-$$

The reaction speed mainly depends on the temperature and the amount of urease. At a temperature lower than 5 °C, the decomposition of urea is very slow, but it would increase greatly with rise in temperature. For example, one unit will liberate 1 μmol of NH_3 from urea per min at pH 7 and 25 °C (Omatete et al. 1991). During the decomposition of urea, gelling of the gelatin solution will occur through hydrogen bonding among the gelatin molecules, and the gelation can take place at room temperature as the gelling point ranges from 15 to 25 °C. Figure 3.24 shows the change in the viscosity with the decomposition time after the addition of urease at 15 °C for the solution containing 2 wt% gelatin and 10 wt% urea. The viscosity increased slowly initially due to less significant urea decomposition and weak hydrogen bonding among the molecules. However, the solution lost fluidity and gelled after about 3 h. This resulted from a 3D network formed by the attraction among the molecular chains.

The time to reach consolidation was clearly influenced by the amount of urease. From Fig. 3.25, it can be seen that about 10 h were required for a urease concentration of less than 5 units, and only 3 h for a urease concentration of 10 units. There was almost no influence of urease content on the time as the urease concentration reached 16 units.

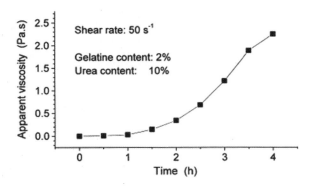

Fig. 3.24 Viscosity of gelatin solution versus time after addition of urease

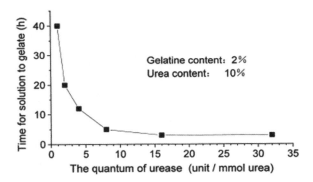

Fig. 3.25 Effect of the urease concentration on the gelation time

3.3.3 Rheology and Zeta Potential of Alumina Suspension Containing Gelatin and Urea

A suspension with 50 vol% alumina and 2 wt% gelatin (based on water) was prepared as shown in the flowchart in Fig. 3.22. The suspension was ball-milled for 24 h. The influence of the amount of urea on rheological properties is illustrated in Fig. 3.26. When the ratio of urea to gelatin (α) was equal to 4, the system showed a relatively large viscosity (the apparent viscosity reached 1,500 MPa s for a shear rate of 100 s^{-1}). In addition, a dilatant flow model was observed in which the viscosity increased with the shear rate. It was implied that there still was attraction among polymer molecules that resulted in a large viscous force during shearing. Under these conditions, agglomerates or flocculation may occur and the homogeneous microstructure of the ceramic body will be destroyed. When $\alpha \geq 5$, the viscosity sharply decreased to lower than 100 MPa s, which facilitated degassing and casting. Moreover the suspension showed a Newtonian behavior. In this case, the viscosity almost did not vary with shear rate, especially for the shear rate above 50 s^{-1}.

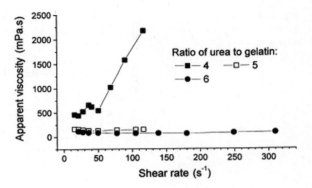

Fig. 3.26 Effect of the urea content on the fluidity of slurry with gelatin and alumina

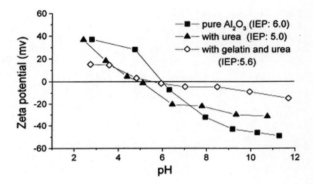

Fig. 3.27 Zeta potential versus pH value

Similar results were also observed for the suspension with 50 vol% alumina and 3 wt% gelatin (based on water). This was helpful to the colloidal process. The rheological behavior was related to the structure of the system, where the attraction between the polymer molecules was limited by the urea content.

The Zeta potential of the alumina suspension was also influenced by gelatin and urea. As shown in Fig. 3.27, the pure alumina suspension had an IEP of 6 and a relatively high zeta potential (absolute value about 40 mV at pH 8–10). For the suspension containing gelatin and urea, the absolute value of the zeta potential was clearly decreased. This may be explained by the fact that the electrostatic double layer was compressed by ions at high gelatin content. It was noted that the suspension still showed a good fluidity when the ratio of urea to gelatin (a) was greater than 5.

3.3.4 Coagulation Forming and Microstructure of Green Body

The apparent viscosity versus time was measured at 15 °C, when the urease (16 units per mmol urea) was added into the 50 vol% alumina suspension with 2 wt% gelatin

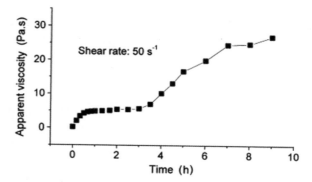

Fig. 3.28 Change in slurry viscosity with time after adding urease

(based on water) and 10 wt% urea (based on water). From Fig. 3.28, the change in apparent viscosity of the slurry was similar to that of a gelatin solution (Fig. 3.24). The viscosity increased relatively slowly in the initial stage and rose considerably after 3 h. This is due to the fact that the coagulation of the slurry resulting from the gel network formed by the gelatin molecules strongly depended on the speed of the enzyme-catalyzed decomposition of urea. The viscosity reached more than 25 Pa s after 7 h and remained constant thereafter. At this point, a 3D gel network was formed, which resulted in complete coagulation of the slurry, and a rigid wet green body was obtained.

The microstructure and pore distribution of the dried green body were examined by an SEM and a mercury porosimeter. As shown in Fig. 3.29, a homogenous structure could be observed, in which large agglomerates and bubble were removed. In addition, the pore distribution was characterized by a narrow single peak while two peaks were observed for the dry pressed green body. These results suggested a well-dispersed slurry, and therefore, good degassing and casting could be performed by the colloidal forming process.

Summary

When the ratio of urea to gelatin (α) was larger than 5, the gelling process was completely restrained when the gelatin solution was cooled down from ~40 to 15 °C (gel point). This was because the hydrogen bonds and van der Walls forces among molecules were limited by the urea, and the viscosity of the 2–3 wt% gelatin solution was less than 10 MPa s. However, the gelling process of the solution was regained when urease was added, due to the enzyme-catalyzed decomposition of urea. The time taken to form the gel strongly depended on the urease concentration. The coagulation speed of the solution increased greatly as the urease concentration reached 10 units. There was almost no change in the gelling time after 16 units of urease.

For the suspension with 50 vol% alumina and 2 wt% gelatin (based on water), the apparent viscosity decreased to 80 MPa s and the suspension showed a Newtonian behavior when the ratio of urea to gelatin α was ≥ 5, which made colloidal processes such as degassing and casting easy to carry out. Moreover, the slurry could be coagulated in a few hours by the gelling of gelatin after the addition of urease (16 unit per

Fig. 3.29 SEM micrograph of dry green body

mmol urea). The dry green body produced by this method showed a homogenous microstructure.

3.4 Alumina Forming Based on Gelation of Sodium Alginate

3.4.1 Research Background

Alginate is a type of gelling polysaccharide, which can be dissolved in water at room temperature and then begin gelling by cross-linking with divalent metal ions at increased temperature after being cast. The mechanism of cross-linking in alginate gels can be considered in terms of an 'egg-box' model involving the cooperative bonding of calcium ions between aligned polyguluronate ribbons (Braccini 1999). Calcium iodate has been applied in this process. It has low solubility at room temperature and high solubility at increased temperature of 60 °C. Thus, the gelcasting

3.4 Alumina Forming Based on Gelation of Sodium Alginate

process can be controlled by the heating rate and the final temperature (Xie and Huang 2001).

A novel gelcasting process by alginate is discussed in this paper. In this process, a solidified agent and chelator are added into the sodium alginate solution simultaneously. Thus, the gelation between calcium ions and sodium alginate is avoided in this stage before casting. Free calcium ions are released when hexanedioic acid is added into the suspension. They react with sodium alginate and form a 3D network. Therefore, the ceramic particles are fixed in this network. The slurry is consolidated and near-net-shape green parts are formed.

(1) Materials and chemicals

The composition of the alumina powder used has been described in Sect. 3.1.1 of this chapter. Controllable casting and solidification are possible through the use of sodium alginate, the solidified agent, the chelator, and the hexanedioic acid ($C_6H_{10}O_4$). Sodium alginate (Na alginate) is used as a gelation reagent, which can dissolve in water at room temperature. Three kinds of solidified agents, namely, calcium carbonate ($CaCO_3$), calcium phosphate secondary ($CaHPO_4$), and calcium phosphate $Ca_3(PO_4)_2$, and two chelators, sodium hexametaphosphate (($NaPO_3)_6$) and tri-sodium citrate ($Na_3C_6H_5O_7$), were employed in the process.

(2) Procedure

The gelcasting process is shown in Fig. 3.30. Sodium alginate was added into the deionized water to form a solution, which had been decomposed at 70 °C for 48 h. Then, the alumina powder, solidified agent, and chelator were added into the solution. The resulting slurry was then ball-milled for 24 h. After degassing in a rotary evaporator under vacuum, hexanedioic acid was added into this slurry before it was cast into a nonporous mold. The slurry consolidated to the shape required, and then the wet green body was demolded and dried at room temperature.

(3) Measurements

The rheological properties of the sodium alginate solution and alumina slurries were measured by a control-rate rotary rheometer (MCR300, Physica, Germany). The measurements were performed within the shear rate range of $0.1-250\,s^{-1}$ at a constant temperature of 25 °C. The gel strength was examined by a simple device, which is usually used in the gum industry. The analysis and measurement of the particle size were done by using a BI-XDC particle analysis device by the method of X-ray centrifugal sedimentation.

The bending strength of both the green body and the sintered samples were examined. A three-point bending method was employed for the test, with a crosshead speed of 0.2 mm/min and a span of 30 mm. The bars of green samples and sintered samples were of dimensions $5 \times 6 \times 36\,mm^3$. The cross-sectional surfaces of the samples were examined by an SEM.

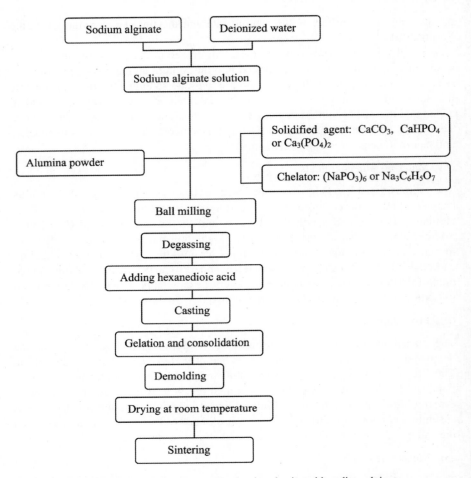

Fig. 3.30 Flowchart of the forming process for alumina slurries with sodium alginate

3.4.2 Gelation Principle of Sodium Alginate

When calcium salts are added into the sodium alginate solution, they form a gel very quickly owing to the reaction between the calcium ions and the sodium alginate, as shown in Eq. (3.1). Gelation occurs during casting due to the fast gelation rate, and therefore results in varying degrees of cross-linking density and heterogeneity within the gel. Thus, controlling the reaction rate by a chemical process is important. In this case, a chelator is used in the process. It forms a chelate complex with the calcium salts (shown in Eq. (3.2)). Therefore, free calcium ions are chelated in the chelate complex, which allows them to be uniformly distributed in the alginate solution and

3.4 Alumina Forming Based on Gelation of Sodium Alginate

does not induce gelation immediately. Then, just before casting, hexanedioic acid is added into the slurry and the calcium ions are released slowly from the chelate complex (shown in Eq. (3.3)). Gelation takes place at this stage, which is shown in Eq. (3.1).

The reactions that occur during the process are as follows:

$$2\text{Na alginate} + \text{Ca}^{2+} \leftrightarrow \text{Ca alginate} + 2\text{Na}^+ \tag{3.1}$$

$$2\text{C}_6\text{H}_5\text{O}_3^{-7} + 3\text{Ca}^{2+} \leftrightarrow \text{Ca}_3(\text{C}_6\text{H}_5\text{O}_7)_2 \tag{3.2}$$

$$\text{Ca}_3(\text{C}_6\text{H}_5\text{O}_7)_2 + 3\text{C}_6\text{H}_{10}\text{O}_4 \leftrightarrow 2\text{C}_6\text{H}_8\text{O}_7 + 3\text{Ca}^{2+} + 3\text{C}_6\text{H}_8\text{O}_4^{2-} \tag{3.3}$$

It has been shown that a slow-release acid is effective in producing homogeneous and relatively slow-setting alginate gels. With the addition of chelator and hexanedioic acid, the time-delayed release of the cross-linking calcium ions allows the calcium alginate suspension to be molded into complex geometry before gelation occurs (Kuo 2001).

(1) Effect of different chelators on gelation

A comparison of samples from Tables 3.3, 3.4 and 3.5 shows that the gelation rate and the gel strength decreased with increase in the amount of the chelator. This might indicate that the conformational rearrangement of the polymer (alginate) is important for cross-linking. A higher chelate complex concentration at the constant condition would result in slower rearrangement of the polymer conformations to form effective cross-linking. Higher content of the chelator yielded higher concentration of the chelate complex, and hence, more hexanedioic acid was required to liberate the calcium ions from the chelate complex in order to initiate gelation. Therefore, for the same content of hexanedioic acid and solidified agent, the calcium ions would be released incompletely with elevated content of chelator; and a gel with lower gel strength was achieved due to this factor.

It could also be seen that the gelation of alginate with various chelators differed considerably. Samples #1 and #6 of Table 3.5 were prepared with the same amount of hexanedioic acid and solidified agent. It was observed that the gel strength with

Table 3.3 Effect of solidified agent ($CaCO_3$) on gelation

No.	Na alginate (ml)[a]	$Na_3C_6H_5O_7$ (g)	$CaCO_3$ (g/ml)	$C_6H_{10}O_4$ (g)	Gelation start time (min)	Gel strength (g/mm^2)
#1	20	0.19	0.01	0.08	7	9.09
#2	20	0.20	0.01	0.08	12	8.92
#3	20	0.20	0.02	0.08	3	17.17
#4	20	0.23	0.02	0.08	8	19.17

[a]The concentration of the Na alginate solution is 1 wt%

Table 3.4 Effect of solidified agent (CaHPO$_3$) on gelation

No.	Na alginate (ml)[a]	Na$_3$C$_6$H$_5$O$_7$ (g)	CaHPO$_3$ (g)	C$_6$H$_{10}$O$_4$ (g)	Gelation start time (min)	Gel strength (g/mm^2)
#1	20	0.19	0.1	0.08	5	11.72
#2	20	0.21	0.1	0.08	10	8.99
#3	20	0.23	0.1	0.08	12	7.16
#4	20	0.23	0.1	0.15	4	26.16
#5	20	0.26	0.1	0.15	4	19.51
#6	20	0.35	0.2	0.15	8	17.29
#7	20	0.40	0.2	0.15	9	10.24

[a]The concentration of Na alginate is 1 wt%

Table 3.5 Effect of solidified agent (Ca$_3$(PO$_4$)$_2$) on gelation

No.	Na alginate (ml)[a]	Na$_3$C$_6$H$_5$O$_7$ (g)	(NaPO$_3$)$_6$ (µl)[b]	Ca$_3$(PO$_4$) (g)	C$_6$H$_{10}$O$_4$ (g)	Gelation time (min)	Gel strength (g/mm^2)
#1	20	0.15	–	0.1	0.08	5	12.86
#2	20	0.18	–	0.1	0.08	8	8.43
#3	20	0.20	–	0.1	0.08	>180	–
#4	20	0.20	–	0.15	0.15	2	29.18
#5	20	–	60	0.1	0.08	2.5	43.9
#6	20	–	120	0.1	0.08	5	37.3
#7	20	–	120	0.3	0.15	1	103.7
#8	20	–	180	0.3	0.15	2	99.33
#9	20	–	240	0.3	0.15	3	70.51

[a]The concentration of the Na alginate is 1 wt%
[b]The concentration of (NaPO3)$_6$ is 20 wt%

the (NaPO$_3$)$_6$ chelator increased much greater than with Na$_3$C$_6$H$_5$O$_7$. In addition, the concentration of the (NaPO$_3$)$_6$ solution was lower than that of Na$_3$C$_6$H$_5$O$_7$, which brought less ions. Therefore, it would have less effect on the suspension and the sintering of green body. Thus, it was concluded that the (NaPO$_3$)$_6$ chelator is suitable for gelcasting.

(2) Rheological property contrast

As shown in Fig. 3.31, there was no obvious effect on the viscosity when CaCO$_3$ and Ca$_3$(PO$_4$)$_2$ were mixed with aqueous alumina suspensions. However, the slurry viscosity increased sharply with the addition of CaHPO$_4$ after ball-milling for 20 h. CaHPO$_4$ ($K_{sp} = 1 \times 10^{-7}$) was found to be slightly soluble in water at room temperature and produced a number of calcium ions after a long stirring time, which initiated calcium gelation and increased the viscosity of the slurry. CaCO$_3$ ($K_{sp} =$

Fig. 3.31 Rheological property of the suspensions with different solidified agent

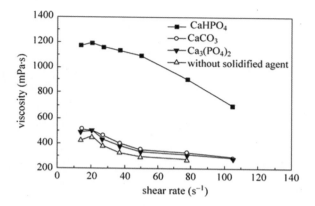

2.8×10^{-9}) and $Ca_3(PO_4)_2$ ($K_{sp} = 2 \times 10^{-29}$) has very low solubility in pure water, allowing its uniform distribution in the alginate solution before the gelation occurred. Therefore, to ensure that the slurry has sufficient fluidity during casting, $CaHPO_4$ should be added into the slurry at a later stage during ball-milling.

3.4.3 Preparation Process of Alumina Green Bodies and Samples by Sodium Alginate

(1) The flowchart of the preparation process of alumina green bodies and sample by sodium alginate is shown in Fig. 3.30.
(2) The properties of the green bodies and sintered samples.

From samples #2 and #3 shown in Table 3.3, it was found that the gelation rate and the gel strength increased with an increase in the amount of the solidified agent. This implies that the availability of calcium ions plays a critical role in the gelation, and a high concentration of calcium ions leads to faster gelation and higher gel strength.

It was found that the microscopic structures of the green body varied greatly with different solidified agents, as shown in Figs. 3.32 and 3.33. $CaCO_3$ could react with acid and release some CO_2 gas. CO_2 could not be degassed completely and formed pores in the green part. These pores functioned as defects and resulted in poor mechanical properties of the ceramic parts after consolidation since they could not be eliminated during sintering. $CaHPO_4$ had to be added into the slurry during ball-milling, which resulted in heterogeneity in the green body. However, no such problem was observed in the case of $Ca_3(PO_4)_2$ and it was found to be a proper solidified agent. To further investigate the difference among the solidified agents, sintered samples with $Ca_3(PO_4)_2$ and $CaHPO_4$ as solidified agents were also prepared. Figure 4 shows the microscopic structure of these samples. It illustrates that the sintered body made from $Ca_3(PO_4)_2$ had a dense and homogeneous structure; while there were several

Fig. 3.32 The SEM micrograph of the cross-section of the green body with different solidified agents **a** $CaCO_3$, **b** $CaHPO_4$, and **c** $Ca_3(PO_3)_2$

3.4 Alumina Forming Based on Gelation of Sodium Alginate

Fig. 3.33 SEM micrograph of the sintered body with different solidified agents **a** $Ca_3(PO_4)_2$ and **b** $CaHPO_4$

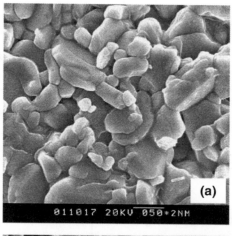

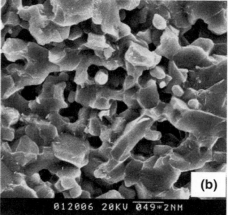

holes and agglomerates in the structure of that made from $CaHPO_4$. This suggests that $Ca_3(PO_4)_2$ is more suitable for the process than the other agents.

(3) Effect of hexanedioic acid on gelation

Samples #3 and #4 described in Table 3.4 showed that the gelation rate and gel strength increased with an increase in the amount of hexanedioic acid. Similar results were observed in the case of samples #3, #4, #6, and #7 shown in Table 3.5. The time taken to form the gel strongly depended on the reaction between the calcium ions and the alginate. As mentioned, hexanedioic acid was employed to release the calcium ions from the chelate complex and to realize the time-delayed gelation. Therefore, the content of the free calcium ions in the solution was mainly determined by adding hexanedioic acid. Higher content of hexanedioic acid resulted in larger concentration of free calcium ions, and correspondingly faster gelation.

Table 3.6 Effect of alumina powder on gelation

No.	Na alginate (ml)[a]	$(NaPO_3)_6$ (ul)[b]	$Ca_3(PO_4)_2$ (g)	$C_6H_{10}O_4$ (g)	Al_2O_3 (g)	Gelation time (min)
#1	20	300	0.3	0.15	72	10
#2	20	400	0.3	0.15	72	13
#3	20	1000	2.3	0.6	72	28
#4	20	400	0.3	0.15	–	7
#5	20	400	0.3	0.15	–	10
#6	20	1000	2.3	0.6	–	26

[a]The concentration of the Na alginate is 1 wt%
[b]The concentration of $(NaPO_3)_6$ is 20 wt%

(4) Consolidation

$Ca_3(PO_4)_2$ was used as the solidified agent and $(NaPO_3)_6$ was used as the chelator. Comparing samples #2 and #4 of Table 3.6, it is clear that the addition of alumina powder decreased the gelation rate. The suspension exhibited greater viscosity due to the addition of alumina, which led to slower rearrangement of the polymer chain conformations to form effective cross-linking. Similar results from the gelation process could also be observed in the consolidation process, for example, the consolidation rate increased with decreasing content of the chelator.

Summary

With the chelator, the time-delayed release of calcium ions was realized, which allowed the slurry to be molded into complex shapes before gelation occurred. Two kinds of chelators, namely, $(NaPO_3)_6$ and $Na_3C_6H_5O_7$, and three kinds of solidified agents, namely, $CaCO_3$, $CaHPO_4$, and $Ca_3(PO_4)_2$ were investigated in the process. Under the same conditions, the gelation rate and gel strength decreased with increasing amount of the chelator. $(NaPO_3)_6$ and $Ca_3(PO_4)_2$ were found to be suitable as chelator and solidifying agent, respectively, for gelcasting. It was also found that with an increasing amount of hexanedioic acid, the gelation rate was higher.

The bending strength of the green body increased with an increasing amount of $Ca_3(PO_4)_2$.

3.5 Gelcasting of Silicon Carbide Based on Gelation of Sodium Alginate

3.5.1 Research Background

(1) Research background

For a long time, silicon carbide ceramics have attracted great interest due to their use in the field of refractories as well as high-temperature structural applications. Silicon carbide ceramics are known to possess a variety of virtues, such as high mechanical strength, high chemical stability, high temperature of decomposition, good thermal conductivity, and semiconductivity. They find specific applications in the petroleum, mechanical, and aerospace industries (Zhou et al. 1998, 2000; Rao 1999a, b).

In this paper, gelcasting of silicon carbide ceramics by using an alginate gelation is discussed. For this process, a solidified agent and a chelator are added into the sodium alginate solution simultaneously. Thus, the gelation between the calcium ions and the sodium alginate is avoided in this stage due to formation of a chelate complex from the chelator and the sodium alginate. When hexanedioic acid is added into the suspension, free calcium ions are released from the chelate complexes and they react with sodium alginate, forming a 3D network. Therefore, the ceramic particles are fixed in this 3D network and the slurry is consolidated.

The starting ceramic powder consisted of coarse silicon carbide and fine silicon carbide powders in order to achieve a high packing density. The mass ratio of the coarse powder to the fine powder was 70:30. the main impurities in the powders included Fe 520 ppm, O 1.17 wt%, Si 0.26 wt%, and free carbon 0.35 wt%. The particle size of the mixture containing coarse and fine SiC powders ranged from 0.2 to 350 μm (Yi et al. 2001).

Controllable casting and consolidation are possible through the use of sodium alginate (Na alginate), sodium hexametaphosphate (($NaPO_3)_6$), tertiary calcium phosphate ($Ca_3(PO_4)_2$), and hexanedioic acid ($C_6H_{10}O_4$). Sodium alginate is used as a gelation reagent, which dissolves in deionized water at room temperature. When the sodium alginate and calcium ions react, the molecular chains attract each other to form a 3D network. Sodium hexametaphosphate, a vitriform solid, is used here as a chelator while tertiary calcium phosphate is used as a solidified agent. Three dispersants, namely, TMAH, DOLARPIX PC33 (PC33), and tri-ammonium citrate are employed.

(2) Research program

The forming process flowchart is shown in Fig. 3.34. Sodium alginate was dissolved in the deionized water to form a solution, which had been decomposed at 70 °C for 48 h to reduce the solution viscosity. Then, the silicon carbide powder, dispersant, solidified agent, and chelator were added into the solution. The resulting slurry was then ball-milled for 24 h. After degassing in a rotary evaporator under vacuum, hexanedioic acid was added into this slurry before it was cast into a nonporous mold, after which the consolidation occurred. After several hours, the wet green body was demolded and dried at room temperature.

(3) Properties measurement

Zeta potentials of the powder were measured by using a zeta potential analyzer (Zeta-Plus, Brookhaven Company, USA). The rheological properties of the solution and SiC slurries were measured by a control-rate rotary rheometer (MCR300, Physics, Germany), and by using a concentric cylinder measurement device with a 1.13 mm gap. The sample volume was 23 ml. The measurements were performed within the shear rate range of $0.1-250$ s^{-1} at a constant temperature of 25 °C. The gel strength

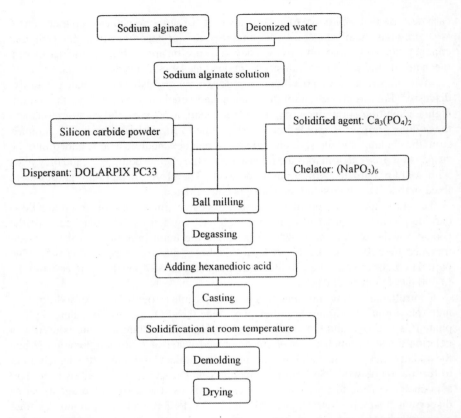

Fig. 3.34 Flowchart of the forming process for silicon carbide slurries with sodium alginate

of the alginate was examined by a simple device, which is usually used in the gum industry (Clare 1993; Plovnick 1997). The room temperature bending strength of the dried green body samples was determined by the three-point bending method, with a crosshead speed of 0.2 mm/min and a span of 30 mm. The bars of green samples were of the dimensions $5 \times 6 \times 36$ mm^3. The cross-section surfaces of the samples were examined with an SEM and the density of the green body was measured by the Hg immersion method based on the Archimedes' principle.

3.5.2 Effect of Dispersant on the Colloidal Behavior of the SiC Suspension

The variations of the zeta potential versus pH for as-received silicon carbide and the suspensions containing TMAH, DOLARPIX PC33, and tri-ammonium citrate as dispersants, respectively, are shown in Fig. 3.35. The addition of sodium alginate

Fig. 3.35 Zeta potential versus pH value of SiC suspension in deionized water (A), in Na alginate solution (B), in Na alginate solution with dispersant TMAH (C), PC 33 (D), and tri-ammonium citrate (E), respectively

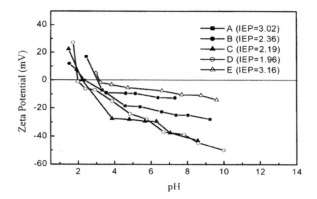

produced a pronounced effect on the electrokinetic behavior of SiC, wherein the IEP of the SiC suspension changed from pH 3.02 to 2.36. This is caused by the absorption of the negatively charged carboxylate alginate chains on the surface of the SiC particles. In the range of pH 7–9, a negative Zeta potential about −10 ~ 15 mV was observed. This may be attributed to the fact that the electrostatic layer was compressed by ions with a high value in sodium alginate (Pierre and Ma 1999; Teng et al. 1997).

TMAH is an organic alkali, PC33 is a kind of polyelectrolyte, and tri-ammonium citrate is an inorganic salt. TMAH and PC33 are alkaline, and result in better dispersion of the SiC powder since SiC powder can be suspended in a basic solution, larger repulsion between ceramic particles and a more stable suspension are achieved with increase in the zeta potential, as shown in Fig. 3.35. In contrast, tri-ammonium citrate is an acid and does not provide good dispersion (Zhou et al. 2000; Yi et al. 2001). DOLARPIX PC33 is the best among the three dispersants due to its high absolute zeta potential value (about 40 mV at pH 7–9) and polyelectrolyte stabilization, which is helpful for the stability of suspensions (Cesarano and Aksay 1988). In this case, DOLARPIX PC33 is a suitable dispersant for the sodium alginate solution compared to the other two.

3.5.3 Rheological Property of SiC Suspension

A suspension with 50 vol% SiC and 1.5 wt% sodium alginate (based on water) was prepared, as shown in Fig. 3.34. The influence of the amount of dispersant (DOLARPIX PC33) on the rheological property of SiC suspension is illustrated in Fig. 3.36, which indicates that the dispersant PC33 (curves B, C, and D) increases the fluidity of the suspensions compared to a suspension without the dispersant (curve A). When the shear rate is $100\ s^{-1}$, the apparent viscosity of the slurry with a volume ratio of 10 μl:1 ml (PC 33 to the sodium alginate solution) system (B) was lower than 1 Pa s, which is the empirical value below which casting process was allowed.

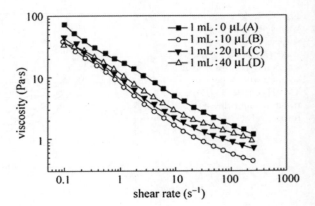

Fig. 3.36 Effect of the dispersant (DOLARPIX PC33) amount on the viscosity of the slurry of a 50 vol% SiC suspension (volume ratio of PC33 to 1.5 wt% Na alginate solution)

The viscosity of the slurry with a volume ratio of 20 μl:1 ml system was about 1 Pa s and was much higher than 1 Pa s for a slurry with a volume ratio of 40 ml:1 ml, which implies that excessive polyelectrolyte (DOLARPIX PC33) is not absorbed on the SiC particles, and therefore, the viscosity of the suspension increases.

On the one hand, it is beneficial to obtain a solids loading as high as possible in order to better control the final geometry of the part, and to achieve higher dimensional tolerances by minimizing the possible complex distortion during drying and sintering shrinkage. On the other hand, the shaping of intricate parts requires the slurry to easily fill molds of possible complex geometry. This in turn means that the viscosity of the slurry has to be kept below some acceptable level. The two-boundary conditions necessitate a well-controlled tailoring of the suspension that can meet both the requirements. Figure 3.37 demonstrates that the suspension has relatively low viscosity and pseudoplastic property when the solids loading is below 55 vol%. When the solids loading increased up to 60 vol%, the viscosity was 7.79 Pa

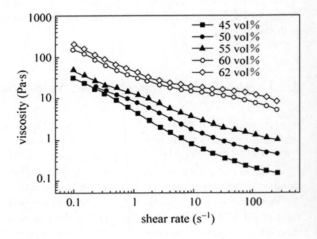

Fig. 3.37 Rheological property of the slurry with various solids loading (based on 1.5 wt% Na alginate)

3.5 Gelcasting of Silicon Carbide Based on Gelation of Sodium …

s under the shear rate of 100 s^{-1}. For the slurry of 62 vol%, the viscosity reached 12.46 Pa s, which does not render it suitable for gelcasting. Therefore, the suspension with solids loading of 45–55 vol% is suitable for this process.

3.5.4 Sedimentation Behavior of the SiC Suspension

When ceramic powder is dispersed in a liquid medium, the particles are influenced by a variety of forces such as force of gravity, drag force from the medium, and Brownian motion of particles. Assuming that the other forces are negligible or of non-varying type, the ceramic particles tend to settle under the force of gravity. The state of dispersion and its stability is measured by the particle settling rate (manifested in terms of an increase of sediment height or sediment volume) versus time (Pierre and Ma 1999; Rao 1999a, b). The particle settling rate manifested in terms of an increase of sediment height or sediment volume as a function of time gives a measure of the state of dispersion and its stability. Relative sedimentation height here refers to the ratio of the mixed suspension height (including the stable suspension and the sedimentation layer) to the whole height of the suspension. The sedimentation height is inversely related to the dispersion of the powder; the more dispersed the powder, the smaller the sediment height.

Sedimentation behavior was conducted in test tubes that had a capacity of 50 ml. Control samples with solids loading of 20, 30, 45, 50, and 55 vol% were included to provide a basis for contrast. Figure 3.38 shows that the sedimentation of the suspensions with low solids loading, such as 20 vol% and 30 vol%, is obvious; while that at high solids loading, such as 45, 50, and 55 vol% cannot even be observed (Rao 1999a, b).

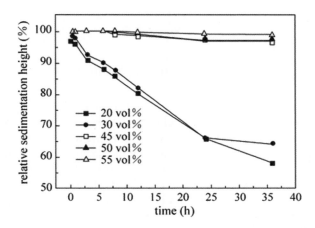

Fig. 3.38 Sedimentation of the SiC slurry with varying solids loading

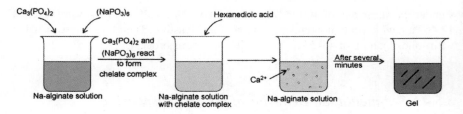

Fig. 3.39 The principle sketch of the gelation process

3.5.5 Gelation Principle and Process of the Alginate Solution

The mechanism of cross-linking in alginate gels can be considered in terms of an 'egg-box' model involving the cooperative bonding of calcium ions between aligned polyguluronate ribbons (Wber and Tomandl 1997; Clare 1993; Cesarano and Aksay 1988). However, it is generally difficult to control the reaction rate, which makes it impossible to complete the casting process at a certain period. Therefore, calcium salts at a controlled reaction rate with alginate were considered in this study. The concentration of the calcium ions released from the salt can be adjusted by the formation and decomposition of the chelate complexes. The principle sketch of the process is shown in Fig. 3.39. Tertiary calcium phosphate (the solidified agent) and sodium hexametaphosphate (the chelator) were added into the alginate solution, and calcium ions were chelated in the chelate complexes. Therefore, calcium ions could not react with alginate and the gelation between calcium ions and alginate was avoided in this stage before ceramic slurry casting. Then, hexanedioic acid was added into the solution; the chelate complexes decomposed and released the calcium ions slowly. The time-delayed release of cross-linking calcium ions allowed the Ca alginate suspension to be molded into complex geometry before gelation occurred. Thus, gelation rate control was realized.

3.5.6 Gelation of the SiC Suspension with Alginate

To study the influence of various Na alginate concentrations, the amount of solidified agent ($Ca_3(PO_4)_2$), and chelator (($NaPO_3)_6$) on the consolidation and the property of the green body, several groups of samples were fabricated, as shown in Table 3.7.

For the same content of solidified agent ($Ca_3(PO_4)_2$) and chelator (($NaPO_3)_6$), sample Nos. 1, 2, and 3 were made from 45 vol% SiC and 1.0, 1.5, and 2.0 wt% Na alginate solutions, respectively. Accordingly, the consolidation time increased from 2 to 5 h, as the Na alginate concentration increased from 1.0 to 2.0 wt%. These results exhibited decrease in the consolidation rate with an increase in the polymer (sodium alginate) concentration, which may be an indication that the conformational rearrangement of the polymer chains is important for cross-linking formation. A higher

Table 3.7 Properties and gelation of the suspension with 45 vol% SiC (400 μl DOLARPIX PC33 in 20 ml Na alginate solution) under various conditions

No.	Na alginate concentration (wt%)	$Ca_3(PO_4)_2$ (g)	$(NaPO_3)_6$ (ml)[a]	pH	Consolidation time (h)[b]	Bending strength (MPa)	Relative green density (%)[c]
1	1.0	4	1.0	8.18	2	2.49	47.86
2	1.5	4	1.0	8.36	3	2.85	52.41
3	2.0	4	1.0	8.66	5	3.09	50.57
4	1.5	6	1.0	8.71	1	3.47	52.16
5	1.5	1.2	1.0	8.04	4	2.54	51.57
6	1.5	4	2.0	8.30	5	3.26	53.99
7	1.5	4	0.4	8.95	0.5	2.67	51.74

[a]Concentration of $(NaPO_3)_6$ is 20 wt% (based on deionized water)
[b]Consolidation time refers to the time of change from slurry to wet green body
[c]The theoretical density of SiC is 3.217 g/cm^3

polymer concentration resulted in higher viscosity, leading to slower rearrangement of the polymer chain conformations to form an effective cross-linked structure. On the other hand, the resulting bending strength and relative green density increased with an increase in the alginate concentration, which indicated that the strength of the green body was mainly determined by the 3D network. It could be concluded that a stronger network and higher strength of the green body can be achieved with increase in the alginate concentration. However, the consolidation rate also increased with the increase of the alginate concentration, and some agglomerates were observed in the consolidation process of the slurry dispersed in the 2.0 wt% Na alginate solution, which probably results in nonuniform structure. In this case, a 1.5 wt% Na alginate solution was employed in the following experiment.

Further characterization of the mechanical properties of the green bodies of samples 4, 2, and 5 were studied, which were fabricated from suspensions with decreasing content of 6, 4, and 1.2 g of $Ca_3(PO_4)_2$ (with 45 vol% SiC and 20 ml 1.5 wt% Na alginate solution), respectively. On comparing samples 5 (2.54 MPa) with 2 (2.85 MPa) and 4 (3.47 MPa), it was found that the strength of the green body increased with the calcium content, presumably due to increase in the cross-linking density. The sodium ions on the Na alginate molecules would be further replaced when more calcium salt was added into the slurry. In this case, the overall contribution of the effective ionic interactions between calcium ions and Na alginate became greater at higher calcium content, leading to better mechanical properties of the green body. Therefore, at low content of $Ca_3(PO_4)_2$ (1.2 g), the sodium ions on the Na alginate molecules would inevitably be replaced incompletely. Thus, the resulting network was not strong, and the density and strength of the green body were reduced. It should also be noted that the suspension with high content of $Ca_3(PO_4)_2$ (6 g), on the other hand, consolidated too quickly, which might have resulted in structural heterogeneity, leading to lower relative density. Therefore, the appropriate content of $Ca_3(PO_4)_2$ was found to be 4 g for this study.

Samples 6, 2, and 7 were produced from 45 vol% SiC slurries and 1.5 wt% Na alginate solution with different contents of 2, 1, and 0.4 ml $(NaPO_3)_6$ (20 wt%), respectively, and the corresponding green bars were also made. When the chelator $((NaPO_3)_6)$ addition is not sufficient, the calcium ions will be chelated incompletely and the redundant calcium ions will react with sodium alginate to form partially gelation, which is detrimental to the fluidity of the slurry. Furthermore, it has a negative effect on the uniformity and density of the green body, as shown in sample 7. Therefore, a high content of $(NaPO_3)_6$ is helpful to the fluidity of the slurry and the ultimate strength of the green body. With an increase in the content of $(NaPO_3)_6$, however, the consolidation rates decrease promptly due to the time-delayed calcium-releasing period. Considering the effect of the chelator addition on both the consolidation rate and the strength of the green body, 1 ml $(NaPO_3)_6$ was found to be appropriate in this study.

Addition of hexanedioic acid plays a critical role in this process, as it can release the calcium ions from the chelate complex and initiate the consolidation. To study the effect of hexanedioic acid on the consolidation, a 45 vol% slurry was prepared, which was composed of 52.4 g SiC powder, 20 ml 1.5 wt% Na alginate solution, 1 ml 20 wt% $(NaPO_3)_6$, 1.2 g $Ca_3(PO_4)_2$, and 1 ml PC33. Disk-shaped green bodies with different hexanedioic acid addition were fabricated. Calcium ions will be released incompletely because of the small amount of hexanedioic acid (<3 g), and therefore, consolidation will not be available. It was observed that a threshold amount of hexanedioic acid is needed to form a fixed shaped green body; below this threshold amount, consolidation is impossible. Table 3.8 demonstrates that the consolidation rate increases with an increasing amount of hexanedioic acid; only when the amount of hexanedioic acid reaches 0.6 g/20 ml, complete consolidation of the Na alginate solution occurs.

Green parts with strength ~3 MPa and perfect surface were obtained in this process. The microstructure of the cross-section of the green body (from the slurry composition of sample No. 2) was studied by an SEM, which shows that the structure of the green body is homogeneous and dense.

Table 3.8 The influence of hexanedioic acid on consolidation

Hexanedioic acid (g)	<0.3	0.5	0.6	0.8	1.0	1.2	>1.5
Consolidation time	>24 h	50 min	50 min	35 min	35 min	20 min	18 min
Consolidating degree	Not consolidated	Fair	Good	Good	Good	Good	Good
The morphology of the green body	Poor	Fair	Good	Good	Good	Good	Good

3.5 Gelcasting of Silicon Carbide Based on Gelation of Sodium Alginate

Summary

Silicon carbide can be suspended in a basic solution. The dispersant DOLARPIX PC33 is better compared to the other two dispersants TMAH and tri-ammonium citrate. A sedimentation experiment showed that the sedimentation of the slurry with solids loading greater than 45 vol% does not clearly occur within 6 h, which indicates that the structure of the green body consolidated within 6 h is homogeneous. The suitable content of sodium alginate for suspensions of 45–55 vol% silicon carbide powder is 1.5 wt%, since such slurries have apparent viscosities lower than 1 Pa s and are pseudoplastic with respect to their rheological properties, which is helpful for casting.

The consolidation rate increases with decreasing concentration of the Na alginate solution, increasing amounts of the solidified agent and hexanedioic acid, and decreasing content of chelator. The green strength is enhanced with the increase in the content of chelator. To obtain the optimal property of the green body, an appropriate concentration of the sodium alginate and the content of the solidified agent should be paid attention. The green strength is mainly determined by the 3D network, which is formed by cross-linking between the sodium alginate and the calcium ions. Considering both the consolidation rate and the green strength, the gelation process of sample No. 2 is suitable for gelcasting. Green bodies of silicon carbide with strength of ~3 MPa, perfect surface, and homogeneous microstructure have been produced by this process.

3.6 Alumina Gelcasting with a Low-Toxicity System of HEMA

3.6.1 Academic Idea and Research Program

HEMA is an innoxious reagent, which is commonly used for soft lens material and as an ingredient in some adhesive resins. HEMA is prone to polymerization due to the vinyl bond while the hydroxy group makes it possible to form a water-compatible polymer. Here, HEMA is selected as a new monomer and it seems to be a successful one. The rheological properties of the ceramic slurries, the microstructure of the green body obtained from the new system, and the pyrolysis property of the system during sintering were carefully investigated.

The composition of the alumina powder used has been described in Sect. 3.1.1 of this chapter.

HEMA produced by Beijing Eastern Acrylic Chemical Co. was used as the monomer and N,N-methylenebisacrylamide (($C_2H_3CONH)_2CH_2$, MBAM) was used as the cross-linker. HEMA and MBAM were dissolved in deionized water to prepare the premix solution. N,N,N',N'-tetramethylethylenediamine (TEMED) and $(NH_4)_2S_2O_8$ were used as the catalyst and initiator, respectively. A commercial product JN281 and TAC were selected as the dispersants. Before casting, the ceramic powder and dispersants were added into the premix solution and thoroughly ball-milled for 24 h to prepare a well-dispersed suspension. After being deaerated for 15 min, polymerization was initiated by adding the initiator and catalyst. Then, the slurry was cast into the molds and heated to 60–80 °C to form the ceramic green bodies of the desired shapes.

The Zeta potentials and rheological properties of the ceramic suspensions were measured by using a ZetaPlus analyzer (Brookhaven Instruments Corporation, USA) and an advanced rheological MCR300 (Physica, Germany) respectively. The detailed pyrolysis of the gelcast polymers was determined by thermogravimetric analysis (TAG) by using a Dupont Thermal Analyst 2000. The room-temperature mechanical strength of the green body was determined by the three-point flexure test.

3.6.2 Colloidal Chemistry and Rheological Property

In order to obtain high-quality ceramic parts by gelcasting, it is important to obtain homogeneously dispersed suspensions with high solids loading. In ceramic suspensions, an intimate relation exists between the rheological property and the suspension structure (i.e., the spatial particle distribution in the liquid), which depends on the zeta potential of the powders. The higher this potential with the same polarity is, the more important the electrostatic repulsion between the particles. The Zeta potentials of different Al_2O_3 suspensions are shown in Fig. 1. The zeta potential of the Al_2O_3 suspensions changes from 7.98 mv at pH = 2.51–16.58 mV at pH = 11.02, with an IEP at about pH = 4.8. Addition of a monomer does not shift the IEP but slightly decreases the absolute value of the Zeta potential. This suggests that the uncharged HEMA molecule may screen the charge developed by the powders in the solution. With the addition of 0.25 wt% TAC and 1 wt% JN281, the IEP of the Al_2O_3 suspensions were shifted to pH = 3.85 and pH = 2.64, respectively, and the zeta potential values became more negative. JN281 can provide a more negative zeta potential, and is likely to be more effective for a well-dispersed suspension (Fig. 3.40).

Figure 3.41 shows the viscosity of the alumina (50 vol%) slurries in the premix solution containing 25 wt% of HEMA. The viscosity curve of the slurry in an AM premix solution is also illustrated for comparison. The high viscosities at the beginning of the curves indicate a 'Bingham' behavior of all the suspensions. Compared with the suspension in the AM solution, which has almost the same viscosity as the suspension in pure water, addition of HEMA slightly increases the viscosity. The slurry in the AM solution shows a shear-thinning behavior while the slurries in the HEMA premix solution show a shear-thinning behavior at low shear rate and a shear-thickening behavior at high shear rate. The shear-thinning behavior reveals that the flow results in a more favorable two-dimensional (2D) structure arrangement rather than a three one of the particles. The 2D arrangement is a layered structure and results in a low resistance of the particle movement between the different layers. Thus, a low viscosity is obtained (Barnes 1989). Due to the damage of the unstable 2D arrangement at very high shear rate, over the critical shearing rate, the viscosities increase

3.6 Alumina Gelcasting with a Low-Toxicity System of HEMA

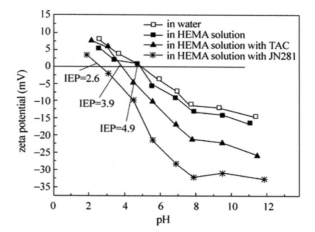

Fig. 3.40 Zeta potential of different Al_2O_3 suspensions

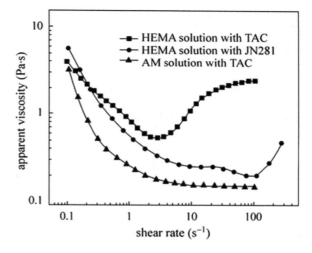

Fig. 3.41 Viscosity versus shear rate for the green strength and microstructure of Al_2O_3 suspensions

again and a shear-thickening behavior is observed in the HEMA system. Although the TAC is a very effective dispersant for the AM system, JN281 seems more appropriate than TAC for the HEMA system. JN281 can provide a much lower viscosity and the critical shearing rate is moved to a higher level than of the TAC system in the HEMA solution. This suggests that JN281 is more effective that TAC in stabilizing the suspension and in decreasing the so-called phase volume fraction (Barnes 1989) in the HEMA solution. These results agree well with the zeta potentials, indicating that the dispersants work by an electrostatic stabilization mechanism.

The strength of the green bodies derived from different HEMA premix solutions with 50 vol% loading is shown in Table 3.9. As expected, the green strength increases with increase in the HEMA concentration. Considering both the mechanical strength and production cost, a HEMA concentration of 25 wt% would be optimal. The

Table 3.9 Strength of the green bodies obtained by different gelcasting systems

	HEMA (20 wt%)	HEMA (25 wt%)	HEMA (30 wt%)	HEMA (40 wt%)	AM (20 wt%)	Gelatin (4.5 wt%)
Green strength (MPa)	14	18	18	20	28	8

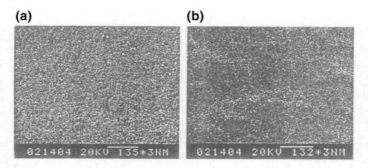

Fig. 3.42 Comparison of the microstructure of the green bodies; **a** HEMA system; **b** AM system

strength of the green bodies derived from AM or gelatin at the same Al_2O_3 loading is also listed. It can be see that although the green strength derived from the HEMA system is lower than that from the AM system, it is much higher than that derived from gelatin (Chen et al. 1999).

Figure 3.42 shows the SEM photographs of the green bodies (50 vol% Al_2O_3) obtained by a HEMA system and an AM system, respectively. A homogeneous structure can be obtained through both systems, suggesting that the HEMA system is suitable for gelcasting.

3.6.3 Binder Burnout and Application of the New System

The TGA curves of the green bodies (50 vol% Al_2O_3) made from a 25 wt% HEMA premix solution and that from a 30 wt% AM solution are shown in Fig. 3.43. The two gelcast bodies show similar pyrolysis behavior. Pyrolysis occurs mainly in the range of 120–680 °C and there are four weight loss peaks for the green body from the AM system. The weight loss in the temperature range of 160–320 °C may be mainly due to the removal of water while those in the range of 320–680 °C may be ascribed to the pyrolysis of the PAM polymer. However, there is only one focused weight loss process in the range of 200–470 °C for the green body from the HEMA system, suggesting that the polymerized HEMA can be removed more easily than the PAM. Based on the above results, complex-shaped alumina parts (Fig. 3.44) were

3.6 Alumina Gelcasting with a Low-Toxicity System of HEMA

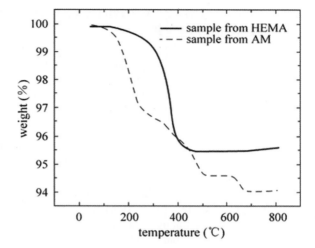

Fig. 3.43 Pyrolysis of the green bodies made from the HEMA system and from the AM system

Fig. 3.44 Ceramic parts fabricated by the new system

successfully fabricated through a premix solution containing 25 wt% HEMA with JN281.

Summary

A new innoxious gelcasting system has been successfully developed by using HEMA. For the Al_2O_3 suspension in the new system, JN281 is more effective than the frequently used TAC. SEM and TGA results indicate that the green bodies made from the new system have a homogeneous microstructure and the organic substance can be removed easily. Dense and homogeneous complex-shaped alumina parts were successfully fabricated.

3.7 A Synergistic Low-Toxicity Gelcasting System by Using HEMA and PVP

3.7.1 Academic Idea and Research Program

(1) PVP overcoming strength distribution in green bodies formed by the HEMA system

We have reported that using HEMA (2-hydroxyethyl methacrylate) as a new monomer. It was exciting that a green strength as high as 18 MPa was obtained from a 25 wt% HEMA aqueous premix solution (Cai et al. 2005). However, a few percent of green bodies might show a much lower strength (~8 MPa as minimum) than the average strength value (18 MPa) because of the heterogeneity of the HEMA hydrogel network. Although the percent is very low, it may constitute a potential problem in cases when narrow strength distribution and high reliability are required. Polyvinyl pyrrolidone (PVP) is a low-toxicity organic reagent widely used in various industries such as cosmetics, pharmaceuticals, and in agricultural formulations as a covering, dispersing, viscosity-enhancement reagent, and also as an adhesive. Owing to the hydrophilic C=O group, it is water-soluble and can form a hydrogel. This paper reports our results of alumina gelcasting with a PVP-modified low-toxicity system. Adoption of PVP as a synergistic gel-forming reagent resolved the problem of wide strength distribution in the green bodies formed by the HEMA system. Besides, the surface exfoliation phenomenon that seems inherent to the AM gelcasting system was also successfully eliminated in the HEMA–PVP system. Analysis of the interaction between HEMA and PVP suggested that the improved microstructure and strength homogeneity, as well as the elimination of surface exfoliation in the new system, were mainly due to their ability to form an intermolecular hydrogen bonding, which greatly improved the homogeneity of the entire poly-HEMA (PHEMA)-PVP co-gel and prevented phase separation and oxygen diffusion.

The alumina powder used was an as-received commercial-grade An-0.5 produced by Xinyuan Aluminum Inc. (Jiyuan, China) with a mean particle size of 2.0 μm. HEMA produced by Eastern Acrylic Chemical Co. (Beijing, China) was used as the monomer. PVP (K-30 type) was an analytically pure reagent with an average molecular weight of ~40,000 provided by the Beijing Chemical Reagent Company (Beijing, China). The cross-linker N,N-methyl-enebisacrylamide (($C_2H_3CONH)_2CH_2$, MBAM), the catalyst N,N,N′,N′–tetramethylethylenediamine (TEMED), and the initiator $(NH_4)_2S_2O_8$ (analytically pure) were the same as those used in the AM gelcasting system, which were produced by Beijing Hongxing Chemical Factory (Beijing, China), Xingfu Fine Chemical Institute (Beijing, China) and Beijing Chemical Reagent Company (Beijing, China), respectively. Tri-ammonium citrate (TAC) provided by Beijing Chemical Reagent Company and a JN281-type polyelectrolyte (ammonium salt of poly (acrylic acid)) provided by Beijing Pinbao Co. were used as the dispersants to achieve a concentrated alumina suspension.

(2) Sample Preparation and Characterization

A 0.05 wt% alumina suspension was used for measurement of zeta potentials (ζ) through a ZetaPlus apparatus produced by Brookhaven Instruments Corporation (Holtsville, NY). The rheological properties of the ceramic slurries were measured by an advanced Physica MCR300 rheometer (Anton Paar Corp., Ostfildern, Germany). The room-temperature mechanical strength of the green bodies was determined by three-point flexural tests. The microstructures of the sample were observed by a HITACHI (Ibaraki, Japan) S-450 SEM.

The procedure for forming green bodies is similar to that of the AM system. HEMA and PVP were dissolved in deionized water, and then the alumina powder and dispersant were added. After thorough ball-milling for about 24 h, the catalyst and initiator were added to the alumina suspension and heated at 80 °C for in situ solidification.

3.7.2 ζ Potentials and Rheological Properties

Development of stable, high solid-loading ceramic suspension with low viscosity is still a prerequisite for the successful application of gelcasting. An effective and necessary route to achieve this is to use a surfactant. TAC is an effective dispersant for alumina and we found that 0.3% TAC (by weight of alumina powder) would be the optimum level for a solution with AM monomer. Its dispersing effect may be mainly due to its electrostatic repulsion. Our results indicated that a polyelectrolyte dispersant, ammonium salt of poly (acrylic acid), had a better effect than TAC for the HEMA monomer solution. Therefore, the commercial polyelectrolyte was also used as the dispersant for the HEMA–PVP system.

The ζ of the particle is defined as the electrical potential between the shear plane surrounding the particle and the bulk solution in a suspension. It is the most important parameter of an electrical double layer and a good measure of the magnitude of the repulsion or attraction between the particles. The ζ s of different suspensions are shown in Fig. 3.45. It can be seen that the alumina particles show small ζ relative values from 8.37 to -14.9 mV in a wide pH range of 3.2–11.6 with an IEP at about pH $= 5.8$. Addition of HEMA or PVP slightly decreases the absolute ζ values of the particle, while they show no obvious effect on the IEP. These may indicate that the neutral HEMA or PVP molecules absorbed on the alumina particle have a charge-screening effect. The dispersant obviously decreases the ζ values under the same pH conditions and moves the IEP to pH $= 3.2$, which can be due to the adsorption of ionized negative chains of the dispersant molecule on the particles. Therefore, it may be expected that the viscosity of the alumina suspension will be increased by adding HEMA and PVP and be decreased by adding JN281 as far as the electrostatic repulsive force is concerned.

Fig. 3.45 Comparison of pH value dependence of Zeta potentials for different suspensions

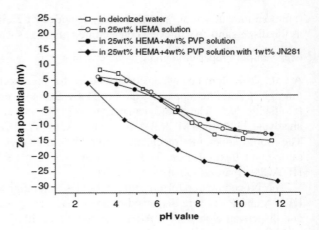

Consistent with the ζ s, we can see from the shear dependence of viscosity in Fig. 3.46. The alumina suspensions with dispersants, especially JN281, show much lower viscosities compared with the suspension in pure water, even with a much smaller volume fraction of 40 vol%. Addition of 4 wt% PVP slightly increases the viscosities of the suspension containing 25 wt% HEMA. Here, suitable contents of HEMA and PVP were selected based on our preliminary tests and the dispersant contents were the optimum values with which the minimum viscosities could be obtained. The data are based on suspensions with pH values of ~7 (without any pH value controlling chemicals), because our previous results indicated that the pH value had a very small role in decreasing the viscosities of the concentrated ceramic slurries (Cai et al. 2005). The data agree with the zeta potential values, indicating that PVP and the effect of the polyelectrolyte on the viscosities can partly be attributed to their

Fig. 3.46 Shear dependence of viscosities of different alumina suspensions

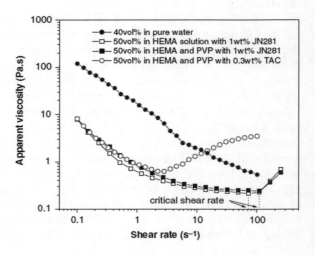

influence on the electrostatic repulsive forces. All three 50 vol% suspensions (with dispersant JN281 or TAC) show shear-thinning behavior at low shear rates and shear-thickening behavior at high shear rates. Shear-thinning behavior generally indicates that the flow can induce a more favorable 2D layered structure arrangement of the particles because a flow-induced layered arrangement can decrease the resistance of particle movement between different layers; thus, a low viscosity can be obtained (Barnes 1989). For all concentrated suspensions of non-aggregating solid particles in a Newton fluid, shear-thickening will occur at a sufficiently high shear rate due to the instability and damage of the 2D arrangement at the high shear stress. The shear rate where viscosity starts increasing is called the critical shear rate (Barnes et al. 1989; Bares 1989). The critical shear rate is related to various conditions such as effective phase volume, particle size distribution, and continuous phase viscosity. Under the same conditions, a lower critical shear rate is undesirable because it means decreased stability of the flow-induced layered structure. Therefore, both the lower viscosities and the higher critical shear rates of the suspension with JN281 compared with those of the solution with TAC indicate that JN281 is better than TAC in stabilizing the suspension of the HEMA system and in decreasing the so-called phase volume fraction of the suspension (Barnes 1989). Besides, it is also apparent from Fig. 3.46 that although the addition of PVP in the HEMA suspension increases the viscosity slightly, it increases the critical shear rate. In fact, we can obtain flowable suspensions with the same maximum solid loading of about 50 vol% in the HEMA solution. This suggests that the addition of PVP to the HEMA system has very little effect on viscosities.

As for the polyelectrolyte effect, we could not attribute its effect only to the increased electrostatic repulsive forces as indicated by the zeta values. It has been reported by Tari et al. (1998) that the increase of the ζ by adding a dispersant may not assure a good dispersion, because each particle may have an 'interaction size' or 'effective diameter' that is approximately its real diameter plus twice the range of total surface forces. The larger effective diameter of particles with higher ζ values may cause a high viscosity in its suspension. However, in our opinion, it is difficult to judge under what circumstances the increase of the 'interaction size' can counteract the effect of the electrical repulsion in preventing particle aggregation, because the 'effective diameter' may be only applicable for a suspension with high solids loading and small particle sizes. However, irrespective of whether the higher ζ value of JN281 means a better electrostatic stabilization or not, it seems reasonable that its better dispersion should also be attributed to its steric stabilization. Generally, steric stabilization requires adsorption of a reasonably dense polymer layer on each particle, and in turn a sufficient supply of polymer in the solution (Horn 1990). This might account for the much higher optimum value of JN281 than that of TAC. An additional level of 1 wt% JN281 can yield a 50 vol% alumina suspension with a viscosity of 8.2 Pa s at a shear rate of about 0.1 s^{-1}, which is generally indicative of a ceramic suspension with good flowability. A further increase of the dispersant content has an inverse effect while a little higher solid loading can greatly increase the viscosity as predicted by the Krieger–Dougherty equation (Barnes 1989; Guo et al. 2003).

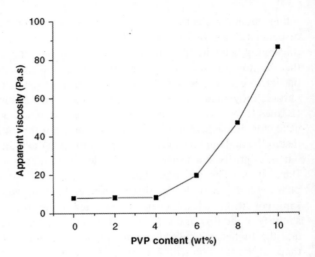

Fig. 3.47 Influence of PVP concentration on the viscosities of 50 vol% alumina suspension at a shear rate of 0.1 s^{-1}

On the other hand, because PVP is a macromolecule (the PVP used here has a molecular weight of ~40,000), it should also have a more or less steric effect that can prevent particles from aggregating. This may be the reason for the small influence of PVP on the suspension viscosity. In fact, PVP is also used as a steric dispersant for inorganic (e.g., CeO_2) and organic materials (e.g., PMMA) (Cao et al. 2000; Yang and Rahaman 1997). However, further increasing the content of PVP above 6 wt% will greatly increase the viscosity of the slurries. The influence of the PVP concentration on the viscosities of a 50 vol% alumina suspension is illustrated in Fig. 3.47. Addition of a polymer to the suspension can have various and subtle effects on the surface force of the particle, depending on the chemical nature of the polymer, the solution, and the particle, as well as on the polymer molecular weight, particle concentration, etc. Bridging flocculation is unlikely to occur in this case for PVP because the polyelectrolyte dispersant is dominant on the particle surface. Therefore, another type of flocculation for soluble polymers, depletion flocculation, should be the reason for the increase of viscosity at a higher PVP concentration, since the osmotic pressure difference in the depletion flocculation mechanism will be enhanced at a higher polymer concentration (Horn 1990; Kiratzis and Luckham 1999). The effect of PVP on increasing the intrinsic viscosity of the liquid phase should also be accounted for.

3.7.3 Activation Energies and Solidification

Different monomers require different suitable cross-linker, initiator, and catalyst concentrations in order to obtain the suitable polymer network structure and strength. Based on our preliminary test, for a 25 wt% HEMA premix solution, the suitable cross-linker, initiator, and catalyst contents are about 8 wt% 0.18 g/L, and 0.3 vol%,

3.7 A Synergistic Low-Toxicity Gelcasting System by Using ...

respectively. The data are similar to those for an AM system. In order to compare the kinetic aspect of the solidification reaction of the HEMA system with that of the AM system, the activation energies E_a of the two systems with the same initiator and catalyst contents were compared. Since the HEMA slurry temperature showed no sharp increase with time after adding the initiator and catalyst, in case of a larger error in measuring the idle time here, the solidification time t_s was tested as a measure of the reaction rate deriving from the E_a of the gels with powders, 1 namely, the time period from addition of the initiator to a solidified state where the slurry just lost its flowability in a flatly placed test tube (beginning of solidification). Due to the inverse proportional relationship of t_s with the reaction rate, its relationship with the temperature T may be expressed by the Arrhenius-type equation:

$$t_s \propto \frac{1}{r} = A \exp^{E_a/RT} \quad (3.4)$$

where r represents the reaction rate, R is the standard gas constant 8.3144 J/K mol, and E_a represents the activation energy of the total reaction. The logarithm plot of t_s with 1/T and the corresponding E_a data are shown in Fig. 4. We can see that with the same initiator and catalyst concentrations of 0.18 g/L and 0.3 vol%, the activation energies of the AM and HEMA systems were about 40.1 and 56.4 kJ/mol, respectively. A small amount of PVP (4 wt%) showed no effect on the E_a value of HEMA because there is no observable change in t_s. Although an error may exist for t_s due to the difficulty in precisely determining the solidification point, the data reveals that HEMA has higher activation energy than AM. Therefore, the gel-forming reaction of HEMA is more sensitive to temperature than that of the AM monomer. The slurry in the AM premix solution can solidify (lost flow ability) in 4–5 min at a temperature of 50 °C, while it will take three times longer for the HEMA-PVP system to solidify at the same temperature. However, at 50 °C, the solidification times are 3 and 5 min for AM and HEMA, respectively. Therefore, we should heat the HEMA-PVP system to accelerate the solidification. Slight heating can substantially shorten the solidification time (Fig. 3.48).

3.7.4 Green Strengths and Microstructures

As noted above, a few percent of green bodies might show a much lower strength than the average strength value. As a result, these samples may fracture under small stress and hence result in large uncertainties in the survival probabilities. Although a wide strength distribution for green parts is not as detrimental as that for sintered ceramics because the sintering process can density the body and remove most pores, it is a serious problem since cracks may still remain in the final ceramics. The effect of the PVP content on the alumina green strength derived from 50 vol% alumina slurries containing 25 wt% HEMA is shown in Fig. 3.49. It is evident that the green strength of the samples increases slightly with the PVP content. It is also evident from

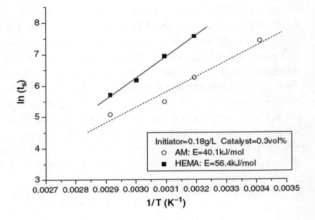

Fig. 3.48 Comparison of activation energy (E_a) data of acrylamide and 2-hydroxyethyl methacrylate gels with 50 vol% alumina powders

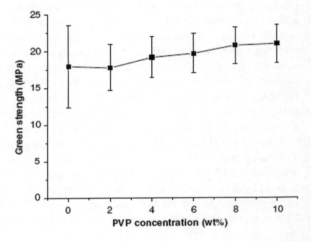

Fig. 3.49 Effect of PVP content on the flexural strength of alumina green bodies; the error bars in the figure represent the standard deviation values of each group of data

the error bars in the figure that a small amount of PVP greatly decreases the standard deviation (where each value is the mean of eight repeats) of the green strength data, indicating that the reagent can increase the strength (microstructural) homogeneity. Although a further increase in the PVP content above 4 wt% can slightly increase the strength, it seems to have no obvious effect on the standard deviation of the data, and the slurry viscosity was found to increase obviously at a higher PVP level as shown in Fig. 3.47. Thus, the optimum PVP content was determined to be 4 wt%. A strength of 19 MPa can be achieved for the sample derived from the solution containing 25 wt% HEMA and 4 wt% PVP. Although it is lower than that derived from AM with the same solid loading (28 MPa), it is much larger than that derived from gelatin (8 MPa) and chitosan (<1 MPa) (Huzzard and Blackburn 1997; Olhero et al. 2000; Chen et al. 1999; Johnson et al. 1991). Such a strength is high enough for further handling and machining.

3.7 A Synergistic Low-Toxicity Gelcasting System by Using …

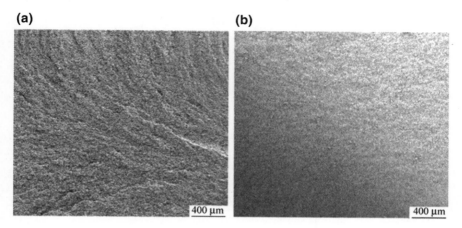

Fig. 3.50 SEM photograph of the fracture surface of alumina green bodies formed from the 2-hydroxyethyl methacrylate premix solution **a** without PVP and **b** with PVP

The SEM photographs of fractured green alumina bodies cast from the HEMA monomer with and without PVP are shown in Fig. 3.50a and b, respectively. The green body without PVP shows many grooves in the fractured surface that are indicative of a heterogeneous packing of the particles, while the addition of PVP obviously improves the homogeneity of the green microstructure. The SEM pictures agree well with the strength data. PVP is a polymer, and a low PVP addition level of 4 wt% has no significant effect on the polymerization of HEMA; therefore, its modification effect should be attributed to reasons other than its effect on the polymerization of HEMA. The reason for this will be discussed in detail in the following section.

3.7.5 Exfoliation Elimination Effect and Analysis of the Interaction Between PVP and HEMA Molecules

In the gelation process of gelcasting, the monomer undergoes a radical-type polymerization after the initiator is added. This chain reaction is strongly inhibited by oxygen, which scavenges very efficiently both the initiating and the polymer radicals to form relatively stable, free radicals of peroxide. Consequently, when polymerization occurred in ambient air, inhibition of gelation due to the presence of oxygen was encountered, and surface exfoliation occurred. A thin layer of about 1 mm will peel off and the geometry of the ceramics will be affected (please refer Fig. 3.51a), especially for small parts. Surface exfoliation also occurs for the HEMA system but not as severely as in the AM system because the peeled layer is thinner. Interestingly, another advantage of the PVP-modified HEMA system is that there is no exfoliation for the bodies that are gelcast from the modified system (please refer Fig. 3.51b). As

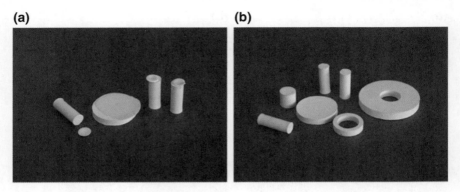

Fig. 3.51 Illustration of the alumina parts gelcast by different systems **a** formed by acrylamide, and **b** formed by PVP improved 2-hydroxyethyl methacrylate

noted above, the diffusion of oxygen from the surface to the inner part should be considered to estimate the exfoliation phenomenon during polymerization. According to Stoke's Law, the relationship for the resistance force of spherical particle movement is given by:

$$F = 6\pi \eta r v \tag{3.5}$$

where η, r, and v are the medium viscosity, particle radius, and movement velocity, respectively. As a result, the diffusion coefficient of oxygen is inversely related to the medium viscosity. Although PVP has little effect in increasing the viscosity of the alumina suspension owing to its steric stabilization, it can increase the intrinsic viscosity of the water phase of the hydrogel. Therefore, the exfoliation suppression effect of the HEMA-PVP system may be partly ascribed to the relatively higher viscosity of the water phase during gelation, which in turn decreases the diffusion coefficient of oxygen and prevents it from diffusing into the system. Dufaud's research on ceramic suspensions for stereolithography and Studer's study on the curing of organic coating drew similar conclusions on oxygen diffusion (Studer et al. 2003; Dufaud and Corbel 2003).

Meanwhile, due to the polar hydroxyl group of PHEMA, the negatively charged oxygen atom of the carbonyl group in PVP can form a hydrogen bond with the active hydrogen atom in PHEMA as shown in Fig. 3.52. The hydrogen-forming ability of the two chemicals is well known. Goh and Lee (1990) have proved that the incorporation of HEMA into EMA (ethyl methacrylate) could increase the compatibility of the copolymer with PVP due to the hydrogen bonding between HEMA and PVP (Goh and Lee 1990). The ability of PVP to form single-phase blends with various polymers containing hydroxyl groups and the existence of hydrogen bonding between hydroxyl and carbonyl groups has been confirmed by FTIR (Moskala et al. 1985; Thyagarajan and Janarthanan 1989; Costa and Vasconcelos 2002). Polymers with hydrogen bonds are very good barriers against permeating gases because the bridging effect of the

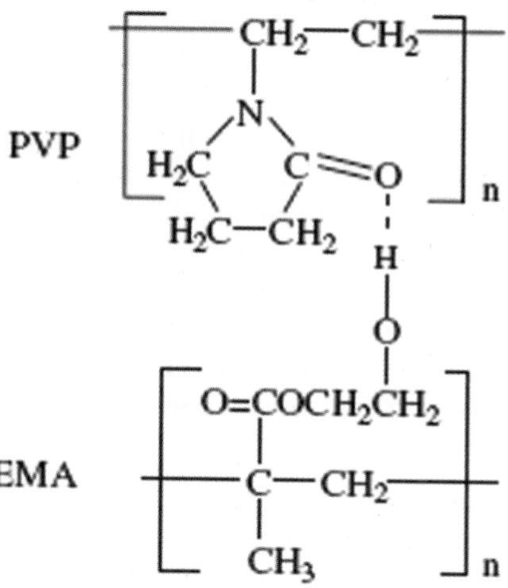

3.7 A Synergistic Low-Toxicity Gelcasting System by Using ...

Fig. 3.52 Illustration of the molecular structures of PVP and poly-2-hydroxyethyl methacrylate and the hydrogen bonding in between

hydrogen bonds could slow down the diffusion of penetrant molecules (Karlsson et al. 2002) Besides, molecular dynamics simulations of the permeation processes of oxygen in PVA hydrogels showed that the processes are highly correlated with the orientational relaxation of the side chains of polymers and the reorganization of the hydrogen bond networks (Tamai and Tanaka 1998). Since the oxygen molecule tends to move along the polymer chain, this movement is sensitive to the number of polymer–polymer hydrogen bonds. Therefore, the diffusivity depended uniquely on the number of the hydrogen bonds (Karlsson et al. 2004). It was also found that the lifetimes of the hydrogen bonds between the polymer chain segments were longer than those between water and polymer chain segments; therefore, the water diffusion in the polymer was also constrained by the hydrogen bonds since the jump of the water molecule from one site to another requires breaking of hydrogen bonds (Karlsson et al. 2002; Tamai et al. 1996). Based on these findings, we consider that another important reason or main reason for the exfoliation suppression effect is the hydrogen bonds formed between PVP and PHEMA.

When PHEMA is obtained by polymerization in water in amounts exceeding the equilibrium water content (EWC 5 (swollen gel weight-dry gel weight)/swollen gel weight), phase separation will result. The EWC of the PHEMA hydrogel is very sensitive to the cross-linker and initiator concentration, and the phase separation can result in heterogeneity, cracks, and low strength in the hydrogel (Perera and Shanks 1996; Dalton et al. 2002; Liu et al. 2000). Sometimes, we found that a very thin layer of moisture was formed on the fractured surface of bodies that were gelcast through the HEMA solution. Therefore, the occasionally formed low-strength samples may have presumably ascribed to the phase separation caused by the decrease of EWC, which may result from the difficulty in precisely controlling the initiator concentration due to its small quantity. It has been reported that incorporation of PVP can greatly enhance the EWC of the HEMA hydrogel and the hydrogen bonding can bind the water more tightly and substantially prevent phase separation (Malak et al.

2003). This is another proof of the effect of PVP in improving the homogeneity through the formation of hydrogen bonding.

Combining the strength data, SEM photographs, and the exfoliation suppression effect, it seems reasonable to conclude that the hydrogen bonding between PVP and PHEMA is the basic reason for the increased gel homogeneity. Thus, we may call the PVP-modified system a synergistic gel-forming system. A combination of HEMA and PVP leads to an obvious improvement that cannot be obtained by either of the reagents. Furthermore, it may be reasonable to propose that the suppression of surface exfoliation and improvement of the gel homogeneity may also be achieved by other gel-forming reagents that can form strong hydrogen bonds with the PAM or PHEMA, provided that we can also achieve high solids loading required for gelcasting with the reagents.

Summary

The addition of PVP greatly increased the homogeneity of the strength data and the microstructure of the green alumina bodies gelcast from a HEMA premix solution. The addition of PVP can also increase the viscosity of the water phase of the PHEMA hydrogel, and thus can suppress the surface exfoliation phenomenon of the HEMA system by preventing oxygen diffusion. On the other hand, the intermolecular hydrogen bonding between PVP and PHEMA has several effects: it can act as a barrier for the diffusion of water and oxygen; it can increase the EWC of the hydrogel, and in turn can prevent phase separation, which is the origin of low-strength samples. Therefore, hydrogen bonding is considered the main reason for the elimination of surface exfoliation and for the increased homogeneity of the strength data and the microstructure of the green alumina bodies. Due to this hydrogen bonding interaction, we may call the HEMA–PVP system a synergistic gelforming system.

References

Alford, N. M., Brichalli, J. D., & Kedall, K. (1987). High-strength ceramics through colloidal control to remove defects, [J]. *Nature 330,* 51.

Amott, S., Fulmer, A., & Scott, W. (1974). The agarose double helix and its function in agarose gel structure, [J]. *Journal of Molecular Biology, 90,* 269–284.

Barnes, H. A. (1989). Solid particles dispersed in Newtonian liquids, [J]. *Journal of Rheology, 33*(2), 329–366.

Barnes, H. A., Hutton, J. F., & Walers, K., (1989). *An introduction to rheology.* Oxford: Elsevier Press [M].

Boersma, W. H., Baets, P. J. M., Laven, J., & Stein, H. N. (1991). Time-dependent behavior and wall slip in concentrated shear thickening dispersions [J]. *Journal of Rheology, 35,* 1093–1119.

Braccini, I. (1999). Conformational and configurational features of acidic polysaccharides and their interactions with calcium ions [J]. *Carbohydrate Research, 317,* 119–130.

Cai, K., Huang, Y., & Yang, J. L. (2005). Alumina gelcasting by using HEMA system, [J]. *Journal of the European Ceramic Society, 25*(7), 1089–1093.

Cao, K., Yu, J., Li, B. G., Li, B. F., & Pan, Z. R. (2000). Micron-size uniform poly (methylmethacrylate) particles by dispersion polymerization in polar media 1, particle size and particle size distribution, [J]. *Chemical Engineering Journal, 78*(2–3), 211–215.

References

Cesarano, J., III, & Aksay, I. A. (1988). Processing of highly concentrated aqueous alpha-alumina suspensions stabilized with poly-electrolytes, [J]. *Journal of the American Ceramic Society, 71*(12), 1062–1067.

Chen, Y. L., Xie, Z. P., Yang, J. L., & Huang, Y. (1999). Alumina casting based on gelation of gelatin, [J]. *Journal of the European Ceramic Society, 19*(2), 271–275.

Clare, K. (1993). Algin. In *Industrial Gums* (3rd Ed.) [M].

Costa, R. O. R., & Vasconcelos, W. L. (2002). Structural modification of poly(2-hydroxyethyl methacrylate)-silica hybrids utilizing 3-methacryloxypropyltrimethoxysilane, [J]. *Journal of Non-Crystalline Solids, 304*, 84–91.

Dalton, P. D., Flynn, L., & Schoichet, M. S. (2002). Manufacture of poly (2-hydroxyethyl methacrylate-co-methyl methacrylate) hydrogel tubes for use as nerve guidance channels, [J]. *Biomaterials, 23*(12), 3843–3851.

Dufaud, O., & Corbel, S. (2003). Oxygen diffusion in ceramic suspensions for stereolithography [J]. *Chemical Engineering Journal, 92*, 55–62.

Fanelli, A. J., Silvers, R. D., Frei, W. S., Burlew, J. V., & Marsh, G. B. (1989). New aqueous injection molding process for ceramic powders [J]. *Journal of the American Ceramic Society, 72*, 1833–1836.

Goh, S. H., & Lee, S. Y. (1990). Compatibility of poly (N-vinyl-2-pyrrolidone) with ethylmethacrylate 2-hydroxyethyl methacrylate copolymers, [J]. *Macromolecular Chemistry and Physics, 191*(7), 3081–3085.

Guo, D., Cai, K., Huang, Y., Li, L. T., & Gui, Z. L. (2003). Water based gelcasting of lead zirconate titanate, [J]. *Materials Research Bulletin, 38*(5), 807–816.

Hoffman, R. L. (1972). Discontinuous and dilatant viscosity behavior in concentrated suspensions. I. Observation of a flow instability, [J]. *Transactions of the Society of Rheology, 16*, 155.

Horn, R. G. (1990). Surface forces and their action in ceramic materials [J]. *Journal of the American Ceramic Society, 73*(5), 1117–1135.

Huzzard, R. J., & Blackburn, S. (1997). A water-based system for ceramic injection molding [J]. *Journal of the European Ceramic Society, 17*(2–3), 211–216.

Jin, J., & Lin, M. (1993). *Utilization and fabrication of algae* [M]. Beijing: Science Press.

Johnson, S. B., Dunstan, D. E., & Franks, G. V. (1991). Rheology of cross-linked chitosan–alumina suspensions used for a new gelcasting process [J]. *Journal of the American Ceramic Society, 85*(7), 1699–1705.

Karlsson, G. E., Johansson, T. S., Gedde, U. W., & Hedenqvist, M. S. (2002). Physical properties of dense amorphous poly(vinyl alcohol) as revealed by molecular dynamics simulation [J]. *Journal of Macromolecular Science, B41*(2), 185–206.

Karlsson, G. E., Gedde, U. W., & Hedenqvist, M. S. (2004). Molecular dynamics simulation of oxygen diffusion in dry and water-containing poly(vinyl alcohol) [J]. *Polymer, 45*, 3893–3900.

Kiratzis, N. E., & Luckham, P. F. (1999). The rheology of aqueous alumina suspensions in the presence of hydroxyethylcellulose as binder [J]. *Journal of the European Ceramic Society, 19*(15), 2605–2612.

Kuo, C. K. (2001). Biomaterials, Ionically cross-linked alginate hydrogels as scaffolds for tissue engineering: Part 1. Structure, gelation rate, and mechanical properties [J]. *Biomaterials, 22*(6), 511–521.

Lange, F. F. (1989). Powder processing science and technology for increasing reliability [J]. *Journal of the American Ceramic Society, 72*(1), 3–21.

Liu, Q., Hedberg, E. L., Liu, Z. W., Bahulekar, R., Meszlenyi, R. K., & Mikos, A. G. (2000). Preparation of macroporous poly(2-hydroxyethyl methacrylate) hydrogels by enhanced phase separation [J]. *Biomaterials, 21*(21), 2163–2169.

Malak, K. M., Hill, D. J., & Whittaker, A. K. (2003). Water sorption into poly[(2-hydroxyethyl methacrylate)-co-(1-vinyl-2-pyrrolidone) [J]. *Polymer International, 52*(11), 1740–1748.

Millan, A. J., Nieto, M. I., Baudin, C., & Moreno, R. (2002a). Thermogelling polysaccharides for aqueous gelcasting—part I: A comparative study of gelling additives [J]. *Journal of the European Ceramic Society, 22*(13), 2217–2222.

Millan, A. J., Nieto, M. I., Baudin, C., & Moreno, R. (2002b). Thermogelling polysaccharides for aqueous gelcasting—part II: Influence of gelling additives on rheological properties and gelcasting of alumina [J]. *Journal of the European Ceramic Society, 22*(13), 2223–2230.

Moskala, E. J., Varnell, D. F., & Coleman, M. M. (1985). Concerning the miscibility of poly (vinyl phenol) blends [J]. *Polymer, 26,* 228–234.

Olhero, S. M., Tari, G., Coimbra, M. A., & Ferreira, J. M. F. (2000). Synergy of polysaccharide mixtures in gelcasting of alumina [J]. *Journal of the European Ceramic Society, 20*(4), 423–442.

Omatete, O. O., Janny, M. A., & Strelow, R. A. (1991). Gelcasting a new ceramic forming process [J]. *Ceramic Bulletin, (AcerS), 70*(10), 1641–1649.

Perera, D. I., & Shanks, R. A. (1996). Swelling and mechanical properties of cross-linked hydrogels containing N-vinyl pyrrolidone [J]. *Polymer International, 39*(2), 121–127.

Pierre, A. C., & Ma, K. (1999). DLVO theory and clay aggregate architectures formed with $AlCl_3$ [J]. *Journal of the European Ceramic Society, 19*(8), 1615–1622.

Plovnick, R. H. (1997). Low-density silicon nitride beads as high-temperature microwave furnace insulation [J]. *Materials Research Bulletin, 32*(6), 749–754.

Rao, R. R., Roopa, H. N., & Kannan, T. S. (1999a). Dispersion, slip casting, and reaction nitridation of silicon–silicon carbide mixtures [J]. *Journal of the European Ceramic Society, 19*(12), 2145–2153.

Rao, R. R., Roopa, H. N., & Kannan, T. S. (1999b). Effect of pH on the dispersibility of silicon carbide powders in aqueous media [J]. *Ceramics International, 25,* 223–230.

Reed, J. S. (1994). *Principles of ceramics processing [M]* (pp. 447–502). New York: Wiley.

Santacruz, I., Nieto, M. I., & Moreno, R. (2002). Rheological characterization of synergistic mixture of carrageenans and locust bean gum for aqueous gelcasting of alumina [J]. *Journal of the American Ceramic Society, 85*(10), 2432–2436.

Studer, K., Decker, C., Beck, E., & Schawlm, R. (2003). Overcoming oxygen inhibition in UV-curing of acrylate coatings by carbon dioxide inerting: Part II [J]. *Progress in Organic Coatings, 48,* 101–111.

Tamai, Y., Tanaka, H., & Nakanishi, K. (1996). Molecular dynamics study of polymer-water interaction in hydrogels, [J]. 1. Hydrogen-Bond Structure. *Macromolecules, 29,* 6750–6760.

Tamai, Y., & Tanaka, H. (1998). Permeation of small penetrants in hydrogels, [J]. *Fluid Phase Equilibrium, 144,* 441–448.

Tari, G., Ferreira, J. M., & Lyckfeldt, O. (1998). Influence of the stabilizing mechanism and solid loading on slip casting of alumina, [J]. *Journal of the European Ceramic Society, 18*(5), 479–486.

Teng, W. D., Edirisinghe, M. J., & Evans, J. R. G. (1997). Optimization of dispersion and viscosity of a ceramic jet printing ink [J]. *Journal of the American Ceramic Society, 80*(2), 486–494.

Thyagarajan, G., & Janarthanan, V. (1989). FTIR and thermal analysis studies of poly(vinyl alcohol)–poly(vinyl pyrrolidone) blends [J]. *Polymer, 30,* 1797–1799.

Ward, A. G., & Courts, A. (1977). *The science and technology of gelatin* (England, Translated by Li W. et al.) [M]. London: Academic Press.

Wber, K., & Tomandl, G. (1997). Preparation of structured ceramics for membranes [J]. *Key Engineering Materials, 132–136,* 1754–1757.

Whistler, R. L. (1973). *Industrial gums* [M] (pp. 45–68). New York: Academic Press.

Xie, Z. P., Huang, Y., Wu, J. G., & Zheng, L. L. (1995). Microwave debinding of a ceramic injection molded-body [J]. *Journal of Material Science Letters, 14,* 794.

Xie, Z. P., Yang, J. L., Chen, Y. L., & Huang, Y. (1998). Gelation forming process of ceramic compacts by agarose, [C]. In *9th CIMTEC-World Ceramics Congress and Forum on New Materials,* Florence, Italy, 14–19 June 1998.

Xie, Z. P., Yang, J. L., Huang, D., Chen, Y. L., & Huang, Y. (1999). Gelation forming of ceramic compacts using agarose, [J]. *British Ceramic Transactions, 98*(2), 58–61.

Xie, Z. P., & Huang, Y. (2001). A new gel casting of ceramics by reaction of sodium alginate and calcium iodate at increased temperature, [J]. *Journal of Material Science Letters, 20,* 1255–1257.

Yang, J. L., Xie, Z. P., Tang, Q., & Huang, Y. (1998). Study on rheological behavior and gelcasting of α-Al_2O_3 suspension, [J]. *Journal of the Chinese Ceramic Society, 26,* 41.

References

Yang, X., & Rahaman, M. N. (1997). Thin films by consolidation and sintering of nanocrystalline powders [J]. *Journal of the European Ceramic Society, 17*(4), 525–535.

Yi, Z. Z., Xie, Z. P., Huang, Y., Ma, J. T., & Cheng, Y. B. (2001). Study on gelcasting and properties of recrystallized silicon carbide, [J]. *Ceramics International, 28*(4), 369–376.

Zhang, J. G., Edirisinghe, M. J., & Evans, J. R. G. (1989). A catalogue of ceramic injection molding defects and their causes, [J]. *Industrial Ceramics, 9,* 72.

Zhang, T., Blackburn, S., & Bridgwater, J. (1994). Properties of ceramic suspensions for injection molding based on agar binders, [J]. *British Ceramics Transactions, 93*(61), 229–233.

Zhou, L. J., Huang, Y., Xie, Z. P., Yang, J. L., & Tang, Q. (1998). Study on gelcasting of silicon carbide. In *Proceedings of the First China International Conference on High-Performance Ceramics, October 1998, Beijing.*

Zhou, L. J., Huang, Y., & Xie, Z. P. (2000). Gelcasting of concentrated aqueous silicon carbide suspension, [J]. *Journal of the European Ceramic Society, 20*(1), 253–257.

Zhou, Z. K., Gu, T. R., & Ma, J. M. (1996). *Foundation of colloidal chemistry* [M] (pp. 318–327). Beijing University Press.

Chapter 4
Generation, Development, Inheritance, and Control of the Defects in the Transformation from Suspension to Solid

Abstract This chapter focuses on the generation, development, inheritance, and control of the defects in colloidal forming process, by which the complex near-net shape ceramic products with a homogeneous microstructure and high reliability were consequentially achieved. The rheological performance of the ceramics suspension is very important to the colloidal forming processes. In our studies, the rheological behaviors of aqueous alumina suspensions were investigated, and the various ways of modifying the rheological behaviors of the suspensions were discussed. A critical solid volume fraction between the shear-thinning and shear-thickening behavior was found in an aqueous alumina suspension. Moreover, the gelation of the suspension with high solid volume fraction of alumina (SVF) got delayed and the strength of the green body decreased, which implies that the fast polymerization of monomers in high SVF alumina suspension was inhibited, and the flexibility of the gelcasting was improved. We found that there was an optimal SVF for the ZTA suspension to obtain the highest bending strength and the highest Weibull modulus for the sintering bodies. Micro mechanisms of agglomerations' generation in ceramic suspensions used in colloidal forming processing were investigated. On the basis of the potential energy between particles in liquid, a microscopic model of agglomerations in ceramic suspensions was presented. Two types of agglomerations, tight and loose, were observed by using an environmental scanning microscope as a quasi-direct method. It was observed that their generation is caused by the deviation of the solid-phase content in the slurry from the stable solid loading. The influences of the idle time on the microstructures and the mechanical properties of the green bodies are also described in this chapter. Vertically laminated cracks were observed in the coagulated green bodies after casting. The cracks were attributed to the shrinkage of the particle network (syneresis) during coagulation. This process in combination with sedimentation during the reaction affected the vertical cracks, which was not observed when a certain idle time before casting was introduced. The idle time diminished the final shrinkage still having a castable slurry of reasonably low viscosity. The study indicated that the solids volume loading in the suspension depended on high valence counterions and ion conductivity constants. Further research revealed that the inner stress in the ceramic green bodies originates from the nonuniform shrinkage during the solidification of the precursor suspension and the drying of the green bodies. The gradients of temperature, the initiator concentration, and the moisture are important original factors causing the inner stress. It is found that a proper amount of a water-soluble polymer, such as hydroxyethyl acrylate (HEA), polyethylene glycol (PEG), and polyvinylpyrrolidone (PVP) added into the concentrated suspension can adjust the polymer network structure and thus reduce the inner stress and cracking in the ceramic green body. The debindering time of large ceramic parts can be significantly shortened by reducing the harmful inner stresses in the green body.

Keywords Defect · Generation · Development and control · Rheological behavior · Counterion · Inner stress · Water-soluble polymer

Property repeatability and product cost are two major considerations in the applications of advanced ceramics and ceramic composites. Both are directly or indirectly related to the microstructural defects occurring during the manufacturing process. The random nature of these defects is often responsible for the poor reliability and performance of ceramic products. It is well known that defects like spallation, delamination, microcracks, or large pores are often present in the green bodies formed by traditional forming processes such as dry pressing, isostatic pressing, and injection molding. These defects can hardly be removed in the subsequent processes and become fracture origins, thereby reducing the properties of the product or leading to higher rejection rate. Since the 1990s, more attention has been paid to colloidal forming processes such as gelcasting, direct coagulation casting, and colloidal vibration forming due to their potential to improve the homogeneity of the ceramic body and reduce harmful microstructural defects.

Among the colloidal forming processes, gelcasting, by which complex ceramic parts can be prepared (Omatete et al. 1991), has been widely studied and applied to various ceramic materials (Nunn and Kirby 1996; Waesche and Steinborn 1997; Ma et al. 2002). Homogeneous green bodies with good strength required for machining can be made through this process. The mechanical properties of the materials after sintering are greatly improved (Bossel et al. 1995; Liu et al. 2002a; Huang et al. 2000).

In this process, the mold temperature is a parameter that can be controlled to induce polymerization in the suspension. However, a temperature gradient exists in the suspension due to the heat transfer from the mold surface to the concentrated suspension. Consequently, the solidification of the suspension is not uniform, resulting in inner stresses in the green bodies. Such stresses are often responsible for the cracking of the green bodies in the subsequent drying or debindering process. As the parts become larger, the inner stresses become more harmful. Therefore, it is of great importance to control the inner stresses in the green bodies while preparing large parts by gelcasting.

The inner stresses in the green bodies originate from differential contraction so that the magnitude of the inner stresses increases with the shrinkage rate and the elastic modulus. There are two sources of shrinkage in the gelcasting process: (1) the forming shrinkage due to slight contraction during the polymerization of the monomers, and (2) the drying shrinkage due to the removal of the moisture. A high solid loading in the suspension is preferred in gelcasting in order to reduce the shrinkage of the green body. However, there is a limit for the solid loading in the suspension so as to achieve the desired viscosity and uniformity. In order to control the shrinkage rate, the wet green body is usually dried slowly under controlled temperature and humidity and then debindered at a very low heating rate. This is not favorable in practice because it leads to a prolonged operation time and high production cost.

The advantages of the fabrication process include dimensional accuracy and complex shaping capabilities, as well as reduction in the manufacturing cost. Fundamental research has been carried out by Young et al. (1991) and Omatete et al. (1991), showing the general feasibility of the process and its advantages in comparison with other liquid forming processes. Omatete et al. (1997) discussed in detail the key aspects of gelcasting: a premix solution, the rheology of gelcasting slurries, the drying process of the gelcast parts, binder burnout, the green strength of dried gelcasting parts, etc. Waesche and Steinborn (1997) investigated the influence of the slip viscosity on the mechanical properties of high-purity alumina by gelcasting. They suggested that the interaction of rheological behavior of the suspension and the generation of flaws in the green body are of crucial importance in view of an optimized microstructure for the gelcasting process.

4.1 Rheological Behaviors of Aqueous Ceramic Suspensions

Ceramic gelcasting is an important colloidal forming process. It has been rapidly developed in the past decade due to its capability of near-net-shape forming of fine ceramic pieces. The advantages of the technique include dimensional accuracy and complex shaping capabilities, as well as reduction in the manufacturing cost. Fundamental research has been carried out by Young et al. (1991) as well as Omatete et al. (1991a, 1991b, 1991c), showing the general feasibility of the process and its advantages in comparison with other wet-forming processes.

Omatete et al. (1997) discussed in detail the key aspects of gelcasting process: premix solution, rheology of gelcasting slurries, drying process of gelcast parts, binder burnout, strength of dried gel-casting green body etc. Waesche and Steinborn (1997) investigated the influence of slip viscosity on the mechanical properties of high purity alumina by gelcasting. They suggested that the interaction of rheological behavior of the suspension and the generation of flaws in the green body are of crucial importance in view of an optimized microstructure for the gelcasting process.

So far, most of the research on gelcasting has been focused on the preparation of suspensions and the process control on gelcasting, such as the effect of dispersants and pH on the viscosity and rheological behaviors of suspensions (Leong and Boger 1991), and the concentration of the monomer, usage of catalysts and initiators, and the degassing time (Young et al. 1991; Omatete et al. 1991a, 1991b, 1991c, 1997). The main objective of these studies on rheological behavior was to prepare suspensions with high solid volume fraction, as it is believed that the advantages of gelcasting over traditional wet-forming processes can only be achieved with high solid loading suspensions (Omatete et al. 1997). An exhaustive relationship between the rheological behavior of the suspension and the mechanical properties of the sintered bodies is little known. In the following content, the rheological behaviors of aqueous ceramic suspensions and their effects on the mechanical properties of the ceramics are expatiated.

4.1.1 Rheological Behaviors of Aqueous Alumina Suspensions

In this section, the rheological behaviors of aqueous alumina suspensions are explained, and several effective ways to modify the rheological properties of the suspension are discussed. There is a critical solid volume fraction for aqueous alumina suspension. If the solid volume fraction is below the critical value, the suspension shows shear-thinning behavior all along. On the other hand, when the solid volume fraction reaches or exceeds the critical value, the rheological behavior changes from shear-shinning to shear-thickening. This is explained in detail below.

A high-purity alumina powder (diameter 3.39 μm) from Xinyuan Alumina Ltd., China, was used in the pertinent study. A series of alumina aqueous suspensions at different solid loadings, ranging from 35 to 55 vol%, were prepared by in polypropylene jars filled with ZrO_2 milling balls for 24 h at a speed of 150 rpm. Solid loading (or solid volume fraction) of alumina in the suspensions were calculated according to the following formula:

$$\phi = \frac{m/\rho_{th}}{m/\rho_{th} + V},$$

where m is the mass of alumina powder (g), ρ_{th} is the theoretical density of alumina, and V is the volume (ml) of the premix solution in the suspension. About 0.25 wt% of tri-ammonium citrate was added as dispersant. The rheological measurements of the suspensions were carried out an hour after the preparation. The operation was performed at 25 °C, using the rotational rheometer Physica MCR 300. Data was obtained logarithmically every 10 s ranging from 0.01 to 250 s^{-1} shear rate (Bergstrom 1998; Yang and Sigmund 2003).

Figure 4.1 shows the rheological curves of aqueous alumina suspensions ranging from 40 to 50 vol% solid volume fraction. Although the viscosity at low shear rate is high, it decreases as the shear rate increases. Moreover, the viscosity decreases faster at low shear rate. When the shear rate increases to a certain value, the viscosity curve becomes smooth. The entire set of curves shows shear-thinning behavior.

Figure 4.2 illustrates the rheological curves of aqueous alumina suspensions ranging from 51 to 55 vol% solid loading. The suspension with 53 vol% solid loading shows shear-thinning behavior in the range of 0.08–100 s^{-1} shear rate. Thereafter, in

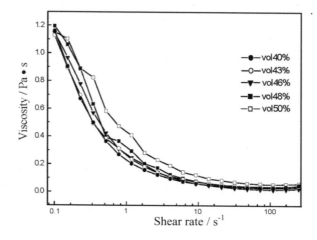

Fig. 4.1 Effects of solid loading on rheological behavior of aqueous alumina suspensions

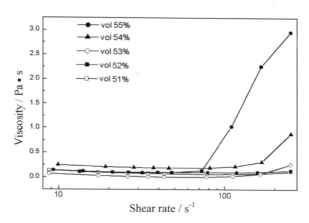

Fig. 4.2 Effects of solid loading of alumina on rheological behavior of aqueous alumina suspension from 51 to 55 vol%

the range of shear rate higher than 100 s^{-1}, the viscosity increases distinctly. Furthermore, it increases to the extent of 0.15 Pa s, which indicates that the whole system presents shear-thickening behavior.

An aqueous alumina suspension with 54 vol% solid loading indicates that at low shear rate, the whole system shows shear-thinning behavior. When the shear rate is in excess of 51.8 s^{-1}, the viscosity increases slightly at the beginning. When in excess of 200 s^{-1}, the curve becomes steeper, and the viscosity increases to the extent of 0.5 Pa s, which suggests more shear-thickening behavior of the whole system.

For the rheological curve of an aqueous alumina suspension with 55 vol% solid loading, the whole system exhibits shear-thinning behavior at low shear rate of 0.08–48.1 s^{-1}. When in excess of 72.7 s^{-1}, the viscosity proliferates and is more than 3 Pa s, which indicates a prominent shear-thickening behavior.

The suspensions exhibit shear-thickening behavior when their solid loading are more than 53, and 55 vol% and show typical shear-thickening characteristics. The viscosity increases significantly with diminutive shear rate increment.

Figure 4.3 shows that the viscosity corresponding to the critical shear rates increases with the increase of solid volume fraction, and the increment speeds up over 50 vol%. Figure 4.4 shows that the critical shear rate decreases with the increase in the solid loading of an aqueous alumina suspension and decreases swiftly over 50 vol%, which suggests that an aqueous alumina suspension at high solid loading undergoes a transition from pseudoplastic to plastic behavior at lower shear rate.

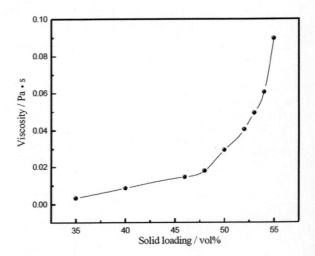

Fig. 4.3 Variation of the viscosity at critical shear rate

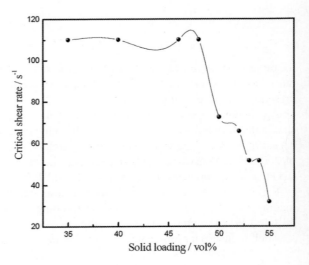

Fig. 4.4 Variation of the critical shear rate at different solid loading

Generally, as the solid loading increases in suspension, better macroscopic performance will be achieved. According to the above explanation, there exists a critical point in the rheological performance of an aqueous alumina suspension, below which the whole system presents shear-thinning behavior all along, and above which the whole system exhibits shear-thinning at the beginning, followed by thickening behavior afterward. The grain packing mode inside the aqueous suspension determines the critical solid loading value. One feasible method to adjust the very critical point is to accommodate the interaction between the grains and modify the grain packing mode.

4.1.2 Effect of Rheological Properties of Suspension on Mechanical Strength of Ceramics

There is no doubt that the rheological properties of the suspension affect the mechanical strength of the final sintered ceramic body. Recent research reveals that the highest flexural strength and the largest Weibull modulus were not achieved at the highest solid loading of the suspensions. The mechanical performance deteriorated when the solid volume fraction reached a certain value. The reason for this phenomenon is that the microstructure of the sintered bodies is to a large extent determined by the structure and the rheological behavior of the suspension, because the colloidal forming is an in situ occurring process.

In the pertinent research, commercially available α-Al_2O_3 (0.44 μm, Ceralox APA-0.5 with MgO, USA) and t-ZrO_2 ($Ce_{0.03}Y_{0.016}Zr_{0.954}O_2$, 0.4 μm, Beijing Founder Ceramics Technology Co. Ltd., China) were used. The polyelectrolytes (ammonium polyacrylate, PAA, Mw = 3,000–5,000, Tianjing Chemical Industry Research Institute, China) was used as a dispersant. The standardized analytical grade $(CH_3)_4NOH$ (TMAH) was used for adjusting the pH value of the suspension.

The essential components in the aqueous ceramic suspension for the forming process are the reactive organic monomers: monofunctional acrylamide, $C_2H_3CONH_2$ (AM), and the difunctional N, N'-methylenebisacrylamide $(C_2H_3CONH)_2CH_2$ (MBAM). These monomers were dissolved in deionized water to obtain a premix solution. The premix solution undergoes free-radical-initiated vinyl polymerization in the presence of an initiator such as ammonium persulfate $(NH_4)_2S_2O_8$. The reaction can be accelerated by using heat or the catalyst N, N, N', N'-tetramethylethylenediamine (TEMED) or both. The resulting cross-linked polymer is an elastic hydrogel (Kulicke and Nottelmann 1989; Tanaka 1981) to be used as the binder.

The premix solution was prepared with AM/MBAM = 23.3:1 (AM: 14.0 wt%, MBAM: 0.6 wt%). After 1 dwb. % (dry-weight basis) dispersant was added, the solution was mixed with alumina powder and zirconia powder, and the resultant suspensions with 40, 45, 50, 53, and 55 vol% solid loadings, respectively, were adjusted to pH = 10 by adding a TMAH aqueous solution. The suspensions were then ball-milled for 48 h to promote dispersion and the admixing process. The suspensions were degassed in a rotary evaporator under vacuum until no further release of air bubbles was observed. Both the initiator, a 5 wt% aqueous solution of ammonium persulfate, and catalyst, TEMED, were then added and stirred slowly to avoid sucking in air. The suspensions were degassed for another 5 min. During the degassing process, the temperature was kept at 0–1 °C to prevent excess evaporation of water and polymerization of the monomers. Each suspension was then cast into an $80 \times 60 \times 8$ mm^3 stainless steel mold, whose temperature was kept at about 70 °C. After the monomers polymerized, the green body was demolded and dried at room temperature under controlled humidity to avoid cracking and nonuniform shrinkage due to rapid drying. The dried green bodies were sintered at 1,580 °C for 2 h. Each sintered body was cut and ground into specimens of about $3 \times 4 \times 36$ mm^3 for the flexural strength measurement and specimens of about $4 \times 6 \times 30$ mm^3 for the single-edge notch beam (SENB) toughness measurement. Before their measurement, each of the specimens for the bending strength was ground by using SiC abrasive of 28, 14, 7, and 3.5 μm, respectively, for 30 min. A notch was cut in each of the specimens at depth of about 0.4–0.5 times the specimen height for the toughness measurements.

Zeta potentials of ceramic powders were determined by a ZetaPlus analyzer (Brookhaven Instruments Corporation, USA). 1 m mol/l NaCl was used as the solvent and the powder concentration was maintained at 0.05 vol%. The pH adjustment was carried out by using HCl and NaOH solutions.

The rheological properties of the concentrated suspensions were measured by a rotation rheometer (MCR 300, a modular compact rheometer, Germany). In measurements of shear stress (t), the thixotropic hysteresis loop (i.e., increasing the shear rate first and then decreasing it between two fixed values) was performed for each sample. Three-point flexural strengths of sintered specimens were determined with 30 mm span using 20 as-ground test pieces of $3 \times 4 \times 36$ mm^3 for each group of samples. Fracture toughness (K_{IC}) was measured using the SENB method with 24 mm span using at least 6 test pieces of $4 \times 6 \times 30$ mm^3 for each group of samples. All measurements were repeated to check for reproducibility and sample alignment. The two-parameter Weibull equation was applied in the characterization of the reliability of composite materials (Quinn 1990). Fracture surfaces of sintered bodies were observed by using an SEM (JSM-6301F, scanning electron microscope, Japan) to estimate the microstructure uniformity and porosity of the specimens.

The relative density of the sintered bodies was determined by Archimedes' principle in water. The formula $(l_0-l)/l_0$ was adopted to calculate the linear sintering shrinkage, where l_0 and l were the length of the specimen before and after sintering, respectively.

The zeta potentials of Al_2O_3 and ZrO_2 were measured individually, as shown in Fig. 4.5. The isoelectric points (pH$_{iep}$-value) were 5.8 and 4.9 for Al_2O_3 and ZrO_2, respectively. At low pH, far from the pH$_{iep}$-value of either powder, only the alumina particles have a high positive zeta potential and are colloidally stable. In the vicinity of the pH$_{iep}$, the particles have a low zeta potential that may be either positive (pH < pH$_{iep}$) or negative (pH > pH$_{iep}$); suspensions prepared within this region are

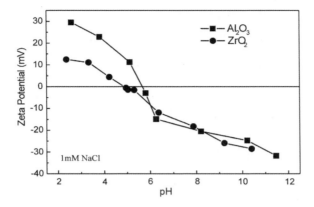

Fig. 4.5 Zeta potentials of Al$_2$O$_3$ and ZrO$_2$ as a function of pH

colloidally unstable and consist of large agglomerates that are likely to lead to porous gelcast ceramics. At high pH, far from the pH$_{iep}$-values of both powders, the particles have a high negative zeta potential and are colloidally stable. The profiles in Fig. 4.5 show that the saturation zeta potential and the breadth of the pH range over which the potential is nearly maximized provide better colloidal stability at high pH for both powders.

Low viscosity and high solid loading are beneficial for both mixing and casting in slurry processing. It is, therefore, important to maintain slurry fluidity while optimizing solid loading. The study was performed on suspensions containing 40–55 vol% of composite powders. Typical plots of apparent viscosity (η) versus the shear rate (γ) after different times of for a 55 vol% suspension are shown in Fig. 4.6. When the milling time was lesser than 48 h, the viscosity of the slurries decreased gradually as the time of milling increased. The results revealed that the absorption

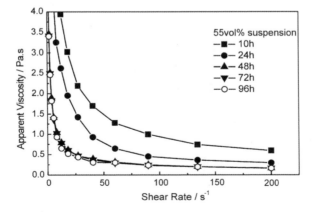

Fig. 4.6 Effect of ball-milling time on rheological behavior of a 55 vol% suspension

of the dispersant on the particles did not reach equilibrium and the suspension was unstable until the time became equal to or more than 48 h. After 48 h of milling, the viscosity of the suspension stabilized. Therefore, the time should be equal to or more than 48 h to obtain a stable suspension at equilibrium.

Figure 4.7a shows the plots of apparent viscosity versus the applied shear rate of stable suspensions after milling for 48 h. It can be seen that all suspensions (40–55 vol% solid loading) exhibited a shear-thinning behavior and relatively low viscosity (of less than 1 Pa s at the shear rate of 10 s^{-1}), which was suitable for casting. Figure 4.7b shows the plots of shear stress versus the applied shear rate of stable suspensions after milling for 48 h. It can be seen that 53 and 55 vol% suspensions possess a thixotropy hysteresis and their yield stresses were 1 and 3.5 Pa, respectively. This phenomenon was associated with the shear-thinning behavior of the suspensions and indicated a flocculated state of particles within the liquid. The decrease in thixotropy, viscosity, and degree of shear-thinning with the decrease of the solid loading of a suspension implied that the degree of powder agglomerate decreased. The thixotropy disappeared, when the solid loading of the suspension was less than 50 vol%.

Concentrated colloidally stable suspensions exhibited shear-thinning behavior in steady shear because of a perturbation of the suspension structure by shear (Bergstrom 1998). At low shear rates, the suspension structure was close to equilibrium because thermal motion dominated over the viscous forces. At higher shear rates, the viscous forces affect the suspension structure and thus shear-thinning occurred. At very high shear rates, the viscous forces dominated and the viscosity plateau measured the resistance to flow of a suspension with a completely hydrodynamically controlled structure. The degree of shear-thinning and the viscosity at high shear rates increased with increase in the volume fraction of solid.

For most systems, the properties of the suspension change drastically at a certain critical particle concentration, Φ_g, which corresponds to the formation of a space-filling particle network (Sigmund et al. 2000). At $\Phi < \Phi_g$, the suspensions have no yield stress. Above Φ_g, the suspensions can sustain a stress before yielding and the elasticity may be significant.

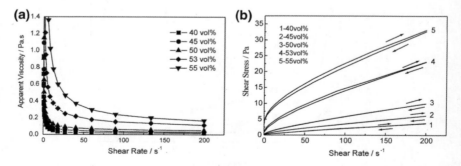

Fig. 4.7 Rheological behavior of the suspension with different solid loadings (ball-milling at 48 h): **a** apparent viscosity versus the shear rate; **b** shear stress versus the shear rate

Fig. 4.8 Bending strength and fracture toughness of sintering bodies prepared from different suspensions

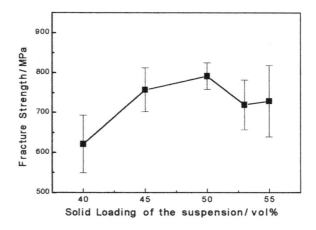

Based on the above description, each of the first three suspensions with solid loadings of 40, 45, and 50 vol%, respectively, had a particle concentration less than Φ_g, and so no yield stress could be observed for them. However, for 53 and 55 vol% suspensions with particle concentration higher than Φ_g, the high solid content resulted in the average separation distance between particles in suspension becoming shorter and formed a space-filling particle network, making flow more difficult. When a stress larger than the yield stress was applied, the gelled structure was broken into smaller units (flocs), which then moved past each other. If the floc attrition was affected by the strength of the hydrodynamic and attractive forces, pseudoplastic behavior prevailed and viscosity decreased with shear rates. The strong shear forces at high shear caused the flow units to be smaller and thereby the flow was facilitated. The destruction of flocs released a constrained solvent, which resulted in a decrease of the effective volume fraction of the flocs. This phenomenon resulted in thixotropic behavior in the system.

Gelcasting is a kind of forming process during which suspensions cure in situ. The structure of a suspension is almost completely kept in the forming bodies after the gelcasting process. This means that the microstructure of the green bodies would be identical with or similar to that of the suspension. In the case of flocs existing in the suspension, the flocs are 'frozen' as aggregates in the green bodies after the gelcasting process. It is well known that dense ceramic materials with good mechanical properties cannot be obtained by sintering green bodies with aggregates. Therefore, the suspensions with 53 and 55 vol% solid loading are not suitable for gelcasting, although they have higher solid loading and their fluidity and low viscosity can meet the requirements of the gelcasting process. On the other hand, the suspensions with solid loadings less than 50 vol% in which no or a few flocs exist are more suitable for use in the gelcasting process. Green bodies of high uniformity can be prepared by gelcasting from these suspensions and ZTA materials with high mechanical properties can be obtained after sintering.

Figure 4.8 shows the relationship between the mechanical properties of sintered samples and the solid loading of the suspension from which the samples were formed by gelcasting. The flexural strength of sintered bodies, which were sintered at 1,580 °C for 2 h, increased gradually from 621.35 ± 71.91 MPa to 791.58 ± 33.42 MPa as the solid loadings of the suspensions changed from 40 to 50 vol%. Their Weibull moduli had the same tendency and reached the highest value of 24.38, as shown in Fig. 4.9. However, as the solid loading exceeded 50 vol%, both bending strength and Weibull modulus of the sintered bodies began to decrease. On the other hand, the fracture toughness of the sintered bodies increases monotonously as the solid loading of the suspension increases from 40 to 55 vol%.

It is well known that the mechanical properties of ZTA ceramics are greatly influenced by the microstructure, including relative density (porosity), grain size, ZrO_2 content, and its distribution, microstructural flaws (Wang and Stevens 1989). From Fig. 4.10, it can be seen that the sintered sample prepared from the suspension with the lowest solid loading, 40 vol%, had the lowest relative density and highest sintering shrinkage. As the solid loading of suspensions increased, the relative densities of

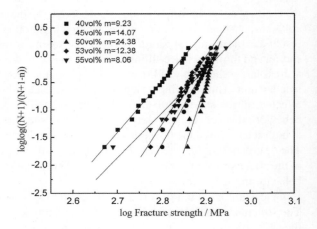

Fig. 4.9 Weibull distribution of sintering bodies prepared from different suspensions

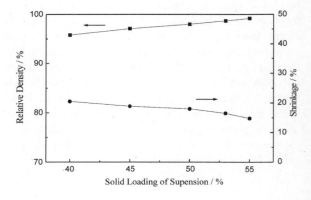

Fig. 4.10 Relative density and shrinkage of sintering specimen as a function of solid loading

4.1 Rheological Behaviors of Aqueous …

the sintered samples increased while the sintering shrinkage decreased. The occurrence of the phenomenon is mainly due to the packing behavior in the suspensions and the gelcasting procedure, as previously discussed. Figure 4.11 also shows that the porosity of the sintered bodies decreased with increase in the solid loading of the suspension.

In general, sintered bodies with higher relative density exhibit better mechanical properties. However, there is also an influence of microstructural uniformity on

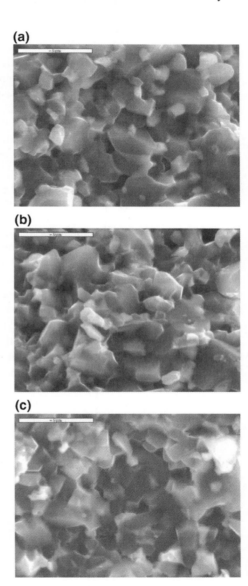

Fig. 4.11 SEM photographs of fracture surfaces of the sintered samples prepared from suspension with **a** 40 vol%; **b** 50 vol%; **c** 55 vol%

Table 4.1 Mechanical properties and calculated flaw sizes of ZTA (Al_2O_3/15 vol% ZrO_2)

Solid loading of the suspension (vol%)	Flexural strength (MPa)	Fracture toughness (MPa m$^{1/2}$)	Calculated flaw sizes (μm)
40	621.35 ± 71.91	5.96 ± 0.19	30.0
45	757.05 ± 55.03	6.40 ± 0.28	22.7
50	791.58 ± 33.42	6.58 ± 0.37	22.0
53	719.44 ± 62.12	6.66 ± 0.33	27.3
55	728.56 ± 89.46	6.87 ± 0.27	28.3

mechanical properties. In the system investigated, the suspensions with solid loading below 50 vol% had uniform microstructure. When the solid loading exceeded 50 vol%, some kind of flocs developed in the suspension, as discussed above. The flocs were maintained in the green bodies after the gelcasting process. These flocs in the green bodies often act as sources of defects that occur during sintering.

The critical flaw size c, which initiates failure, can be estimated from the relation (Kingery et al. 1976),

$$c = \frac{K_{IC}^2}{\pi \sigma_f^2}$$

where K_{IC} is the fracture toughness of the material and σ_f is the flexural strength.

Table 4.1 shows the calculated critical flaw size for each composition. The critical flaw size decreased gradually from 30 to 22 μm as the solid loadings of the suspensions changed from 40 to 50 vol%. However, as the solid loading exceeded 50 vol%, the critical size began to increase. This emphasizes the fact that a suspension with uniform microstructure will result in sintered compacts with relatively small flaws, thus improving the mechanical strength and reliability of the ZTA.

Based on the above discussions, the following can be concluded. (1) Flexural strength and Weibull modulus increase with increase in solid loading up to a maximum of 791.58 ± 33.42 MPa and, 24.38 at 50 vol% solid loading respectively. (2) All suspensions exhibit a shear-thinning behavior and relatively low viscosity, but suspensions with more than 50 vol% solid loading possess thixotropy hysteresis and yield stresses. (3) Although suspensions with good casting behavior could also be prepared for higher solid concentrations, a decrease of strength as well as of Weibull modulus were observed beyond a 50 vol% solid volume fraction. This was a result of flocculation occurring in the suspension: the flocs may act as sources of flaws in the sintered compact.

4.1.3 Effects of Solid Volume Fraction on Colloidal Forming

Water-based gelcasting of ceramics, since its invention by Young et al. (1991), developed rapidly in the past decades. Besides the polymerization of monomers, some new gelation mechanisms such as coordinative chemistry (Morissette and Lewis 1999) and protein chemistry (Lyckfeldt et al. 1999) have been introduced in the system of ceramics gelcasting in recent years. Generally, researchers focus on the preparation and gelation of dense ceramic suspensions, and most of them seek an optimized circumstance of dispersants, pH, and something else to increase the solid volume fraction of the suspensions, which is important to obtain ceramic parts with good properties and high performance. However, little attention is paid to how the solid volume fraction affects the technical process, which is also important to ensure good product quality. The effects of solid volume fraction in alumina suspensions on the rheological behavior and gelation of the suspensions, and debindering and sintering of green bodies prepared by colloidal forming of alumina have been investigated, and the results suggest that a moderate solid volume fraction is advantageous to both process control and the strength of the final products.

In this study, alumina powders, CT3000SG from Alcoa Qingdao Ltd., were used as received. The powders have a size $D_{50} = 0.3$ μm and a specific surface area 5.55 m^2/g. Tri-Ammonium Citrate (TAC) was used as dispersant. 15 wt% acrylamide (AM) premix solution was prepared with some MBAM (N, N'-methylenebisacrylamide, 2 wt% to AM) added. Empirically, TAC added into the suspension was 0.2 wt% of alumina powders. The mixture, consisting of alumina, TAC, and premix solution with a certain ratio, was ball-milled for 48 h to prepare the suspension.

The rheological curve of the suspension was measured with an MCR300 rheometer. To determine the gelation idle time by measuring the temperature (Young et al. 1991), the suspension (30 ml) was poured into a 25 °C thermostatic bath, and then an initiator (4 wt% $NH_4S_2O_8$ solution) and a catalyst (tetramethylethylenediamine) were added into the desired suspension in the ratio initiator:catalyst:premix solution $= 4$ μl:2 μl:1 ml. The temperature of the suspension was collected and recorded by a computer linked with the temperature detector (Yang et al. 2002). TGA and DTA tests of green body were performed with a SETARAM 92 thermo-analyzer. The density of the sintered body was measured according to Archimedes' method; the strength of samples was measured by three-point flexural tests; and the microstructure was analyzed with an S-450 SEM.

The effect of solid volume fraction on the gelation of alumina suspensions was studied. Since the polymerization of acrylamide monomers is an exothermic reaction, the rise in the temperature of the suspension denotes that the polymerization has begun. The period from adding the initiator and catalyst into the alumina suspension to the commencement of the rise in temperature of the suspension is called idle time. From Fig. 4.12, it can be seen that the idle time prolongs as the solid volume fraction of alumina suspension increases. After the initiator and catalyst were added into alumina suspension, the initiator decomposed and yielded free radicals gradually. The polymerization of acrylamide monomers was initiated by free radicals. When the

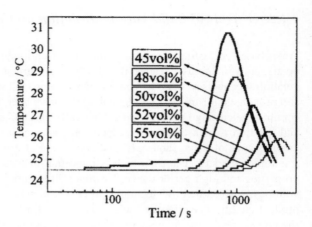

Fig. 4.12 Temperature rise of Al$_2$O$_3$ suspensions with different solid volume fractions during gelation

free radical met with an interface or another free radical, they combined with each other and lost the ability to initiate the polymerization. As the solid volume fraction of alumina suspension increased, the total surface of alumina particles per unit volume suspension increased, and the probability of the free radicals interacting with the particle surface increased. At the beginning of polymerization, the free radicals interacting with the particle surface are unable to initiate the polymerization of acrylamide monomers. As a result, the gelation of the alumina suspension was delayed, and the idle time was prolonged for the high solid volume fraction suspension.

Effect of solid volume fraction on the debindering of alumina green bodies: The polymerization of acrylamide monomers in the suspensions terminated when the growing end of the polyacrylamide chain collided with the surface of alumina particles. Therefore, the length and the cross–linkage of the polyacrylamide chain decreased when the solid volume fraction of the suspension increased. This resulted in the strength of the green bodies prepared by gelcasting decreasing with the solid volume fraction rising (Fig. 4.13), because the strength of the green bodies depends on the polyacrylamide chains formed, and the C–C bond in polyacrylamide chains with longer length or higher cross–linkage have a higher strength.

Since the strength of the C–C bond between the branches and principal chain is lower than that in the principal chain, the branches of polyacrylamide chains separated from the principal chains and burned out first during the debindering of the green bodies. This produced the first exothermal peak in the DTA pattern (Fig. 4.14a). When the temperature was increased continuously, the C–C bonds in the principal chains began to break and burn, and this resulted in the second exothermal peak in the DTA pattern (Fig. 4.14a). Since the length of the principal chain affects the C–C bonds in the principal chains more than that in the branches, the position of the exothermal peaks in the DTA pattern of the green body prepared with higher solid volume fraction suspension shifts to a lower temperature, and the shift is more obvious for the second exothermal peak (Fig. 4.14b). Correspondingly, the weight

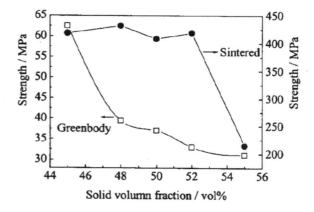

Fig. 4.13 The effects of suspension solid volume fraction on the strength of green bodies and sintered bodies

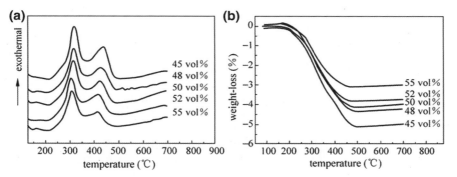

Fig. 4.14 a DTA and **b** TGA patterns of green bodies prepared by colloidal forming of Al$_2$O$_3$ suspensions with different solid volume fraction

loss ended at lower temperature for the green bodies from higher solid volume fraction suspensions (Fig. 4.14b).

Effect of solid volume fraction on the sintering of green bodies: Figure 4.15 shows that all alumina samples had similar density after being sintered at 1,580 °C. However, a steep fall of the strength for alumina samples prepared from 55 vol% suspensions can be seen from Fig. 4.13. The steep fall of the strength may result from the microstructure of the sintered bodies, corresponding to coarse crystal grains, which represent the characteristics of overfiring (Fig. 4.17d). From the rheological curves of alumina suspensions (Fig. 4.16), it can be seen that the yield stress of the suspensions increased with increasing solid volume fraction, and 55 vol% suspension showed obvious shear-thickening behavior. Since shear-thinning is the result of breakage of particle aggregates in suspensions, a higher yield stress of the alumina suspension corresponded to a higher strength of the particle aggregates in it. When sheared, particle aggregates in a 55 vol% suspension could not be broken entirely before shear-thickening occurred, which indicated the formation of hydrodynamic

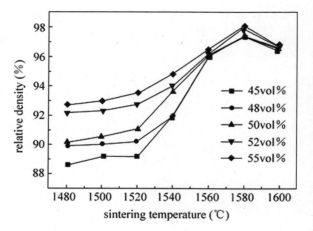

Fig. 4.15 The effects of sintering temperature on the relative density of Al_2O_3 samples from suspensions with different solid volume fraction

aggregates. Therefore, the microstructures of 55 vol% suspension and the green body prepared from it were far from homogeneous when compared to that of suspensions with less solid volume fraction. Like the pores, the aggregates in the green bodies also hindered the densification during sintering; and more disadvantageously, the aggregates resulted in the overgrowth of local crystal grains and then a coarse microstructure of the sintered body. As a result, the strength of the alumina sample from 55 vol% suspension decreased, although it had the highest density. In addition, the results showed that the microstructure rather than the relative density played a more important role in affecting the strength of the sintered bodies. Choosing a better dispersant may overcome the aggregates affecting the sintering of the green bodies, and this aspect should be paid attention to in further research studies.

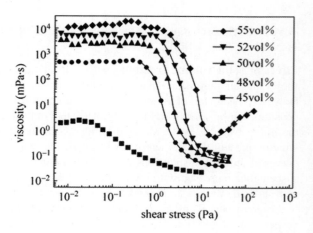

Fig. 4.16 The rheological curves of Al_2O_3 suspensions with different solid volume fraction

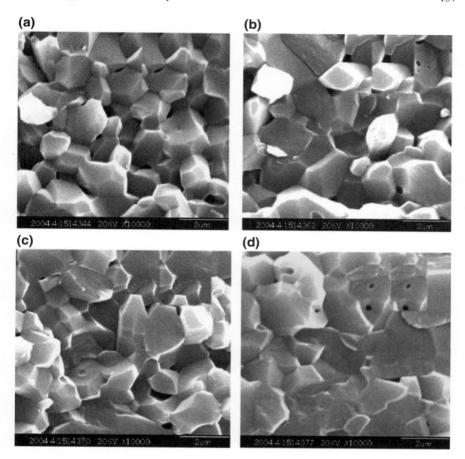

Fig. 4.17 SEM micrographs of fracture surface of sintered samples (1580 °C × 2 h) from Al_2O_3 suspensions with different solid volume fraction, **a** 45 vol%; **b** 50 vol%; **c** 52 vol%; **d** 55 vol%

The results indicated that the polymerization of acrylamide monomers was delayed and hindered in high solid volume fraction alumina suspensions for gelcasting, because alumina particles provided a surface to adsorb free radicals. Therefore, the idle time prolonged for the high solid volume fraction suspensions during gelation; and when heated, the exothermal peaks shifts to lower temperature for the green bodies from high solid volume fraction suspensions. Although the strength from high solid volume fraction suspensions decreased, it is high enough to sustain the configuration of the green bodies. Therefore, the flexibility of gelcasting could be improved with high solid volume fraction suspensions. However, excessive solid volume fraction suspensions had aggregates, resulting in a coarse microstructure and decreased strength of sintered samples, which should be avoided.

4.2 Generation and Development of Defects

4.2.1 Generation Mechanisms of Agglomerations in Ceramic Suspensions

A novel colloidal forming technology with in situ solidifying and near-net-size forming can significantly improve the microstructure and uniformity of the ceramic green body and reduce the production cost of advanced ceramics (Huang et al. 2007).

The characteristics of the new colloidal forming technology are based on a highly dispersed ceramic suspension that can be converted into uniform structural green bodies and products through in situ curing yet retain the original structures of the suspension (Giuliano 2003). However, if there are agglomerations in the suspension, they can cause uneven microstructure in the ceramic products, leading to a decrease in the reliability of their performance (Lang 1989). In addition, agglomerations also reduce the fluidity of the suspension and restrict the solid-phase content and so the realization of the net size molding is affected. Therefore, the eradication and control of agglomerations is quite necessary to optimize the colloidal forming process and its applications.

Tight aggregates often exist in ceramic raw materials, and are formed due to the chemical bonding of powder particles, while soft aggregates are formed by van der Waals force between particles (Sigmund et al. 2000). By selecting high-quality powder or effectively screening the powder prior to the colloidal forming processes, the formation of aggregates could be avoided. Soft aggregates are usually decomposed in the preparation process of suspension, because in most liquid mediums the attraction between the powder particles in water is far weaker than that in air. For example, the Hamaker constant of alumina is 3.67×10^{-20} J in water, but 15.2×10^{-20} J in air. The van der Waals attraction potential between the particles is directly proportional to the Hamaker constant.(Bergstrom 1997). The soft aggregates in the suspension can be further eliminated by adding a dispersant and adjusting the pH values. Even then, many studies (Rahaman 2003; Liu 1998) found that the grain clusters or unusual large-size grains can often be found in actual sintered ceramics; this phenomenon is more observable in products made with nano- or micro-powders (Lim et al. 2000). The reason for this is that different degrees of aggregates in suspension still exist.

At present, research on agglomerations in suspension focuses on two aspects: the effects of reunion on the rheological characteristics of suspension (Zhang et al. 2004; Bender and Wagner 1996) and the dispersibility of the powder, which can be effectively improved by using dispersing agents and by other methods (Tari et al. 1998; Takai et al. 2006). However, a few studies on the causes of agglomerations and on the effect of microscopic characteristics of ceramic have been conducted. Thus, the categories and generation mechanisms of agglomerations have been investigated by analyzing the basic role of the potential energy between ceramic powder particles, observing agglomerations in suspension, and studying the effect of agglomerations on the ultimate properties of materials. The objectives of these studies are to control

4.2 Generation and Development of Defects

and eradicate agglomerations, to improve colloidal processes, as well as to obtain a more complete theoretical basis on the subject.

(1) Pertinent Experimental Procedures

The ceramic powder used in this research was fine alumina (Al_2O_3 powder, type CT3000, Alcoa Chemical) with a mean particle size of 0.2 mm. Deionized water was used as solvent and ammonium citrate (type TAC, AR, the Beijing Chemical Reagents Company) was used as the dispersing agent. A premixed solution of monomers was prepared in acrylamide amide (AM, $C_2H_3CONH_2$), N, N′-methylenebisacrylamide (MBAM, $(C_2H_3CONH)_2CH_2$), and deionized water in accordance with the mass ratios of 14.5:0.5:85. The pH value of the suspension was adjusted using ammonia and hydrochloric acid in the course of preparation. Slurry consisting of alumina powders and premixed solution as well as dispersant was mixed by for 24 h. The slurry, with 0.02% in mass (the same as below) of initiator (ammonium persulfate, $(NH_4)_2S_2O_4$,) and 0.01% organic catalyst (N, N, N′, N′-tetramethylethylenediamine, TEMED) in total suspension was stirred thoroughly before gelling, and thereafter was solidified at 30 °C. Alumina slurries with alumina volume fractions of 20, 45, 50, and 55% were prepared with 0.2 wt% TAC dispersant (relative to alumina powder), and the adjusted pH = 10. The unmolded wet green body was heat–treated to dry and the organic contents were removed. Subsequently, the dried bodies were sintered into ceramic products at 1,500 °C for 2 h.

The distribution and morphology of agglomerations were directly observed by using an environmental scanning electron microscope (ESEM, Model Quanta 200, FEI Company, the Netherlands). The microstructures of the wet green bodies after colloidal forming and solidifying in situ displayed a true microstructure of the suspension, when observed under an ESEM in a low vacuum, high humidity mode with a pressure of 600 Pa and a humidity of 98% on the dissociation plane of specimens. The microstructures of the sintered bodies were observed using a SEM (Model JSM-6460, JEOL Ltd., Japan). The bending strengths of the sintered bodies were measured by three-point bending test on samples of size 4 mm × 3 mm × 40 mm with a loading speed of 0.5 mm/min. Fracture toughness K_{Ic} was measured by a four-point bending test with an SENB, on an universal test machine (Model AG-2000G, Shimadzu, Japan) with a sample size of 4 mm × 6 mm × 30 mm and a loading speed of 0.1 mm/min. Each set of data involves 4–5 samples.

(2) Generation Mechanisms of Agglomerations in Ceramic Suspensions

According to DLVO theory (Zongqi et al. 2001; Reed 1988), the motivating force for coagulation is the van der Waals attraction force, which is a function of the dielectric constant of the medium and the mass and distance between the particles. For two small spherical particles of radius a, the potential energy of attraction is V_{vdw}; the repulsive force is provided by the interaction of two electrical double layers. The magnitude of the repulsive force depends on the size and shape of the particles, the distance D between their surfaces, thickness of the double layer κ^{-1}, and the dielectric constant of the liquid medium. The total potential energy V_{total} is the algebraic sum of the attractive van der Waals potential energy V_{vdw}, and the repulsive energy V_{elect}:

Fig. 4.18 Sketch of potential energy between two particles in medium (Zongqi et al. 2001)

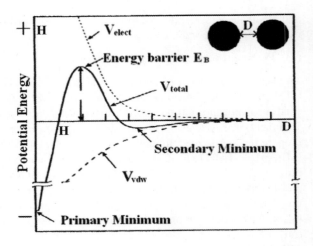

$$V_{\text{total}} = V_{\text{vdW}} + V_{\text{elect}}$$
$$= -\frac{Aa}{12D} + 2\pi\varepsilon_0\varepsilon_r a\Psi_0^2 \exp(-\kappa h)$$

where A represents the Hamaker constant, ε_0 is the vacuum dielectric constant, ε_r is the relative dielectric constant of the solvent, and Ψ_0 is the particle surface potential energy. The potential role between particles changes with the change in the distance between particle surfaces, as shown in Fig. 4.18 (Zongqi et al. 2001).

From Fig. 4.18, it can be observed that there are minima at two locations on the particles' potential energy curve, i.e., the 'Primary Minimum' corresponding to very short distances between the particles and the 'Secondary Minimum' corresponding to the relatively bigger distances between the particles. When the surface distances between powder particles correspond to the trap of potential energy (i.e., the primary minimum), the system's energy is in its lowest energy level. In still or slow flow suspensions, Brownian movement is the main movement form of the powder particles, and its $K_b(T)$ energy is 4×10^{-21} J at $T = 300$ K. The energy difference between the potential E_B of the powder particles and the 'secondary minimum' is usually about 10^{-19} J (Zongqi et al. 2001). Thus, it is very difficult for particles of the suspension to cross over the barrier E_B to get into the 'first trap'. Thus, in general terms, the 'secondary minimum' represents that the system is in its minimum energy state. Therefore, the interparticle separation corresponding to the secondary minimum can be called 'stable separation D_0'. For a given powder particle size, the solid-phase content in the suspension is determined directly by the distance between the particles. The solid volume fraction corresponding to the 'stable interval D_0' of the powder particles is defined as the stable solid-phase volume fraction φ_0 of this suspension. When the solid-phase content is lower than φ_0 in suspension, the surface interval of the powder particles is greater than D_0 and the potential energy

4.2 Generation and Development of Defects

between the particles is not at the lowest level, deviating from the secondary potential well. Under such circumstances, the particles spontaneously close up to each other and restore stability at pitch D_0. As a result, aggregate structures are formed in the local area and reduce the system energy to a minimum. Corresponding to the depth of the secondary potential well, the attraction force in aggregates is usually relatively weaker. However, in this process, a 'loose reunion' forms because of the attraction due to the large gap between the particles. Large gaps are also left between aggregates. The average distance between particles in the suspension is equal to D_0 when the solid-phase content reaches φ_0. Despite the surface distance between particles deviating from space-balanced positions because of Brownian motion, the system spontaneously restores the particle distance to D_0 again on its own to maintain the minimum energy. In this manner, the suspension takes on a stable and uniform structure. If the solid-phase content of the suspension exceeds φ_0 while the average distance between particles is shorter than D_0, the energy of the system increases rapidly, leading to the potential energy of particles crossing over the barrier E_B. Therefore, the probability for the potential energy level falling into the 'first trap' increases significantly due to Brownian motion. As a result, some particles strongly close up to form the reunion structure and fall into the 'first trap'. On the other hand, the rest particles then restore the 'stable interval D_0', and thereby the system returns to its lowest energy state. The particle reunion structure formed is defined as a tight agglomeration because of its short particle distances and the strong mutual attraction potential among particles. These three conditions are displayed in Fig. 4.19 (Ruifeng 2007).

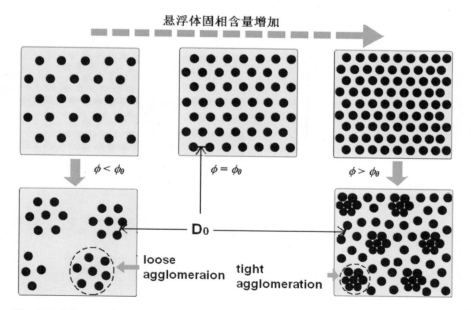

Fig. 4.19 Microscopic model of agglomerations in suspension

(3) **Effect of Solid-Phase Content on the Behavior of Alumina Suspension**

The solid volume fractions in the alumina slurries selected are 20%, 50%, and 55%. The observation results of the agglomerations in alumina suspensions with different solid-phase contents are shown in Fig. 4.20.

Fig. 4.20 Agglomerations in alumina suspensions with different solid phase contents.

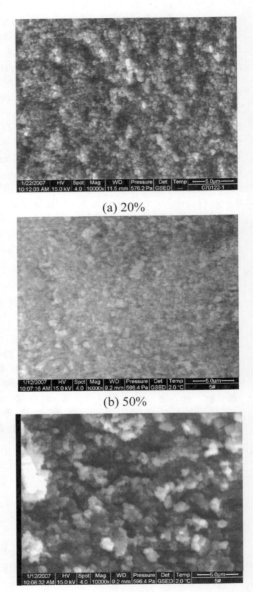

(a) 20%

(b) 50%

(c) 55%

4.2 Generation and Development of Defects

From Fig. 4.20a, it can be seen that the ceramic powders at the submicron level in the suspension with very low solid-phase content (20%) did not distribute evenly, despite the addition of an appropriate proportion of dispersant. Instead, a partial reunion of loose states could exist. When the solid-phase content was increased up to 50%, a suspension with a highly uniform structure was observed, as shown in Fig. 4.20b. This can be explained as the powder particles in suspension achieved a stable distribution. When the solid-phase content was increased to 55% (Fig. 4.20c), large agglomerations with diameters of 1–2 μm could be clearly observed. This phenomenon is because of the formation of tight aggregates in the suspension due to the powder particles fully closing up to each other. Based on these observations, the optional alumina solid-phase content is about 50% to form a stable uniform dispersion in suspension.

The agglomeration generation mechanism in an alumina ceramic suspension can be summarized as follows: when the solid-phase loadings are less than φ_0, a loose agglomeration is produced. When the solid-phase content is higher than φ_0, a tight agglomeration is formed. When the solid-phase content is equal to φ_0, the particles in suspension take on a uniform and stable dispersion, but no reunions appear. If the solid-phase content is too low in a ceramic suspension used for colloidal molding, it causes not only a heavy shrinkage but also an uneven shrinkage of the ware during latter heat treatment.

(4) Effect of Agglomerations on Microstructures of Ceramic Materials

The microscopic structures of the sintered ceramics are shown in Fig. 4.21. It can be seen from Fig. 4.21a that there were locally uneven cluster structures of 5–10 μm sizes in the sintered alumina ceramics, which were made of alumina slurry with a solid-phase content of 20%. Since the solid-phase volume fraction in suspension was less than φ_0 for a stable suspension, looser reunions were formed. When the phase content of the suspension increased to 45%, the structure of ceramics sintered became more uniform, but the stocks of grains formed clusters in the local region with larger gaps between the grain clusters (see locations specified by the white arrows in Fig. 4.21b). When the solid fraction was increased to 50%, the grains and pores of the ceramics were evenly distributed with grain sizes mostly in the range of 1.0–2.0 μm as shown in Fig. 4.21c. It is apparent that the solid-phase content in this suspension was consistent with stable solid loading φ_0 on which the powder particles were distributed evenly. When the solid loading of suspension continued to increase up to 55 vol%, large-size grains of more than 5 μm appeared again (black arrows in Fig. 4.21d) in the sintered alumina. This is because solid loading in the suspension exceeded φ_0 and resulted in formation of reunions (see Figs. 4.19 and 4.20).

(5) Effect of Agglomerations on Mechanical Properties of Alumina Ceramics

The strength and toughness of sintered alumina ceramics shown in Fig. 4.21b–d are measured and listed in Table 4.2. The data in Table 4.2 corresponding to Fig. 4.21 show that when the solid-phase volume fraction in suspension is 50%, the prepared ceramic displays the highest strength with the best particle distribution uniformity in the sample. This can be explained that the solid phase content is approaching stable

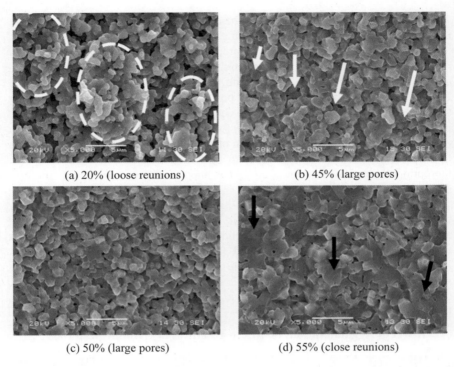

Fig. 4.21 Microstructures of alumina ceramics sintered at 1,500°C for 2 h and using suspensions with different solid-phase volume fractions.

Table 4.2 Influence of solid-phase content on the properties of sintered alumina ceramics

Solid phase content in slurry (%)	Relative density of sintered body/%	Bending strength/MPa	K_{Ic}/(MPa m$^{1/2}$)
45	96.6 ± 0.5	429 ± 28	4.1 ± 0.6
50	97.5 ± 0.3	483 ± 16	4.2 ± 0.2
55	98.2 ± 0.3	456 ± 32	4.1 ± 0.5

solid loading φ_0 and the powder particles in suspension are optimally dispersed. Owing to this, the ceramic obtains the most uniform microstructure and its strength is significantly improved. When the solid-phase volume fraction in the suspension is reduced to 45%, both the strength and density of the prepared ceramics decrease. This is due to the formation of loose reunions, which is because the solid-phase content in the suspension is lower than φ_0. Although the internal particle distributions of aggregates maintain stable states, the distances between loose reunions are relatively great.

4.2 Generation and Development of Defects

The existence of loose reunions affects the mass transferring and sintering densification due to the great distance between loose reunions. The ceramic characteristics of lower density and less uniformity are the direct results of large pores and uneven microstructure.

When using a suspension with a solid-phase volume fraction of 55%, the density of the sintered ceramic increases by 0.7%, while the breaking strength decreases. This is mainly because tight reunions are formed as the solid-phase content in the suspension is greater than φ_0. The particles within the reunion regions are very close to each other, and so sintering takes place among them first and thereafter the particles grow into large grains when the sintering temperature increases.

The toughness K_{Ic} values are not significantly different among the ceramics prepared by slurries with three different solid phase contents, but the bending strengths are relative lower for those prepared by suspensions with solid-phase content of 45 and 55%.

This shows that the aggregates generated in suspension can reduce the microstructure uniformity and performance reliability of the final ceramic material. Therefore, the suspensions used in colloidal molding fabrication must reach the stable solid-phase volume fraction φ_0 in order to avoid formations of agglomeration, which can ultimately affect the performance of ceramic materials.

4.2.2 Influences of Idle Time on Microstructures and Mechanical Properties of Green Bodies by Direct Coagulation Casting

Ceramic green bodies can be produced by enzyme-catalyzed reactions from suspensions with high solids loading (generally, >50 vol%). This forming process is known as direct coagulation casting (DCC) (Gaucker et al. 1999a, b). It is a near-net-shape forming method, which is suitable for ceramic parts having complex geometry. In this paper, we will focus on discussing the coagulation method, i.e., increasing the ionic strength. This method can produce much higher wet strength of the green bodies by aging for 1–3 days than the pH-shift method and may be suitable for other ceramic systems, such as zirconia or silicon carbide (Si et al. 1999). The ionic products ammonium and carbonate of the internal hydrolysis lead to coagulation by compressing the electrostatic double layer (Gaucker et al. 1999a, b; Si et al. 1999).

The microstructures of the green bodies produced by pH-shift to the isoelectric point (IEP) are always very dense and homogeneous without cracks in the green body as well as in the sintered state. The consolidation time of the suspensions is usually about 0.5–2 h, depending on the temperature and urease concentration. In this coagulation method, no syneresis or sedimentation occurs. This means that the particle network is not contracted during coagulation, although the separation distance between the neighboring particles is decreased (Helbig et al. 2000).

The coagulation time by increasing the ionic strength is longer than that of pH-shift. In general, several h or up to 3 days are necessary to get a maximum strength of the wet green bodies. Its compressive wet strength is higher by a factor of up to 10 compared to that of the pH-shift method but it strongly depends on the coagulation time. This is explained by the gel network aging, during which the number of particle contacts is increased, thereby improving the strength of the gel network (Balzer et al. 1999). In our study, the microstructure of wet or dry ceramic green bodies produced by increasing the ionic strength always showed cracks on the surface and in the interior of the green bodies.

In this section, the author investigated the microstructures of coagulated alumina suspensions by increasing the ionic strength in aqueous suspensions by the enzymatic urea/urease reaction. We observed three stages of the coagulation process by increasing the ionic strength: the initial stage where no shrinking was found, followed by a stage of syneresis in combination with sedimentation, and finally the aging stage, where the compressive strength of the wet green body increased. We discovered that cracks were formed during the first two stages of the coagulation. They can be reduced or even avoided by introducing an idle time before casting the slurries or by adding salts during preparation of the slurries. By increasing the idle time or the ionic strength prior to casting, the microstructure of the green bodies gradually changes from an inhomogeneous structure with many macroscopic cracks to a homogeneous microstructure without cracks.

(1) **Materials Used in the Pertinent Research**

High purity alumina powder (HPA-0.5 MgO, CONDEA Vista Company, USA) containing 0.5% MgO in the form of $MgAl_2O_4$, was used. The powder has a d_{50} value of 0.5 μm and a surface area of 9.0 m^2/g. The other chemicals were deionized water (>18 MΩ/cm), urea, p.a. (Sigma), 2 N HCl (Titrisol, Merck), ammonium carbonate $CH_8N_2O_3$ (Fluka, FLUKA Chemie AG, Switzerland), diammonium citrate, purum (Alfa, Johnson & Mathey Brandenberger AG, Switzerland), and urease (activity: 71,000 units/g, Lyo. Special Quality, Roche, Diagnostics Mannheim, Germany). The solids loading of slurries, if not explicitly mentioned, was 57 vol%.

(2) **Preparation of the Slurries: Methods of Coagulating Slurries**

The slurries for the pH-shift method.

The slurries for the pH-shift method were prepared as follows: first, urea and HCl were added to deionized water, and then the alumina powder was slowly introduced during stirring. After mixing homogeneously, the slurries were ball-milled for 24 h with alumina milling balls. The pH of the ball-milled shurries was measured and adjusted to 3.5–4.5 with 2 N HCl. A urease water solution was dispersed by ultrasonic treatment, and the slurry was cooled to around 5 °C and degassed. The urease solution was then added to the cooled slurry. After mixing for about 5 min, the slurries were cast into the mold. After casting for 2 h, the coagulated slurries were demolded.

After careful drying, the microstructure was examined. The following parameters were used in the pH-shift experiment: 0.3 mol/l urea (referred to liquid), urease of 1 unit/g (to alumina).

4.2 Generation and Development of Defects

The pH-shift slurry is prepared at a pH of around 4 and destabilized by the urea hydrolysis reaction to shift the pH of 9. The pH of the slurry buffered by the produced ammonium carbonate is pH 9.

The slurries for the ionic strength shift method.
Due to the adsorption of diammonium citrate on the particles, the surface charge on the alumina particles changed. The IEP shifted from 9.2 to 3.4 (Helbig et al. 2000). Well-dispersed ball-milled slurries can be prepared within a pH range from neutral to basic. The suspension is stable at a pH of around 9 and at low ionic strength. Then, it is destabilized at a constant pH of 9 by increasing the ionic strength. The preparation process of the ionic-strength slurries before casting is similar to the pH-shift slurry except for replacing HCl by diammonium citrate at the beginning. After adding urease, the degassed and cooled slurry is idled at different times. Before casting, the idle slurries were shaken for 5' by hand. One day after casting, the slurries were demolded and carefully dried. Then, the microstructures were examined. The following parameters were used during the preparation of the slurries: 2.2 mol/l urea (referred to liquid), 0.45 wt% diammonium citrate (referred to alumina), and 10 unit/g urease (referred to alumina). Diammonium citrate is about 150% of the amount needed for a monolayer on a powder with 9.0 m^2/g; the ionic strength after completion of the hydrolysis reaction is adjusted by the amount of urea added.

(3) Characterization of Coagulating Slurries and Green Bodies

A light microscope and stereo lens (from Leica co., Germany) was used to observe the microstructure of the green bodies. An electronic image capturing system was used to store the pictures.

In order to determine the volume shrinkage of the liquid phase during the hydrolysis of urea, the prepared slurry was degassed and cooled down to about 5 °C and was cast into two 100 ml graduated cylinders with height 25 cm and diameter 2.5 cm. One was used for measuring the temperature of the slurry. The other one was used to monitor the volume change versus the time during the reaction.

The rheological behavior of the slurries and suspensions was measured with a controlled–stress rheometer by using a vane tool. The drying of the sample was prevented by the application of a silicon oil layer on top of the slurry surface.

(4) Crack Formation During ΔI Coagulation

The coagulation of ceramic suspensions with high solids loading (>50 vol%) was investigated. The coagulation experiments were carried out using a special type of DCC procedure. This method uses the enzyme-catalyzed hydrolysis of urea to increase the ionic strength (ΔI) in the suspension at a fixed pH of 9. The microstructures of the achieved green bodies were examined by photographing the fracture surfaces of the dried green bodies. The volume shrinkage (syneresis) and the concentrated suspension were monitored during coagulation.

The fracture microstructure of all of the dried green bodies coagulation by ΔI showed vertically oriented cracks in the carefully dried green body. Samples coagulated by ΔpH are dense and crack-free. The crack-to-crack distance of the laminated cracks in the ΔI coagulated samples was about 0.3–2 mm.

Table 4.3 Experimental details for ΔI coagulation

No	Experimental conditions	Fracture microstructure	Notes
1	Different ratios of urea/urease	Cracks	See Table 4.2
2	Mixing the slurry with urea/urease in low-temperature ball-milled jar	Cracks	Distribution of urease
3	Using high-power ultrasonic to distribute urease in the ball-milled slurry	Cracks	Distribution of urease
4	Replace diammonium citrate by PAA	Cracks	Influence of dispersant
5	Casting slurry onto the surface of liquid with high specific density (2.93 g/ml)	Cracks	Avoiding wall effects

The influence of several experiment parameters on the crack formation was investigated (Table 4.3). First, the coagulation kinetics was changed. The amount of added urea was varied from 0.55 to 2.2 mol/l urea, and the used amount of urease was varied from 2.5 to 40 unit/g urease (2.2 mol/l urea) in slurries of 53–62 vol% alumina (see Table 4.4). All samples showed curved lines on their surface as demonstrated in Fig. 4.22a. After three days, the surface of the slurry could still be plastically deformed through touching (Fig. 4.22b). As all samples showed these cracks, we conclude that the cracks are not related to any of the different conditions and ratios of urea/urease and volume loading of the slurries.

Inhomogeneously distributed urease as a cause for the cracks could also be ruled out. Using high-power ultrasonic treatment in the suspension after adding urease

Table 4.4 Effect of different ratios of urea/urease on microstructure of green bodies

Sample no.	Urea (mol/l)	Urease (unit/g)	Fracture microstructure	Volume loading (%)	Notes
JL008	2.2	10	Cracks	57	–
JL064	2.2	5	Cracks	57	–
JL066	2.2	2.5	Cracks	57	–
JL065	1.1	5	Cracks	57	–
JL067	1.1	2.5	–	57	No coagulation
JL068	0.55	2.5	–	57	No coagulation
JL025	2.2	15	Cracks	55	–
JL026	2.2	15	Cracks	53	–
JL084	2.2	20	Cracks	57	–
JL088	2.2	40	Cracks	57	–
JL184	2.2	10	Cracks	62	–

4.2 Generation and Development of Defects

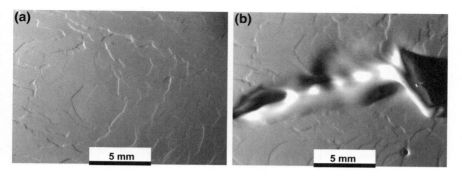

Fig. 4.22 Surface pattern (**a**) and surface after touching (**b**) of noncoagulated alumina slurry after 3 days by increasing ionic strength

as well as an experiment where urea and urease were added to the slurry and milled in a low-temperature mixing container for 15′–30′ did not influence the crack microstructures (see No. 2 and No. 3 in Table 4.3).

One might speculate that the crack formation is caused by the ammonium citrate used for electrical stabilization. Therefore, the alumina particles were stabilized by polyacrylic acid (Aldrich 2000 g/mol), which stabilizes by electrostatic and steric reasons. However, no difference was observed and an influence by the dispersant can be ruled out (see No. 4 in Table 4.3).

We also wanted to check the mold surface on the crack formation. We cast the slurry onto the surface of a hydrophobic liquid of high density (1, 1, 2, 2-tetrabromoethane; $\rho = 2.693$ g/mm^3) (see No. 5 in Table 4.3). The fracture surface of these two wet green bodies did not differ from the usually observed ones after coagulation. Therefore, cracks were not generated by the contact of the weak slurry sticking to the wall.

(5) Influence of Idle Time on the Microstructure of Green Bodies

To follow the crack development and to observe in which stage cracks are formed, we introduced an idle time before casting. The idle time is defined as the period of time starting with the addition of urease and ending with the casting of the slurry. After the idle time before casting, the slurry was shaken well by hand in order to destroy any larger coagulated structures. The microstructures of the fracture surface of the dried green bodies with different idle times are shown in Fig. 4.22. It is very surprising that the idle time strongly influences the formation of cracks in the green bodies coagulated by increasing the ionic strength. The results clearly show that no cracks occur in samples that were cast after 150 min of idle time.

The sample shown in Fig. 4.23a was directly cast after the dispersion of the urease in the slurry, which took about 5 min (the minimum idle time tested). A vertically laminated crack structure in the green body was observed. With increase in the idle time, the distance between the vertically laminated cracks became considerably larger (Fig. 4.23a, g). On the other hand, the number of cracks in the green body decreased gradually (Fig. 4.23b, g), until finally the cracks completely disappeared (Fig. 4.23h).

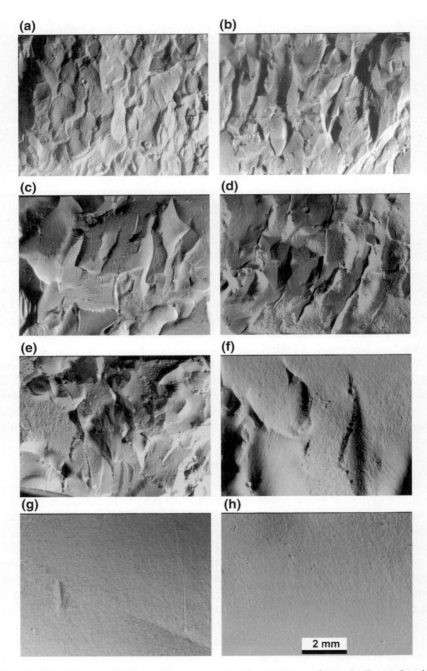

Fig. 4.23 The influence of different idle times on the microstructures of green bodies produced by increasing ionic strength: **a** 5 min; **b** 20 min; **c** 40 min; **d** 60 min; **e** 80 min; **f** 90 min; **g** 120 min; **h** 150 min

4.2 Generation and Development of Defects

The structure became dense and homogeneous on a millimeter-to-centimeter length scale when the idle time reached about 150 min (Fig. 4.24h). The shortest necessary idle time to avoid cracks is defined as the critical idle time.

When the idle time increases, the ionic strength and the viscosity in the slurry also increase. Thus, the factors that determine the critical idle time include the starting temperature, ionic strength in the suspension, the amount of added urea and urease, as well as the solids loading of the suspension.

(5) The Influence of Addition of Different Salts on the Microstructures of the Green Bodies

Increasing the idle time means not only increasing the ionic strength within the suspensions, but also shearing the slurry before casting. In order to simulate the effect of increasing the idle time on the microstructure of green bodies, we introduced different amounts of salt into the suspensions prior to . For simulating the real reaction system, an ammonium carbonate ($CH_8N_2O_3$) was chosen. It was added to the initial slurry containing the urea to result in salt concentrations of 0.001–0.4 mol/l at a pH of 9. The microstructures of the green bodies with varying amounts of salt are shown in Fig. 4.24. The slurries were cast right after the addition of urease without any idle time. The results clearly indicate that the crack development is connected to the salt concentration in the suspensions. The resulting microstructures of the green bodies are the same at salt concentrations up to 0.05 mol/l of added salt (Fig. 4.24c). The micrograph of the sample with a salt addition of 0.1 mol/l is comparable to a sample with 40 min of idle time (Fig. 4.24d). A crack-free structure was achieved by the addition of 0.4 mol/l salt (Fig. 4.24h). This is comparable to the microstructure of the sample that was cast after 150 min of idle time. This experiment documents that increasing the idle time means increasing the ionic strength inside the suspension. It is reasonable to derive that the crack formation is connected to the salt concentration within the slurry prior to casting.

To identify the reasons of crack formation, we investigated the volume shrinkage of the slurry during coagulation.

(6) Syneresis During Coagulation

In order to characterize the network formed during coagulation after casting, we measured the volume change of the suspension. The shrinkage (syneresis) of the particle network as well as a decrease of the water phase due to the urea hydrolysis were monitored. The volume shrinkage of the coagulating alumina particle network as a function of the reaction time at different volume loadings is shown in Fig. 4.25. There are three stages that are observed in our experiment that lead to the final coagulated particle network structure. It has to be mentioned that the different time stages depend on the urease/urea content as well as the slurry temperature.

The volume shrinkage versus the reaction time clearly shows three stages.

Stage I: The temperature of the slurry increases to 26.5 °C (room temperature) and the urea decomposition starts. The volume of the slurry remains constant during the first 60 min. In this stage, the particles gradually start to agglomerate as the ionic strength

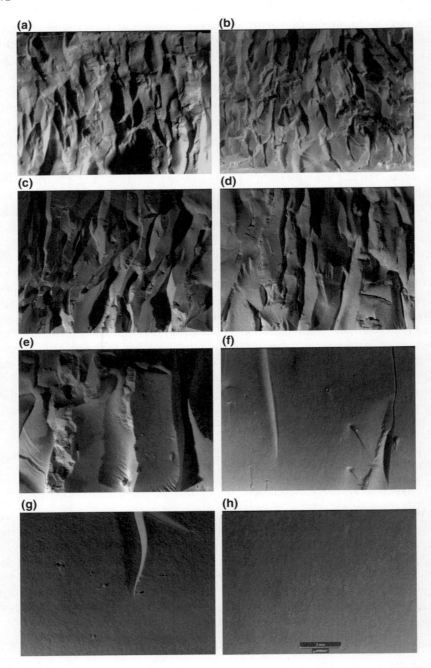

Fig. 4.24 The influence of different amounts of hydrated ammonium carbonate on microstructures of the green body by increasing ionic strength: **a** 0.001 mol/l; **b** 0.01 mol/l; **c** 0.05 mol/l; **d** 0.1 mol/l; **e** 0.2 mol/l; **f** 0.3 mol/l; **g** 0.35 mol/l; **h** 0.4 mol/l

4.2 Generation and Development of Defects

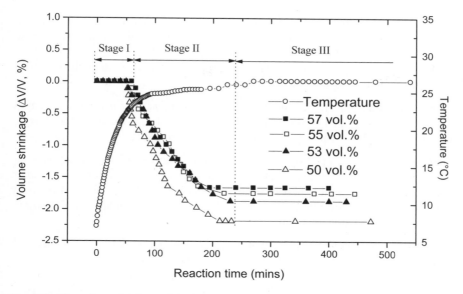

Fig. 4.25 The volume shrinkage of slurries versus the coagulation time at different volume loading

increases. The gel network formed by the particle agglomerates is very weak. The viscosity of the slurry remains low and no syneresis is observable.

Stage II: In this stage, a remarkable shrinkage of the coagulating particle network up to 240 min is observed (Fig. 4.25). The volume shrinkage is a consequence of syneresis and sedimentation. Syneresis results from the spontaneous contraction of the gel without evaporation of the constituent pore fluid (Brinker and Scherer 1990; Scherer 1989). In our experiments, we only observed one-dimensional compaction along gravity direction and no lateral shrinkage. This resulted from the combination of both sedimentation and syneresis of the destabilized suspension.

Stage III: The syneresis of the suspension stops. The viscosity is high and the particle network starts to age. During this period, the strength of the gel network is still increasing as shown in previous studies (Balzer et al. 1999; Dewhurst et al. 1999).

The Idle Time is defined as the period of time starting with the addition of urease and ending with casting of the slurry. Different idle times actually indicate different ionic strengths in the suspension. Before casting at every idle time, the slurry must be shaken homogeneously in order to destroy its formed microstructure and then rebuild its new structure. By increasing the idle time or the ionic strength prior to casting, the microstructure of the green bodies gradually changes from an inhomogeneous structure with many macroscopic cracks to a homogeneous microstructure without cracks. The microstructure of the green body becomes homogeneous and crack-free when the idle time is about 150 min, which corresponds to 0.4 mol/l of added ammonium carbonate in this study.

The influence of the idle time on the microstructure is related to the ionic strength in the slurry, the starting temperature of the slurry, the amount of urea and urease, and the solid volume loading of the suspension.

The coagulation process of increasing the ionic strength consists of the following three stages as described below.

Stage I: Liquid stage: Particles start to agglomerate with increase in the ionic strength. The gel network formed by the particles is very weak. The viscosity of the slurry is low. There is lack of syneresis and sedimentation during this stage.

Stage II: Sedimentation and syneresis stage: The one-dimensional compaction along the direction of gravity is dominant and no lateral shrinkage is observed. The main parameter influencing the syneresis strain rate is gel viscosity. The high gel viscosity will eventually shut down the process of syneresis, but not until relatively high strains have been attained after destabilization.

Stage III: Aging stage: This stage involves improving the strength of the gel network by increasing the number of contacts in the contracted structure.

4.3 Effect of Ionic Conductance on Preparation of Highly Concentrated Suspension

4.3.1 Academic Idea and Research Program

With the development of the in situ consolidation forming processing such as DCC (Graule et al. 1994), gelcasting (Omatete et al. 1991a), temperature-induced flocculation (Bergstrom 1994), a highly concentrated ceramic suspension with low viscosity becomes the prerequisite to achieve high density and good homogeneity of green bodies for their application.

Polyelectrolytes are widely used as dispersants to prepare highly concentrated ceramic suspensions in industrial application. A 62 vol% aqueous alumina suspension could be prepared with polymethacrylic acid and polyacrylic acid (Cesarano and Askay 1988). However, the use of polyelectrolytes (or dispersants) makes a great difference for the same materials due to different preparation processes. The selection of dispersants is very important for a variety of commercial powders. On the other hand, a suitable graduation of fine and coarse particles is usually an effective method to achieve a highly concentrated suspension (Chong et al. 1971), but it will lead to an increase of processing cost for commercial powders.

According to rheological theory, macrorheological behaviors of a suspension depend on the micro-interaction forces between particles. In an aqueous medium, the interaction forces between particles usually include the van der Waals attractive force and the electrostatic repulsive force, which depends on ionic valence number and concentration that affect the thickness of the electrical double layer and the critical

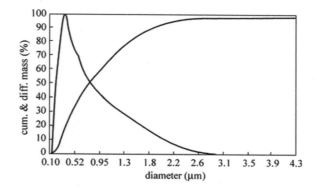

Fig. 4.26 Particle size cumulative and differential distribution of α-Al_2O_3

coagulation concentration. We studied the effects of ion valence and ion concentration on the solids volume loading in alumina suspension. For the first time, an ion chromatogram technique was introduced to determine a variety of ions in the ceramic suspension and an ion conductivity constant related to solids volume loading in the suspension was used to characterize the quality of powders from different parts of the world.

The starting materials used in our study were α-Al_2O_3 powder (produced by a special ceramic factory in Wu county, Jiangsu, China) of high purity (>99.99%) of submicrometer size and relatively regular shape, which was prepared by pyrolysis of aluminum-ammonium phosphide. The chemical reactions are as follows:

$$Al_2(SO_4)_3 + (NH_4)SO_4 + 24H_2O \longrightarrow 2NH_4Al(SO_4)_2 \cdot 12H_2O$$

$$2NH_4Al(SO_4)_2 \cdot 12H_2O \longrightarrow 1100 - 1300\,°C \alpha - Al_2O_3 + 2NH_3 \uparrow + 4SO_3 \uparrow + 25H_2O$$

The particle size distribution measured by an X-ray centrifugal sedimentation technique (instrument type: BI-XDC, made by Brookhaven Instruments Corporation, New York, USA) is shown in Fig. 4.26.

Viscosity of the suspensions was measured by a rotary viscometer with a small sample chamber. The rheological measurement method of suspension was as follows: (1) All of the samples measured were milled to disperse particles in an alumina jar with alumina balls at pH = 4 for 10 h. (2) The shear rate of each suspension used was fixed to 60 s^{-1} for comparison.

The Zeta potential was calculated from the measured electrophoretic mobility by using electrophoretic light scattering (ELS) with a 5 mW laser source. The design of a novel electrode assembly to avoid electroosmosis phenomenon due to no need to focus at any stationary plane is shown in Fig. 4.27 The suspension of 0.05 vol% concentration was chosen.

To avoid the effect of particles on ion conductivity measurements, suspensions with different solid volume loading were first centrifuged to obtain samples of the supernatant mother liquor. Then, its conductivities were measured. These measured values, which were the ionic conductivities of the mother liquor (namely free ion

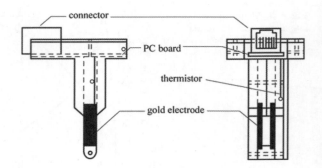

Fig. 4.27 The design of zeta plus electrode assembly

conductivity of suspension), were assumed to be ion conductivities of the suspensions since there was no way to measure the actual suspension in this study.

An ion chromatogram technique (Type: DX-100, made by Dionex Corporation, USA) was used to determine ions in the suspension. An AS4A-SC analytical column and AG4A guard column were used to determine anions with an AMM-2 anion micromembrane suppressor. Typically, a 1.8 mM Na_2CO_3/1.7 mM $NaHCO_3$ eluent with flow rate of 1 ml/min was used. The injection volume used was 25 µl. A CS12 analytical column and CG12 guard column were designed for cations with a CMMS-II cation micromembrane suppressor. The CS12 analytical column should have generated a system back pressure of approximately 800 psi when operated with a 20 mM HCl eluent at a typical flow rate of 0.5 ml/min. 50 mM KOH was used as a regenerant. The injection volume used was 25 µl. The anions and cations in the 5 vol% alumina suspension are shown in Fig. 4.28.

4.3.2 The Relationship Between Ion Conductivity Constants and Solids Volume Loading

In this section, we first introduced the result of Zeta potential versus the pH. The variation of Zeta potential with pH of raw materials and hydrated alumina that had been aged at 160 h in the acid medium is shown in Fig. 4.29. This figure implies that the IEPs of alumina increase from 5.6 to 9.2 with hydration time. When hydration time reached 160 h, the IEP of alumina became a constant. This is because an aluminum monohydrate layer on alumina particle was formed in an aqueous medium (Carniglia 1981). This hydration process is due to the thermodynamical instability of alumina in water.

By attrition during the wet-milling process, the aluminum monolayer is continuously removed from the surface and aluminum ions are released from the particles into the water. The unhydrolyzed Al^{3+} is stable below pH = 3 in the acid medium. With increasing pH, the Al^{3+} ion hydrolyzes and condenses, forming a variety of charged mononuclear and polynuclear complexes (Baes and Mesmer 1976). Many of these large polynuclear hydroxides and oxohydroxides are unstable. This is why

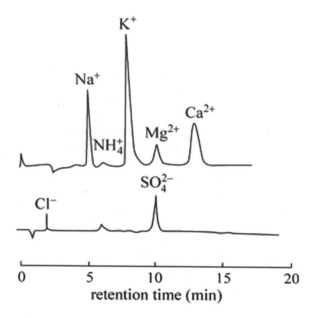

Fig. 4.28 The ion chromatogram of anions and cations in the alumina suspension

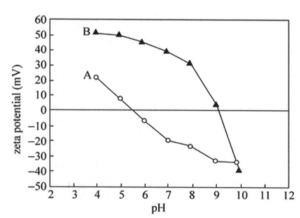

Fig. 4.29 Zeta potential versus the pH of alumina

the IEP of alumina increases with the hydration time. When the hydration process reaches equilibrium, the IEP of the alumina gradually becomes stable. Therefore, all of the alumina used to prepare the highly concentrated suspension was aged at 160 h in the aqueous acid medium.

Ion concentration in the suspension: A ceramic suspension consists of the suspension medium and ceramic powder. Ion concentration (n_t) in the suspension consists of ion concentrations (n_m) of suspension mediums and the soluble ion concentration (n_p) in the ceramic powders as follows:

$$n_t = n_m + n_p$$
$$= (n_{m1} + n_{m2} + \ldots n_{mi} + \ldots) + (n_{p1} + n_{p2} + \ldots n_{pi} + \ldots)$$

where n_{mi} is the ion concentration of the ith suspension medium; n_{pi} is the soluble ion concentration of the ith ceramic ceramic powder. Two or more kinds of ceramic starting materials are usually doped as sintering aids or composition of ceramics in the research and production of ceramic materials. A variety of ceramic powders have many of different soluble ions. Sometimes, the suspension media consists of manifold mediums, and so the effects of ions and ionic concentration on the highly concentrated suspension become very complex. Better understanding of ions and ionic concentrations in every powder and medium is essential in the preparation of a highly concentrated suspension.

High-valence counterions: Counterions are the reverse charged ions to particle charge. Highly valence counterions mean counterions whose charge number is larger than or equal to 2. The critical coagulation concentration depends inversely on the sixth power of the charge numbers (ze) according to:

$$n_c = \frac{107\varepsilon^3 (kT)^5 \gamma^4}{A^2 (ze)^6}$$

where ε is a dielectric constant, k is the Boltzmann constant, T is the absolute temperature, γ is a interfacial energy, A is a Hamaker constant, z is an electrovalent number, and e is an electrical charge. If there were soluble high-valence counterions in the ceramic powder, they would increase with increase in the solid content of the suspension, and the critical coagulation concentration would drop dramatically. In other words, the van der Waals attractive potential would increase and the electrostatic repulsion between the particles would reduce, and to some extent the particles would occur to agglomerate, the viscosity of the suspension would rise rapidly. Thus, a highly concentrated suspension cannot be prepared. If high-valence counterions were removed from the ceramic suspension, the critical coagulation concentration would not drop rapidly with increase in the volume loading. Thus, there is a possibility of preparing a concentrated suspension. Figure 4.30 shows an anion chromatogram curve of 5 vol% suspension after removing SO_4^{2-} from the alumina powder.

Ion conductivity constants: With increase in the volume loading, the ion concentration in the suspension would increase. How does one characterize the change of ion concentration in the suspension? Figure 4.31 shows that ion conductivities increase linearly with solids volume loading of seven kinds of alumina powders containing different ion concentrations (A, B, C, D, E, F, and G) after removing SO_4^{2-}. For a concentrated suspension, these lines have a greater deviation. This is because the free ions in the suspension are constrained by the overlapping of the electrical double layers around the particles. The inclination slopes of these lines were given according to:

$$K = \frac{\Delta C}{\Delta \Phi_V}$$

Fig. 4.30 Ion chromatogram of 5 vol% alumina suspension after removing SO_4^{2-}

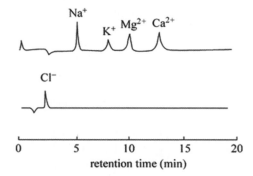

Fig. 4.31 Ion conductivity versus the solids volume loading

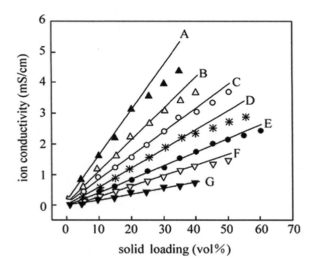

where K is a constant, whose unit is ms cm^{-1}. It is defined as the ion conductivity constant. For a given ceramic powder, because the ion concentration of the suspension medium does not change with solids volume loading, the ion conductivity constant never varies even though the suspension medium used is different. For a fixed starting material with the same particle size, distribution, shape, specific surface area, etc., if there were different ion concentrations, the ion conductivity constants would be different. Thus, it can be used to characterize the soluble ion concentration in ceramic powders. It is a very important parameter related to the successful preparation of a concentrated suspension.

The relationship between ion conductivity constants and solid volume loading: After removing high-valence counterions, with the increase in volume loading, the ion concentration in the suspension would not immediately cause the reduction of critical coagulation concentration. If the ion concentration in the suspension is very low, the electrical double layer would be very thick, resulting in the increase of the viscosity of the suspension due to overlapping between the electrical double layers. If the ion concentration is very high, the electrical double layer would be very thin, resulting in particle agglomeration due to the increase of attractive potential and reduction of electrostatic repulsion. The above two situations cannot lead to the successful preparation of a highly concentrated suspension with low viscosity. Only with a suitable ion concentration leading to a thin electrical double layer that would maintain enough repulsion to disperse particles, a highly concentrated suspension with low viscosity can be achieved.

At pH = 4, seven alumina powders with different ion conductivity constants as stated above (A, B, C, D, E, F, and G) were used to prepare aqueous suspensions with varied solids volume loading after for 10 h. The viscosity of each suspension was measured. The apparent viscosities of the suspensions at shear rate = 60 s^{-1} (as the Y-axis) and solid volume loading (as the X-axis) are shown in Fig. 4.32. These ion conductivity constants as a function of solid volume loading of alumina were discovered after removing SO_4^{2-} from alumina. If viscosity were 1 Pa s at shear rate 60 s^{-1}, these results were apparently shown in Fig. 4.33. With $K = 2.7–8.1$ ms cm^{-1}, a highly concentrated suspension of more than 50 vol% with apparent viscosity of

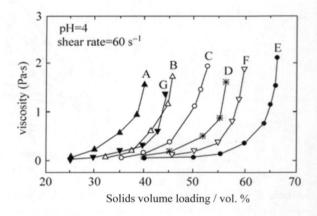

Fig. 4.32 Viscosity versus the solids volume loading of alumina suspensions with different ion conductivity constants

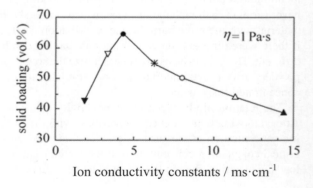

Fig. 4.33 Solids volume loading versus the ion conductivity constants (k) of alumina suspension

1 Pa s was successfully prepared and a maximum volume loading of alumina was achieved when $K = 4.5$ ms cm^{-1}.

Based on the above results, we can conclude the following:

(1) IEPs of alumina are a function of hydration time of the particle surface. In an acidic medium, IEPs of alumina changed gradually from 5.6 to 9.2 and then stabilized when hydration time reached 160 h.
(2) Ion conductivities increase linearly with solid volume loading of suspensions. Its slopes is defined as ion conductivity constant, which is an important parameter to characterize the quality of ceramic powder directly related to the preparation of concentrated suspension.
(3) At pH = 4, the particle surface of alumina is charged positively; a highly concentrated suspension (>50 vol%) cannot be prepared due to the presence of high-valence counterion of SO_4^{2-}.
(4) When the ion conductivity constant $K = 4.5$ ms cm^{-1}, a concentrated suspension of 65 vol% with a viscosity of 1 Pa s at a shear rate of 60 s^{-1} was successfully prepared.
(5) A novel way to prepare a highly concentrated suspension of ceramics is to control the ion conductivity constants suitably to reduce the thickness of the electrical double layer and to maintain repulsion between particles, after removing high-valence counterions.

4.4 Control of Inner Stress in Green Body

4.4.1 Origin, Transformation and Control of Inner Stress in Green Body

The colloidal forming technique (Gaucker et al. 1999a; Young et al. 1991; Novak et al. 2002), which is characterized by in situ solidification of the concentrated ceramic suspension, is a novel process to form ceramic green bodies. The typical route of the colloidal process is preparing a suspension with high solid loading and low viscosity, and then solidifying the suspension cast in a pore-free mold through various solidification mechanisms such as DCC (Gaucker et al. 1999a) and gelcasting (Young et al. 1991). The green body prepared by the colloidal process has a homogeneous microstructure similar to the precursor suspensions, and therefore the structure homogeneity and reliability of the ceramics are improved. However, asynchronous solidification of the suspensions caused by various factors, such as the temperature gradient or initiator distribution, has always been found for most colloidal processes. The asynchronous solidification makes the suspension shrink nonuniformly, and then inner stress develops in the ceramic green body, which is liable to be an origin for cracks during subsequent handling. In this section, the occurrence and transformation of inner stress in ceramic green bodies prepared by colloidal forming were expatiated, and some methods to control and eliminate inner stress in ceramic gelcasting green bodies were proposed.

In the pertinent study, alumina powders with a size $D_{50} = 1.86$ μm from Henan Xiyuan Ltd., were used as received. Tri-Ammonium Citrate (TAC) was used as dispersant. 15 wt% acrylamide (AM) premix solution was prepared with some MBAM (N, N′-methylenebisacrylamide, 2 wt% to AM) added. Alumina powders and TAC were added into the premix solution to prepare suspensions. 4 wt% $NH_4S_2O_8$ (AP) solution and TEMED were used as initiator and catalyst, respectively (Young et al. 1991). The time of the polymerization onset is characterized by the idle time (Young et al. 1991). The temperature of the suspension during gelation was recorded by a computer linked with the temperature detector (Yang et al. 2002). The pictures of the fracture surface of the green bodies were taken with a digital camera.

Inner stress caused by the concentration gradient of the initiator during solidification: The gradient of initiator in the suspension results in an uneven distribution of free radicals, and then the asynchronous solidification occurs. A previous study (Zhao 2002) showed that Cu^+ ions could accelerate the decomposition of ammonium persulfate (AP) due to the following reaction:

$$S_2O_8^{2-} + Cu^+ \rightarrow SO_4^{2-} + SO_4^{-\bullet} + Cu^{2+}$$

Thus, a copper beaker was used to provide a gradient of free radicals in the Al_2O_3 suspension (Fig. 4.34). After the suspension containing the initiator was cast into the copper beaker, Cu^+ diffused from the beaker wall to the center of the suspension. As a result, a gradient of Cu^+ ions and then a gradient of the free radicals appeared. To avoid the influence of axial diffusion of Cu^+ ions, a nylon bottom was used. During the polymerization, the temperature changes at three points (A, B, and C) were

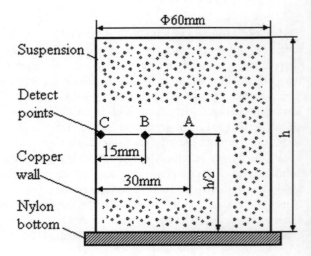

Fig. 4.34 Schematic device to test the effects of the free radicals gradient on the solidification

4.4 Control of Inner Stress in Green Body

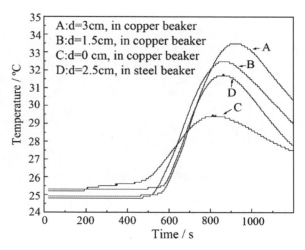

Fig. 4.35 The effects of the free radicals gradient on the solidification of the Al_2O_3 suspension

detected and collected, and the corresponding results are shown in Fig. 4.35. From the temperature rise in Fig. 4.35, it is found that the suspension near the beaker wall solidifies within a short time of about 200 s. This is a result of the effects of high Cu^+ concentration on the polymerization of AM monomers. However, the high Cu^+ concentration near the beaker wall has very little effect on the solidification of the suspension in the center. Therefore, the temperature rise of the center point in the copper beaker has no obvious difference from that in the stainless steel beaker. Since the diffusion of ions is slow in a dense Al_2O_3 suspension, the enrichment zones of initiator or free radicals, once they appear, will not be easy to eliminate. The suspensions in the enrichment zones solidify in the first instance and become contraction centers, so that the shrinkage of subsequent solidifying zones is constrained. As a result, a complex distribution of inner stress occurs. To analyze the effect of initiator enrichment in the suspension on the green body, AP and TEMED were added into a 300 ml suspension, and then the suspension was cast without sufficient stirring to prepare a green body., The macrocracks that appeared in the green body during debinding are shown in Fig. 4.36a. From the orientation of the cracks, it could be determined that there exist at least two origins in the green body. The reason for the random cracks is the inhomogeneous distribution of the initiator or free radicals, which yields multiple centers of contraction and causes stress concentration.

Inner stress caused by the temperature gradient during solidification: The effect of the temperature gradient on the solidification of Al_2O_3 suspension was studied using a double-layer beaker (Fig. 4.37). The outer ring of the beaker was filled with 60 °C water, and a 100 ml 25 °C suspension was cast into the core of the beaker. Then, the temperatures at three points (A, B, C in Fig. 4.37) were detected and collected. The temperature rise at points A, B, C during solidification is shown in Fig. 4.38. After the suspension was cast into the beaker, the 60 °C water in the

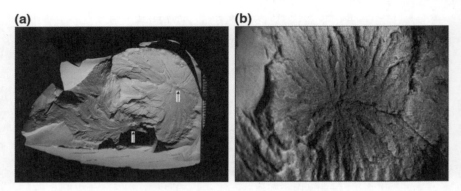

Fig. 4.36 The cracks in the green bodies caused by the gradient of free radicals (**a**) and temperature (**b**)

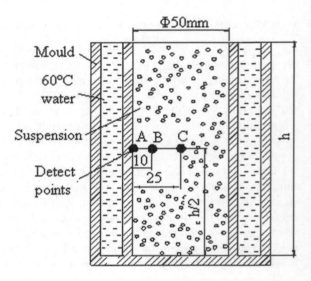

Fig. 4.37 Schematic device to test the effects of temperature gradient on the solidification

outer ring of the beaker produced an annular temperature gradient in the suspension. Thus, the solidification proceeded radially from the beaker wall to the center of the suspension. This constrained the solidification of the suspension near the center by the outer solidified ring. As a result, the center part of the green body endured radial and circumferential tensile stress, and the circumferential tensile stress was dominant for a sample with a small radius of curvature. Therefore, radial cracks in the green body were observed (Fig. 4.36b) after the polymer chains, which integrate with the ceramic powders in the green body, broke during debinding.

Generally, the magnitude of the inner stress is determined by the elastic modulus and the shrinkage ratio of the green body. The suspensions for colloidal forming usually have high solid loading, and therefore the wet green bodies have a small

4.4 Control of Inner Stress in Green Body

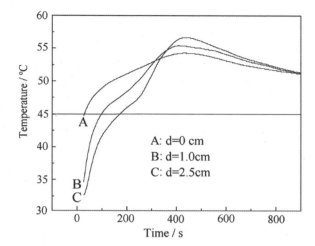

Fig. 4.38 The effects of temperature gradient on the solidification of Al_2O_3 suspension

shrinkage ratio (<0.5%) when demolded. On the other hand, the elastic modulus of the wet green body is very small (10^{-1} MPa) (Ruisong et al. 1994), and hence, the inner stress in the wet green bodies is also very small. However, the wet green body may shrink more than 3–4% and the elastic modulus may rise to the magnitude of 1 GPa during drying. This results in a considerable magnitude of inner stress. Due to the binding of the polymer chains, the inner stress does not cause macrocracks during drying, but could result in damage during debinding and sintering.

Inner stress caused by moisture gradient in the green body during drying: Although the moisture distribution in the wet green body is homogeneous, a moisture gradient exists in the green body during drying, because the moisture drains at different rates at the surface and the center of the wet green body. The moisture gradient causes the green body to shrink unevenly, and then the inner stress occurs. Figure 4.39 shows the drying rate and the shrinkage ratio of an Al_2O_3 disk ($\Phi 10.5$ cm $\times$ 3.3 cm) prepared by gelcasting of a 50 vol% Al_2O_3 suspension. As shown in Fig. 4.39, the

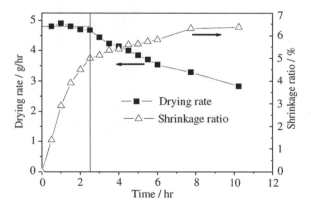

Fig. 4.39 The shrinkage ratio and drying rate of Al_2O_3 disk prepared by gelcasting during drying

shrinkage finished about 80% during the constant-rate drying period (Abbas et al. 1999), which varied from 0 to 2.5 h, but the moisture drained only 7 wt% during the same period. The rearrangement of particles on the surface and the shrinkage of the surface are complete at the end of this period. After the constant-rate drying period was complete, the interface between the solid and the liquid water withdrew from the surface to the interior of the green body. This means that the remaining 18 wt% moisture in the green body would be removed by diffusion through a capillary. The 'dried' superficial layer constrains the shrinkage of the interior part of the green body, and therefore inner stress appears in the drying green body. During debinding, radial and annular cracks appear in the Al_2O_3 disk (Fig. 4.40). The influence of the temperature gradient can be ignored because the Al_2O_3 suspension solidifies at room temperature. The nonuniform shrinkage during drying is likely to be an origin of inner stress in the green body. For the disk sample, radial and circumferential tensile stresses were introduced into the green body at the interface between the solid and the liquid water during drying. The radial cracks in the green body are caused by circumferential stress, and the annular cracks are caused by radial stress.

The inner stress in the green body prepared by gelcasting mainly results from the asynchronous solidification of the precursor suspension, and remains in the green body. The gradients of temperature and initiator concentration in the ceramic suspensions and the moisture in the green bodies are the major origins of the inner stress. This stress could result in macrocracks during debindering. To eliminate or reduce the inner stress in the green bodies, temperature-induced solidification should be avoided, and drying the green body at controlled moisture and temperature is required.

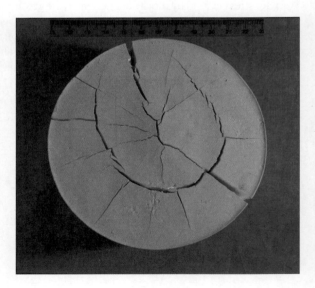

Fig. 4.40 Cracks in Al_2O_3 disk prepared by gelcasting after debinding

4.4.2 Release and Control of Inner Stresses in Ceramic Green Body

Among the colloidal forming processes, gelcasting, by which complex ceramic parts can be prepared (Omatete et al. 1991a), has been widely studied and applied to various ceramic materials (Nunn and Kirby 1996; Waesche and Steinborn 1997; Ma et al. 2002). Homogeneous green bodies with good strength required for machining can be made through this process. The mechanical properties of the materials after sintering are greatly improved (Bossel et al. 1995; Liu et al. 2002b; Huang et al. 2000). The main steps in the gelcasting process are as follows (Nunn and Kirby 1996; Waesche and Steinborn 1997; Ma et al. 2002; Young et al. 1991). First, the ceramic powder (e.g., alumina or silicon nitride) is thoroughly mixed with small quantities of gel initiators, catalysts, monomers, crosslinkers, and sintering additives to form a homogeneous suspension with high solid loading and low viscosity. Next, gelation is initiated by casting the suspension into a non-porous mould at elevated temperature. A green body is then formed by in situ polymerization with a three-dimensional gel network holding all the ceramic particles together. The green bodies are then demolded, dried, debindered, and finally sintered at a high temperature (This paragraph is probably repeating previous chapters).

In recent years, the authors developed a pressure-assisted gelcasting process using injection molding of colloidal suspension (Huang et al. 2004). The mold temperature and hence the temperature gradient can be reduced owing to the pressure-induced solidification mechanism. The reliability of the products is improved due to more homogeneous solidification but involves extra cost. The authors found that the cracking of the green bodies could be reduced by the addition of a plasticizer or a moderator in the suspension. The inner stresses are correlated with the elastic modulus of the green body that can be controlled by modifying the gel network by using moderators. This mechanism has been explored by using lab simulations. The strategies to minimize the inner stresses will be reported and the implications on the cracking of green bodies and the efficiency of the debindering process will be discussed.

In this research, a commercial Al_2O_3 powder with mean particle size of 3.84 μm, produced by the Xin-Yuan Al_2O_3 plant in the Henan Province of China, was used. The other chemical reagents used were: deionized water with conductivity of 1.02 μS cm^{-1}, acrylamide (AM) as monomer, hydroxyethyl acrylate (HEA) as moderator, methylenebisacrylamide (MBAM) as crosslinker, $(NH_4)_2S_2O_8$ as initiator, ammonium citrate as dispersant, and N, N, N′, N′-tetraethylmethylenediamine (TEMED) as catalyst.

Premix solutions were prepared by dissolving MBAM, AM, and HEA in deionized water. Since the moderator (HEA) is a kind of monomer, the total concentration of HEA and AM was kept constant at 14 wt%, while the ratio of HEA to AM varied. The concentration of the crosslinker, MBAM, was kept at 0.6 wt%. The concentrations of the initiator and catalyst are referred to reference (Huang et al. 2004). Six premix solutions with decreasing HEA concentration, M1, a, b, c, d, and e, were prepared, as summarized in Table 4.5. Al_2O_3 powder was dispersed into the premix solutions

Table 4.5 Weight percent of HEA in the monomers in different premix solutions

Premix solution	HEA/(HEA + AM)
M1	1
A	1/3
B	1/5
C	1/9
D	1/11
E	0

Table 4.6 The molds used to form green bodies

Number	Name	Dimensions
1	Simulation mold	Shown in Fig. 4.41
2	A glass beaker of 150 ml	Φ5.6 cm
3	A plastic beaker of 1000 ml	Φ11.3 cm
4	An annular stainless steel mold	External: Φ12.0 cm Inner: Φ5.3 cm
5	A spherical nylon mold	Φ7.0 cm

at a solid loading of 50 vol%. In order to obtain a low viscosity, 1 wt% of ammonium citrate of the ceramic powder was added into the suspensions as dispersant. The suspensions were subjected to ball-milling for 24 h and then poured into a mold to form ceramic green bodies. The details of the molds are given in Table 4.6. The green bodies were dried for 48 h at room temperature and then put into an oven at 80 °C until constant weight was obtained. The dried green bodies were debindered for 40–50 h by heating slowly from the room temperature to 600 °C and then sintered in a furnace at 1,560 °C for 2 h.

The viscosity of the suspensions was measured using a viscometer (MCR-300 mode, Physica Corporation, Germany). The images of the cracks in the ceramic green bodies were taken at a short focal distance ratio with a digit camera. Test bars with dimensions of 6.5 mm × 5.5 mm × 42 mm were prepared for evaluating the flexural strength and elastic modulus of the dry green bodies by using three-point bending tests on a testing machine.

To simulate the asynchronous solidification of ceramic suspensions, an organic glass mold was made. The mold was divided into three slots (L, M, and R) using two dummy plates, as shown in Fig. 4.41. The suspensions with different solidification rates were cast into different slots, and then merged by withdrawing the two dummy plates immediately after casting. In order to facilitate demolding, a trapezoidal section of the mold was used.

Initiation of inner stresses: The solidification rate of a ceramic suspension increases with the temperature and the concentrations of the solidification agents, including the initiator and the catalyst (Huang et al. 2004). Using the equipment shown in Fig. 4.41, a series of experiments were designed, as summarized in Table 4.7,

4.4 Control of Inner Stress in Green Body

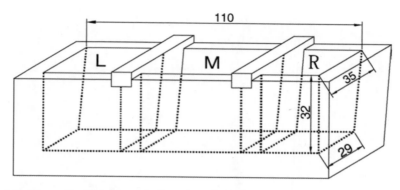

Fig. 4.41 The mold to simulate nonuniform solidification of the ceramic suspension

Table 4.7 Experiments to simulate nonuniform solidification in the ceramic suspension[a]

Test number		Temperature (°C)	Catalyst (μl)	Initiator (μl)
#1		18	150	150
	L	18	50	150
	M	18	150	150
	R			
#2		18	150	150
	L	4	150	150
	M	18	150	150
	R			
#3		18	150	150
	L	18	100	150
	M	18	50	150
	R	18	150	150

[a]Room temperature is 18 °C each slot with 50 ml of 50 vol% Al_2O_3 suspension, which was made from premix solution 'e'

to examine the effect of nonuniform solidification rate on the cracking of ceramic green bodies. In experiment #1, the dose of the catalyst in the middle slot was less than that in the left one and the right one. In experiment #3, the dose of the catalyst decreased in turn from the left to right. In experiment #2, the initial temperature in the middle slot is lower than that in the other two. The dose of the initiator is the same in all suspensions as shown in Table 4.7.

Under the conditions in experiments #1 and #2, the suspension in the right and the left slots solidified faster than that in the middle slot. In experiment #3, the suspension in the left slot solidified first, then the middle one, and finally the right one. No cracks were observed in all three samples after drying and debindering. However, there were cracks in samples #1 and #2 after sintering, as shown in Fig. 4.42. The cracking of

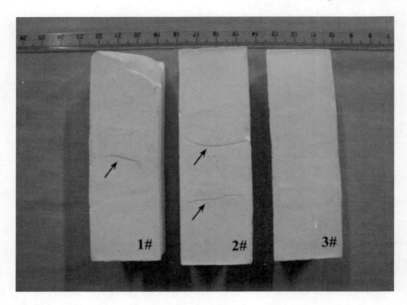

Fig. 4.42 Samples to simulate the nonuniform solidification of ceramic suspensions

samples #1 and #2 is related to the development of inner stresses in the green bodies. Since the suspensions in the left and right slots solidified earlier, the contraction of the green body in the middle slot was restricted. Therefore, a tensile stress developed in the vertical direction in the middle layer. The maximum tensile stress occurred in half depth where the cracking was observed in sample #1. The temperature distribution in experiment #2 is complicated by the heat transfer from the suspensions on both sides and from the atmosphere to the suspension in the middle. The solidification rate is not uniform in each layer. This leads to a complex distribution of inner stresses and thus the crack occurred at the one-third regions in the length direction. No cracks were found in sample #3 because the inner stresses are significantly reduced by a gradual transition in the solidification rate from the left to the right. These simulations confirm that the inner stresses can be traced back to the forming stage due to the gradient of initial temperature or the concentrations of solidification agents. In the subsequent drying stage, the inner stresses are inherited and magnified due to further contraction. Control of the inner stresses developed during the forming process is critical to the success of the subsequent drying and debindering processes after gelcasting.

Control of inner stresses: Fig. 4.43 shows the effect of the amount of the moderator on the strength of the ceramic green goodies. It is found that the strength of the green bodies decreases almost linearly with increasing concentration of the moderator. The strength of the green bodies (M_1) is less than 2 MPa in which the moderator replaces the original acrylamide (AM) monomer entirely. While the strength of the green bodies with a concentration of moderator of 20–33 wt% (samples 'b' and 'a') is about 20 MPa, which is slightly less than that of those produced without any

4.4 Control of Inner Stress in Green Body

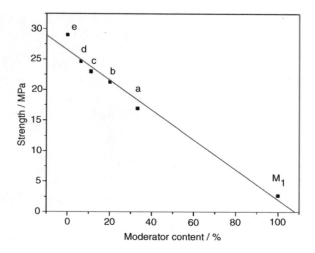

Fig. 4.43 Strength of ceramic green bodies versus the amount of HEA

moderator (sample 'e'). The elastic modulus of the green bodies versus different concentrations of the moderator is shown in Table 4.8. This indicates that the elastic modulus of the green bodies also decreases with the concentration of moderator. The elastic modulus of the green body 'e' (prepared from an original AM premix solution) is 8.63 GPa while the green body 'M_1' has an elastic modulus of 0.84 GPa in which the modulator replaces the original acrylamide (AM) monomer entirely. This is because the elastic modulus of a green body is determined by the strength of the gel network structure. The results indicate that the structure of the polymer network can be adjusted by the copolymerization of the acrylamide (AM) and moderator (HEA) monomers.

Five Al_2O_3 green bodies a, b, c, d, and e were produced by casting 100 ml of the suspensions into a 150 ml beaker respectively. The amount of moderator was varied from 33 wt% to 0 in the total weight of monomers as summarized in Table 4.5. The surfaces of these Al_2O_3 green bodies after debindering are shown in Fig. 4.44. The green bodies c, d, and e cracked when the amount of the moderator was not sufficient to soften the polymer network. However, the inner stresses were effectively reduced and cracking was suppressed when the concentration of the moderator was more than 20 wt%. By choosing a proper amount of the moderator, cracking of the green bodies can be avoided and sufficient strength of the green bodies can be retained.

Table 4.8 Elastic modulus of green bodies versus the concentration of moderator[a]

Test sample	M1	a	c	e
Content of the modulator (%)	100	33.3	11	0
Elastic modulus (GPa)	0.84	3.57	5.00	8.63

[a]The concentration of the moderator can be referred from Table 3.1

Fig. 4.44 Surface patterns of the Al_2O_3 green bodies after debindering

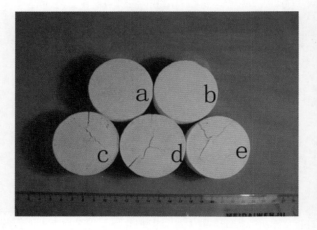

Figure 4.45 indicates the influence of HEA on the viscosity of ceramic suspension with 50 vol% solid loading. It was found that the viscosity increased slightly with the amount of HEA. Nevertheless, the viscosity was less than 1 Pa s, even if the amount of HEA was one-third of the total weight of HEA and AM (sample 'a'). The suspension was still suitable for gelcasting.

Effect on cracking on large ceramic green bodies: Two annular Al_2O_3 green bodies with diameter as large as 12 cm were obtained by casting suspensions, one containing 33 wt% moderator while the other had no moderator, into hot molds (the annular type, as shown in Table 4.6) respectively. From Fig. 4.46, it can be seen that the green body without HEA cracked annularly, while the other containing HEA did not crack at all. Disks were also prepared from these suspensions using a plastic beaker with a diameter of 11.3 cm. The disk with moderator was free of cracks, as shown in Fig. 4.47. The results confirm that relatively large ceramic green bodies can be made successfully by gelcasting with the addition of a proper moderator.

Fig. 4.45 Effect of HEA on the viscosity of the ceramic suspension

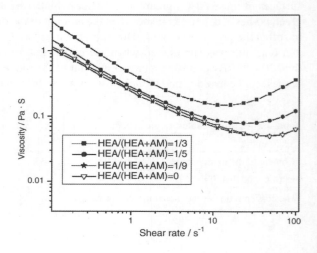

4.4 Control of Inner Stress in Green Body

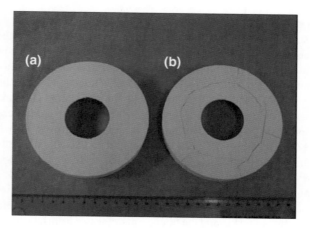

Fig. 4.46 Annular alumina green bodies: **a** with 33 wt% moderator and **b** no moderator

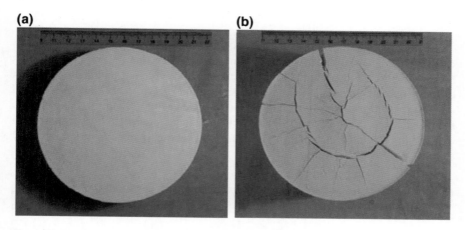

Fig. 4.47 Green bodies of alumina disks: **a** with 33 wt% moderator and **b** no moderator

Table 4.9 Results of large-size ceramic balls

Number	1	2	3	4
Gel system	Without HEA	Without HEA	Adding HEA	Adding HEA
Casting method	Hot mold	Microwave	Hot mold	Microwave
Result	Cracked after debindering	Cracked after debindering	Not cracked after debindering, cracked after sintering	Good

Improvement in the efficiency of debindering: As shown in Table 4.9, a series of large-size ceramic balls were prepared by using a spherical nylon mold under various conditions. The green bodies were dried, debindered, and sintered following the procedure described in Sect. 2.2. It is found that both samples without HEA (sample #1 and #2) cracked after debindering even if solidification was induced by microwave heating to minimize the gradient of temperature. It is inferred that the gel network with no moderator is very stiff, which results in significant inner stresses in the green body. The inner stresses are significantly reduced with the addition of HEA and this is why sample #3 survived the debindering process. However, the sample cracked during the sintering process. This indicated that a certain level of inner stresses existed in the green body due to the temperature gradient and increased with further shrinkage during sintering. In experiment #4, both strategies were explored: the gradient of temperature was minimized by using microwave and the gel network was softened by the addition of HEA. Ceramic balls prepared using this process were successfully debindered and sintered without cracking, as shown in Fig. 4.48 (the linear shrinkage rate was 16.4% after sintering and the final diameter was 5.6 cm). Further experiments showed that the debindering time could be reduced to 9 h (from room temperature to 600 °C) without cracking, compared to 40–50 h by using conventional heating process (Young et al. 1991). This significantly increases productivity and reduces the production cost.

The initiation and evolution of inner stresses in the green body formed by gel-casting have been discussed above. It can be concluded that the important factors responsible for the inner stresses are the temperature gradient and the distribution of the solidification agents. Adding proper amount of HEA into the concentrated suspension improves the flexibility of the polymer network and thus reduces the inner stresses in the green body. By reducing the inner stresses in the green body at the forming stage, the debindering time of large ceramic parts can be significantly shortened.

Fig. 4.48 Sintered alumina balls (the linear shrinkage rate was 16.4% after sintering and the final diameter was 5.6 cm)

4.5 Suppression of Surface Exfoliation with the Addition of Organic Agents

4.5.1 Suppression of Surface Exfoliation by Introducing Polyacrylamide (PAM) into a Monomer System in Suspension

Gelcasting involves the casting of the slurry that consists of ceramic powder, water, and water-soluble organic monomers. After casting, the monomer in the slurry is polymerized to form a gelled network which holds the ceramic powders together in a mold. Compared to other casting methods, gelcasting produces results in much more homogeneous green bodies with little density variation over the parts and higher flexural strength. However, the gelcasting technique has a main problem, surface spallation.

The Oak Ridge National Laboratory (USA) (Young et al. 1991) first reported flaws and spallation on the surface of dried green bodies prepared by gelcasting in air. Their study concluded that the surface spallation phenomenon is due to the presence of oxygen, which hinders polymerization and gelation of the monomer in the system (Odian 1991; Landham et al. 1987). Surface spallation could result in green bodies with a powdery surface that can be a few millimeters thick. This will reduce the strength and make it difficult to control the dimension of the final products, and thus it must be avoided in production. It was found that gelcasting in a nitrogen atmosphere could suppress the effect of oxygen inhibition and hence prevent the occurrence of surface spallation (Janney and Omatete 1992; Gilissen et al. 2000). In industrial production, however, a technical operation under nitrogen introduces complexity with significant increase in the production cost.

In our research, we attempted to resolve the surface spallation problem by introducing a water-soluble polymer, polyacrylamide (PAM), into the acrylamide-based system. The effects of the polymer on the gelcasting of alumina and the surface spallation phenomenon of green bodies were investigated.

Materials in the experiments: The starting ceramic powder used in this experiment with an average particle size of 3.47 μm was a commercial high-purity α-Al_2O_3 powder. Calcium carbonate, silicon dioxide, and kaolin were used as its sintering aids. For gelcasting, the following were used: acrylamide [$C_2H_3CONH_2$] (AM) as the monomer, N, N′-methylenebisacrylamide [$(C_2H_3CONH)_2CH_2$] (MBAM) as a coupling agent, N, N, N′, N′-tetramethylethylenediamine (TEMED) as a catalyst, ammonium persulfate as an initiator, and ammonium citrate as a dispersant. Polyacrylamide (Aldrich Chemical Company, Inc. USA) used in the study had an average molecular weight of 10 000. All reagents were chemically pure.

Determination of PAM amount: A proper amount of PAM may eliminate the surface spallation phenomenon of green bodies cast in air, but decrease the fluidity of the suspension. To determine the optimum amount of PAM, which is enough to eliminate the surface spallation phenomenon of green bodies cast in air, the following experiments were performed. Suspensions with different amounts of PAM were prepared. The amount of PAM and the degree of surface spallation are shown in Table 4.10. The suspension with 2.8 wt% PAM (based on alumina) has a better fluidity and has no surface spallation phenomenon. Thus, 2.8 wt% PAM was chosen as the additive for the suspensions in the subsequent experiments.

Process of gelcasting with PAM: The preparation of suspensions is similar to that of reference (Omatete et al. 1991). The 2.8 wt% PAM (based on alumina) and 0.25 wt% dispersant (based on alumina) were first completely dissolved in a premix solution, which was prepared by dissolving proper amounts of AM and MBAM in deionized water (the mass ratio is 50:1:315), to produce a complex solution. Alumina powder and its sintering aids were then added into the solution. Suspensions with solid loading of 50 vol% were mixed manually first and then milled for 24 h in a nylon resin jar using alumina balls as ball-milling media to break down the agglomerates formed during solvent evaporation and to achieve a good homogeneity. The prepared suspensions were subjected to measurements for the rheological property. 100 μl initiator and 50 μl catalyst were added to the suspension (30 ml) to form a slurry, which was used for measurement of gelation characteristics and to fabricate cast samples for microstructure and mechanical characterization.

Binder burnout and subsequent sintering were carried out in stationary air and conducted separately. Binder burnout was done at 700 °C for 2 h, with a heating rate of 2 °C/min and a natural cooling. Sintering was carried out at 1,550 °C for 2 h with a heating rate of 1 °C/min from room temperature to 1,250 °C and a rate of 0.5 °C/min from 1,250 to 1,550 °C and a cooling rate of 1 °C/min to 1,000 °C, followed by natural cooling.

Measurements: The shear viscosity is measured using an MCR (MCR300, Paar Physica, Germany) with a concentric cylinder having a diameter of 27 mm. Steady-state shear flow curves are measured at a shear rate range between 0.1 and 250 s^{-1}; the duration between the measuring points is 10 s and each measurement lasts 3 min.

Table 4.10 Relationship between sample surface and polymer amount

Amount of PAM (wt%)[a]	Surface of sample	Viscosity (Pa s)	
		Shear rate 1.18 s^{-1}	Shear rate 48.1 s^{-1}
0	Spallation	0.45	0.10
1.0	Spallation	0.65	0.26
2.0	Spallation	1.68	2.81
2.5	Spallation	4.78	4.24
2.8	Without spallation	5.91	5.43
3.0	Without spallation	7.30	9.42
4.0	Without spallation	33.5	37.8

[a]PAM content is based on alumina

4.5 Suppression of Surface Exfoliation ...

The temperature ramp rotation mode is used to measure the temperature dependence of the slurry viscosity.

A suspension was prepared at a loading of 5 vol% solids. This suspension was allowed to age for 24 h with continuous stirring. 0.4 ml of the above suspension was then diluted to 200 ml to prepare a final working solution with loading of 0.01 vol% solids for analysis. Tests were performed on alumina in two different solutions: 10^{-2} mol/l potassium nitrate (KNO_3) without PAM, and 2.8 wt% PAM in 10^{-2} mol/l potassium nitrate (KNO_3). The pH was adjusted with HCl and NaOH solutions of appropriate concentration. Zeta potential was calculated from the measured electrophoretic mobility in the pH interval from 3 to 12 for alumina using the ELS method (Brookhaven Instruments Corporation, USA).

Bulk densities of dried green and sintered bodies were determined by the Hg immersion method based on Archimedes' principle. The flexural strengths of bars of green and sintered bodies were examined by the three-point bending test with a span of 30 mm at a loading rate of 0.5 mm/min. The bars of green and sintered bodies were of dimensions 0.5 cm × 0.6 cm × 4.2 cm and 0.3 cm × 0.4 cm × 3.6 cm, respectively. Fractal surfaces of the green and sintered specimens were examined by an SEM (JSM-6460LV, JEOL Ltd., Japan).

Dispersion of Al_2O_3 powder: Fig. 4.49 shows the zeta potential of the Al_2O_3 powder in water with and without PAM as a function of pH. Error bars of ±2 mV indicate the reproducibility of the values. Adding PAM affects the surface charge of the particles in dispersion and their zeta potential. With 2.8 wt% PAM added to the slurry, the zeta potential–pH curve shifts toward a lower pH region, and the isoelectric point of the slurry moves from pH 7.4 to pH 3.6. Increased zeta potential values (higher than −30 mV) were observed at pH > 10.5, and the slurry with PAM could be well dispersed in a much wider pH range compared to that without PAM. The shift of the isoelectric point of the oxide powders toward lower pH values is typical of the PAM function and similar behavior has been observed for Al_2O_3 slurries with the addition of NH_4-PMA (Christos et al. 2000).

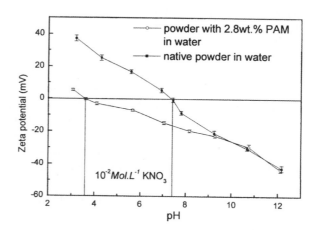

Fig. 4.49 Zeta potential of the Al_2O_3 powder in water with and without 2.8 wt% PAM as a function of pH

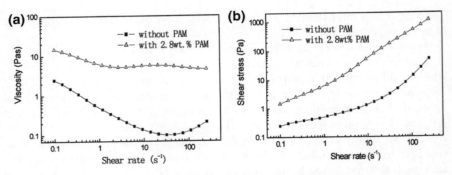

Fig. 4.50 The curves of the rheological properties for the suspensions with and without 2.8 wt% PAM

Rheological properties of Al$_2$O$_3$ suspensions with PAM: In order to further examine the colloidal behavior, we present in this section the curves of the rheological behavior. Figure 4.50a shows the variation of steady-state viscosity as a function of shear rate for the suspensions with and without PAM. It can be seen that the suspension without the addition of PAM displays relatively strong shear-thinning behavior. The degree of shear-thinning decreases significantly after the addition of PAM. Using PAM, highly concentrated suspensions can be prepared but their fluidity is likely to be compromised. Concentrated stable suspensions usually exhibit shear-thinning behavior due to perturbation of the suspension structure by shearing (De Kruif 1990). Another characteristic shown in Fig. 3.5.2a for the suspension with the addition of PAM is the shear-thickening (dilatancy). At high shear rates (>4 s^{-1}, depending on solid loading), the viscosity increases as the shear rate increases. Hoffman (1972) suggested that shear-thickening is a consequence of an order-to-disorder transition of the particle microstructure.

Figure 4.50b shows the corresponding variation of shear stress as a function of shear rate for the suspensions with and without PAM of Fig. 4.50a. The rheological behavior depicted in Fig. 4.50b has been analyzed using the Herschel–Bulkley model (Herschel and Buckley 1926).

$$\tau = \tau_y + k\gamma^n \quad (1)$$

where τ_y is the yield stress, τ is the shear stress, γ is the shear rate, n is the shear rate exponent, and k is a constant. The yield stress values presented in Table 4.11 are calculated from this model. The suspension with the addition of PAM exhibits a higher yield stress compared to the suspension without PAM.

Shih et al. (1999) recently reported that the suspension yield stress (τ_y) scales with the solids concentration (Φ) in the following form:

$$\tau_y \sim \left(1 - 1.5\alpha\zeta^2\right)\left(A/24_{So}^{3/2}\right)\left(1/R^{d-3/2}\right)\Phi^m \quad (2)$$

4.5 Suppression of Surface Exfoliation ...

Table 4.11 Hershel–Bulkley parameters τ_y, and n of alumina suspensions (50 vol%, 0.25 wt% dispersant, 25 °C) for 2.8 wt% PAM

Sample	Monomer content[a] (wt%)	Polymer content (wt%)	Yield stress τ_y (Pa)	Constant k	Shear rate exponent n	Correlation ratio R_xy
A	3.2	0	0.3238	0.0863	0.9698	0.9214
B	3.2	2.8	1.3252	4.7269	0.9973	0.9987

[a]Contents of the monomer and polymer are based on alumina

where α is a constant related to the Debye thickness (k^{-1}) and the surface separation (s_o) between the particles; ζ the zeta potential; A the Hamaker constant; R the particles radius; d the Euclidean dimension; and $m = (d + X)/(d - D_f)$, with D_f and X the fractal dimension of the clusters and the backbone of the clusters, respectively. It can be concluded that the shear stress (τ_y) increases with the decrease in the zeta potential at given the solids concentration from Eq. (2). The result of the present research is fit to this equation.

Gelation characteristics of Al_2O_3 slurries with PAM: It is well known that the reaction of free radical polymerization is exothermal. When the polymerization starts, the temperature of the system is enhanced. Thus, the gelation characteristics of the suspension may be described by the variation of temperature with time.

Figure 4.51 shows how the temperature of the alumina slurry with and without PAM varies with time. It can be seen that the increase in the temperature of the slurry without PAM is faster and reaches a higher level at a shorter time, compared to the slurries with PAM. The initial increase in temperature (i.e., the consolidation starting point) occurs at different times between the PAM-containing system (about 800 s) and the one without PAM (about 700 s). Above 700 s, the monomers start to polymerize, resulting in a temperature rise because of exothermal reaction. Besides, the increase in temperature terminates at about 1,300 s (i.e., the consolidation completing point) for

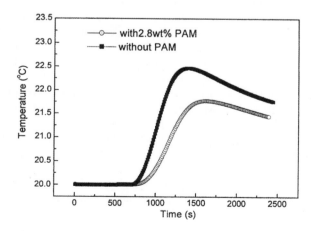

Fig. 4.51 The relationship between temperature and time

Fig. 4.52 Alumina green body surfaces gelcast in air from slurries **a** without PAM and **b** with 2.8 wt% PAM

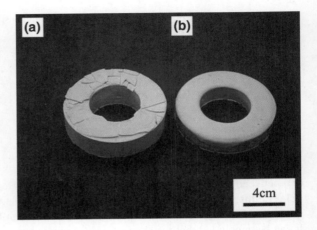

the PAM-free system, but for the PAM-containing system the consolidation continues at least at 1,650 s. A possible reason for the suspension with PAM requiring a longer time for the initial rise in temperature compared to the PAM-free system is that the contact among monomers AM would be interfered and hindered by PAM in the solution. Hence, longer time would be required for the AM to become a hydrogel in the presence of PAM. Further studies would be required to confirm this.

Surface of gelcast alumina green bodies: Fig. 4.52 shows annular cylinders of alumina obtained by gelcasting. Figure 4.52a, b are green bodies gelcast from slurry without the addition of PAM and slurry with addition of 2.8 wt% PAM, respectively. The surface of the green body gelcast in air from the slurry without addition of the polymer shows surface spallation phenomenon, but a surface without spallation is observed in the green bodies gelcast in air from the slurry containing 2.8 wt% PAM. The mechanism of eliminating surface spallation by PAM is similar to the one in our previous study (Ma et al. 2003).

Properties of alumina green and sintered bodies: Fig. 4.53a presents the SEM micrograph of an alumina green body obtained from the system by adding PAM, while Fig. 4.53b shows the SEM micrograph of a green body without PAM. It can be seen that a satisfactory microstructure with very few pores has been achieved for both systems and a polymer network can be observed in both green bodies. The variation in density and green strength of the dried bodies with and without 2.8 wt% PAM are shown in Table 4.12. The density of the green body prepared from the suspension with PAM increases compared to that without PAM. The green strength, on the other hand, shows a significant increase after the addition of PAM. After adding 2.8 wt% PAM, the density and strength of the green body are approximately 58.6% of the theoretical density (3.85 g/cm^3, the mixed density of alumina and sintering aids) and 43.38 MPa, respectively.

The previous studies indicated that the strength of a green body formed by gelcasting was provided by the polymer gel (Asad et al. 2000; Bauer et al. 1999).

4.5 Suppression of Surface Exfoliation ...

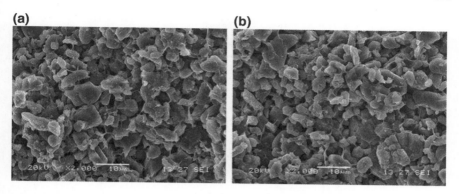

Fig. 4.53 The microstructures of green bodies prepared from suspension **a** without PAM and **b** with 2.8 wt% PAM

Table 4.12 The bulk density and flexural strength of dried green and sintered bodies with and without 2.8 wt% PAM

PAM content	Flexural strength of green body (MPa)	Density of green body (g/cm^3)	Flexural strength of sintered body (MPa)	Density of sintered body (g cm^{-3})
0	37.11 ± 1.67	2.176	351.85 ± 19.10	3.71
2.8 wt%	43.38 ± 4.58	2.323	347.63 ± 24.72	3.68

However, for the monomer solution with 2.8 wt% PAM, as the polymerization proceeds, the native PAM with free amino and keto groups are likely to form a more complicated network structure through hydrogen bonding with the keto groups of the PAM polymerized by the monomers, resulting in an intimate mix of polymer chains (as shown in Fig. 4.54). The relatively high gel strength obtained for the monomer-polymer system with respect to a pure monomer system is attributed to the enhanced polymer network due to this cross-linking through hydrogen bonding. Hence, the flexural strength of green bodies prepared from the suspension with addition of PAM increases significantly compared to those without PAM.

Figure 4.55 illustrates the SEM micrograph of alumina ceramics. It should be noted that the ceramics have a high density and uniform microstructure without developing a huge grain growth. The flexural strengths are about 350 MPa for both ceramics obtained from the sole AM system and ones from the mixed PAM and AM system (as shown in Table 4.12).

Based on the above expatiation, we can conclude that the surface spallation phenomenon of green bodies gelcast in air can be eliminated by adding a small amount of PAM to the monomer solution. It is found that by using PAM, highly concentrated suspensions (~50 vol% solids loading) can be prepared with a small reduction in fluidity of the slurries (0.1 Pa s, at shear rate of 48.1 s^{-1}).

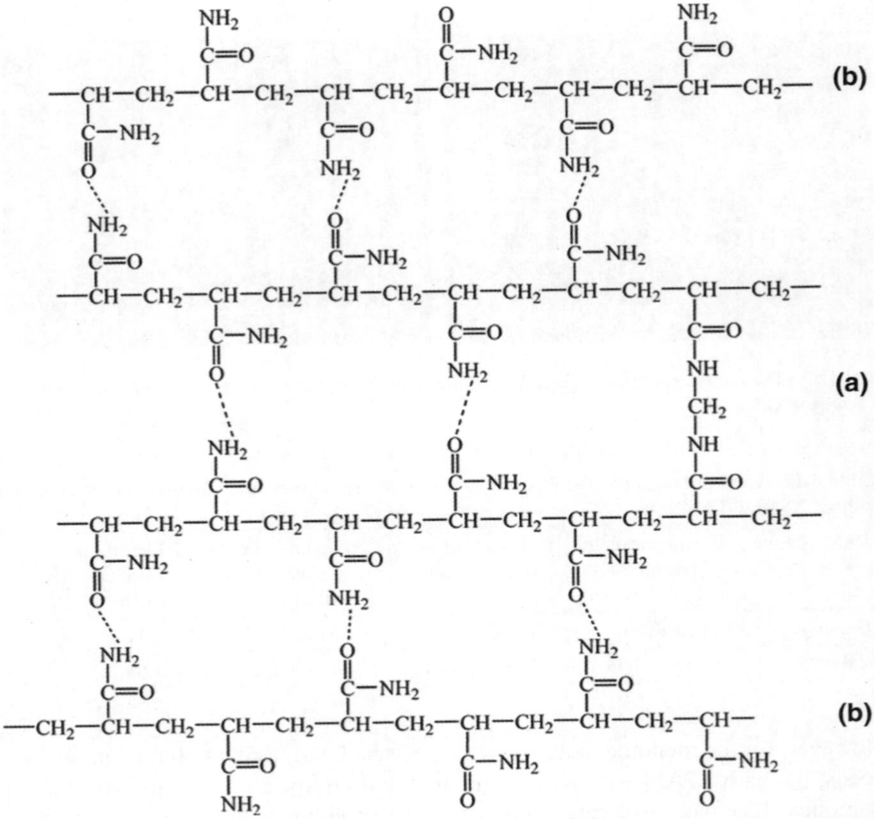

Fig. 4.54 The proposed structure of cross-linked PAM (**a**) gel after an addition of native PAM (**b**)

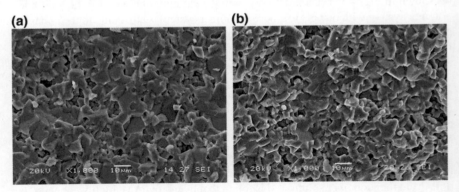

Fig. 4.55 The microstructures of sintered bodies prepared from suspension **a** without PAM and **b** with 2.8 wt% PAM

With 2.8 wt% PAM added to the slurry, the zeta potential–pH curve shifts toward a lower pH region, and the isoelectric point of the slurry shifts to pH 3.5 as the added PAM results in a reduction of the positive shear-plane potential.

After adding PAM to the AM-containing slurry, the suspension exhibits shear-thinning behavior with yield stress. The increase of the temperature becomes more rapid and reaches a higher level at a lower period for the sole AM system compared to the PAM/AM mixed system.

The flexural strength of green bodies prepared from the PAM/AM mixed slurry is higher than that from the AM-only system. The relatively high green body strength 43 MPa obtained for the PAM/AM system is attributed to the cross-linking between the native PAM and polymerized PAM chains through hydrogen bonding and polymer gel. A dense and homogeneous microstructure is observed for the alumina samples prepared from the slurries containing mixed PAM and AM. The flexural strengths are about 350 MPa for both ceramics.

4.5.2 Suppression of Surface Exfoliation by Introducing Polyethylene Glycol (PEG) into Monomer System in Suspension

Our study was intended to resolve the problem of oxygen inhibition by chemical methods. In an attempt to develop components of an aqueous system, we investigated gelcasting by adding nonionized water-soluble polymers and noticed that the surface exfoliation phenomenon of green bodies gelcast in air was eliminated by the addition of a proper amount of polymers to the monomer solution. A new introduced component is not toxic, which provides the possibility for further industrialization.

Polyethylene glycol (PEG) has the simplest structure of the nonionized water-soluble polymers. Its physical conformation with 10,000 molecular weight is a white wax-state solid. PEG has good stability, adhesional wetting, and low toxicity, and it burns out completely, resulting in wide application in the casting of ceramics.

The purpose of this study is to investigate and discuss the effect of the water-soluble polymer PEG with a molecular weight of 10,000 on the gelcasting of alumina, based on the precondition that the surface exfoliation of green bodies cast in air is inhibited.

The starting ceramic powder used in this study was a commercial high-purity α-Al_2O_3 powder with an average particle size of 3.47 μm. For gelcasting, acrylamide [$C_2H_3CONH_2$ (AM)] was used as the monomer; N, N′-methylenebisacrylamide [$(C_2H_3CONH)_2CH_2$(MBAM)] was used as a coupling agent; N, N, N′, N′-tetramethylethylenediamine [TEMED] was used as a catalyst; and ammonium persulphate was used as an initiator. To improve the dispersion of the powder and fluidity of the suspension, an organic base, ammonium citrate, is used as a dispersant. PEG with a molecular weight of 10,000 was purchased from Tianjin Chemical Company, China.

Table 4.13 Relationship between the amount of PEG and the degree of surface exfoliation of green body

Amount of PEG (wt%)	Surface of sample	Viscosity (Pa·s)	
		Shear rate 11.8 s^{-1}	Shear rate 21.1 s^{-1}
0	Exfoliation	0.447	0.105
0.5	Exfoliation	0.684	0.874
1.0	Exfoliation	0.709	0.981
1.3	No exfoliation	0.730	1.09
1.5	No exfoliation	0.882	1.55
2.0	No exfoliation	1.6	3.57

Preparation of suspensions: A proper amount of PEG may eliminate the surface exfoliation phenomenon of green bodies cast in air, but decreases the fluidity of the suspension. To determine the optimum amount of PEG, which is enough to eliminate the surface exfoliation phenomenon of green bodies cast in air, the following experiments were performed. Suspensions with various amounts of PEG were gelcast. The amount of PEG and the degree of surface exfoliation are shown in Table 4.13. The suspension with 1.3 wt% PEG (based on alumina) has a better fluidity and exhibits no surface exfoliation phenomenon. Thus, 1.3 wt% PEG was chosen as the additive for the suspensions in subsequent experiments.

The PEG (1.3 wt%, based on alumina) and dispersant (0.25 wt%, based on alumina) were first completely dissolved using mechanical stirring for 5 min in a premix solution, which was prepared by dissolving AM and MBAM in demonized water. The premix solution served as a dispersing media for the ceramic powder. Next, the alumina powder was added. The suspensions, with solids loading of 50 vol% were mixed manually and then milled for 24 h in a nylon resin jar by using alumina balls as a milling media to break down the agglomerates and to achieve a good homogeneity. The prepared suspensions were used to evaluate their rheological properties and viscoelastic properties. The prepared suspension (25 ml) with 100 μl initiator and 50 μl catalyst was used to evaluate the gelling characteristics and to fabricate samples. The surfaces of the samples were inspected and their flexural strength was measured.

Zeta potential is a very important physical parameter, which can affect the stability of the aqueous ceramic slurry; the higher the zeta potential, the higher the repulsive energy and the more stable the slurries. Figure 4.56 presents the measured results of zeta potential with or without PEG in a 0.01 wt% alumina suspension. The zeta potential for alumina powder only ranged from 38 mV at pH 3 to ~40 mV at pH 12 with an IEP of pH 7.3, while that in the case of the suspension containing 1.3 wt% PEG ranged from 24 mV at pH 3 to ~38 mV at pH 12 with an IEP of pH 7.1.

It was found that the PEG did not dissociate and had hardly any effect on the alumina surface charges (Fig. 4.57).

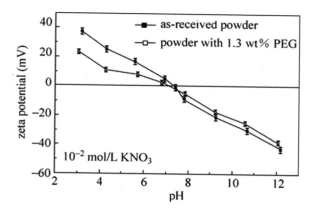

Fig. 4.56 Zeta potential of the Al$_2$O$_3$ powder in water without and with 1.3 wt% PEG

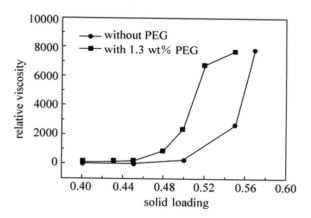

Fig. 4.57 Relative viscosity of Al$_2$O$_3$ suspensions without and with 1.3 wt% PEG versus the solids loading at a shear rate of 110 s^{-1}

The viscosity of a suspension is strongly dependent on the solid volume fraction, with the viscosity approaching infinity at a maximum volume fraction, Φ_m, where Φ_m relates to the particle concentration at which the average separation distance between the particles approaches zero and the particles pack together, making flow impossible.

The relative viscosity is the ratio between the viscosity of the suspension and the viscosity of the monomer solution (1.45 mPa s). The experimental points are fitted to a modified Krieger–Dougherty equation (Lyckfeldt and Ferreira 1998):

$$\eta_r = (1 - \Phi/\Phi_m)^{-n},$$

where Φ_m and n are used as fitting parameters. The best fit of the experimental data to the above equation shows that Φ_m is lower for the suspension with PEG ($\Phi_m = 0.52$) compared with the Al$_2$O$_3$ suspension ($\Phi_m = 0.58$). The difference in the value of Φ_m illustrates that the addition of PEG has an influence on the packing behavior of particles, but this influence is not significant.

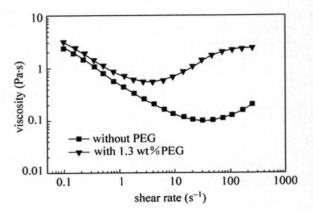

Fig. 4.58 Viscosity of the suspensions without and with 1.3 wt% PEG

The steady shear data of suspensions without and with PEG are shown in Fig. 3.5.10. Both f these suspensions display the same type of rheological behavior. Shear-thinning occurs at low shear rates and shear-thickening occurs at very high shear rates. However, these suspensions show significant differences in viscosities at the same volume fraction of solids as seen in Fig. 4.58. The suspension without PEG has lower viscosity and the critical point from shear-thinning to shear-thickening occurs at higher shear rate ($\gamma = 32$ s^{-1}) while the suspension with PEG exhibits higher viscosity and the critical point from shear-thinning to shear-thickening occurs at lower shear rate ($\gamma = 4$ s^{-1}).

Concentrated, colloidally stable suspensions display shear-thinning because of a perturbation of the suspension structure by shear (Lyckfeldt and Ferreira 1998). At low shear rates, the suspension structure is close to equilibrium because thermal motion dominates over the viscous forces. At higher shear rates, viscous forces affect the suspension structure, and shear-thinning occurs. At very high shear rates, the viscosity increases as the shear rate increases.

With the addition of polymers such as PEG into the monomer solution, the system becomes a mixture of organic monomer and polymer. It is important that the PEG is absorbed on the ceramic particle surface as homogeneously as possible. For this reason, ball-milling of ceramics with PEG has been commonly used. Steady shear properties of Al$_2$O$_3$ suspensions with PEG at various ball-milling periods are shown in Fig. 4.59. It can be seen that the optimum ball-milling period is 24 h for an alumina suspension to which PEG has been added. All of the suspensions are characterized by a shear-thinning behavior at low shear rates and then a thickening behavior at high shear rates except the suspension with 0 h ball-milling period. These results imply that the PEG molecule would be absorbed on the alumina particles, and influence the interaction between the particles during ball-milling.

Figure 4.60 shows the log–log plot of the variation of G' and G'' with the solids loading of alumina suspension with 1.3 wt% PEG, where G' is the storage modulus and G'' is the loss modulus. G', which is proportional to the elastic modulus of the gel, characterizes the elastic response of the gel. G'', which is proportional to the viscous

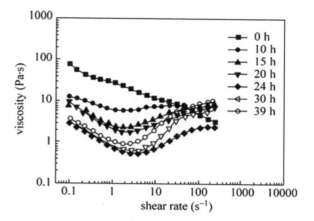

Fig. 4.59 Effect of ball-milling time on the viscosity of suspensions without and with adding 1.3 wt% PEG

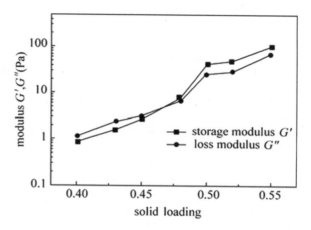

Fig. 4.60 Variation of G' and G'' with the solids loading of alumina suspension with 1.3 wt% PEG at constant frequency 1.27 Hz

modulus of the gel, characterizes the viscous response of the gel. The modulus of the suspension shows a gradual change from more viscous (G' < G'') to more elastic (G' > G'') with increasing volume fraction. At relatively low solids loading, there is a weak interaction between the particles or/and PEG, which are only slightly compressed. Under these conditions, the suspension behaves as a viscous fluid with little elastic contribution. When the solid loading is increased, G'' becomes relatively large until a critical solid loading ($\Phi_m = 0.47$) is reached where G' ≈ G'', above which greater inter-particle interaction occurs. Above the critical solids loading, G' continues to increase until it becomes close to G* (complex modulus, $G^* = G' + i\,G''$, where $i = (-1)^{1/2}$). Under these conditions, the system becomes predominantly elastic, with significant interpenetration and compression of the chains occurring. This is also reflected in a rapid increase in the dynamic viscosity η' and the whole suspension behaves like a gel.

Fig. 4.61 Relationship of the storage modulus of alumina suspension versus solidification time

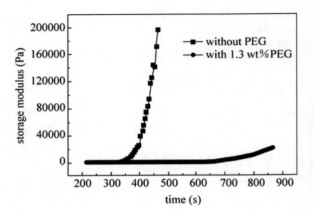

For gelcasting, the polymerization is induced by an initiator and the reaction can be accelerated by a catalyst and temperature. By using an initiator and at the same time fixing the amount of catalyst and the temperature, the structural change of the suspension versus time could be observed (Lyckfeldt and Ferreira 1998).

Figure 4.61 shows how the storage modulus of alumina/PEG slurry varies with time. It can be seen that the increase of the storage modulus is more rapid and reaches a higher level at the same temperature without PEG, compared to those of slurry with PEG. The storage modulus of the slurry with PEG is lower than that of the slurry without PEG, which indicates the strength of green body with PEG is less than that of the green body without PEG.

The reason that the storage modulus of the suspension with PEG varies earlier and higher than that of the suspension without PEG is that the contact among monomers AM might be interfered and hindered by the PEG in the solution. Hence, longer time would be required for AM to become a hydrogel in the presence of PEG. Above 350 s, the monomer starts to polymerize, resulting in an increased viscosity and an elastic viscoelastic response (Fig. 4.60).

The physical properties of the gelled sample from the alumina suspension with PEG were studied. An acrylamide/PEG-based gelcasting system is a viable alternative to one based on in situ polymerization of monomeric systems. This study highlights the importance of suspension composition (i.e., polymer, monomer, cross-linking agent, and solid content) and rheological and gelatin behavior. A minimum PEG concentration in the solution is required to inhibit the surface exfoliation phenomenon. By a series of experiments, it is concluded that 1.3 wt% PEG is the optimum concentration. Figure 4.62 shows the surface of an annular cylinder of alumina obtained by using the present method. Figure 4.62a, b show green bodies gelcast from the slurry with additions of 1.3 wt% PEG and the slurry without PEG, respectively. It can be seen that the surface of the green bodies without PEG exhibits a surface exfoliation phenomenon. However, the surface exfoliation phenomenon of green bodies gelcast in air was eliminated by adding 1.3 wt% PEG to the monomer solution. The

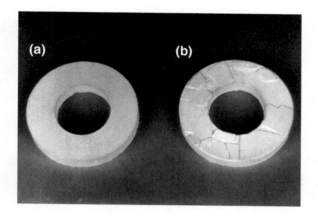

Fig. 4.62 Surface of alumina green body prepared from the suspension **a** with 1.3 wt% PEG and **b** without PEG

mechanism that inhibits surface exfoliation phenomenon of green bodies gelcast in air by the addition of a polymer is referred to our further study.

An advantage of gelcasting over slip casting is the high strength of the dried green body. In this study, after adding 1.3 wt% PEG, the density and strength measured were approximately 53.1% of the theoretical density and 9.7 Mpa, as seen in Fig. 4.63. The addition of 1.3 wt% PEG decreases the green density and the green strength of the dried bodies compared to the strength of the green body without PEG (37 MPa). However, the reduced strength is enough to enable inexpensive green body machining into even more complex shapes. The decrease in strength can be explained by the change in microstructure of the green body.

Figure 4.64a, b show the microstructures obtained by the gel formed by polymerization of monomer without and with PEG mixed with alumina. There is a good microstructure with very few pores in the green bodies without PEG and also in those with PEG. However, it is to be noted that there is an obvious network structure in the green bodies without PEG compared to those with PEG.

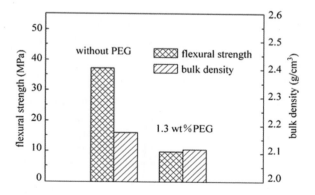

Fig. 4.63 Bulk density and flexural strength of dried green bodies without and with 1.3 wt% PEG

Fig. 4.64 Microstructures of green bodies prepared from suspension **a** without PEG and **b** with 1.3 wt% PEG

Thus, a new gelcasting system based on aqueous Al_2O_3-PEG-AM suspensions cross-linked with a coupling agent MBAM was successfully developed. An advantage of this system over the monomer-based system is that it eliminates the surface exfoliation phenomenon of green bodies gelcast in air and makes gelcasting an easy process. Addition of 1.3 wt% PEG has little influence on the dispersion of the alumina slurry and on its rheological properties. A marked decrease in both gelation rates and flexural strength of ceramic parts is observed for green bodies prepared from suspensions with PEG. However, the flexural strength of ceramic parts (9.7 MPa) is high enough for them to be machined inexpensively into more complex shapes.

4.5.2.1 Suppression of Surface Exfoliation by Introducing Polyvinylpyrrolidone (PVP) into Monomer System in Suspension

Surface spallation of gelcast alumina green bodies due to the exposure of the surface to air during gelation was successfully eliminated by adding a proper amount of polyvinylpyrrolidone (PVP) into the aqueous acrylamide premix solution. The influences of PVP on the colloidal characteristics of alumina in aqueous solutions, the rheological properties of alumina slurries, the gelation process powders, and the properties and microstructures of the gelcast green bodies and the sintered samples were systematically investigated.

In this research, the as-received alumina powder used in the experiments was a commercial high-purity α-Al_2O_3 powder (Zhengzhou Advanced Ceramics Factory, China), with a mean particle size of 3.47 μm. Its chemical composition is shown in

Table 4.14. Calcium carbonate, silicon dioxide, and kaolin were used as sintering aids. For the colloidal system for gelcasting, acrylamide ($C_2H_3CONH_2$) (AM) was used as the monomer; N, N′-methylenebisacrylamide (($C_2H_3CONH)_2CH_2$) (MBAM) was used as a coupling agent; ammonium persulfate was used as an initiator; N, N, N′, N′-tetramethylethylenediamine was used as a catalyst; and ammonium citrate was used as a dispersant. The poly (vinylpyrrolidone) (K15, Aldrich Chemical Co., Inc.) used in the study had an average molecular weight of 10,000. All these reagents were chemically pure.

The flowchart for the gelcasting process with the addition of PVP is shown in Fig. 4.65. The procedural details can be found elsewhere (Ma et al. 2003).

For the measurement of zeta potentials, suspensions were prepared at solids loading of 5 vol%. The suspensions were allowed to age for 24 h with continuous stirring;

Table 4.14 Chemical compositions of commercial alumina powder

Composition (wt%)						Density (g cm^{-3})	α-Al$_2$O$_3$ (wt%)
Al$_2$O$_3$	SiO$_2$	Fe$_2$O$_3$	Na$_2$O	B$_2$O	Other		
>99.7	0.05	0.03	0.05	0.04	0.1	>3.96	>95

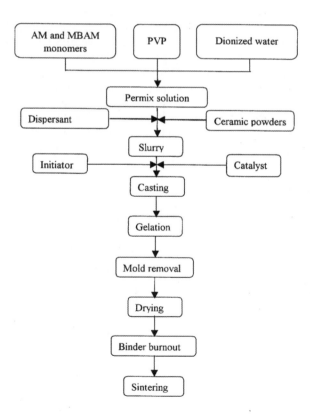

Fig. 4.65 Flowchart of the gelcasting process with addition of PVP

0.4 ml of the above suspensions was then diluted to 200 ml to obtain a final working solution with solids loading of 0.01 vol% for analysis. Tests were performed on the alumina powder, in two kinds of aqueous solutions: deionized water solution, with 0 and 15 wt% of PVP; and a premix solution with 1 wt% of dispersant and also with 0, 15, and 20 wt% of PVP. All these aqueous solutions contained potassium nitrate (KNO_3) with its concentration remaining the same at 10^{-2} M. The pH value of the solutions during measurement was adjusted with dilute HCl and NaOH solutions. Zeta potentials were calculated from the measured electrophoretic mobility in the pH interval from 3 to 12 by using ELS based on Henry's equations. The measurement was conducted at 25 ± 0.1 °C by using ZetaPlus (Brookhaven Instruments Corporation, USA).

The analysis of rheological properties of slurries was accomplished by using a rotary rheometer (MCR-300, Physica Instrument Corp., Germany). A cylinder with a volume of about 23 ml was used. The static rheological measurements were performed in the shear rate range of 0.1–250 s^{-1} at a constant temperature of 25 °C. Measurement of the storage modulus of slurries was conducted at 25 °C and 1 Hz.

Infrared spectra of dried green bodies were recorded at room temperature using an FT-IR spectrometer (FTIR-560, Nicolet Instrument Corp.). The bulk density of the specimens at different stages, namely, after being dried and sintered, was determined by the Hg immersion method based on the Archimedes' principle.

The flexural strength of bars of green and sintered bodies was examined by the three-point bending method, with a span of 30 mm. The dimensions of the former were 0.5 cm × 0.6 cm × 4.2 cm, while those of the latter were 0.4 cm × 0.3 cm × 3.6 cm. Fractal surfaces of the green and sintered specimens were examined by an SEM (S-450, Hitachi Instrument Corp., Japan).

Influence of PVP on colloidal characteristic of alumina powder in aqueous solutions: Zeta potential curves of alumina powder versus the pH values in deionized and premix solutions without and with PVP are shown in Fig. 4.66. It can be seen that the IEPs and shapes of curves of alumina powder in the premix solutions with dispersant

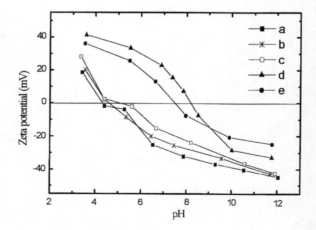

Fig. 4.66 Effect of PVP on the zeta potentials of alumina powder in different aqueous solutions: **a** premix with 0 wt% PVP and 1 wt% of dispersant; **b** premix with 15 wt% PVP and 1 wt% of dispersant; **c** premix with 20 wt% PVP and 1 wt% of dispersant; **d** deionized water without PVP; **e** deionized water with 15 wt%. All solutions with 10^{-2} M KNO_3

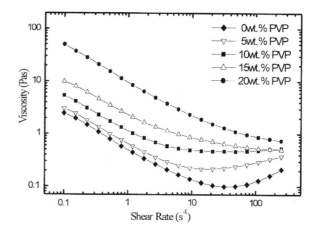

Fig. 4.67 Static rheological properties of alumina slurries with solid loading of 50 vol% and various contents of PVP in the premix solution

are different from those of the powder in deionized water. The differences can be attributed to the absorption of dispersant and monomers on the powder. However, in the same solution, either deionized water or premix solution, curve shapes of zeta potentials for the powder almost remain the same with or without PVP, with little differences in IEPs. The IEP changed from 8.2 to 7.5 when 15 wt% of PVP was added to the deionized solution. However, the IEP of the powder did not change so much in the premix solution with the content of PVP increasing up to 20 wt%. On the whole, PVP had little influence on the colloidal characteristics of alumina powder in aqueous solutions.

Influence of PVP on rheological properties of alumina slurries and their gelation: The static rheological properties of alumina slurries with solid loading of 50 vol% and various contents of PVP in the premix solution are given in Fig. 4.67. It can be seen that, just as in other colloidal systems, the viscosity, on the whole, increases with an increase in the content of polymers. The viscosity of slurries with the same solid loading versus the PVP content at a shear rate of 48.1 s^{-1} is illustrated in Fig. 4.68. With an increase in PVP content in the premix solution from 0 to 20 wt%, the viscosity increases from 0.101 to 1.11 Pa s, correspondingly, which is attributed to the adsorption of PVP on the surface of the alumina particles.

Interestingly, the slurry without PVP first exhibits shear-thinning and then relatively strong shear-thickening. However, with an increase in the content of PVP in the system, the shear thickening behavior becomes less and less obvious. For the suspension prepared from a premix solution with a PVP content of 20 wt%, the slurry becomes nearly only shear-thinning and belongs to almost pure pseudoplastic fluid within the measurement range.

Stable, concentrated suspensions usually exhibit shear-thinning behavior because of a perturbation of the suspension structure by shear (De Kruif 1990). At low shear rates, the suspension structure is similar to the equilibrium structure at rest, because the effect of thermal motion dominates over the viscous forces. At high shear rates, the viscous forces affect the suspension structure more and shear-thinning occurs.

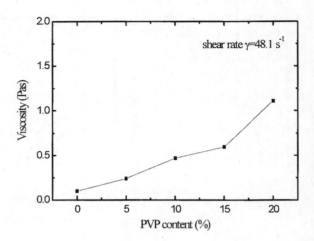

Fig. 4.68 Effect of PVP content in premix solution on viscosity of 50 vol% alumina slurries

However, if solid loading in the suspension is too high, the slurry may exhibit shear-thickening when the shear rate is above a critical value. Below the critical shear rate, well-stabilized particles have a tendency to form a layered structure with close packing in the layers but higher separation between layers as a result of laminar flow (Song and Evans 1996). At high shear rates, the layered arrangement becomes disrupted into a random three-dimensional structure (Russel 1987). This structural change requires greater packing space and presents an increase in the apparent viscosity.

From the effect of adding PVP into the system on the rheological behavior of the alumina slurry for gelcasting, it can be concluded that the addition of PVP increases the stability of the suspension, especially at relatively high shear rate. The increase in differences in the static rheological properties of the alumina slurries with increasing PVP from 0 to 20 wt% of premix solution reflects the interactions between ceramic particles and PVP. Since PVP is the only polymer in the slurries, the interaction between the polymer and alumina particles is stronger than that between the particles and other organic substances with the same content as PVP. Thus, the addition of PVP has an obvious influence on the viscosity of the slurries. As content of PVP increases, the interactions between the particles and PVP become the stronger. When the content of PVP in the premix solution is up to 20 wt%, the interactions between particles and PVP play such an important role in the stability of the slurry that an increase in shear stress results in destruction of polymer chains. This decreases the number of tangling points and the extended state of the polymer, but the layered structure of the slurry cannot be disrupted. Even though the above speculative explanation sounds reasonable, we still need further investigations to prove it.

In the gelcasting process, the rheological properties of concentrated ceramic suspensions play an important role in controlling the shape-forming behavior and the green body properties. The results of our experiments indicate that alumina slurries with high solid loading and good fluidity can be prepared when a proper amount of PVP is added. It can be seen from Figs. 4.67 and 4.68 that when the PVP content in

4.5 Suppression of Surface Exfoliation ...

the premix solution is 15 wt%, the viscosity of the slurry with solids loading of 50 vol% is only 0.6 Pa s at a shear rate of 48.1 s^{-1}, which is ideal for gelcasting.

Concentrated colloidal suspensions usually display viscoelastic behavior, which can be characterized by dynamic rheological measurements, namely oscillation measurements. During the measurements, a frequency-dependent shear stress or strain is applied to a suspension, and the shear modulus is obtained. The shear modulus $G*$ can be expressed in complex form in terms of a storage modulus G' and a loss modulus G'' as

$$G* = G' + i G'' \tag{1}$$

where $G' = G* \cos\delta$, $G'' = G* \sin\delta$, $i = (-1)1/2$, and δ is the loss angle (Collyer and Clegg 1988).

The storage modulus G' represents the ratio of the in-phase stress to the strain and gives a measure of the elastic properties. The loss modulus G'' represents the ratio of the out-of-phase stress to the strain and gives a measure of the viscous properties. $G*$, G', and G'' are only a function of the angular frequency.

The variation of the storage modulus versus time of the 50 vol% alumina slurries gelated from premix solutions with varying amounts of PVP is shown in Fig. 4.69. The figure shows that the starting time of gelation gelated with an increase in PVP content. Furthermore, the rate of increase in the storage modulus decreases with an increase in PVP content when gelation starts. According to the reactive mechanism of the system described in the literature (Molyneux 1982), in the presence of ammonium persulfate and MBAM, PVP can be coupled to form hydrogel at 110 °C. However, after the addition of TEMED, PVP can yield hydrogel through the coupling reaction even at room temperature. Therefore, it is thought that the presence of PVP consumes initiator, catalyst, and coupling agent in the ceramic slurries; as a result, longer time is required for AM to become a hydrogel and the rate of gelation also decreases. In

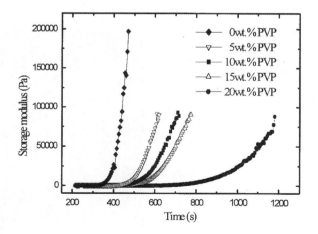

Fig. 4.69 Time-dependent storage modulus of alumina slurries prepared from a premix solution with different PVP content during gelation at 25 °C

addition, the contact among AM monomers would be and hindered by an increase in the amount of PVP in the solution.

Elimination of surface spallation of green bodies with the addition of PVP and the mechanism: Cylinders of alumina green bodies (Fig. 4.70) with diameter 2.0 cm and height 1.5 cm were prepared from 50 vol% alumina slurries with their premix solutions containing 0, 5, 15, and 20 wt% PVP, respectively. The upper faces of the samples, corresponding to the upper face of the mold, were exposed to air during the gelation of their slurries. The side and bottom faces, were isolated from air during the whole gelation process. It can be seen from the figure that, for all of the samples, the side and lower faces show no surface spallation, but for the green body without PVP, the upper face exhibits severe surface spallation (Fig. 4.70a). This indicates that inhibition of gelation due to the presence of oxygen can take place in the area that is exposed to air during the gelation process. If the area is isolated from air during gelation, there will be no surface spallation. Comparing the upper faces of the green bodies, it can be found that with an increase in PVP content, the surface spallation is gradually suppressed. When the PVP content is up to 15 wt% of the premix solution, the spallation phenomenon is completely eliminated (Fig. 4.70c). Considering both the rheological properties of the slurries and the elimination of surface spallation of green bodies, it is found that 15 wt% PVP content in the premix solution is optimal for the preparation of alumina slurries with low viscosity and green bodies without surface spallation. Therefore, a premix solution with 15 wt% PVP was selected for subsequent investigations. To further explore the mechanism of eliminating surface spallation by PVP, infrared spectroscopy was used to analyze the surface and bulk of alumina green bodies prepared from the premix solutions with 15 wt% PVP and without PVP.

Figure 4.71 shows the results of our investigation by IR spectra to explore the mechanism for elimination of surface spallation. The IR curves for AM, PAM, PVP, and alumina powder are curves a, b, c, and d, respectively, in the figure. IR spectra for the bulk and the spalled surface layer of the sample without PVP are shown as

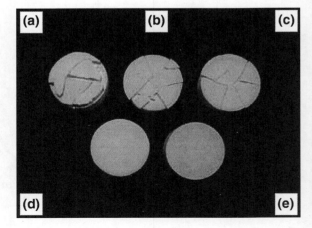

Fig. 4.70 Alumina green bodies prepared from premix solution with various PVP content (the upper face of the samples exposed to air during gelation): **a** without PVP, **b** 5 wt% PVP, **c** 10 wt% PVP, **d** 15 wt% PVP, **e** 20 wt% PVP

4.5 Suppression of Surface Exfoliation ... 217

Fig. 4.71 Infrared spectra for **a** AM, **b** PAM, **c** PVP, **d** alumina, **e** bulk of green body without PVP, **f** surface layer of green body without PVP, **g** bulk of green body with 15 wt% PVP, and **h** surface layer of green body with 15 wt% PVP

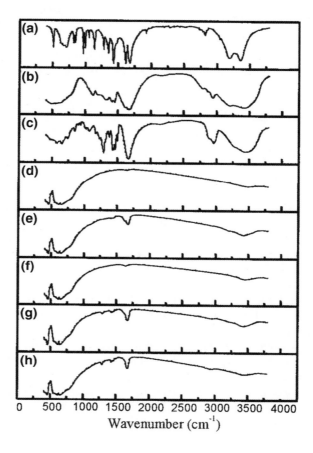

curves e and f. IR spectra of the bulk and the surface layer of the sample with PVP are also shown as curves g and h.

There are a number of absorption peaks in the spectrum for AM (curve a). The absorption peak at 1,667 cm^{-1} is the characteristic C=O stretching vibration and the absorption peak at 1,610 cm^{-1} is the characteristic C=C stretching vibration for the monomers. The IR spectrum of PAM (curve (b)) differs in many aspects from that of AM (curve a). The major absorption peaks around 1,667 cm^{-1} become much broader for the PAM sample than for AM and there is a shoulder at about 1,610 cm^{-1}. The absorption peak for PAM at 1,667 cm^{-1} represents the same characteristic of the C=O stretching vibration as seen in the AM sample. However, the shoulder at 1,610 cm^{-1} belongs to the C=C stretching vibration of the remaining monomer AM, which has not completely polymerized. In addition, the absorption peak at 3,330 cm^{-1} in curves a and b corresponds to the N–H stretching vibration for both the monomer AM and polymer PAM. PVP is only an acrylic resin without N–H groups. It does not show an IR absorption at 3,330 cm^{-1} due to lack of the N–H group, but has a strong characteristic absorption peak of C=O at 1,667 cm^{-1} (curve c). Furthermore, both

IR absorptions at 1,450 and 1,260 cm^{-1} make PVP distinct from AM and PAM. The IR analyses for AM, PAM, and PVP are in a good agreement with those reported in the literature (Santhiya et al. 2000; Lang 1974; Kumar et al. 1999). The broad band in the vicinity of 600–800 cm^{-1} as well as the strong band at 447 cm^{-1} for the α-alumina powder (curved) are due to Al-O stretching vibrations. The absorption band at 1,630 cm^{-1} is due to the presence of moisture in the sample.

Shapes of the IR spectra of the samples from the green bodies are similar to that of pure alumina, with weak characteristic peaks of the organic substances if they were in the samples, as shown in curves e, f, g, and h. The reason is that the dried green bodies contained more than 92 wt% alumina, but less than 8 wt% of organic substances.

Comparing curve e with curves a, b, and d, it is found that curve d has the characteristic peaks of AM and PAM, i.e., the bulk of the green body without PVP contains either PAM or AM. From the high strength of the green bodies we will discuss below, it can be concluded that the main organic substance in the bulk should be PAM. However, by comparison of curve f with curves a, b, and d, it can be seen that there are no characteristic peaks of AM and PAM at 1,667 cm^{-1} except a minor adsorption at 1,630 cm^{-1} due to moisture in curve f. In other words, there is no AM or PAM in the spalled surface layer of the alumina green bodies without PVP according to the resolution of the IR analyzer. It could be considered that there was no PAM on the surface layer of green bodies exposed to air during its gelation without PVP. A possible explanation for the difference in the content of PAM and AM on the spalled surface layer and in the bulk of the green body is that AM on the surface of the green body did not polymerize when it was exposed to oxygen in air, and hence it may vaporize or migrate to the bulk of the sample during consolidation of the slurry.

For the green body with PVP, it is found that both the bulk (curve g) and the surface layer (curve h) demonstrate characteristics of PVP with three characteristic IR absorption peaks at 1,450, 1,260, and 1,667 cm^{-1}. Since the intensity of the PVP absorption peak at 1,667 cm^{-1} is much stronger than that of PAM and AM and there is heavy overlap between the two IR spectra, it is not certain from the IR study whether PAM is also present on the surface. However, by comparison of the curves f and h of the surface layers of green bodies with and without PVP, we could still assume that the interaction of PVP chains with the alumina particles in the surface layer was responsible for the elimination of surface spallation of the green body with PVP.

From the above analysis, it can be concluded that the absence of PAM in the surface layer of a green body with the surface exposed to air during gelation causes surface spallation. However, this problem can be eliminated by the addition of a proper amount of PVP to the premix solution. The improvement is mainly due to PVP and its coupled substance, both of which serve as binders for ceramic particles.

Influence of PVP on properties and microstructure of alumina green and sintered bodies: To examine the influence of the addition of PVP on the properties and microstructures of gelcast alumina green and sintered bodies, premix solutions with

4.5 Suppression of Surface Exfoliation ...

0, 5, 10, 15, and 20 wt% PVP were used for the whole process of gelcasting of alumina. The PVP content based on the whole solid content in the slurries, namely, PVP content in the dried green bodies, should be 0, 1.3, 2.8, 4.4, and 6.3 wt%, respectively.

The microstructures of the alumina green bodies prepared from 50 vol% slurries containing PVP of 0 and 15 wt% after drying for 12 h at 100 °C are shown in Fig. 4.72. It can be found that there is no obvious difference in the microstructures between the green bodies with and without PVP. Both green bodies have relatively high uniformity and density. The density and flexural strength of the green bodies prepared from 50 vol% slurries containing various amounts of PVP after drying for 12 h at 100 °C are shown in Fig. 4.73. The addition of PVP has little effect on the density of the green bodies, but has a significant effect on their flexural strength. The strength of the green bodies remains almost the same at 37 MPa when the PVP content in the premix varies from 0 to 10 wt%, and then declines when the PVP content is more than 10 wt%. A previous investigation has shown that the strength of alumina green bodies is determined by hydrogels that form by the polymerization of acrylamide (Khan et al. 2000; Bauer et al. 1999). When the PVP content is relatively low, it does not significantly affect the polymerization of acrylamide. Meanwhile, PVP may improve a certain amount of strength by coupling and forming hydrogels. With further increase of PVP in the premix to more than 10 wt%, the coupling of PVP can impair the polymerization of acrylamide. The strength of a hydrogel of PVP is lower than that of acrylamide; hence, the strength of alumina green bodies continuously decreases with increasing PVP in the systems. However, the strength of the green bodies is still high enough for any further process even with PVP content up to 20 wt%.

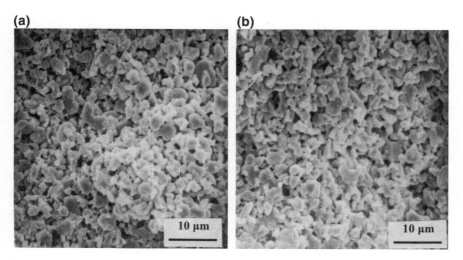

Fig. 4.72 SEM of fracture surface of alumina green bodies prepared from premix solution with various PVP content: **a** without PVP, **b** with 15 wt% PVP

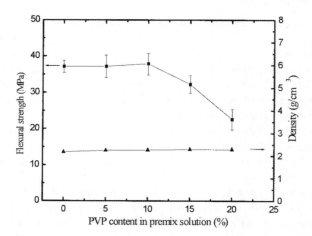

Fig. 4.73 Effect of PVP content in premix solution on the density and flexural strength of alumina green bodies

After binder burnout, green bodies with different amounts of PVP were sintered at 1,550 °C for 2 h, all of which had a density of about 3.6 g cm^{-3}, with a flexural strength of about 305 MPa. Microstructures of fractal surfaces of the materials prepared from 50 vol% slurries containing 0 and 15 wt% PVP are shown in Fig. 4.74. There is also no obvious difference between the uniform microstructures. The results above indicate that the addition of PVP, with its content up to 20 wt% in premix solutions

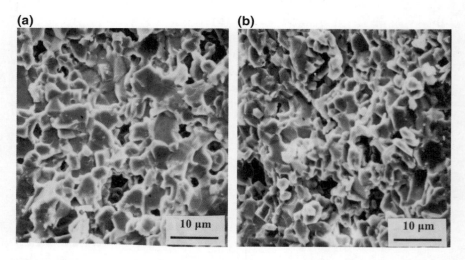

Fig. 4.74 SEM of fracture surface of alumina sintered bodies prepared from premix solution with various PVP contents: **a** without PVP, **b** with 15 wt% PVP

and 6.3 wt% in the dried green bodies, has no significant effect on the final properties and microstructure of gelcast alumina ceramics.

We can conclude that the surface spallation of gelcast alumina green bodies due to the exposure of the surface to air during gelation can be successfully eliminated by adding a proper amount of PVP in the acrylamide premix solution during gelcasting. PVP serves as a substitute binder for polyacrylamide for the particles in the green bodies. This is important to their surface layer where the gelation is inhibited due to the presence of oxygen.

The addition of PVP to the premix solution has no obvious influence on the colloidal behavior of the alumina powder in the solution, but has a slightly negative effect on the rheological properties of the slurry and its gelation process. When the amount of PVP in the premix solution is up to 15 wt%, the fluidity of the alumina slurry with 50 vol% solids loading and the time required for its gelation are still suitable for gelcasting. The addition of PVP also has a slightly negative influence on the strength of dried green bodies when the PVP content in the premix solution is more than 10 wt%, but the strength is still high enough to allow the green bodies to undergo further processing.

It is found that for the range of PVP concentrations used in the premix solution in our study, the addition of PVP had no obvious influence on the final properties and microstructure of the alumina materials.

Overall, PVP is effective in eliminating the surface spallation of gelcast alumina green bodies with their surfaces exposed to air during gelation. The suggested mechanism implies that it should be possible to extend the use of PVP to other ceramic and monomer systems for gelcasting.

References

Abbas, E. N., Harn, Y. P., Draskovich, B. S., & Pollinger, J. P. (1999). *Journal of the American Ceramic Society, 82,* 513.

Asad, U. K., Brian, J. B., & Paul, F. L. (2000). Interaction of binders with dispersant-stabilized alumina suspensions. *Colloid Surface A, 161,* 243–257.

Bossel, C., Dutta, J., Houiet, R., Hilgorn, J., & Hofmann, H. (1995). Processing of nano-scaled silicon powders to prepare slip cast structural ceramics. *Materials Science and Engineering A, 204*(1/2), 107–112.

Bender, J., & Wagner, N. J. (1996). Reversible shear thickening in monodisperse and bidisperse colloidal dispersions. *Journal of Rheology, 40,* 899–916.

Bergstrom, L. (1997). Hamaker constants of inorganic materials. *Advances in Colloid and Interface Science, 70,* 125–169.

Baes, C. F., & Mesmer, R. E. (1976). *The hydrolysis of cations* (pp. 112–123). New York: Wiley.

Bergstrom, L. (1994, August 23). Method for forming ceramic powders by temperature-induced flocculation, US Pat. No. 5340532.

Bergstrom, L. (1998). Shear thinning and shear thickening of concentrated ceramic suspensions. *Colloids and Surfaces A: Physicochemical and Engineering Aspects, 133,* 151–155.

Balzer, B., Hruschka, M. K. M., & Gauckler, L. J. (1999). *Journal of Colloid and Interface Science, 216,* 379.

Brinker, C. J., & Scherer, G. W. (1990). *Sol–gel science.* San Diego, CA: Academic Press.

Bauer, W., Ritzhaupt-Kleissl, H.-J., & Hausselt, J. (1999). Micropatterning of ceramics by slip pressing. *Ceramic International, 25,* 201–205.

Collyer, A. A., & Clegg, D. W. (1988). *Rheological Measurement.* London, UK: Elsevier Applied Science.

Zongqi, C., Guangxin, W., & Guiying, X. (2001). *Colloid and interface chemistry (in Chinese)* (pp. 183–189). Beijing: Higher Education Press.

Ruifeng, C. (2007). *Generation and evolution of defects in ceramics by gelcasting: Mechanisms and applications (in Chinese, dissertation)* (pp. 57–58). Beijing: Tsinghua University.

Carniglia, S. C. (1981). Thermochemistry of the aluminas and aluminum trihalides. *Journal of the American Ceramic Society, 64,* C-62–C-63.

Cesarano, J., & Askay, I. A. (1988). Processing of highly concentrated aqueous α-alumina suspension stabilized with polyelectrolytes. *Journal of the American Ceramic Society, 71*(12), 1061–1067.

Chong, J. S., Christiansen, E. B., & Bear, A. D. (1971). Rheology of concentrated suspensions. *Journal of Applied Polymer Science, 15,* 2007–2021.

Christos, A., Athena, T., & Ioanna, L. (2000). Effect of slurry rheological properties on the coating of ceramic honeycombs with yttria-stabilized-zirconia washcoats. *Journal of the American Ceramic Society, 83,* 1033–1038.

Dewhurst, D. N., Cartwright, J. A., & Longergan, L. (1999). *Marine & Petroleum Geology, 16,* 793.

De Kruif, C. G. (1990). In J. P. Hulin, A. M. Cazabat, E. Guyon, & F. Carmona (Eds.), *Hydrodynamics of dispersed media* (pp. 79–85). Amsterdam: Elsevier Applied Science.

Giuliano, T. (2003). Gelcasting ceramics: A review. *American Ceramic Society Bulletin, 82,* 43–47.

Graule, T. J., Baader, F. H., Gauckler, L. J. (1994). Shaping of ceramic green body compacts direct from suspensions by enzyme-catalyzed reactions, In *cfi/Ber.* DKG (Vol. 71, No. 6).

Gaucker, L. J., Graule, T. J., & Baader, F. H. (1999a). *Materials Chemistry and Physics, 61,* 78.

Gaucker, L. J., Graule, T. J., Baader, F. H., & Will, J. (1999b). *Key Engineering Materials, 159–160,* 135.

Ruisong, G., Yirui, C., Zhengfang, Y., & Qiming, Y. (1994). *Journal of Tianjin University, 27,* 665.

Gilissen, R., Erauw, J. P., Smolders, A., Vanswijgenhoven, E., & Luyten, J. (2000). Gelcasting, a near net shape technique. *Materials & Design, 21,* 251–257.

Huang, Y., Ma, L. G., Tang, Q., Yang, J. L., Xie, Z. P., & Xu, X. L. (2000). Surface oxidation to improve water-based gelcasting of silicon nitride. *Journal Materials Science, 35*(14), 3519–3524.

Huang, Y., Zhang, L., Yang, J., et al. (2007). Research progress of new colloidal forming processes for advanced ceramics (in English). *Journal Chinese Ceramic Society, 35*(2), 1–10.

Huang, Y., Ma, L. G., Le, H. R., & Yang, J. L. (2004). Improving the homogeneity and reliability of ceramic parts with complex shapes by pressure assisted gelcasting. *Materials Letters, 58,* 3893–3897.

Helbig, M. H., Hutter, M., & Schonholzer, U. P. (2000). *Journal of Colloid and Interface Science, 222,* 46.

Hoffman, R. L. (1972). Discontinuous and dilatant viscosity behavior in concentrated suspensions I. Observation of a flow instability. *Transactions of the Society of Rheology, 16,* 155–173.

Herschel, H., & Buckley, R. (1926). *Proceedings American Society for Testing Materials, 26,* 621–623.

Reed, J. S. (1988). *Principles of ceramics processing* (pp. 162–163), New York: A Wiley-Interscience Publication, Wiley.

Janney, M. A., & Omatete, O. O. (1992). Method for molding ceramic powder using a water-based gelcasting process, US Patent, 4145908.

Kumar, V., Yang, T., & Yang, Y. (1999). Interpolymer complexation: I, Preparation and characterization of a poly(vinyl acetate phthalate)–poly(vinylpyrrolidone) (PVAP–PVP) complex. *International Journal of Pharmaceutics, 188,* 221–232.

Kulicke, W. M., & Nottelmann, H. (1989). Structure and swelling of some synthetic, semisynthetic, and biopolymer hydrogels. In J. E. Glass (Ed.), *Polymers in aqueous media* (pp. 15–44). Washington, DC: American Chemical Society.

References

Khan, A. U., Briscoe, B. J., & Luckham, P. F. (2000). Interaction of binders with dispersant stabilized alumina suspensions. *Colloids and Surfaces A: Physicochemical and Engineering Aspects, 161,* 243–257.
Kingery, W. D., Bowen, H. K., & Uhlmann, D. R. (1976). *Introduction to ceramics* (2nd ed., pp. 785–787). New York: Wiley.
Leong, Y. K., & Boger, D. V. (1991). Surface chemistry and rheological properties of zirconia suspensions. *Journal of Rheology, 35*(10), 149–165.
Liu, X. L., Huang, Y., Yang, J. L., & Su, L. (2002a). Mechanical properties of ZTA formed by gelcasting. *Key Engineering Materials, 224*(2), 295–299.
Liu, X., Huang, Y., & Yang, J. (2002b). *Ceramics International, 28,* 159–164.
Lang, L. (1974). *Adsorption spectra in the infrared region* (Vol. 4). Budapest, Hungary: Akademiai Kiado.
Liu, D.-M. (1998). Dispersion characteristic of nano-sized ceramic powder in an aqueous medium. *Journal of Material Science Letters, 17,* 207–210.
Lim, L. C., Wong, P. M., & Jan, M. (2000). Microstructural evolution during sintering of near-monosized agglomerate-free submicron alumina powder compacts. *Acta Materialia, 48,* 2263–2275.
Lang, F. F. (1989). Powder processing science and technology for increased reliability. *Journal of the American Ceramic Society, 72*(1), 3–15.
Lyckfeldt, O., Brandt, J., & Lesca, S. (1999). *Journal of the European Ceramic Society, 20,* 2551.
Landham, R. R., Nahass, P., Leung, D. K., Ungureit, M., & Bowen, W. E. (1987). Potential use of polymerizable solvents and dispersants for tape casting of ceramics. *American Ceramic Society Bulletin, 66,* 1513–1516.
Lyckfeldt, O., & Ferreira, J. M. F. (1998). Processing of porous ceramics by 'starch consolidation'. *Journal of the European Ceramic Society, 18,* 131–140.
Ma, L. G., Yang, J. L., Zhao, L., & Huang, Y. (2002). Gelcasting of a rutile mixture applied to extrusion forming. *Journal of the European Ceramic Society, 22*(13), 2291–2296.
Morissette, S. L., & Lewis, J. A. (1999). *Journal of the American Ceramic Society, 82,* 521.
Ma, J. T., Xie, Z. P., Miao, H. Z., Zhou, L. J., & Huang, Y. (2003). Elimination of surface spallation of alumina green bodies prepared by acrylamide-based gelcasting via poly (vinylpyrrolidone). *Journal of the American Ceramic Society, 86*(2), 266–272.
Nunn, S. D., & Kirby, G. H. (1996). Green machining of gelcast ceramic materials. *Ceramic Engineering and Science Proceedings, 17*(3), 209–213.
Omatete, O. O., Janney, M. A., & Strehlow, R. A. (1991a). Gelcasting—A new ceramic forming process. *Ceramic Bulletin, 70*(10), 1641–1649.
Omatete, O. O., Janney, M. A., & Nunn, S. D. (1997). Gelcasting from laboratory development toward industrial production. *Journal of the European Ceramic Society, 17,* 407–413.
Omatete, O. O., Tieggs, T. N., & Young, A. C. (1991b). Gelcast reaction bonded silicon nitride composites. *Ceramic Engineering and Science Proceedings, 12*(7–8), 1257–1264.
Omatete, O. O., Bleier, A., Westmoreland, C. G., & Young, A. C. (1991c). Gelcast zirconia-alumina composites. *Ceramic Engineering and Science Proceedings, 12*(9–10), 2084–2094.
Odian, G. G. (1991). *Principles of polymerization* (pp. 262–266). New York: Wiley.
Molyneux, P. (1982). *Water-soluble synthetic polymers: Properties and behavior* (pp. 172–175). Boca Raton, FL: CRC Press.
Quinn, G. D. (1990). Flexure strength of advanced structural ceramics: A round robin. *Journal of the American Ceramic Society, 73,* 2374–2384.
Rahaman, M. N. (2003). *Ceramic processing and sintering* (pp. 567–573). New York: M Dekker.
Russel, W. B. (1987). *The dynamics of colloidal systems* (pp. 8–10). Madison, WI: University of Wisconsin Press.
Sigmund, W. M., Bell, N. S., & Bergstrom, L. (2000). Novel powder-processing method for advanced ceramics. *Journal of the American Ceramic Society, 83*(7), 1557–1574.
Si, W. J., Graule, T. J., Baader, F. H., & Gaucker, L. J. (1999). *Journal of the American Ceramic Society, 82,* 1129.

Scherer, G. W. (1989). *Journal of Non-Crystalline Solids, 108,* 18.
Novak, S., Kosmac, T., Krnel, K., & Drazic, G. (2002). *Journal of the European Ceramic Society, 22,* 289.
Shih, W. Y., Shih, W.-H., & Aksay, I. A. (1999). *Journal of the American Ceramic Society, 82,* 616–624.
Song, J. H., & Evans, J. R. G. (1996). Ultrafine ceramic powder injection molding: The role of dispersants. *Journal of Rheology (NY), 40,* 131–152.
Santhiya, D., Subramanian, S., Natarajan, K. A., & Malghan, S. G. (2000). Surface chemical studies on alumina suspensions using ammonium polymethacrylate. *Colloids and Surfaces A: Physicochemical and Engineering Aspects, 164,* 143–154.
Tari, G., Ferreira, J. M., Fonseca, A. T., et al. (1998). Influence of particle size distribution on colloidal processing of alumina. *Journal of the European Ceramic Society, 18,* 249–253.
Takai, C., Tsukamoto, M., Fuji, M., et al. (2006). Control of high solid content yttria slurry with low viscosity for gelcasting. *Journal of Alloys and Compounds, 408–412,* 533–537.
Tanaka, T. (1981). Gels. *Scientific American, 244*(1), 124–138. [3].
Waesche, B., & Steinborn, G. (1997). Influence of slip viscosity on the mechanical properties of high purity alumina made by gelcasting. *Key Engineering Materials, 132–136,* 374–377.
Wang, J., & Stevens, R. (1989). Zirconia-toughened alumina (ZTA) ceramics. *Journal Materials Science, 24,* 3421–3440.
Young, A. C., Omatete, O. O., & Janney, M. A. (1991). Gelcasting of alumina. *Journal of the American Ceramic Society, 74*(3), 612–618.
Yang, Y., & Sigmund, W. M. (2003). *Journal of the European Ceramic Society, 23,* 253–261.
Yang, J., Su, L., Ma, L., & Huang, Y. (2002). *Key Engineering Materials, 224–226,* 667.
Zhang, L., Ma, T., Yang, J., et al. (2004). Rheological behavior of alumina suspensions. *Journal of Inorganic Materials (in Chinese), 19,* 1145–1150.
Zhao, L., Yang, J.-L., Ma, L.-G., & Huang, Y. (2002). *Materials Letters, 56,* 990.

Chapter 5
Gelcasting of Non-oxide Ceramics

Abstract In this chapter, the non-oxide powder surface characteristics before and after treatment, and the effect of surface modification on the solids loading of its aqueous suspension were studied. For Si_3N_4 powders, a universal method to prepare highly concentrated suspensions has been established. The results show that Si_3N_4 with high reliability was obtained via processing strategy, for example, using a combination of coating, oxidizing, gelcasting, cold isostatic pressing, and gas-pressure sintering (GPS).

Keywords Non-oxide ceramics · Concentrated suspensions · Si_3N_4 · SiC · Surface modification · Coating · Oxidizing

Non-oxide ceramics, such as SiC, Si_3N_4, and Si_3N_4-bonded SiC, are important and promising advanced ceramics. For example, SiC ceramic materials are indispensable advanced materials in a broad range of applications, including medical biomaterials, high-temperature semiconductors, synchrotron optical elements, and lightweight/high-strength structural materials. However, the application of non-oxide ceramics has still been limited mainly owing to their low reliability and high cost.

An approach to overcome these problems is to prepare homogeneous and near-net-shaped green bodies through colloid-forming in situ processes, such as temperature-induced flocculation (Bergstrom 1994), direct coagulation casting (Graule 1996; Baader 1996), and gelcasting (Young 1991; Omatete 1997). These techniques can be used not only to prepare complicated shapes, but can also be employed to decrease or eliminate defects, such as hard agglomerates, pores, and cracks, which would result in relatively low reliability of the structural components. Among these colloid-forming in situ processes, gelcasting is one of the most potential technique for industrial applications because it is capable of producing uniform green bodies with high strength to machine them in green state.

In recent years, gelcasting has received increasing attention for its simple process, low binder content, and other characteristics mentioned earlier. Although there have been many previous studies focusing mainly on gelcasting of oxides (Carisey 1995; Wang 1997; Rolf 1997; Sun 1996; Zhou 1980a), gelcasting of non-oxides still needs further investigation.

5.1 Effects of Powder Surface Modification on Concentrated Suspension Properties of Silicon Nitride

5.1.1 Contributing Factor and Elimination of Macropores in Silicon Nitride Green Bodies

In the forming process of water-based gelcasting, a complex suspension system is composed of deionized water, monomer, crosslinker, dispersant, catalyst, initiator, and Si_3N_4 powder. The gas-discharging reactions, particularly those that occur during the gel-forming process, may form large pores in the green body if the amount of the discharged gas is immense. Thus, not only the homogeneity of the green body is destroyed, but densification becomes difficult during sintering.

The surface characteristics of Si_3N_4 powders strongly depend on the different production methods (Mezzasalma 1996; Sanchez 1996; Greil 1991). In the process of nitridation of Si powder, some free Si and other impurities remain in the nitridated product before further purification, and their contents in the final powders are determined by the method of purification.

In this study, several possible gas-discharging reactions were investigated. The contributing factor of large pores in the green body was studied, and the homogeneous green body without large pores was obtained from alkali-cleaned Si_3N_4 powder.

Two kinds of Si_3N_4 powders were used in this study, and were termed as powder A (Beijing Founder High Tech. Ceram. Corp., Ltd.) and powder B (H. C. Starck, Berlin, FRG). The characteristics of these two powders are shown in Table 5.1.

In the water-based gelcasting system, the medium employed was deionized water. The monomer used was acrylamide (AM) [$CH_2CHCONH_2$], and the crosslinker was N, N'-methylene-bisacrylamide (MBAM) [$(CH_2CHCONH)_2CH_2$]. N, N, N', N'-tetramethylethylenediamine (TEMED) [$NH_2(CH_3)_2C(CH_3)_2CNH_2$] and ammonium persulfate [$(NH_4)_2S_2O_8$] were used as the catalyst and initiator, respectively. Tetramethyl ammonium hydroxide [$(CH_3)_4NOH$, TMAH] was employed as a dispersant to break up the agglomerate of the powders.

Zeta potentials were measured by ZetaPlus (Brookhaven Instrument Corp., USA). The particle-size distributions were measured by BI-XDC particle-size analyzer

Table 5.1 Characteristics of powder A and powder B

	d_{50} (μm)	α-phase (%)	O (%)	Fe (ppm)	Al (ppm)	Ca (ppm)	Si + SiO_2
Powder A	0.98	93.1	2.2	7	1300	190	1.38
Powder B	0.45	95	1.58	130	420	45	0.76
Powder A after alkali-cleaning	2.55	93.1	0.83				0.09

(Brookhaven Instrument Corp., USA). The apparent viscosity of Si_3N_4 suspensions was investigated by a rotating viscometer (model NSX-11, Chengdu Instrument Plant, China). The fracture morphologies of the green bodies were examined by scanning electron microscopy (SEM). Gas chromatography analyzer (model 3400, Beijing Analytical Instrument Plant, China) was used to examine the gas composition discharged during the colloidal process.

Concentrated Si_3N_4 suspension with 45 vol% solids loading was prepared by thoroughly mixing the monomer, crosslinker, dispersant, and Si_3N_4 powders in deionized water. After ball-milling, the catalyst and initiator were added into the slurry. After being degassed through vibration and/or vacuum treatment, the suspension was cast into non-pore molds. When heated, the slurry solidified in situ, because of the occurrence of gelling between the monomer and crosslinker. After demolding and drying, the green body was examined by SEM. A large amount of macropores with millimeter level were found in the green bodies gelcasted with powder A, as shown in Fig. 5.1a. The suspension was vibrated and vacuumed sufficiently to remove the bubbles caused by air. Thus, it can be deduced that the large pores were formed in the gel-forming process.

In the gelcasting system of powder A, there are three possible gas-discharging reactions:

1. Hydrolysis of Si_3N_4:

$$Si_3N_4 + 6H_2O \Leftrightarrow 3SiO_2 + 4NH_3 \uparrow \tag{5.1}$$

2. Oxidation of Si_3N_4 by $(NH_4)_2S_2O_8$:

$$Si_3N_4 + 6S_2O_8^{2-} + 12OH^- \Leftrightarrow 3SiO_2 + 2N_2 \uparrow + 12SO_4^{2-} + 6H_2O \tag{5.2}$$

3. Reaction between free Si and H_2O:

$$Si + 4H_2O \Leftrightarrow H_4SiO_4 + 2H_2 \uparrow \tag{5.3}$$

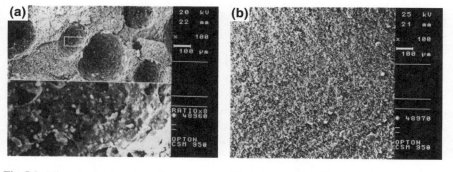

Fig. 5.1 Micrographs of green bodies gelcasted with **a** as-received and **b** alkali-cleaned powder A

Fig. 5.2 Device draft of gas-gathering experiments

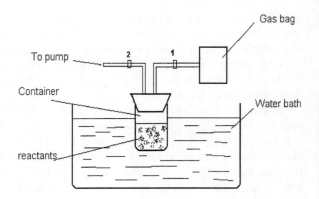

A series of gas-gathering experiments were designed to examine the contributing factor of pores formed during the gel-forming process. The gas-gathering set is shown in Fig. 5.2. First, the container and gas-collecting bag (0.5 L) were vacuumed after valves 1 and 2 were opened. Subsequently, valve 2 was closed and the container was put into water bath at 80 °C for 2 h. If some amount of gas was collected in the gas bag, valve 1 was closed and the gas bag was taken off for composition analysis. The arrangement and results of the experiments are shown in Table 5.2.

It can be seen from Table 5.2 that ammonia was not detected in all experiments. This is owing to the slow hydrolysis speed of Si_3N_4 and greater solubility of ammonia in water. The solubility of ammonia has been known to be 89.9 ml per 100 g of H_2O at 0 °C and 7.4 ml per 100 g of H_2O at 96.6 °C. However, for practical gelcasting process, the Si_3N_4 suspension was ball-milled for a specific period of time to increase the degree of hydrolysis. Some bubbles were found in the slurry and the ammonia smell was strong enough when the ball-milling time reached 48 h. This indicates that the solubility of ammonia is saturated in the suspension. Thus, if the slurry was cast directly without vacuum-degassing, large pores might be found in green bogy because of supersaturation of ammonia in the suspension at the temperature of solidification (70–80 °C). During the vacuum-treatment process of the slurry, the pH value of the atmosphere (measured by wet pH test paper) decreased from ~11 to ~9, which was close to that of the slurry without ball-milling. This suggests that the ammonia yielded during ball-mill process was removed through vacuum treatment. Thus, it can be concluded that the ammonia formed by the hydrolysis of Si_3N_4 cannot cause large amount of pores in the green body used for the present study.

It was confirmed by Experiment No. 6 (Table 5.2) that Reaction 2 occurred and discharged large amount of N_2 when the amount of $(NH_4)_2S_2O_8$ was greater and pH value was higher (pH = 13.9). The small amount of H_2 detected at the same time will be discussed later. In the practical suspension, the initiator solution [water solution of $(NH_4)_2S_2O_8$] was diluted and a small amount was added into the slurry. For 100 ml of slurry with solid volume loading of 45 vol% Si_3N_4, the amount of N_2 produced was less than 0.1 ml, as $(NH_4)_2S_2O_8$ was exhausted completely. Hence, this reaction cannot cause large amount of pores in the green body. This was confirmed

5.1 Effects of Powder Surface Modification … 229

Table 5.2 Arrangement and results of gas-gathering experiments

Experiment No.	Si_3N_4 powder (g)		Reactants					Gas composition		
			Deionized water (ml)	Dispersant (ml)	NaOH (g)	Initiator (g)		H_2 (vol%)	N_2 (vo.%)	NH_3 (vol%)
1	Powder A	30	20	2	0	0.01		97.54	2.46	0
2	Powder A after alkali-cleaning	30	20	2	0	0.01				
3	Si powder	4	40	0	0	0		84.12	15.88	0
4	Si powder	2	40	4	0	0		90.56	9.44	0
5	Powder B	30	20	2	0	0.01				
6	Powder B	10	40	0	10	10		11.52	88.48	0

by Experiment No. 5 (Table 5.2), in which the amount of dispersant and initiator was about the same as that in practical suspension, and no gas was collected in the gas bag. For other experiments, the little amount of N_2 detected may be owing to incomplete vacuum treatment and/or leakage of air into the gas bag.

To examine Reaction (5.3), Si powder was used. It can be seen from Experiment No. 3 and 4 (see Table 5.2) that Si can react with water, and H_2 is obtained at 80 °C. The reaction becomes stronger and more gas is collected during the same period when the basic dispersant is added. The reason for the absence of gas production in Experiment No. 5 (see Table 5.2) is because of the fact that powder B had little element Si (which was observed during its purification process introduced earlier). When compared with Experiment No. 1, the Si_3N_4 powder used in Experiment No. 2 was powder A that was alkali-cleaned to eliminate free Si, and hence, no gas was detected (see Table 5.2). The reason for the detection of little amount of H_2 in Experiment No. 6 is that only a limited quantity of element Si in powder B is exposed to the outside layer of Si_3N_4, which is consumed by $(NH_4)_2S_2O_8$, or the presence of a little amount of free Si on the powder surface.

According to the results and analysis presented in Table 5.2, it can be concluded that the contributing factor for the large amount of pores in the green body is H_2 produced by the reaction of free Si with the basic media.

As discussed earlier, large pores in the green body prepared using powder A were formed by H_2 produced during the reaction between free Si and water. To avoid these pores in the green body, it is natural to eliminate free Si by treating the as-received powder. In this study, alkali-cleaning was used to treat the powder, and the following process was employed: Powder A was suspended in NaOH solution at a concentration of 5 wt% and stirred for 6 h in a water bath at 80 °C. Subsequently, the centrifuged powder was washed using deionized water until a pH value below 10 was reached. The modified powder A was tested in a gas-collecting experiment and no gas was found to be discharged (Experiment No. 2 in Table 5.2). Furthermore, its content of free Si was observed to be 0.09%.

The particle-size distributions of the as-received and alkali-cleaned powder A are shown in Fig. 5.3. It can be observed from the figure that the range of alkali-cleaned particle size becomes wider, because the alkali-cleaning process causes some irreversible agglomeration of powder A and some small particles might be carried off by the deionized water. From the zeta potential curves presented in Fig. 5.4, it can be observed that the isoelectric point (IEP) of alkali-cleaned powder A slightly shifts to the right and the zeta potentials of the two powders are close to each other when the pH value is beyond 10. The slight shift of IEP may be owing to the reduction of silica-like surface layer of powder A (Greil 1989a). As shown in Fig. 5.5, the apparent viscosity of the alkali-cleaned powder A suspension was lower than that of the suspension of the original powder A, and both of them were observed to behave as pseudo-plastic fluids. However, lower viscosity of alkali-cleaned powder A suspension is considered to be advantageous for gelcasting. The fracture morphology of the green body gelcasted with alkali-cleaned powder A is shown in Fig. 5.5, and it can be seen that no large pores existed and the green body was homogeneous.

5.1 Effects of Powder Surface Modification ...

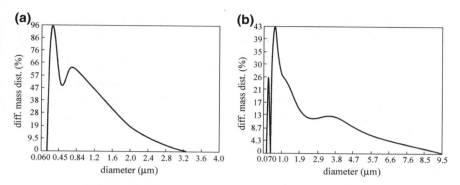

Fig. 5.3 Particle-size distributions of **a** as-received and **b** alkali-cleaned powder A

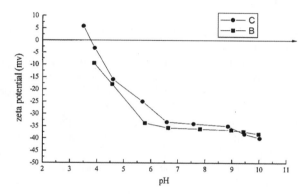

Fig. 5.4 Zeta potentials of **b** as-received and **c** alkali-cleaned powder A

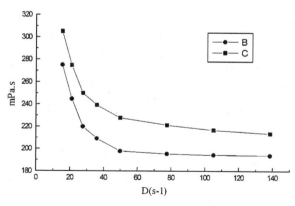

Fig. 5.5 Apparent viscosity of suspensions (45 vol%) of **b** alkali-cleaned and **c** as-received powder A

Thus, it can be concluded from the earlier discussion that (1) there are three possible gas-discharging reactions in the Si_3N_4 suspension of water-based gelcasting: hydrolysis of Si_3N_4 discharging ammonia, oxidation of Si_3N_4 by $(NH_4)_2S_2O_8$ discharging N_2, and the basic reaction of free Si with water discharging H_2; (2) the contributing factor of large amount of pores in the green body of powder A is H_2 discharged during the reaction between free Si and water; and (3) homogeneous green body without large pores is obtained by the use of alkali-cleaned powder A in which free Si has been eliminated.

5.1.2 Effect of Foreign Ions on Concentrated Suspension of Silicon Nitride

In this study, the influence of the concentration of ions on the colloidal behavior of concentrated suspensions of Si_3N_4 was investigated. The foreign ions in the powders of Si_3N_4 mainly come from raw materials and preparation processes. For example, the Si_3N_4 powders of Founder Ceramics Corp. are prepared by nitridation of Si, which is followed by attriting, wet ball-milling, and iron removing by hydrochloric acid. After washing with running water, the powders are dried for reservation. In the light of this process, we can observe that the ions in Si, a little remnant Fe^{2+}, and some amount of Fe^{3+}, Ca^{2+}, and Mg^{2+} in the running water still remain in the final powder of Si_3N_4.

The Si_3N_4 powders used in this study were obtained from Shanghai Materials Research Institute with d_{50} of 2.57 μm (hereafter referred to as S0) and Founder Ceramics Corp. with d_{50} of 2.88 μm (hereafter referred to as F0). Their particle-size distributions are shown in Figs. 5.6 and 5.7. The Na^+ ion-exchange resin (Model Molselect CM-25) was supplied by Roanal Corp., Hungary. The deionized water was supplied by the water station of Microelectronics Institute, Tsinghua University.

The powder F0 and S0 were deionized by washing with deionized water and reacting with ion-exchange resin, respectively. The zeta potentials of the powders

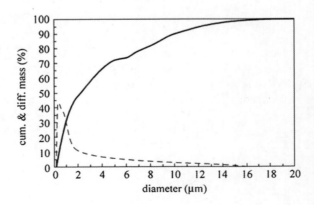

Fig. 5.6 Particle-size cumulative and differential mass distribution of silicon nitride S0

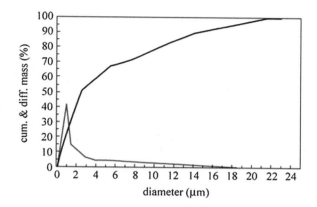

Fig. 5.7 Particle-size cumulative and differential mass distribution of silicon nitride F0

before and after deionization were measured, and the conductivity and concentration of the ions of the supernatant of the centrifuged concentrated suspension were measured. Furthermore, the viscosity of the suspensions before and after deionization was compared.

The washing process employed in this study is as follows [3]:

(1) About 50 g of Si_3N_4 powder was mixed with 200 ml of deionized water to obtain a homogeneous suspension.
(2) The suspension was centrifuged, the supernatant was decanted, and 200 ml of deionized water was added again. This operation was repeated for five times.
(3) The final powder was obtained after drying, grinding, and sieving.

The ion-exchange process employed in this study is as follows [3]:

(1) The washed powder was mixed with an appropriate amount of deionized water and ion-exchange resin, and was subsequently ball-milled for 24 h.
(2) The ball-milled suspension was sieved to remove most of the resin.
(3) The powder was calcined at 500 °C to remove the residual resin.

The calcined powder was processed using two different methods, referred to as method a and method b:

Method a: The calcined powder is directly ground and sieved to obtain the final powder.

Method b: The calcined powder is washed to obtain the final powder.

The particle-size distribution and zeta potential of the powders were measured by BI-XDC and ZetaPlus of Brookhaven Instruments Corp., U.S.A. The conductivity of the supernatant was measured by digital conductimeter (model DDS—IIAt) of Shanghai Leici Instruments Corp., and the ion concentrations were measured by ionographimeter (model IONPACSX. 100) of Dionex, U.S.A. The apparent viscosity was measured by rotating viscometer (model NSX—11) of Chengdu Instruments Plant. The centrifuge (model TGL-5) was supplied by Shanghai Anting Scientific Instruments Plant.

The interaction between the particles in the slurry mainly relates to particle surface adsorption and surface charge determined by the properties of the solid–liquid interface where the electric double layer is located (Levine 1996; James 1980; Levine 1971). One important factor influencing the electric double layer is the surface charge of the particles. According to Mezzasalma and Baldovino, the reason for the Si_3N_4–H_2O interface charge is the presence of $SiNH^{3+}$, $SiOH^{2+}$, and SiO^- under acidic conditions and $Si(OH)_{n+1}^{n-1}$ and $Si(OH)_n O^{(4-n)-}$ under alkaline conditions. Furthermore, the location and density of these surface groups change along with the variation of the pH value, which determines the charge density and charge character of the surface of Si_3N_4 (Mezzasalma 1996). The impurities, such as C, O, etc., introduced during the production process normally produce immense difference in the charging mechanism of different Si_3N_4 powder. Another important factor influencing the electric double layer is the species and concentration of ions in the slurry (Kitahara 1992), which affects the height of potential barrier and distance of interaction between the particles. The abovementioned two factors completely determine the interaction between the particles.

The conductivity of the supernatants of the suspensions prepared by powder F0 and powder S0 before and after washing and ion exchange is shown in Table 5.3. Washing with deionized water greatly decreases the concentration of the ions in the powder, but it is difficult to remove the foreign ions thoroughly because there is an adsorb–resolve equilibrium of ions at the particle surface through which the ions continuously resolve into water from the particle surface. After washing for five times, the conductivity of the supernatant has been found to remain at 20 $\mu S\ cm^{-1}$ or so. The advantage of ion exchange is that the ion-exchange resin can exchange the high-valence counter-ions, such as high-valence cations for Na^+. This can greatly decrease the harmfulness of the counter-ions according to the abovementioned Schulze–Hardy rule. Furthermore, this is also the main reason for the low conductivity of the ion-exchange method b. On the contrary, the high conductivity of the ion-exchange method a is owing to the introduction of Na^+ by calcinations of the remnant Na^+ ion-exchange resin. The effect of the two deionization methods can also be observed from Table 5.4 that presents the ionography data of the supernatants.

Deionization decreased the concentration of the ions in the slurry, and thus, changed the density of the ions at the particle surface through adsorb–resolve equilibrium. As a result, the zeta potential of the powder was charged. Figures 5.8 and 5.9 show the zeta potentials of powder F0 and powder S0 before and after deionization. It can be seen that the pH_{iep} of both powder F0 and powder S0 slightly dropped

Table 5.3 Conductivity of the supernatants of the suspensions ($\mu S\ cm^{-1}$)

Raw materials	Unprocessed (powder: H_2O = 1:2 wt)	Washed with deionized water (five times)	Ion-exchange method a	Ion-exchange method b
F0	175.6	23	620	10
S0	180.4	23	620	10

5.1 Effects of Powder Surface Modification ...

Table 5.4 The ion concentration of different samples (ppm)

Sample	Na^+	NH_4^+	K^+	Ca^+
1	1.8	16.6	–	30.4
2	5.8	1.4	2.2	12.9
3	0.1	0.7	–	8.2
4	0.7	0.4	–	2
5	46.4	–	–	–

Notes: Sample 1: The supernatant of the suspension was obtained by mixing 20 g of unprocessed powder F0 with 40 ml of deionized water
Sample 2: The supernatant of the suspension was obtained by mixing 20 g of unprocessed powder S0 with 40 ml of deionized water
Sample 3: The supernatant of the suspension was obtained by mixing 20 g of washed powder F0 with 40 ml of deionized water
Sample 4: The supernatant of the suspension was obtained by mixing 20 g of washed powder S0 with 40 ml of deionized water
Sample 5: The supernatant of the suspension was obtained by mixing 55 g of powder F0 processed by ion-exchange method a with 300 ml of deionized water.

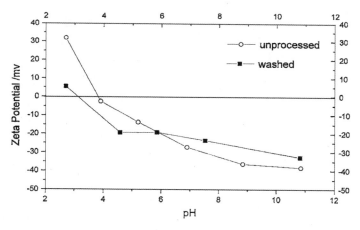

Fig. 5.8 Zeta potential of powder F0 before and after deionization

after deionization. This is owing to the decrease in the density of the counter-ions cations at the surface of the particles. As shown in Fig. 5.10, the specific adsorbed cations exist in the Stern layer, and when the concentration of OH^- in the diffusion electronic double layer increases along with the increase in the pH value, it results in a pH_{iep} when the potential at the slide-plane zeta potential becomes zero. When the density of the cations in the Stern layer decreases, the OH^- concentration at which the zeta potential is zero, or the pH_{iep}, also decreases.

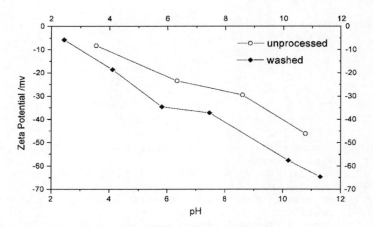

Fig. 5.9 Zeta potential of powder S0 before and after deionization

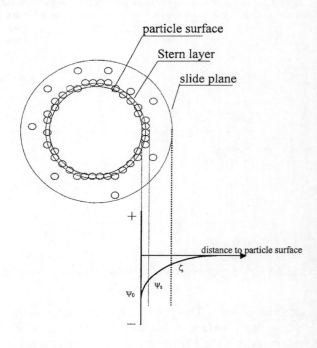

Fig. 5.10 Schematic diagram of the electric double layers at the surface of a particle

Thus, according to the formula

$$1/\kappa = \left(\varepsilon kT/8\pi ne^2 z^2\right)^{1/2}$$

where κ—Debye parameter; k—Boltzmann constant; T—absolute temperature; n—ion concentration; and z—ion valence.

5.1 Effects of Powder Surface Modification ...

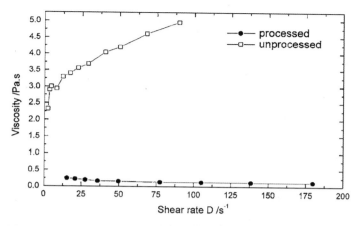

Fig. 5.11 Viscosity comparison of powder F0 suspensions with solid volume concentration of 49% before and after processing with ion-exchange method b

When ion concentration decreases, the electric double layer becomes thicker and increases the radius of the slide plane. Very low ion concentration results in larger radius of the slide plane and increases the resistance of the particles to rearrange and slide against each other in concentrated suspensions, and thus, increases the viscosity. However, when dispersant J (TMAH) is added to the suspension, the viscosity of the suspension of deionized powder F0 becomes less than one-tenth of that of the unprocessed powder F0 with the same solid volume concentration of 49%, as shown in Fig. 5.11. This may be owing to the fact that more dispersant molecules are adsorbed and repulsion between the particles is improved when more room is obtained in the Stern layer after deionization.

In the suspensions with relatively low solid volume concentration, the distance between the particles is large when compared with the particle size, and enlargement of the radius of the slide plane may not lead to evident increase in the resistance of the particles to rearrange and slide against each other. Moreover, the repulsion between the particles increases after deionization; thus, the dispersity is improved and viscosity is decreased. Figure 5.12 shows the viscosity comparison of the suspensions of powder S0 before and after processing with ion-exchange method b. It can be seen that because of the raise in the zeta potential (Fig. 5.9) and the consequent increase in the repulsion between the particles, the apparent viscosity of the suspension dropped over 90%.

(1) Both washing with deionized water and ion-exchange method b can effectively decrease the ion concentration in the suspensions and on the surface of the Si_3N_4 particles. Furthermore, ion-exchange method b can particularly decrease the concentration of high-valence counter-ions—high-valence cations.
(2) The pH_{iep} of the deionized powder decreased owing to the decrease in the counter-ions—cations adsorbed onto the particle surface.

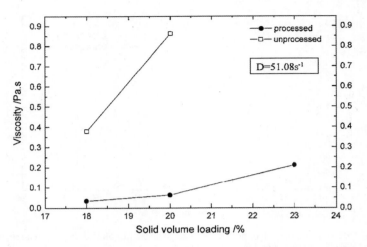

Fig. 5.12 Viscosity comparison of powder S0 suspensions before and after processing with ion-exchange method b

(3) The repulsion between the particles in the suspension increased after deionization. Thus, the dispersity of the suspension improved and the apparent viscosity of the suspension of powder S0 with solid volume concentration of 20% decreased below 10% after deionization.

(4) Deionization can facilitate the adsorption of dispersant molecule on the particle surface and improve suspension dispersity. When dispersant J was added to the suspensions, the viscosity of the suspension of deionized powder F0 decreased below one-tenth of that of the unprocessed powder F0 with the same solid volume concentration of 49%.

5.1.3 Effects of Acid Cleaning and Calcinations on the Suspension Properties of Silicon Nitride

A universal method for preparing highly concentrated suspensions of Si_3N_4 powders has not yet been developed. The purpose of this study has been to investigate the particle surface characteristics before and after treatment, and the effect of surface modification on the solids loading of its aqueous suspension.

The powder characteristics of the two types of Si_3N_4 (referred to as FD1 and FD2) as well as the rheological properties of their aqueous suspensions were studied. Both the Si_3N_4 powders used in this study (Beijing Founder High Tech. Ceramic Corp., Ltd., Beijing, China) were obtained by direct nitridation of Si. The only difference between them was in the pulverization process. One powder was pulverized through vibration with an aqueous medium (FD1), and the other was pulverized through

5.1 Effects of Powder Surface Modification ...

ball-milling with an ethanol medium (FD2). The final products were obtained after drying without further purification.

The FD1 and FD2 were acid-cleaned as follows: the mixture of Si_3N_4 powder with 5% HCl (100 g: 200 ml) was put into a ball-milling jar (without milling balls) and trundled for the first 24 h. After the mixture was centrifuged, the supernatant was removed and fresh acid was added to the jar. Subsequently, the suspension was trundled for another 24 h. The process was repeated for four times. Finally, the acid-cleaned powder was leached with de-ionized water for four or five times to remove the residual acid, and the powder was subsequently dried.

To form a silica layer on the particle surface, the Si_3N_4 powder was calcined in air at 600 °C for 6 h. A standard muffle furnace with a temperature controller was used with a static air atmosphere. The depth of the powder layer was ~10 mm.

(1) Powder Characteristics

Figure 5.13 shows the particle-size distribution of the FD1 and FD2 powders. Both the powders have a wide distribution and a size at the micrometer level. It can be observed that the specific surface area (SSA) of the FD1 powder is larger than that of the FD2 powder, and that the SSAs of the treated samples are almost the same as the corresponding as-received powders (Table 5.5). This difference in the size distribution may be owing to the pulverization process. The introduced energy of vibration milling for FD1 was larger than that of ball-milling for FD2 powder. Both bulk and surface pulverization mechanisms had an equivalent impact during the vibration-milling process, while the surface pulverization mechanism played an important role during ball-milling.

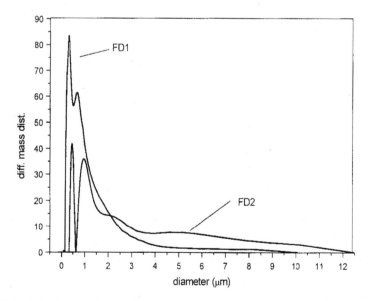

Fig. 5.13 Particle-size distribution of the FD1 and FD2 powders

Table 5.5 Surface oxygen content and specific surface area of FD1 and FD2 powders

Powder	Oxygen content (mg m^{-2})	Specific surface area (m^2 g^{-1})
FD1 (as-received)	2.98	12.4
FD1 (acid-leached)	2.91	12.5
FD1 (acid-leached and calcined)	5.97	12.2
FD2 (as-received)	0.56	8.2
FD2 (acid-leached)	0.53	8.0
FD2 (calcined)	11.93	8.3

Table 5.5 shows the surface oxygen content of the samples. For the calcined samples, the surface oxygen content increased remarkably when compared with the as-received powders, which implies a silica-like layer formed on the particle surface after calcination. Furthermore, Table 5.6 shows the ion conductivity and (Mg^{2+} + Ca^{2+}) concentration of the supernatants, and Table 5.7 shows the chemical composition of the two as-received powders. It can be observed that the ion conductivity and (Mg^{2+} + Ca^{2+}) concentration of the as-received FD1 powder are much higher than those of the as-received FD2 powder. This considerable difference may be primarily owing to the pulverization process. For the FD1 powder, the liquid medium used during the pulverization process was tap water that contains many kinds of

Table 5.6 Ion conductivity and (Ca^{2+} + Mg^{2+}) concentration of the supernatants of FD1 and FD2 powders

Powder	Ion conductivity (mS cm^{-1})	Ca^{2+} + Mg^{2+} concentration (ppm)
FD1 (as-received)	0.716	74.7
FD1 (acid-cleaned)	0.186	11.8
FD1 (calcined)	0.015	6.8
FD2 (as-received)	0.136	21.1
FD2 (acid-cleaned)	0.180	6.8
FD2 (calcined)	0.047	4.2

Table 5.7 Characteristics of the as-received powders

Powder	Silicon	Nitrogen	Chemical composition (%)[†]			Calcium	Chlorine
			Oxygen	Iron	Aluminum		
FD1	48.4	37.3	6.4	0.58	0.78	0.27	0.14
FD2	60	40	2.8	0.50	0.28	0.13	

[†]The chemical composition was obtained by sequential X-ray fluorescence (XRF), and the SSAs were measured by the BET method. The oxygen content determined by XRF agreed well with the data measured by the inert gas fusion method (shown in Table 5.5), which were 6.43% and 2.28% for the as-received FD1 and FD2 powders, respectively

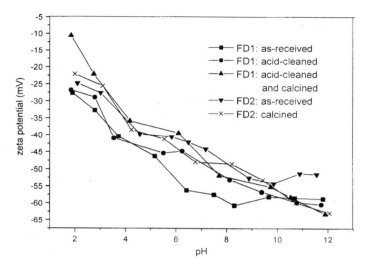

Fig. 5.14 Zeta potential versus pH of FD1 and FD2 powders

high-valence cations, such as Mg^{2+}, Ca^{2+}, and $Fe^{3+/2+}$. As further purification after pulverization was not carried out, both the ions from the tap water and soluble ions of the powder remained. However, for the FD2 powder, the liquid medium was ethanol and no foreign ions were introduced during pulverization, which can be confirmed from the chemical composition of the two as-received powders (Table 5.7). For the acid-cleaned FD1 sample with small ion conductivity and ($Mg^{2+} + Ca^{2+}$) concentration, the majority of the soluble ions were eliminated during acid-cleaning and subsequent leaching by de-ionized water. For the calcined FD1 powder, ion conductivity and ($Mg^{2+} + Ca^{2+}$) concentration were even lower than the acid-cleaned sample, which may be because of the formation of nonsoluble compounds on the powder surface and/or diffusion of the soluble ions from the particle surface inwards.

Figure 5.14 presents the zeta potential versus pH curves of the FD1 and FD2 powders. According to Fig. 5.14, the as-received FD1 powder has a higher diffuse-layer potential than the as-received FD2 powder, which indicates that it would be easier to disperse FD1 powder than FD2 powder. However, for the treated samples, the difference in the zeta potentials is not distinctive at a high pH. These powders also seem to have extremely low IEPs (<pH 2), which conflict with the results presented in the literature (Greil 1991; Bergstrom 1989). Furthermore, in this investigation, FD1 and FD2 powders were observed to have more surface oxygen (Table 5.5), which, according to Hackley et al, indicates low IEPs (Hackley 1993).

The diffuse reflectance infrared Fourier transform (DRIFT) spectroscopy findings, from 2400 to 4000 cm^{-1} of the as-received and treated samples, are shown in Figs. 5.15, 5.16, and 5.17. According to the literature (Nilsen 1987; Ramis 1989; Busca 1986), the peak ranging from 3400 to 3200 cm^{-1} corresponds to the amine structure, and the one at ~3000 cm^{-1} is caused by the stretching vibration of the

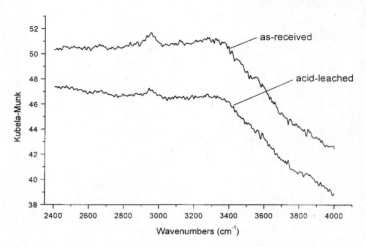

Fig. 5.15 DRIFT spectra of the as-received and acid-leached FD1 powder

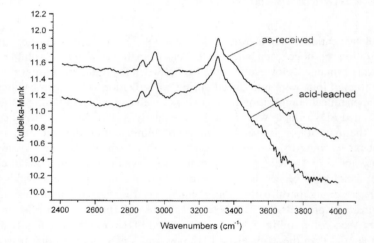

Fig. 5.16 DRIFT spectra of the as-received and acid-leached FD2 powder

C–H bonding. The spectrum of the as-received FD1 powder (Fig. 5.15) has a peak at ~3000 cm^{-1}, but it is difficult to discern the peak from 3400 to 3200 cm^{-1}. The spectrum of the as-received FD2 powder (Fig. 5.16) demonstrates the existence of the amine structure and C–H bonding, and the peak at ~2900 cm^{-1} is also caused by the C–H stretching. The two peaks at 3400–3200 cm^{-1} and 3000 cm^{-1} of the as-received FD2 powder are more pronounced than those of the as-received FD1 powder, which implies that the amount of the amine structure and C–H bonding in FD2 powder are more developed than those in FD1 powder. Furthermore, the spectra of the acid-cleaned FD1 and FD2 powders are similar to those of their respective

5.1 Effects of Powder Surface Modification … 243

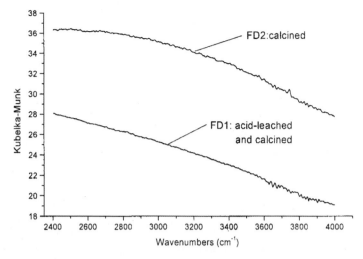

Fig. 5.17 DRIFT spectra of the calcined FD1 and FD2 powders

as-received powders. As shown in Fig. 5.17, the patterns of calcined FD1 and FD2 powders are very similar and considerably different from their respective as-received samples.

(2) Rheological Properties of Concentrated Silicon Nitride Slurries

Rheological Properties of Acid-Cleaned Silicon Nitride Powders: For the as-received and treated powders, the viscosity versus shear rate curves of the FD1 and FD2 powders are shown in Figs. 5.18 and 5.19, respectively. It is clear that the concentrated suspensions cannot be directly prepared from the as-received powders. As shown in

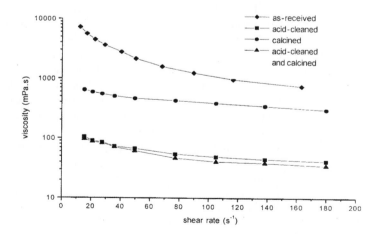

Fig. 5.18 Rheological curve of 40 vol% suspension of FD1 powder (pH 11.5)

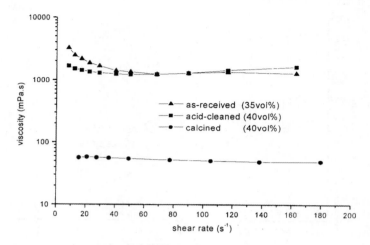

Fig. 5.19 Rheological properties of FD2 suspensions (pH 11.5)

Fig. 5.15, acid-cleaning can improve the aqueous dispersibility of FD1 powder dramatically, and the slurry viscosity of the acid-leached FD1 powder decreased greatly when compared with its respective as-received sample. Although acid-leaching can also improve the dispersibility of FD2 powder (Fig. 5.16), the apparent viscosity of a 40 vol% suspension of acid-leached FD2 powder is still beyond 1500 mPa s. The viscosity curve for the slurry with 40 vol% solids loading of acid-leached FD2 powder can be divided into two different rheological regions. At a low shear rate, the slurry behaves as a pseudo-plastic fluid, while it shows shear thickening behavior when the shear rate is beyond 50 s^{-1}.

Rheological Properties of Calcined Powders: The rheological curves of calcined FD1 and FD2 powders are shown in Figs. 5.18 and 5.19, respectively. It can be observed that the viscosity of 40 vol% slurry for the direct-calcined FD1 powder decreased greatly when compared with the as-received sample, but it is still much higher than that of the acid-cleaned powder (Fig. 5.18). This implies that although calcination can improve the aqueous dispersibility of FD1 powder, the degree of improvement is not as pronounced as with acid-cleaning. The rheological curves of 40 and 50 vol% slurries for the acid-cleaned and calcined FD1 samples are shown in Figs. 5.18 and 5.20, respectively. It can be observed that the aqueous dispersibility of this sample is better than that of the acid-leached one, and the viscosity of the 50 vol% slurry is lower than 300 mPa s at low shear rates.

Figures 5.19 and 5.20 show the rheological curves of the slurries for the calcined FD2 powder with solids loading of 40 and 50 vol%, respectively. It can be observed that the slurry with solids loading of 40 vol% behaves almost as a Newtonian liquid and its viscosity remains at ~50 mPa s. Furthermore, the suspension with solids loading of 50 vol% can be found to show pseudo-plastic characteristics at low shear rates and shear-thickening when the shear rate is beyond 35 s^{-1}. From the rheological curves of the as-received and treated FD2 slurries, it can be concluded that calcination

5.1 Effects of Powder Surface Modification ...

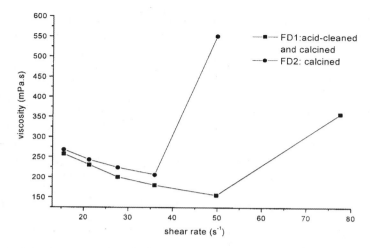

Fig. 5.20 Rheological properties of 50 vol% suspension of treated FD1 and FD2 powders (pH 11.5)

could enhance the solids loading dramatically to 50 vol%, while the improvement from acid-leaching is inferior.

High-concentration slurries (≥50 vol%) cannot be prepared from the as-received powders of FD1 and FD2. Both acid-leaching and calcination can improve their aqueous dispersibility, but the effect of the same method varies with different powders. The solids loading of the FD1 slurry can reach 50 vol% through acid-cleaning and calcination, while the effect of direct calcinations is not similar to acid-cleaning. However, calcination of the FD2 slurry can enhance the solids loading to 50 vol%, while acid-leaching does not produce a similar effect. With regard to the slurries of FD1 and FD2 powders with 50 vol% solids loading (Fig. 5.20), although both exhibit similar rheological behaviors, FD2 powder is observed to be a stronger dilatant than FD1 power at high shear rates. The different dilatancies of the slurries may be owing to the differences in the particle size and size distribution.

(3) Discussion

The flocculation of the slurry is affected by the concentration of the electrolytes. According to the Schulze–Hardy rule, in an aqueous suspension at 25 °C (Kitahara 1992),

$$n_c = 8 \times 10^{-22} \left(\frac{\gamma^4}{A^2 z^6} \right) \quad (5.4)$$

where n_c is the critical flocculation concentration, $\gamma = [\exp(Z/2) - 1]/[\exp(Z/2) + 1]$ ($Z = ze\varphi_0/(kT)$, where z is the valency of the counter-ion, e is the electron charge, k is the Boltzmann constant, and T is the absolute temperature), and A is the Hamaker constant. The critical flocculation concentration n_c is inversely proportional to the sixth power of the counter-ion valency. Furthermore, if the counter-ion valency is higher, then the critical flocculation concentration becomes lower. When compared with the mono-valence counter-ions, high-valence counter-ions are observed to lead to coagulation at lower molar concentrations. In addition, the presence of specific surface impurities on dry Si_3N_4 powder, which are primarily imparted from the processing steps (milling, leaching, etc.) has also been reported (Bergstrom 1992b; Peuckert 1987; Rahaman 1986). According to Malghan (Malghan 1992), cationic (Na^+, Al^{3+}, Ca^{2+}, Fe^{3+} etc.) and anionic (Cl^-, F^-, etc.) species can get dissolved from the surface impurities and form surface complexes by reacting with the surface silanols. Thus, the soluble, especially the counter-ions with high valency, may affect the slurry rheology and may be the limiting factor in the preparation of concentrated suspensions.

The ion conductivity and ($Mg^{2+} + Ca^{2+}$) concentration of the supernatant of the as-received FD1 powder were higher when compared with other samples, while the rheological property of its 40 vol% slurry was very poor. On the other hand, the ion conductivity and ($Mg^{2+} + Ca^{2+}$) concentration of the acid-leached FD1 powder were far lesser than those of the as-received sample, indicating that most of the soluble species in the as-received powder as well as the soluble high-valence counter-ions have been eliminated. The DRIFT spectrum of the acid-leached FD1 powder was very much similar to that of the as-received sample, and the difference in the zeta potentials was indistinct at the discussed high pH. According to Hackley et al. (Hackley 1994), reduction in these ions should also affect the zeta potential; however, this was not observed in this study. Hence, it can be concluded that the dramatic improvement in the dispersibility was caused by the elimination of the soluble high-valence counter-ions. Furthermore, the zeta potential of the FD1 powder, which was acid-leached and calcined, was almost the same as that of the acid-cleaned sample, while the DRIFT spectrum was different. The viscosity of the 40 vol% slurry showed a slight decrease when compared with the acid-leached powder, implying that oxidation had an indistinctive effect on the aqueous dispersibility of the acid-leached sample. However, the ion conductivity and ($Mg^{2+} + Ca^{2+}$) concentration of the direct-calcined FD1 powder were lesser than those of the acid-leached sample. The remarkable reduction in the soluble species may be owing to the formation of insoluble species or the diffusion of these species from the particle surface inwards. From Fig. 5.18, it is clear that the effect of direct calcination of FD1 powder on dispersibility is inferior to acid-leaching. These insoluble species may be unfavorable for dispersibility, and hence, the improvement is restricted when compared with acid-leaching. Thus, it can be concluded that for the as-received FD1 powder, the crucial factor limiting the aqueous dispersibility is the presence of soluble species, especially the high-valence counter-ions. The results of acid-leaching correlate with those obtained by (Laarz 1997), while the effects of oxidation are conflicting. Laarz et al. attributed the improvement in dispersibility to a characteristic physical change in the powder, and presumed that the effect of oxidation was unexpected.

The dispersibility of the as-received FD2 powder was very poor, such that the viscosity of a slurry with 35 vol% solids loading was beyond 1000 mPa s (Fig. 5.20). Soluble species from the powder may not be responsible for the poor dispersibility because the ion conductivity and (Mg^{2+} + Ca^{2+}) concentration of the as-received powder were very small. According to Fig. 2, the as-received FD1 powder has a higher zeta potential than that of the as-received FD2 powder, indicating that it would be easier to disperse FD1 powder than FD2 powder, which was confirmed in the rheological study (Figs. 5.18 and 5.19). For the acid-leached, leached, and then calcined FD1 and FD2 powders, the zeta potentials were almost the same at the discussed high pH, which implies that the dispersibility of these three samples should be similar, and this was also confirmed (Figs. 5.18 and 5.19).

The dispersant used in this study was an aqueous solution of $(CH_3)_4NOH$, which is an organic alkali, and thus, the slurry of Si_3N_4 was stabilized via electrostatic repulsion at pH 11. Surface charging of Si_3N_4 particles may contribute to the dissolution of the surface groups as follows:

$$[SiOH_2]^+ \xleftarrow{H^+} [SiOH] \xrightarrow{OH^-} [SiO]^- + H_2O \tag{5.5}$$

$$[Si_2NH_2]^+ \xleftarrow{H^+} [Si_2NH] \xrightarrow{OH^-} [Si_2N]^- + H_2O \tag{5.6}$$

$$[SiNH_3]^+ \xleftarrow{H^+} [SiNH_2] \xrightarrow{OH^+} [SiNH]^- + H_2O \tag{5.7}$$

These reactions are pH-dependent and essentially equivalent to the specific adsorption of protons or hydroxyls. Although the charging mechanism of Si_3N_4 in basic solution has not yet been well-established (Kulig 1991; Harame 1984; Greil 1989a), there has been a similar viewpoint that the silanol group, Si–OH, is favorable for surface charging. If the silanols are more, then the electrostatic repulsion between the particles becomes greater, with other conditions remaining constant. From the viewpoint of preparing highly concentrated slurries, the substitution of Si–OH for amine structures is beneficial to improve the dispersibility of Si_3N_4 powder. On the other hand, it is difficult to prepare highly concentrated aqueous slurries if the amount of amine structures or other hydrophobic groups is dominant on the particle surface. According to Schwelm et al. (Schwelm 1993), stable esters that are hydrophobic with the configuration Si–O–C–R would be formed on the Si_3N_4 surface by the reaction of alcohol with the silanol groups. The CH_3 groups of the alcohol are believed to shield the negative charge of the electron pairs at the oxygen ion, which are the active sites for deflocculant coupling. Temperatures up to 500 °C can remove the esters from the powder surface.

The DRIFT spectrum of the as-received FD2 powder showed that the dominant surface groups were the amine structures (Si_2NH and $SiNH_2$) and C–H bonds, which were indistinct in the FD1 sample. From the pulverization process of FD2 powder, it can be concluded that the Si_3N_4 powder would react with alcohol to form esters on the particle surface, and that the distinctive C–H bonding is caused by the ester, Si–O–C–R. As the hydrophobic groups are dominant on the particle surface of the as-received FD2 powder, it is easy to understand the poor aqueous dispersibility. The DRIFT pattern of the acid-leached FD2 powder was similar to that of the as-received sample, and the rheological improvement of the slurry was limited. However, for the calcined FD2 powder, both the amine structures and C–H bonding disappeared completely, and the DRIFT spectrum of the calcined FD2 powder was similar to that of the acid-cleaned and calcined FD1 sample. The solids loading of the slurries for these two samples reached 50 vol%, while the viscosity was below 300 mPa s at low shear rates. For the as-received FD2 powder, the results indicated that the crucial factor limiting the preparation of the highly concentrated slurry was the hydrophobic groups on the particle surface.

(4) Summary

In this study, two types of commercial Si_3N_4 powders were studied, and only their pulverization processes were found to be different. Several differences in the size distribution, soluble species, and rheological properties of the aqueous slurries were found. Both acid-leaching and calcination were found to improve their aqueous dispersibility, but the degree for the same method varied with different powders. For the as-received FD1 powder, the crucial factor limiting dispersibility was the high-valence counter-ions, which could be eliminated almost completely through acid-leaching. Because of the diffusion or reaction of the soluble species, improvement occurred with the use of direct calcination, but was inferior to acid-leaching. In the as-received FD2 powder, the surface hydrophobic groups of Si–O–C–R were found to limit solids loading. In the FD2 powder, the effect of calcination on dispersibility was remarkable, while the effect of acid-cleaning was not similar to that of calcination. Furthermore, the solids loading of the aqueous suspensions of acid-cleaned and calcined FD1 and FD2 powders reached 50 vol% with a viscosity below 300 mPa s.

Particle-size distribution was determined with a particle-size analyzer (Model BI-XDC, Brookhaven Instruments Corp., Holtsville, NY), and the zeta potentials were measured using ZetaPlus (Brookhaven Instruments Corp.). The chemical composition was determined with a sequential X-ray fluorescence spectrometer (Model XRF-1700, Shimadzu Corp., Tokyo, Japan) and the specific surface area was measured by the BET method. The total oxygen content of the powders was determined with the inert gas fusion method (Natansohn 1991). To determine the bulk oxygen content, the powders were leached in 5% HF for 1 h to remove the surface oxygen (Natansohn 1993).

Slip rheological properties were determined using a concentric cylinder viscometer (Model NSX-11, Chengdu Instrument Plant, Sichuan, China) at 25 °C. The medium used to prepare concentrated suspensions was de-ionized water. TMAH $[(CH_3)_4NOH]$ was added as a dispersant to break up the agglomerates. For the rheological measurements, all the slurries were ball-milled for 4 h to break up any network structure, and the pH of the slurries was fixed at 11.5 ± 0.1 by adding $(CH_3)_4NOH$ and HCl. The experiments were performed by stepping up to higher shear rates, and the increasing curves were measured. At each shear rate, the samples were presheared for 1 min to reach equilibrium.

To examine the soluble ions in the powder, both digital conductivity gauge (Model DDS-11AT, Rex Instruments Factory, Shanghai, China) and ion chromatographic analyzer (Model IONPAC DX-100, Dionex Corp., Sunnyvale, CA) were used.

A suspension with 5 g of Si_3N_4 and 20 ml of de-ionized water was stirred for 30 min and then centrifuged; the supernatant was used to measure the ion conductivity and concentration of the cations.

The DRIFT spectroscopy (Model 750 FTIR, Nicolet Instrument Corp., Madison, WI) was used to collect information about the surface chemistry of the as-received and treated Si_3N_4 powders. The samples were dried in an oven for 24 h at 150 °C before being examined. Dry air was used in the Fourier transform infrared (FTIR) spectroscopy sample chamber, until no free-water peaks were observed.

5.1.4 Effects of Liquid Medium and Surface Group on Dispersibility of Silicon Nitride Powder

The purpose of this study was to investigate the effects of liquid media and ball-milling process on the particle surface group and aqueous dispersibility of Si_3N_4 powders.

The commercial α-Si_3N_4 powder used in this study was prepared by direct nitridation of Si powder (grade M11, H. C. Starck, Berlin, FRG). The powder properties of the as-received product is shown in Table 5.8. The liquid media used to modify the particle surface group of the Si_3N_4 powder were de-ionized water (about 18×10^6 Ω), analytically pure ethanol, and isopropanol.

To study the influence of liquid media on particle surface group and aqueous dispersibility, the Si_3N_4 powder was treated as follows: mixtures of the as-received Si_3N_4 powders with different liquids (100 g: 200 ml) were ball-milled for 3 days at a rotary speed of 100 rev min^{-1}, and then dried at 100 °C for 5 days. The weight ratio of ZrO_2 milling balls to Si_3N_4 powder was 1:1.

To compare the influence of the ball-milling process on the particle surface group and aqueous dispersibility of the mixture with ethanol as the liquid medium, the weight ratio of the milling balls to Si_3N_4 powder and the rotary speed were increased to 3:1 and 400 rev min^{-1}, respectively. Other parameters such as the ratio of the powder weight to liquid volume and drying process were invariant.

All the treated powders were divided into two portions: one part was used directly for rheology measurement and DRIFT spectra test, and the other portion was first calcined at 600 °C for 6 h and then was used for the same examinations. A standard muffle furnace with a temperature controller, having a hearth size of $25 \times 15 \times$

Table 5.8 The composition and properties of the as-received silicon nitride powder

Chemical composition						Specific surface area (m^2 g^{-1})	α Phase (%)
N (%)	O (%)	C (%)	Fe (ppm)	Al (ppm)	Ca (ppm)		
38.69	1.44	0.20	15	500	61	12.9	91.7

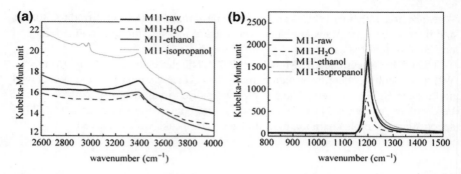

Fig. 5.21 DRIFT spectra of the silicon nitride powders ball-milled in different liquids: **a** 4000–2600 cm^{-1}; **b** 1500–800 cm^{-1}

12 cm, was used with an air static atmosphere, and with the powder layer of about 10 mm.

The liquid medium to prepare the aqueous suspensions was de-ionized water, and $(CH_3)_4NOH$ was added as a dispersant to break up the agglomerates of Si_3N_4 powders. After ball-milling for 24 h, the slurries with a pH of 11.5 ± 0.3 were measured using a rotating viscometer (model NXS-11, ChengDu Instruments Plant, China) to determine their rheological properties.

The DRIFT spectra were recorded using a Nicolet 750 Fourier transformation spectrometer (8 cm^{-1} resolution, 64 scans, mirror face of the specimen cup as background). All the Si_3N_4 powders were measured in their undiluted state.

(1) Effect of the Liquid Media—Water, Ethanol, and Isopropanol

The DRIFT spectra of the as-received and liquid-treated Si_3N_4 powders are shown in Fig. 5.21. According to the literature (Stadelmann 1989; Nilsen 1987; Ramis 1989; Busca 1986), the wide peak from 3200 to 3500 cm^{-1} corresponds to the amine structures (Si_2–NH and Si–NH_2) on the particle surface, while the peak at 3740 cm^{-1} results from the isolated Si–OH surface group. Earlier investigations of the DRIFT spectra of the undiluted Si_3N_4 samples in the range of 800–1500 cm^{-1} have shown (Busca 1986; Dai 2001) that the strong sharp peak at about 1200 cm^{-1} is owing to the anti-symmetric stretching vibration of the siloxane surface group, Si–O–Si. It can be seen from Fig. 5.21a that the amine structures (3200–3500 cm^{-1}) have obscure changes, and that the weak peak corresponding to isolated Si–OH group disappears after liquid treatment. With regard to powders ball-milled in ethanol and isopropanol (referred to as powder M11–ethanol and powder M11–isopropanol, respectively), three peaks are observed to emerge at 2800–3000 cm^{-1}, which correspond to the stretching vibration of the C–H bonding, while these peaks are nonexistent in the DRIFT spectrum of the sample treated in de-ionized water (referred to as powder M11–H_2O). When compared with powder M11–ethanol, the C–H peak of powder M11–isopropanol is found to be more intensive. At the range of 800–1500 cm^{-1} (Fig. 5.21b), the peak intensity of Si–O–Si group of powder M11–H_2O is observed to decrease and that of powder M11–isopropanol is found to increase, when compared with the as-received powder (referred to as powder M11–raw), while the peak intensity of M11–ethanol remained invariant.

As previously investigated (Dai 2001), the surface oxygen content of the as-received Si_3N_4 powder was 0.72 wt% and there were excess of Si–O–Si groups on the particle surface, which could be hydrolyzed gradually in the aqueous solution as follows:

$$Si-O-Si + 2H_2O \rightarrow 2Si-OH \tag{5.8}$$

Thus, the number of Si–O–Si surface groups will decrease owing to the hydrolysis reaction, which is consistent with the observations in the DRIFT spectrum of powder M11–H_2O.

Both the powders ball-milled in ethanol and isopropanol are quite volatilizable, and the emergence of C–H bonding implies reaction of the liquids with the particle surface. The reaction may occur as follows:

$$Si-O-H + HO-CH_2CH_3 \rightarrow Si-O-CH_2CH_3 + H_2O \tag{5.9}$$

$$Si-OH + HO-CH(CH_3)_2 \rightarrow Si-O-CH(CH_3)_2 + H_2O \tag{5.10}$$

The number of Si–OH groups (isolated and adjacent) is observed to decrease and that of the Si–O–Si surface species is found to remain invariant according to these two reaction equations. Furthermore, the emergence of the C–H bonding and disappearance of the isolated Si–OH groups in their DRIFT spectra lend support to the postulate. This finding is also consistent with the results obtained by Schwelm et al. (Schwelm 1993), who observed that the Si–OH groups at the Si_3N_4 particle surface react with isopropanol and form esters of Si–O–C–R. Their results also demonstrated that the Si–O–C–R surface groups are very stable and can be burnt out at temperatures above 450 °C. The peaks of the amine structures of the powders M11–ethanol and M11–isopropanol are observed to be similar to those of the as-received sample, implying that there is no reaction between the amine structures and the liquid media, ethanol and isopropanol. With regard to powder M11–ethanol, the number of the Si–O–Si surface groups remained similar to that of the as-received powder, which demonstrates that the Si–O–Si groups do not react with ethanol. However, with regard to powder M11–isopropanol, the obvious increase in the Si–O–Si amount indicates the occurrence of other reactions of isopropanol with Si_3N_4 particle surface. It has been proposed in an earlier paper (Dai 2001) that there is another kind of Si–O bonding (neither Si–O–Si nor Si–OH), which is unstable and easily forms Si–OH groups in aqueous solution, at the amorphous oxygen-rich layer on the Si_3N_4 powder surface. The long-term ball-milling of Si_3N_4 powders in isopropanol may cause the following reaction, which increases the amount of Si–O–Si surface groups:

$$2Si-O + Si-H + HO-CH-(CH_3)_2 \rightarrow Si-O-Si + Si-O-CH-(CH_3)_2 + H_2O \tag{5.11}$$

The rheological properties of 40 vol% aqueous slurries of the as-received and liquid-treated powders are shown in Fig. 5.22a. In contrast to the rheological properties of slurry M11–raw, the apparent viscosity of slurry M11–H$_2$O is reduced, while that of slurries M11–ethanol and M11–isopropanol are increased. As mentioned earlier, hydrophobic Si–O–C–R groups are formed on the particle surface of the powders ball-milled in ethanol and isopropanol. This hydrophobic surface group deteriorates the aqueous dispersibility (Dai 2001). Furthermore, the 40 vol% suspensions of the liquid-treated powders after thermal oxidation at 600 °C for 6 h, as exemplified in Fig. 5.22b, show similar rheological behaviors. It can be clearly seen from Fig. 5.23 that the DRIFT spectra of these liquid-treated samples are very similar after calcination, which indicates that the hydrophobic Si–O–C–R groups are removed, and

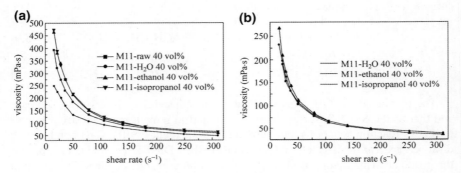

Fig. 5.22 Rheological properties of the slurries of the liquid-treated silicon nitride powders: **a** before and **b** after calcination

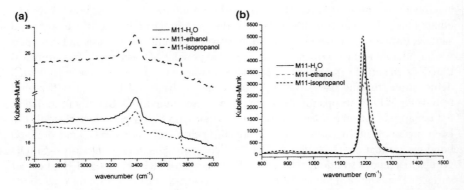

Fig. 5.23 DRIFT spectra of the liquid-treated silicon nitride powders after calcination: **a** 4000–2600 cm^{-1} and **b** 1500–800 cm^{-1}

5.1 Effects of Powder Surface Modification ...

the surface characteristics of the liquid-treated powders are recovered, which are identical after calcination.

(2) Effect of Ball-Milling Process

Though the aqueous dispersibility of the liquid-treated powders was modified, the change was not evident, which may be owing to the weak reaction of the liquid media with the particle surface of the Si_3N_4 powder. To study the influence of ball-milling process on the aqueous dispersibility and particle surface group, the weight ratio of the milling balls to the powders ball-milled in ethanol was increased from 1:1 to 3:1, and the rotary speed was simultaneously increased from 100 to 400 rev min^{-1}. The treated samples were referred to as powders M11-11 and M11-34. As shown in Fig. 5.24, the DRIFT spectra of powder M11-34 exhibited remarkable difference when compared with those of powder M11-11. First, the amount of the amine structures and C–H bonding increased dramatically, and the peak positions also demonstrated deviation (Fig. 5.24a). Second, the peak of the adjacent Si–OH surface groups at 3500–3600 cm^{-1} emerged. Third, the amount of the Si–O–Si surface groups decreased obviously and the peak position shifted visibly to lower frequency. Besides the condensation reaction of ethanol with the Si–OH surface species to form Si–O–C–R groups, the results also demonstrated that other reactions occur at the particle surface, which is related to the ball-milling process. During the ball-milling process of powder M11-34, the particle surface endured intensive mechanical impact, which broke the Si–O–Si bonding and caused the reaction of ethanol with the ruptured Si–O–Si groups:

$$Si-O-Si + HO-CH_2CH_3 \rightarrow Si-O-CH_2CH_3 + Si-OH \quad (5.12)$$

According to this reaction, the amount of the Si–O–Si groups decreases, while that of Si–O–C–R and Si–OH group simultaneously increases, which is confirmed by the DRIFT spectra of powder M11-34. Furthermore, the peak shift of the Si–O–Si groups from 1190 to 1175 cm^{-1} also demonstrates the presence of large tensile stress in the cyclic Si–O–Si structure owing to the intensive impact. The DRIFT spectra

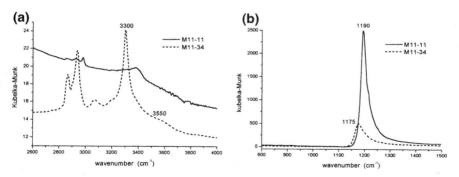

Fig. 5.24 DRIFT spectra of the silicon nitride powders after different ball-milling processes: **a** 4000–2600 cm^{-1} and **b** 1500–800 cm^{-1}

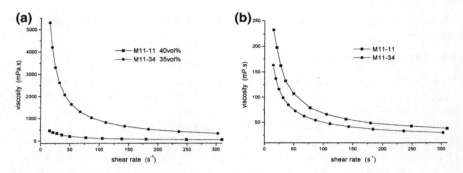

Fig. 5.25 Rheological properties of the slurries of powders M11–11 and M11–34: **a** before calcination and **b** after calcination

of powder M11–34 conclusively show that the dominant process during ball-milling is the reaction of ethanol with the ruptured Si–O–Si groups, which generates large amount of Si–O–C–R groups and obviously consumes the Si–O–Si structures.

Figure 5.25 illustrates the rheological properties of the slurries of powders M11–11 and M11–34. With regard to the ball-milled powders before calcination (Fig. 5.25a), the aqueous dispersibility of powder M11–34 is observed to be much inferior to that of powder M11–11, and the presence of excess hydrophobic Si–O–C–R groups on the particle surface is found to deteriorate the dispersibility. However, with regard to the thermal-oxidized powders (Fig. 5.25b), the apparent viscosity of slurry M11–34 is found to be lower than that of slurry M11–11, owing to the small amount of Si–O–Si groups of powder M11–34 after calcination, as shown in Fig. 5.26.

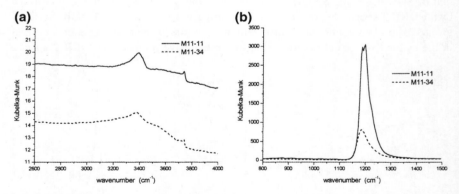

Fig. 5.26 DRIFT spectra of powders M11–11 and M11–34 after calcination: **a** 4000–2600 cm^{-1} and **b** 1500–800 cm^{-1}

(3) Summary

After ball-milling in different liquids, the particle surface group and aqueous dispersibility of Si_3N_4 powders had different alternations. The aqueous dispersibility of the powder treated in de-ionized water showed some improvement because of the decrease in the Si–O–Si surface groups. On the other hand, in the samples treated in ethanol and isopropanol, hydrophobic Si–O–C–R groups were formed on the particle surface, which deteriorated the aqueous dispersibility. After calcination in static air, the liquid-treated samples with identical ball-milling process demonstrated similar particle surface characteristics, and thus, similar aqueous dispersibility.

Ball-milling process had significant effect on the particle surface group and its aqueous dispersibility. Under the conditions of small weight ratio of milling balls to powders (1:1) and low rotary speed (100 rev min^{-1}) of powder M11–11, the Si–OH surface groups reacted with ethanol and formed little amount of hydrophobic Si–O–C–R groups. By increasing the weight ratio of the milling balls to powders and the rotary speed to 3:1 and 400 rev min^{-1} (powder M11–34), respectively, the reaction of the ruptured Si–O–Si surface groups with ethanol was dominant and the formation of large amount of Si–O–C–R structures remarkably depressed its aqueous dispersibility. With regard to the two samples ball-milled in ethanol, the subsequent calcination eliminated the hydrophobic Si–O–C–R groups and improved their aqueous dispersibility to a greater extent. With regard to powder M11–34, most of the Si–O–Si groups were broken during the ball-milling process and were not restored after subsequent thermal oxidation, thus, resulting in the superior aqueous dispersibility in contrast to that of powder M11–11 after calcination.

5.2 Gelcasting of Silicon Nitride Ceramics

5.2.1 Preparation of Silicon Nitride Ceramics with Surface-Coated Silicon Nitride Powder

One of the key steps of gelcasting is the preparation of a highly concentrated ceramic suspension with low viscosity, a characteristic which is strongly dependent on the surface chemistry of the ceramic powder. The surface status and characteristics of the Si_3N_4 particles are relatively complex (Mezzasalma 1996; Greil 1991; Sanchez 1996). To obtain consistent rheological behavior in the slurries from different batches of Si_3N_4 powders, the technical parameters must be continuously modified, which makes the process very difficult to control during industrial production and increases the production costs.

By coating Si_3N_4 with Al_2O_3 and Y_2O_3, we expected different Si_3N_4 powders to exhibit similar colloidal characteristics, resulting in a more consistent forming process. In addition, we also expected an improvement in the dispersion of the powder, because the surface of the coated Si_3N_4 powder takes up some characteristics of the oxide coatings, which have good dispersibility in aqueous media (Fagerholm 1994;

Shih 1996; Fagerholm 1996). A coating layer of the powders can separate the surface of the powders from the aqueous media, which improves the stability of the powders. Additionally, the sintering aids in the coating layer are presumed to cause uniform dispersion of the Si_3N_4 particles. Uniform dispersion can reduce the diffusion distance and improve the sintering rate, and thus, can be especially advantageous for the densification of the covalent-bonded compounds.

5.2.1.1 Preparation Procedure

(1) Surface Coating

The compounds Y_2O_3 and Al_2O_3 are usually coated onto the Si_3N_4 powder as sintering aids.

The two main methods for coating Al_2O_3 and Y_2O_3 onto the Si_3N_4 powder are the hydrolysis of the metal alkoxides (Joshi 1994) and the precipitation of inorganic salts (Kim 1997; Garg 1990; Yang 1999).

For inorganic salt precipitation, $Al(NO_3)_3$ and $Y(NO_3)_3$ are made to react with $(NH_2)_2CO$ to produce insoluble basic salts. The basic salts are then physically or chemically adsorbed onto the surface of the Si_3N_4 particles, forming a coating layer. After calcination, the basic salts decompose, and the oxide coating forms.

The advantages of the precipitation process are the low price of the salt precursor and the feasibility of upscaling the technique. Owing to these advantages, inorganic salt precipitation was used as the coating method in this study.

The Si_3N_4 used for coating was a commercially available powder (Founder High Technology Ceramics Co., Ltd., Beijing, China). All the Si_3N_4 powders mentioned in this chapter were obtained from this source, unless otherwise cited. Table 5.9 shows the physical and chemical characteristics of the powder, according to the analysis by the Shanghai Institute of Ceramics and Tsinghua University.

The process of coating Si_3N_4 particles with Y_2O_3–Al_2O_3 is illustrated in Fig. 5.27. An aqueous solution containing 0.200 mol L^{-1} of $Al(NO_3)_3$ (purity 98.0%, Yili Fine Chemical Corp., Yili, China), 0.276 mol L^{-1} of $Y(NO_3)_3$ (purity 99.0%, Lanzhouhuaxin Chemical Corp., Lanzhou, China), and 50 g L^{-1} of urea (purity 99.0%, Beijing Chemical Reagents Co., Beijing, China) was heated at 95 °C for 100 min and then cooled quickly to room temperature. The resulting clear liquid was termed as the coating concentrate. An aqueous dispersion containing 73.4 g of Si_3N_4 powder, etched previously by a solution of 5 wt% HF, 33 wt% HNO_3, and 62 wt% H_2O to remove surface SiO_2 in 800 ml of deionized water, was agitated ultrasonically for 10 min and then added to 200 ml of the coating concentrate. The mixture, with a pH value of 5.5 ± 0.2, was kept at 50 °C and vigorously stirred in a

Table 5.9 Physical and chemical characteristics of silicon nitride powder used for gelcasting

d_{50} (μm)	α-Phase (%)	O (%)	Fe (ppm)	Al (ppm)	Ca (ppm)	Si + SiO_2 (%)
0.98	93.1	2.2	7	1300	190	1.38

5.2 Gelcasting of Silicon Nitride Ceramics

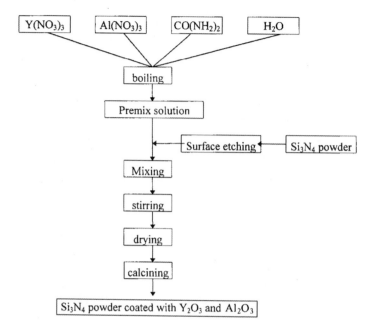

Fig. 5.27 Flowchart of the Y_2O_3–Al_2O_3 coating process

closed flask for up to 48 h, and subsequently dried on a hot plate to obtain it in the form of a powder. The Si_3N_4 powder coated with yttrium–aluminum salts was passed through a sieve (100 mesh) and calcined at 1000 °C in N_2 gas for 2 h, to produce the Si_3N_4 powder coated with 7.5% Y_2O_3 and 2.5% Al_2O_3. A pure precursor of the Y_2O_3–Al_2O_3 coating was also prepared similarly for comparison.

The nature of the coating layers was observed by transmission electron microscopy (TEM).

The chemical compositions of the coated powder and the pure coating material were examined by X-ray diffractometry (XRD; Model D/Max IIIB, Rigaku Co., Ltd., Tokyo, Japan).

The zeta potentials of the powders in a 1 mol m^{-3} KNO_3 solution were measured (Model ZetaPlus, Brookhaven Instruments Corp., Ltd., Worcestershire, U.K.), and diluted HCl and NaOH solutions were used for pH adjustment.

The original or coated powder was dispersed using TMAH [(CH$_3$)$_4$NOH] as a dispersant in a premixed solution prepared by dissolving AM (C$_2$H$_3$CONH$_2$) and MBAM [(C$_2$H$_3$CONH$_2$)$_2$CH$_2$] in deionized water.

The viscosities of the suspensions at pH 10.5 ± 0.2 after 24 h of rolling in polyethylene bottles using alumina balls were examined using a rotary viscometer (Model NXS-11, Chengdu Instrument Factory, Chengdu, China).

The suspensions were rolled to create a good dispersion of the ceramic powders, rather than for milling purposes. Our previous study showed no obvious change in the surface coating of Si_3N_4 whiskers after 24 h of rolling, and hence, we will not discuss the effect of rolling on surface coating in this study.

(2) Gelcasting

The AM and MBAM were applied during gelcasting as monomers for polymerization, initiated by an initiator, ammonium persulfate [$(NH_4)_2S_2O_8$] and accelerated by a catalyst, TEMED. All these reagents were chemically pure. The gelcasting process was similar to that of the previous studies (Young 1991; Omatete 1997). First, the original Si_3N_4 with 7.5 wt% Y_2O_3 and 2.5 wt% Al_2O_3 powder (Jiangsu Wuxian Special Ceramics Plant, Wuxian, China), or a coated Si_3N_4 powder was suspended with TMAH as a dispersant in the premixed solution. The slurry with a solids loading of 45 vol% was degassed for 10 min after rolling for 24 h in polyethylene bottles using alumina balls. The suspension was degassed for another 3 min when the initiator and catalyst were added. All the abovementioned operations were conducted at room temperature. Subsequently, the slurry was cast into a nonporous mold, which was then kept at 70 °C. After the monomers had polymerized, the green bodies were demolded and dried at room temperature under controlled humidity, to avoid cracking and nonuniform shrinkage caused by rapid drying. The microscopic morphology of the green bodies was observed by SEM (Model CSM950, OPTON Co., Ltd., Munchen, Germany).

(3) Gas Chromatography

The Si_3N_4 suspensions with the same contents of coated or mechanically mixed sintering aids for gelcasting used for comparison, were all at a solids loading of 45 vol%. After the gelcasting process, large amounts of macropores were observed in the original Si_3N_4 green bodies, which might be associated with the gas evolved from the reaction of the powders with the dispersant and the monomer-containing premixed solution. To verify this assumption, a laboratory-made gas-gathering device, shown in Fig. 5.28, was designed to collect the gas formed during the reactions. The reagents (ceramic powders, premixed solution, and dispersant) were contained in a gas-tight

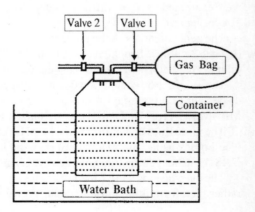

Fig. 5.28 Schematic representation of the gas-gathering device

container. Valves 1 and 2 of the device were initially opened, and the gas bag and the container were vacuumed before valve 2 was closed. The container was then put into a water bath at 80 °C, and the reagents were agitated continuously by a magnetic stirrer. After 30 min, water was breathed into the container via valve 2, until all of the gas in the container had been squeezed into the gas bag. Valve 1 was subsequently closed, and the gas bag was removed from the instrument for the analysis by gas chromatography (Model 3400, Beijing Analytical Instrument Plant, Beijing, China). For comparison, Si_3N_4 powder (SN-E10, UBE Industries, Ltd., Yamaguchi, Japan) was also tested for its gas-discharging reaction.

(4) Sintering and Testing

The coated Si_3N_4 green bodies formed by gelcasting were subjected to binder out and gas-pressure sintering (GPS) in a furnace (Model FPW 180/250-2200-100-SP, KCE Corp., Ro¨dental, Germany). The GPS process was conducted at 1750 °C for 2.5 h under a N_2 pressure of 3 atm (1 atm = 9.8066 × 10^4 Pa), and subsequently at 1900 °C for 2.5 h under a N_2 pressure of 60 atm. The heating and cooling rates were 10 °C min^{-1}.

As the original Si_3N_4 green bodies prepared by gelcasting contained numerous macropores, investigating the effect of the surface coating on the mechanical properties of the Si_3N_4 ceramics by comparing the ceramics gelcasted from the original and the coated Si_3N_4 powders was meaningless. The original Si_3N_4 samples were prepared through cold isostatic pressing, dry-pressed at 1 atm, and then were cold isostatically pressed at 20 atm, for comparison with the gelcasting samples and for the evaluation of the effect of surface coating and gelcasting on the performances of the Si_3N_4 ceramics. The original Si_3N_4 green bodies, with densities of 55% of the theoretical value, were sintered by the same procedure used for the coated and gelcasted bodies.

The bulk densities of the resulting ceramic bodies were measured by the Archimedes' method.

Specimens for the bending test were ground and polished longitudinally to measure 3 × 4×36 mm. Bend tests were performed on 20 specimens using the three-point bend method with a 30-mm span at a cross-head speed of 0.5 mm min^{-1}.

Furthermore, Weibull modulus values were estimated using the two-parameter model.

The fracture surfaces and polished/etched surfaces of the sintering bodies were also observed using SEM.

5.2.1.2 Characteristics of the Coated Silicon Nitride Powder

The micrographs of Si_3N_4 particles coated with Al_2O_3 by the same coating process used in this work have already been illustrated in a previous paper (Tang 1998).

The TEM micrographs of the Y_2O_3–Al_2O_3-coated Si_3N_4 particles in the present work are shown in Fig. 5.29. A layer of precipitates, with a thickness varying from ~5 to 40 nm, is closely attached to the particle surfaces. However, a more careful

Fig. 5.29 TEM micrographs of yttrium oxide–aluminum oxide-coated silicon nitride particles

observation of the figure shows that the surface of the Si_3N_4 powder is not completely covered by the precipitates. In addition, it can be observed that some precipitates make no contact with the Si_3N_4 particles.

The XRD curves (Fig. 5.30) demonstrate that the coated layer is a Y_2O_3–Al_2O_3 layer consisting of Y_2O_3, δ-Al_2O_3, $Y_3Al_2(AlO_4)_3$, and $Y_4Al_2O_9$ resulting from the decomposition of the Y_2O_3–Al_2O_3 precursor.

The zeta potentials of the coated Si_3N_4 particles are shown in Fig. 5.31. Unlike any curve of the pH-dependent zeta potential for Al_2O_3, Y_2O_3, or Si_3N_4, the zeta-potential curve for the coated Si_3N_4 powder exhibited three IEPs, i.e., three "charge" reversals in the pH range. A previous study (Hackley 1994) demonstrated that up to three reversals of the electrokinetic potential can occur when polyvalent metal ions (Al^{3+} and Y^{3+} ions in our work) are present during the acid–base titration of the oxides. Furthermore, the study attributed these three "charge" reversals with an increase in pH to the native substrate (the surface of the Si_3N_4 in the present work),

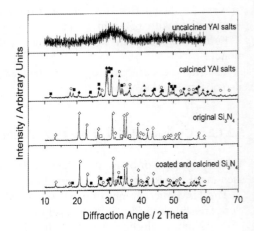

Fig. 5.30 XRD curves of yttrium oxide–aluminum oxide precursor salts and uncoated/coated silicon nitride powders ((♦) Y_2O_3, (▲) δ-Al_2O_3, (○) $Y_3Al_2(AlO_4)_3$, (◊) α-Si_3N_4, (□) β-Si_3N_4, (■) $Y_4Al_2O_9$)

5.2 Gelcasting of Silicon Nitride Ceramics

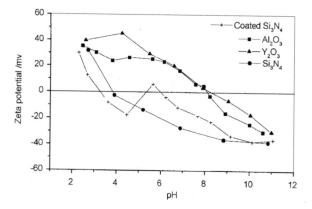

Fig. 5.31 Zeta potentials of some ceramic powders

the onset of surface hydroxide precipitation (Y_2O_3–Al_2O_3 in the present work), and the mixed substrate/precipitate surface, and theorized that the appearance of the three IEPs indicates inadequate coverage of Y_2O_3–Al_2O_3 on the surface of the Si_3N_4 particles, as observed using TEM.

At any rate, when the original or the coated Si_3N_4 was suspended in an aqueous solution, the absolute values of its zeta potentials were higher at a high pH, inducing higher electrostatic interparticle repulsion, which is important for establishing a colloidally stable suspension. Thus, both the original and the coated powders easily got dispersed in an alkaline solution. Figure 5.32 shows the flow curves of the slurries prepared from the two powders. At pH 10.5 ± 0.2, the viscosity of the coated Si_3N_4 slurries with the same solids loading as that of the uncoated Si_3N_4 slurry decreased significantly. When the solids loading were increased to 50.6 vol%, the viscosity of the coated Si_3N_4 slurry was still much lower than that of the uncoated Si_3N_4 slurry with a solids loading of 45 vol%.

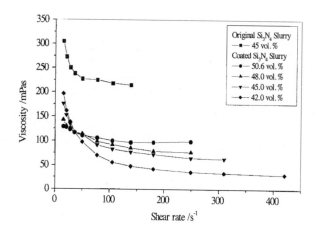

Fig. 5.32 Viscosity comparison of the suspensions of original and coated silicon nitride powders

However, the increase in the rheological properties of the coated Si_3N_4 slurries has not been adequately explained by the zeta-potential data presented in Fig. 5.31, where the zeta-potential values for the coated and original Si_3N_4 powders are almost identical at pH 10.5 ± 0.2. The electrokinetic behavior of a powder is only one of the contributors to the colloidal characteristics of the powder. In the present work, changes in the particle-size distribution and shape of the Si_3N_4 powder may contribute to the increase in the rheological properties of the coated Si_3N_4 slurry; however, confirmation of the exact contributors requires further investigation.

5.2.1.3 Structure and Characteristics of Gelcasted Silicon Nitride Green Bodies

Following the solidification of the slurry prepared from the uncoated Si_3N_4 powder, numerous millimeter-sized macropores were found in the green bodies, as shown in Fig. 5.33a. As the slurry had been vibrated and vacuumed before it got solidified, there were no macroscopic bubbles. Therefore, it can be concluded that the macropores in the green bodies were not introduced by the bubbles in the slurry. Thus, the macropores must have formed during the solidification of the slurry. The relative quantity of the ingredients in the Si_3N_4 slurry indicates that the following three gas-discharging reactions may have occurred in the system:

Hydrolysis of Si_3N_4:

$$Si_3N_4 + 6H_2O = 3SiO_2 + 4NH_3 \uparrow \tag{5.13}$$

Oxidation of Si_3N_4 by the initiator:

$$Si_3N_4 + 6S_2O_8^{2-} + 12OH^- = 3SiO_2 + 2N_2 \uparrow + 12SO_4^{2-} + 6H_2O \tag{5.14}$$

Hydrolysis of free Si:

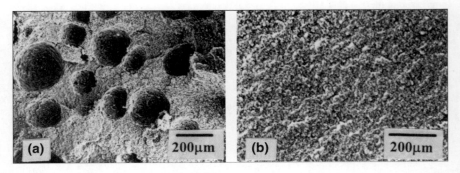

Fig. 5.33 SEM micrographs of the green bodies produced by gelcasting, as prepared from **a** original silicon nitride powder and **b** coated silicon nitride powder

5.2 Gelcasting of Silicon Nitride Ceramics

$$Si + 4OH^- = SiO_4^{4-} + 2H_2 \uparrow \tag{5.15}$$

All these reactions are strongly enhanced in the aqueous solutions with high pH values.

During gelcasting of the original Si_3N_4, a high content of the basic dispersant, TMAH, in the Si_3N_4 suspensions produced large quantities of bubbles in the slurries, even after milling only for several hours. An earlier study (Raider 1936) reported that Si_3N_4 can be hydrolyzed in an aqueous solution even at room temperature and neutral pH values, which confirms Reaction (5.13). However, this hydrolysis was found to pose an intrinsic difficulty for gelcasting Si_3N_4-containing aqueous suspensions. During our experiments, we continuously vacuumed the slurry-containing flask to remove the bubbles from the suspensions before casting; however, this exercise failed to create a macropore-free original Si_3N_4 green body after solidification. To determine the macropore-forming reaction during the solidification of the slurry system, we conducted a series of gas-gathering experiments. The experimental arrangement and results are shown in Table 5.10.

Ammonia was not detected in any of the experiments, possibly because of its relatively high solubility in water (Zhou 1980a) and the limited resolution of the analytical instrument used; however, the production of ammonia during the experiments was confirmed by its smell.

Table 5.10 Results of gas chromatography experiments

Experiment No.	Reactant (reacting at 80 °C for 30 min)		pH	H_2 (ml)	N_2 (ml)
	Powder	Solution			
1	Si_3N_4/40 g	H_2O/250 ml, $(NH_4)_2S_2O_8$/5 g	3.36	0	30.8
2	Si_3N_4/40 g	H_2O/250 ml, $(NH_4)_2S_2O_8$/0.04 g	6.60	0	0
3	Si_3N_4/40 g	H_2O/250 ml, TMAH[a]/12.5 ml	12.50	96.5	0
4	Si/0.25 g	Premix[b]/250 ml	4.02	20.2	0
5	Si/0.25 g	H_2O/250 ml, TMAH/6.0 ml	12.20	146.4	0
6	UBE Si_3N_4/40 g	H_2O/250 ml, TMAH/12.5 ml	12.50	0	0
7	Oxidized $Si_3N_4^c$/40 g	H_2O/250 ml, TMAH/12.5 ml	12.50	0	0
8	Coated $Si_3N_4^d$/40 g	H_2O/250 ml, TMAH/12.5 ml	12.50	0	0

[a]TMAH: Tetramethyl ammonium hydroxide [$(CH_3)_4NOH$], 10 wt%
[b]Premix solution: $C_2H_3CONH_2$ (AM), $(C_2H_3CONH)_2CH_2$ (MBAM), and deionized water
[c]Oxidized at 800 °C for 2 h
[d]Coated Si_3N_4, coated using Al_2O_3 and Y_2O_3

Large amounts of N_2 gas were detected in Experiment No. 1, suggesting that Reaction (5.14) could occur in the system at a high concentration of the initiator $(NH_4)_2S_2O_8$. However, during the practical gelcasting process, the actual amount of the initiator added to the slurry was much lower than that used in Experiment No. 1, which could generate a negligible amount of discharged N_2 gas, as certified by Experiment No. 2. Thus, Reaction (5.14) cannot be considered as a major cause for the high amount of macropores in the gelcasted Si_3N_4 green bodies. Moreover, Experiment No. 3 illustrates that Si_3N_4 powder can also react with TMAH [$(CH_3)_4NOH$] water solution and discharge H_2 gas at 80 °C. A small amount of free Si was present in the Si_3N_4 powder used in our experiments. Therefore, the free Si is expected to hydrolyze the subsequent Reaction (5.14), releasing H_2 gas.

Experiment No. 4 indicates that Si powder could get hydrolyzed in an aqueous solution at 80 °C, discharging a small amount of H_2 gas. Experiment No. 5 further indicates that more H_2 gas was produced in the system when a basic dispersant was added. The Si_3N_4 powder used for gelcasting was prepared by nitrification of the Si powder. The extent of the reaction did not reach 100%, and thus, more or less free Si remained in the Si_3N_4 powder. The content of free Si contained in the powder provided by the manufacturer was more than 1 wt%. The UBE Si_3N_4 powder was prepared by the decomposition of imide, and thus, contained no free Si. The absence of H_2 gas in Experiment No. 6 certified the difference between the two powders. Finally, no H_2 gas was detected in Experiment No. 7 and 8, when the Y_2O_3–Al_2O_3-coated powder or oxidized Si_3N_4 powders were heated (800 °C for 2 h in air) in the basic aqueous solution, indicating that surface modification caused by the coating and surface oxidation of Si_3N_4 could effectively restrain the deleterious gas-discharging reactions in the slurry system.

Several early investigations (Hackley 1997; Popper 1960; Ezis Ezis 1974) have reported that Si has a tendency to get hydrolyzed in water, which is an exothermic reaction, and that difficulties are associated with Si processing in an aqueous environment. Evolution of H_2 gas was observed when Si was mixed with an aqueous solution. The degree of reactivity increased as the particle size decreased and the alkalinity increased, and severe gassing was reported for pH values greater than 10. According to the present authors' knowledge, no earlier study has reported the reaction between free Si in the Si_3N_4 powder and an alkaline solution, partly because of the fact that free Si in Si_3N_4 powder is normally much lesser, and thus, the reaction is slow and inconspicuous, and also partly because Si_3N_4 slurries with pH values greater than 10 were not heated to temperatures as high as 70–80 °C in the early researches.

The results presented in Table 5.10 indicate that the hydrolysis of free Si in Si_3N_4 powder in the basic dispersant-containing aqueous solution is one of the major causes of macropores formation in the Si_3N_4 green bodies obtained in this study. However, it is not clear whether the reaction between Si_3N_4 and TMAH in the solution was the other major cause responsible for macropores formation, because the amount of ammonia formed in our aqueous system was difficult to detect, owing to its high solubility in water.

5.2 Gelcasting of Silicon Nitride Ceramics

As the coating layer of Y_2O_3–Al_2O_3 effectively restrained the deleterious gas-discharging reactions in the slurry system, macropore-free green bodies were obtained via gelcasting, as shown in Fig. 5.33b. According to our analysis, three mechanisms contributed to the elimination of macropores in the green bodies:

(a) The coated Y_2O_3–Al_2O_3 layer acted as a barrier, retarding contact and reaction between the basic solution and free Si in the Si_3N_4 powder.
(b) The content of free Si was decreased by hydrolysis before gelcasting, when the Si_3N_4 powder was treated in the aqueous coating concentrate at 50 °C for 48 h.
(c) The content of free Si was decreased by nitridation before gelcasting, when the yttrium–aluminum salt-coated Si_3N_4 powder was calcined in N_2 gas at 1000 °C for 2 h.

Further investigation is necessary to identify the exact roles of these three mechanisms.

5.2.1.4 Improvement of Mechanical Properties

Both Al_2O_3 and Y_2O_3 can be simultaneously coated onto the surface of Si_3N_4 particles in the same proportion and quantity as used for sintering aids. With this type of coated powder, complex-shaped green bodies can be gelcasted and then sintered into the final components.

The relative densities of the sintering samples prepared by either gelcasting the coated Si_3N_4 or by cold isostatic pressing of the original Si_3N_4 were 98% of the theoretical density. The mechanical characteristics of the Si_3N_4 ceramic bodies prepared by gelcasting the coated Si_3N_4 powder were much better than those of the bodies prepared by cold isostatic pressing of the original powder. The room-temperature bending strength and Weibull modulus of the GPS gelcast samples produced from the Y_2O_3–Al_2O_3-coated Si_3N_4 powder via gelcasting were 840.57 MPa and 15.55, respectively, whereas those of the materials prepared from the uncoated powder by conventional cold isostatic pressing were 582.70 MPa and 6.04, respectively.

Micrographs of the polished and etched surfaces of the two samples, shown in Fig. 5.34, may explain the discrepancies in their mechanical performances. Figure 5.34 shows that the grain sizes of the cold isostatically pressed samples were nonuniform and that some of the particles were very large and obviously spherical, which may be the main reason for its low bending strength. On the other hand, the grains in the coated and gelcast sample were spindle-shaped, indicating that the coating promoted liquid formation and grain rearrangement but not excessive grain growth. Apparently, homogeneous distribution of the sintering aids is considered to be beneficial for the formation of green bodies with uniform microstructures by coating and gelcasting, and may improve the mechanical properties of the sintered bodies.

Differences in the grain shape can also influence the rupture mode of the samples, as shown in Fig. 5.35. The large grains in Fig. 5.35a can be easily removed, because they have large interface areas and high interfacial strengths. Thus, when the angle

Fig. 5.34 Micrographs of the polished and etched planes of different GPS silicon nitride samples prepared from **a** uncoated powder, by dry and cold isostatic pressing, and **b** coated powder, by gelcasting

Fig. 5.35 Micrographs of the fracture surfaces of different GPS silicon nitride samples: **a** uncoated powder, produced by dry and cold isostatic pressing and **b** coated powder, produced by gelcasting

between the interface and the crack plane is relatively large, the crack will propagate mostly through the interface. When the angle is small, the crack will spread mostly along the grain boundary, and as a result, the grains will flake off. The grains in Fig. 5.35b are more spindle-shaped, and thus, more grains are found to flake off and pull out. All these phenomena increase the distance and energy dissipation during crack spreading, thus increasing the toughness of the sample. Therefore, the spindle shape of the grains may be another reason, besides grain uniformity, for the high Weibull modulus of the coated samples.

5.2.1.5 Conclusions

The following conclusions can be drawn from the abovementioned investigations.

(1) A layer of Y_2O_3–Al_2O_3 was coated onto the surface of the Si_3N_4 particles by the precipitation of inorganic salts. The electrokinetic and colloidal characteristics of the Si_3N_4 powder were significantly changed by the coating layer, and the dispersion of the Y_2O_3–Al_2O_3-coated Si_3N_4 powder was significantly increased over that of the original powder.

(2) For the first time, we found that free Si in the Si_3N_4 powder may pose a severe problem to the colloidal-forming process. In our present investigation, numerous millimeter-sized macropores were found in the green bodies of the original Si_3N_4 powder. The macropores were observed to have been formed by the H_2 gas discharged as a result of a reaction between the free Si in the Si_3N_4 powder and the heated basic solution, and this reaction was constrained by the Y_2O_3–Al_2O_3 coating layer, resulting in macropore-free and uniform Si_3N_4 green bodies prepared by gelcasting. Thus, the Y_2O_3–Al_2O_3 layer coating also improved the stability of the Si_3N_4 powder in an aqueous solution.

(3) As the Y_2O_3–Al_2O_3 coating also acted as a sintering aid, the surface coating as well as the gelcasting process significantly improved the mechanical properties of the Si_3N_4 ceramics. The room-temperature bending strength and Weibull modulus of the GPS gelcast samples produced from the Y_2O_3–Al_2O_3-coated Si_3N_4 powder were 840.57 MPa and 15.55, respectively, which were much higher than those of the cold isostatically pressed samples prepared from the original Si_3N_4 powder. Surface coating yielded a uniform distribution of the sintering aids; gelcasting ensured uniformity of the green bodies, and at the same time, surface contact between the Si_3N_4 powder and the sintering aids seemed to improve the sinterability of the powders.

5.2.2 Preparation of Silicon Nitride Ceramics with Surface-Oxidized Silicon Nitride Powder

A method of oxidation to improve water-based gelcasting of Si_3N_4 was studied. This method was found to be not only effective but also easy to utilize, because an effective coating layer of silica (SiO_2) was formed on the surface of the Si_3N_4 powder.

The Si_3N_4 powder (Beijing Founder High Technology Ceramics Corporation) with an average size d_{50} of 0.6 μm was used in this study (Fig. 5.36). In addition, deionized water with a conductivity of 1.02 μS cm^{-1}, AM as the monomer, MBAM as the crosslinker, $(NH_4)_2S_2O_8$ as the initiator, and TEMED as the catalyst were also employed.

Photoelectron spectra were obtained using an ESCALAB220i-XL electron spectrometer with a base pressure of 1×10^{-9} mbar and an EP300 electron spectrometer

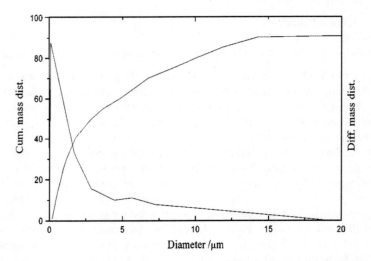

Fig. 5.36 Particle-size cumulative and differential mass distribution of the silicon nitride powder

with a base pressure of 1×10^{-8} mbar. The apparent viscosity of the Si_3N_4 suspensions was measured by rotating viscosimeter (model NSX-11, Chengdu Instrument Plant, China). The ion concentration of the suspension was measured by IONPACDX-100 ion chromatography instrument, and the ion conductivity of the suspensions was measured by the ion conductivity meter (model DDS-II A_T, Shanghai Leici Instrument Plant, China). The bending strength of the samples was measured through three-point bending method. The micrograph of the gelcast green bodies was observed using a CSM950 SEM from OPTON Corporation of Germany.

5.2.2.1 The influence of Surface Oxidation on the Composition and Surface Status of the Silicon Nitride Powder

Under the conditions of high temperature and high oxygen pressure, the following reactions were observed to take place:

$$Si_3N_4 + 3O_2 = 3SiO_2 + 2N_2 \tag{5.16}$$

After the formation of SiO_2, the reaction on the interface between the SiO_2 layer and Si_3N_4 powder was controlled by the diffusion of oxygen across the SiO_2 layer. In the meantime, the following reactions occurred:

$$Si_3N_4 + SiO_2 = 2Si_2N_2O \tag{5.17}$$

$$2Si_2N_2O + 3O_2 = 4SiO_2 + 2N_2 \tag{5.18}$$

5.2 Gelcasting of Silicon Nitride Ceramics

The Si_3N_4 powders were oxidized in the molybdenum silicide furnace, by maintaining the temperature at 600 °C, 800 °C, 850 °C, and 900 °C, respectively, for 2 h. The weights of the oxidized powders were compared with those of the raw powders, and the results are shown in Fig. 5.37. Thermogravimetric analysis (TGA) curves of Si_3N_4 at 600 °C, 800 °C, 850 °C, and 900 °C, respectively, are shown in Fig. 5.38.

It can be observed through Fig. 5.37 that the oxygen content from 600 to 850 °C increases slowly, but it rises sharply when the oxidizing temperature reaches 900 °C. Furthermore, Fig. 5.38 indicates that the gain in the oxidation speed at 900 °C is relatively high. This is because the transformation from α-quartz ($\rho = 2.533$ g cm^{-3}) to α-tridymite ($\rho = 2.228$ g cm^{-3}) at 870 °C takes place in the SiO_2 layer. Owing to the change in the density, the SiO_2 film layer cracks, and the oxygen infiltrates through the crack, leading to more active reactions. At 900 °C, the sharp oxidation reaction results in a large increase in the SiO_2 content in the Si_3N_4 powder. In fact, the actual increase in the SiO_2 content may be more than that measured in our study, because the weight loss during the experiment, which can be attributed to the combustion of oxygen and carbon in the Si_3N_4 powder (Castanho 1997a), is not considered. As is well known, too much of glass phase derived from excess of SiO_2 will remarkably degrade the mechanical properties of Si_3N_4 at elevated temperature. Hence, it is necessary that the Si_3N_4 powder be oxidized below 900 °C.

Figure 5.39 presents the X-ray photoelectron spectrometer (XPS) results of the oxidation powder at 850 °C. In Fig. 5.39a, it can be observed that the Si (2p) peak divides into two, one at 102.30 eV representing Si_2N_2O, and the other at 103.49 eV representing SiO_2. In Fig. 5.39b, N (1 s) peak at 397.91 eV represents Si_2N_2O

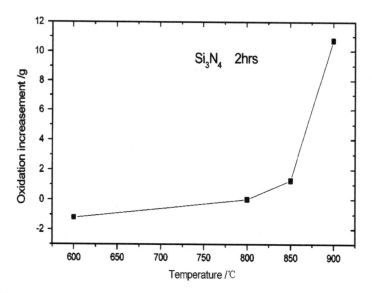

Fig. 5.37 Increase in the oxidation of silicon nitride powder oxidized in the molybdenum silicide furnace for 2 h

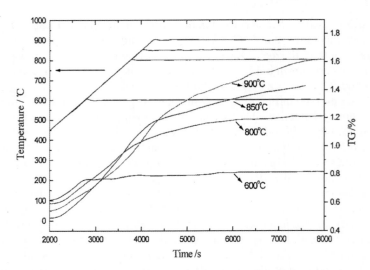

Fig. 5.38 TGA curves of the oxidizing silicon nitride powder at different temperatures

(Castanho 1997a, b; Brow 1986). Thus, it can be concluded that the powder surface mainly consists of Si_2N_2O and SiO_2, and that the surface modification based on oxidation can yield a layer of coating.

5.2.2.2 Influence of Surface Oxidation on the Colloidal Characteristics of the Slurry and Structure of the Green Body

In the gelcasting process of Si_3N_4, the presence of some cations in the slurry, namely anti-ions, are very much disadvantageous to rheologies of the slurries (Kitahara 1992). Impurity ions in the Si_3N_4 powder can be removed by washing and ion-exchange resin, but these methods are expensive and time-consuming. It has been found in our study that oxidizing at high temperature could easily and effectively reduce the ion concentration in the slurry. In our study, the original Si_3N_4 powder and several kinds of oxidized Si_3N_4 powders were dispersed in deionized water, respectively, and the corresponding homogenous suspensions were acquired. After a few days, they were centrifuged and the ion conductivity of the watery solution on top of the suspensions was measured. Figure 5.40 shows the curves of the ion conductivity versus time. With the rise in the oxidizing temperature, the ion conductivity of the watery solutions decreased significantly, indicating that the ion concentration in the slurry decreased remarkably. The decrease may be attributed to the isolation of the SiO_2 layer and the change in the formation of the ions on the surface of Si_3N_4 powder. Table 5.11 presents the variations in the K^+ and NH_4^+ concentration versus time. It can be seen from the table that the concentration of both K^+ and NH_4^+ decreased after oxidation of the powder. The decrease in the ion concentration was greater with

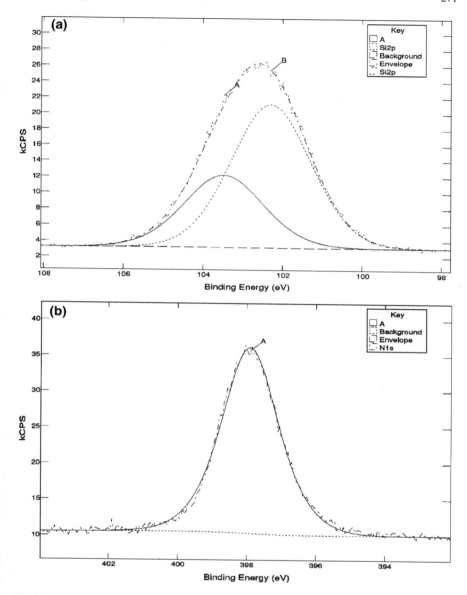

Fig. 5.39 XPS spectrum of the silicon nitride powder oxidized at 850 °C. **a** Si (2p) and **b** N (1 s) peak

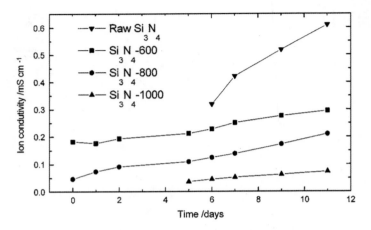

Fig. 5.40 Curves of the ion conductivity of the silicon nitride suspensions versus time

Table 5.11 Variation of K^+ and NH_4^+ concentration versus time

	Original Si_3N_4 powder (0 day)	Original Si_3N_4 powder (7 days)	Si_3N_4 powder oxidized at 600 °C (11 days)	Si_3N_4 powder oxidized at 800 °C (11 days)
K^+(ppm)	3.47	3.43	1.13	0.59
NH_4^+(ppm)	24.8	92.73	10.57	1.21

the increasing oxidizing temperature. The decrease in the K^+ concentration may be owing to the volatilization of the potassium salts and the formation of insoluble solid solution. The concentration of NH_4^+ decreased obviously because of the decomposition of the ammonium salts and the isolating function of the oxide layer on the Si_3N_4 particles. In addition, Fig. 5.40 indicates that the ion conductivity of the oxidized powder varies relatively smoothly, while that of the raw powder varies rather sharply. Thus, it can be concluded that a homogeneous coating was formed on the surface of the oxidized powder, effectively retraining the deleterious reaction on the surface of the Si_3N_4 powder, and that after oxidizing, the improvements in the rheologies of the Si_3N_4 slurries can be partly attributed to the decrease in the impurity ions.

Figure 5.41 shows the viscosity at different shear rates of the slurries of the Si_3N_4 powders oxidized at different temperatures and with the same solids loading of 42 vol% (the solid density was considered as 3.18 g cm^{-3}). It can be observed from the figure that below 850 °C, the suspension viscosity decreased along with the increase in the oxidizing temperature, and that above 850 °C, the suspension viscosity increased again, because of the sharp increase in the SiO_2 content and the resulting increase in the solids volume loading. An oxidizing temperature of 850 °C was observed to be ideal for obtaining slurries with low viscosity.

As stated earlier, some gas-discharging reactions between the Si_3N_4 particles and the medium occur when the raw Si_3N_4 powder comes into contact with the medium.

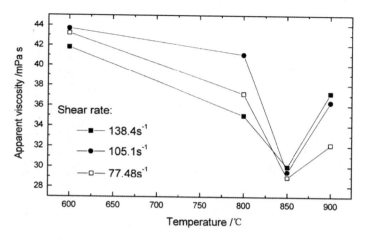

Fig. 5.41 Viscosity of the suspensions of oxidized silicon nitride powders at different temperatures

In the gas-gathering experiment (Dai 1999), harmful reactions were detected in the slurry of the unprocessed Si_3N_4 powder. However, no reaction was detected between the oxidized Si_3N_4 powder and the basic medium, indicating that surface oxidation could effectively restrain the deleterious gas-discharging reactions in the slurry system. Figure 5.42 shows the SEM micrographs of the green body gelcasted from (a) the original powder and (b) the oxidized powder. It can be seen from Fig. 5.42 that the former had notable bubbles, while the latter was homogeneous without any bubbles. Thus, the strength of the green bodies after surface modification was greatly improved to about twice as much as that without surface modification, reaching about 20 MPa.

5.2.2.3 Influence of Surface Oxidation on Sintering

With the increase in the oxygen content in the Si_3N_4 powder after oxidation, the liquid phase in the process of sintering also increased. This influenced the mechanical property and microstructure of the samples. The mechanical properties of the gelcast samples of the oxidized powder as well as the dry-pressed and then isostatically-pressed samples are shown in Table 5.12, and their microstructures are shown in Fig. 5.42.

It can be observed from Table 5.12 that the mechanical property of the gelcast samples is much better than that of the dry-pressed and then isostatically-pressed samples. However, in Fig. 5.43, no obvious difference in the microstructure can be observed. This indicates that the increase in the liquid phase during sintering did not lead to excessive growth of the grains. The difference in this mechanical property may be owing to the defects caused by inhomogeneity in the dry-pressed and then isostatically-pressed samples, as well as their different grain-boundary phases.

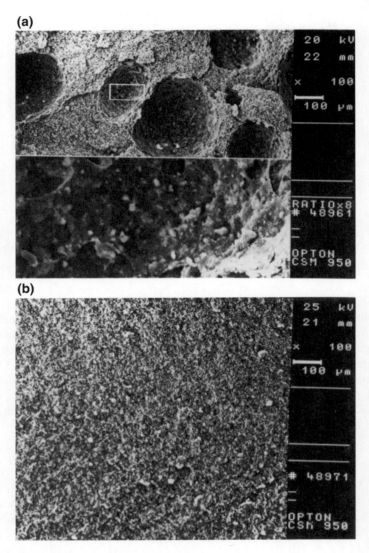

Fig. 5.42 Micrographs of the gelcast green bodies from the **a** raw silicon nitride powder and **b** oxidized powder

5.2.2.4 Summary

(a) After oxidizing in the air, a layer of SiO_2 and Si_2N_2O was formed on the surface of the Si_3N_4 particles and their oxygen content increased significantly, which increased the liquid phase in sintering. However, the mechanical properties of

5.2 Gelcasting of Silicon Nitride Ceramics

Table 5.12 Comparison of the mechanical property of different samples

	GPS	
	Bending strength at room temperature $\sigma_{f,\ RT}$(MPa)	Weibull modulus m
Gelcasting of oxidized powder	785.16	11.78
Dry-pressed and isostatically–pressed sample	582.7	6.04

Fig. 5.43 Micrographs of different samples: **a** Oxidation + gelcasting + GPS and **b** Dry press + isostatic press + GPS

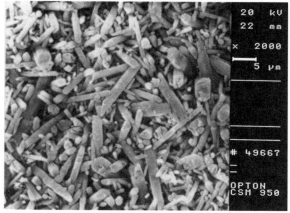

(a) Oxidation+gelcasting+GPS

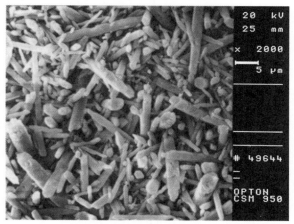

(b) Dry press+isostatic press+GPS

the gelcast samples with the oxidized powder were found to be much better than those of the dry-pressed and then isostatically-pressed samples.
(b) After oxidizing, the impurity ions in the Si_3N_4 suspensions decreased. The effect was more obvious when the oxidizing temperature was higher.
(c) After surface oxidation, the viscosity of the Si_3N_4 slurry sharply decreased and the lowest viscosity was obtained when the Si_3N_4 powder was oxidized at 850 °C for 2 h.
(d) After surface oxidation, the oxide layer on the particle surface restrained the reaction between the free Si in the Si_3N_4 powder and basic solution, and as a result, homogeneous gelcast green bodies without macroscopic pores were obtained.

5.2.3 Preparation of Silicon Nitride Ceramics Using Combination Processing

This study aims to produce high-reliable silicon nitride ceramics via a processing strategy using a combination of coating, oxidizing, gelcasting, cold isostatic pressing, and GPS.

5.2.3.1 Experimental Procedure

(1) Materials and Reagents

The Si_3N_4 powder used in this study (H. C. Starck, M-11, Germany) was 91.7% β-phase. It exhibited a specific surface area of 12.85 $m^2\ g^{-1}$ and a mean particle size of 430 nm. The major impurities in the powder were C and O with 0.20 and 1.44 wt%, respectively, and a total metal content (Fe + Al + Ca) of 0.058 wt%. The compounds, $Y(NO_3)_3.6H_2O$ and $AlCl_3.6H_2O$, coprecipitated with $NH_3.H_2O$, were applied as the precursors of Y_2O_3–Al_2O_3 sintering aids coated onto the surface of the Si_3N_4 particles. Deionized water was employed as a reacting medium, followed by alcohol as a washing liquid in the coating process. Powders of yttria and alumina (Jiangsu Wuxian Special Ceramics Plant, China) with a purity of 99% were used as the sintering aids for the uncoated Si_3N_4 powder. Furthermore, AM ($C_2H_3CONH_2$) and MBAM [$(C_2H_3CONH)_2CH_2$] were used in the gelcasting process as monomers for polymerization, which was initiated by an initiator, ammonium persulfate [$(NH_4)_2S_2O_8$], and accelerated by the catalyst, TEMED. All the materials and reagents used were chemically pure.

(2) Surface Modification of the Silicon Nitride Powder

To adjust the colloidal characteristics of the Si_3N_4 powder and the properties of the resultant ceramics, the as-received powder was coated with its sintering aids and then oxidized in air.

5.2 Gelcasting of Silicon Nitride Ceramics

The procedure for coating the Si_3N_4 particles with Y_2O_3–Al_2O_3 sintering aids is presented in Fig. 5.44. The precursors $Y(NO_3)_3 \cdot 6H_2O$ and $AlCl_3 \cdot 6H_2O$ were first dissolved in deionized water at a concentration of 0.4 mol L^{-1}, followed by 2 h of stirring. The resulting clear liquid was called the "coating solution". Second, the as-received Si_3N_4 powder was dispersed in deionized water to prepare an aqueous Si_3N_4 suspension with solids loading of 10 wt% and a basic pH of 10, adjusted by the addition of $NH_3 \cdot H_2O$. The suspension was then stirred and ultrasonically agitated for 4 h to achieve high homogeneity. Subsequently, a certain amount of coating solution was dripped into the suspension at a rate of 10 drops min^{-1}, which was vigorously stirred and kept constant at a pH of 10 by the simultaneous addition of

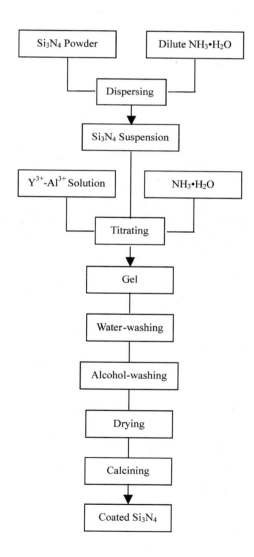

Fig. 5.44 Procedure for coating silicon nitride powder with sintering aids

concentrated $NH_3 \cdot H_2O$. The mixture was stirred for another 30 min after the coating solution was completely added. Nitrate and chlorine ions as well as deionized water were totally eliminated by washing and filtering for four times with deionized water, and for another four times with alcohol. The resultant Si_3N_4 powder coated with Y–Al hydroxides was then dried at 80 °C for 24 h and passed through a sieve (100 mesh) followed by calcination in N_2 atmosphere for 2 h. Pure Y–Al hydroxides were also prepared, analyzed by differential thermal analysis (DTA), calcined at different temperatures, and detected using XRD, to determine the proper calcination temperature for the Si_3N_4 powder. The content of the sintering aids was adjusted to 7.5 wt% Y_2O_3 and 2.5 wt% Al_2O_3. After calcination, the coated Si_3N_4 powder was sieved with a 100 mesh and oxidized at 600 °C for 3 or 6 h in air, to reach a certain oxygen content on the surface of the powder which corresponds to a silica layer. For comparison, the as-received Si_3N_4 powder was also oxidized at 600 °C for 3 h in air.

(3) Forming and Sintering

Gelcasting was employed to obtain Si_3N_4 green bodies with high homogeneity, which can be further improved by subsequent cold isostatic pressing. The gelcasting process used was similar to that used in previous studies (Young 1991; Omatete 1997). First, the Si_3N_4 powder with 7.5 wt% Y_2O_3 and 2.5 wt% Al_2O_3 was suspended in the premixed solution of AM and MBAM with TMAH as the dispersant. The suspensions with a pH value of 10.5 ± 0.2 were degassed for 5 min after rolling for 24 h in polyethylene bottles using alumina balls as the milling media. The slurry was degassed for another 3 min when the initiator and catalyst were added. All the abovementioned operations were carried out at room temperature. Subsequently, the slurry was cast into a nonporous mold, which was then kept at 60–80 °C. After consolidation, the green bodies were demolded and dried under controlled humidity to avoid cracking and nonuniform shrinkage owing to rapid drying. Subsequent binder burnout was carried out at 600 °C for 3 h in air, followed, if necessary, by cold isostatic pressing for 1 min at a pressure of 200 MPa.

GPS of the green bodies was carried out in a KCE furnace (FPW 180/250-2200-100-SP, KCE Corp., Germany). The GPS schemes were as follows: 1750 °C for 1.5 h under the N_2 pressure of 0.3 MPa and then 1900 °C for 1.0 h (sample SN5), 1.5 h (samples SN1–SN4), 2.0 h (sample SN6), or 2.5 h (sample SN7), under the N_2 pressure of 6 MPa with both heating and cooling rates of 10 °C min^{-1}.

(4) Characterization and Reliability Evaluation

The transformation temperature of yttrium–aluminum hydroxides to their oxides was tested using TGA/DTA (Setaram PC92, France). XRD (D/Max IIIB, Rigaku, Japan) was employed to determine the completeness of the transformation. The nature of the coating layers was observed by TEM (JEM-200CX, Japan). The oxygen content was measured by inert gas pulse, while the viscosity of the Si_3N_4 suspensions with different solids loading was examined by a rotational rheometer (MCR300, Physica Corp., Germany).

5.2 Gelcasting of Silicon Nitride Ceramics

Bulk densities of the resulting green bodies and ceramic bodies were measured using the Archimedes' principle. The microscopic morphology of the green bodies and the fracture surfaces were also observed using SEM (JSM-6301F, Japan). Specimens for bend tests with the final dimensions of 3 × 4×36 mm were ground and polished longitudinally with diamond pastes to a finish of 0.5 μm and lightly beveled on the long edges of the tensile surface. Bending tests were performed on 20 specimens per batch using the three-point bending method with a 30-mm span at a cross-head speed of 0.5 mm min^{-1}. Weibull moduli were estimated based on the two-parametric Weibull distribution.

5.2.3.2 Influence of Surface Modification on the Colloidal Properties of Silicon Nitride Powder

Coating, followed by oxidation, is applied to modify the surface characteristics of the Si_3N_4 powder, and thus, its rheological performance. For the coating process, a proper calcination procedure should be determined for the transformation of Y–Al hydroxide to Y–Al oxide on the surface of the powder. For this purpose, DTA and XRD were selected. The DTA curve of pure Y–Al hydroxides revealed an exothermic peak at 892 °C, which corresponded to the transformation temperature (Fig. 5.45). However, calcination of the hydroxides at 900 °C for 2 h in N_2 atmosphere failed to produce crystalline oxides; only when calcined at 1000 °C for 2 h, the resulting products were clearly crystalline (Fig. 5.46).

The TEM micrographs of Y–Al oxide-coated Si_3N_4 particles are shown in Fig. 5.47. It can be observed from this figure that a layer of nano-sized precipitates is closely attached to the surface of large Si_3N_4 particles and the thickness of the precipitates is varied from about 5 to 40 nm. A more careful observation of the figure shows that the surface of small Si_3N_4 particles is not completely covered by precipitates. This is a typical feature for the coating of sub-micron powder with nano-sized precipitates, as the smallest Si_3N_4 particles are also in the nano-size regime.

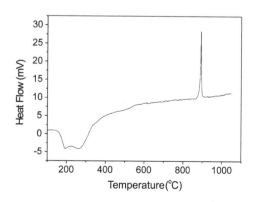

Fig. 5.45 TGA curve of transformation from yttrium–aluminum hydroxide to yttrium–aluminum oxide

Fig. 5.46 XRD patterns of yttrium–aluminum hydroxides calcined at different temperatures

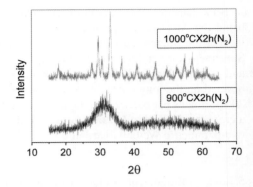

Fig. 5.47 TEM micrograph of the coated silicon nitride particles

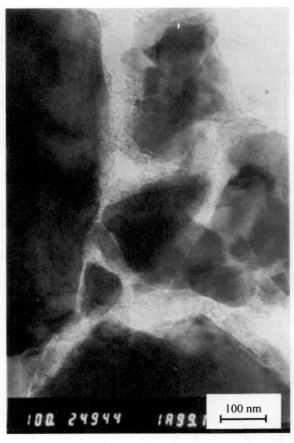

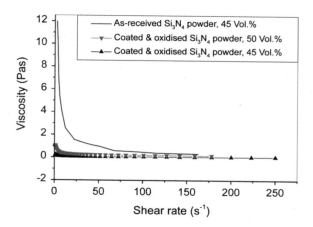

Fig. 5.48 Rheological curves for the suspensions of as-received and surface-modified (coated and oxidized) silicon nitride powders

However, colloidal characteristics and rheological properties of the surface-modified powder are still obviously adjusted.

As reported in previous studies (Huang 2001; Tang 1998), the zeta potentials of the coated Si_3N_4 particles are rather complex. Unlike any curve of the pH-dependent zeta potentials for Al_2O_3, Y_2O_3, or Si_3N_4, the zeta potential for the coated Si_3N_4 powder possesses three IEPs, i.e., three "charge" reversals in the pH range, indicating an inadequate coverage of Y–Al oxides on the surface of Si_3N_4 particles, as observed by TEM. However, the dispersibility of the powder could be improved by coating. Earlier studies also demonstrated that the dispersion of Si_3N_4 powder is enhanced after being oxidized below 850 °C in air. In the current study, it was found that the dispersibility of the coated and oxidized (600 °C × 3 h in air) powder is improved when compared with that of the as-received powder (Fig. 5.48). Furthermore, at a pH of 10.5 ± 0.2, the viscosity of the slurries of the coated and oxidized Si_3N_4 powder with the same solids loading was greatly decreased when compared with the original Si_3N_4 slurry. When the solids loading was increased up to 50 vol%, the viscosity of the coated and oxidized Si_3N_4 slurry was still much lower than that of the as-received Si_3N_4 slurry, although it had a lower solids loading of 45 vol%. The reason for this observation is complex, but it is commonly believed that the oxides on the surface of the non-oxides such as Si_3N_4 lead to surface properties similar to those of the oxides that exhibit good dispersibility in the aqueous media (Fagerholm 1996).

5.2.3.3 Influences of Coating, CIP, and Oxidation on the Properties of Silicon Nitride Ceramics

It was demonstrated in earlier studies that relative to dry pressing of the original Si_3N_4 powder with mixed sintering aids, both gelcasting of Si_3N_4 powder with coated sintering aids and gelcasting of oxidized powder with mixed sintering aids were able to produce Si_3N_4 ceramics with higher bending strength and reliability. Hence, it was concluded that coating and oxidizing as well as gelcasting are beneficial for the resulting properties of the material. In this study, we combined coating, oxidation, gelcasting, and CIP to prepare Si_3N_4 ceramics and to investigate their influences on the final properties of the material. There were four batches of Si_3N_4 specimens, each corresponding to a processing strategy listed as follows:

SN1: Mixed sintering aids + Oxidation (600 °C × 3 h) + Gelcasting + GPS
SN2: Coated sintering aids + Oxidation (600 °C × 3 h) + Gelcasting + GPS
SN3: Coated sintering aids + Oxidation (600 °C × 3 h) + Gelcasting + CIP + GPS
SN4: Coated sintering aids + Oxidation (600 °C × 6 h) + Gelcasting + CIP + GPS

The GPS scheme was 1750 °C for 1.5 h under the N_2 pressure of 0.3 MPa, and then 1900 °C for 1.5 h under the N_2 pressure of 6 MPa. The processing strategies and their results are shown in Fig. 5.49 and Table 5.13.

The surface oxygen contents of the Si_3N_4 powders presented in Table 5.13 do not include those of Y–Al oxides coated onto the surface of the powder. The original Si_3N_4 powder contains 1.44 wt% oxygen, with 0.72 wt% in the bulk and 0.72 wt% on the surface. When oxidized at 600 °C for 3 h in air, the oxygen content on the surface of the original powder is up to 2.94 wt%, while that of the coated powder is up to 2.73 wt%, which is slightly lesser, indicating that the coated layer on the surface of the Si_3N_4 powder is inhibited to a certain extent by the oxidation of the powder. The restraint of the existing layers on further oxidation of the Si_3N_4 powder is also illustrated by the influence of the oxidation time at 600 °C: at 3 and 6 h, the increments were not proportional to the oxidation times.

Even though there remains a small difference in the oxygen contents of the SN1 and SN2 samples, it can be assumed that the main contributor to higher reliability of the SN2 sample relative to SN1 sample is the addition of the sintering aids via coating. There are three advantages of coating Si_3N_4 powder with the sintering aids in

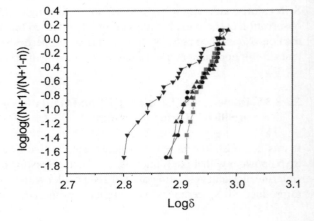

Fig. 5.49 Properties of silicon nitride ceramics with different processing schedules. ▼: SN1, ▲: SN2, ●: SN3, ■: SN4

5.2 Gelcasting of Silicon Nitride Ceramics

Table 5.13 Results of the influences of coating, oxidizing, and CIP on the properties of silicon nitride ceramics prepared by gelcasting

No.	Surface oxygen content of Si_3N_4 powder (wt%)	Adding sintering aids by	CIP	Relative density (%)		Bending strength (MPa)	Weibull modulus
				Green bodies	Sintering bodies		
SN1	2.94 (600 °C × 3 h)	Mixing	No	47.7	99.0	773.5 ± 95.1	8.6
SN2	2.73 (600 °C × 3 h)	Coating	No	47.7	98.9	814.5 ± 68.5	12.4
SN3	2.73 (600 °C × 3 h)	Coating	Yes	64.2	98.8	868.9 ± 59.8	15.1
SN4	3.64 (600 °C × 6 h)	Coating	Yes	64.2	99.2	877.2 ± 40.5	22.8

aqueous gelcasting: (1) modification of the surface characteristics of Si_3N_4 particles to improve their dispersibility, and thus, the homogeneity of its slurry; (2) uniform distribution of the sintering aids in green bodies to promote the sintering processes; and (3) partial separation of the surface of the Si_3N_4 powder from the aqueous media with high basicity, to avoid the formation of macropores in the green bodies. These advantages result in uniform Si_3N_4 green bodies, homogeneous sintered bodies, and consequently, high reliability of the material.

The influence of CIP on the bending strength and reliability of Si_3N_4 ceramics gelcasted from coated and oxidized Si_3N_4 powders becomes evident if we compare the corresponding values of SN2 and SN3 samples (Table 5.13). The bending strength and Weibull modulus of the former were 814.5 ± 68.5 MPa and 12.6, respectively, while those of the latter were 868.9 ± 59.8 MPa and 15.1, respectively, indicating that CIP could significantly improve the reliability of the material. This can be explained by analyzing the microstructures of the green bodies obtained with and without CIP, as shown in Fig. 5.50. The Si_3N_4 powder was observed to be chemically unstable in the aqueous media, especially in the alkaline solution. If the processing condition was not properly controlled during gelcasting, then millimeter-sized macropores formed in the Si_3N_4 green bodies. However, by decreasing the pH value of the slurry and coating the Si_3N_4 powder with its sintering aids, green bodies without macropores could be prepared (Zhou 2000) Nevertheless, there were still some micron-sized pores in the green bodies prepared, as shown in Fig. 5.50a. After CIP, not only the micropores in the bodies disappeared, but the relative density of the green bodies also increased, as can be seen from Fig. 5.50b, both of which further optimized the homogeneity of the green bodies and the final ceramics. Accordingly, the reliability of the material was increased.

Comparison of the SN3 and SN4 samples revealed the influence of the oxygen content of the Si_3N_4 powder on the bending strength and reliability of the Si_3N_4 ceramics, as shown in Table 5.14. The bending strengths of the SN3 and SN4 samples

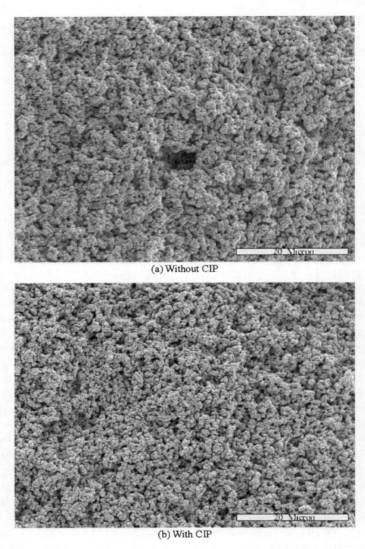

Fig. 5.50 Micrographs of the coated and oxidized silicon nitride green bodies obtained by gelcasting

were observed to be almost similar, while the Weibull modulus of the latter was 22.8, which is much greater than that of the former (with a value of 15.1). Similar to coating and CIP, oxidation was found to have more influence on the reliability rather than on the strength of the Si_3N_4-based ceramics. The coated Si_3N_4 powders of SN3 and SN4 samples were oxidized at 600 °C for 3 and 6 h in air, which led to a surface oxygen content of 2.73 and 3.64 wt%, respectively. This surface oxygen content of Si_3N_4 powder was owing to silica, which also acted as a sintering aid and played an

Table 5.14 Properties of silicon nitride ceramics with various second holding times in two-stage GPS

No.	The second holding time (h)	Relative density (%)	Bending strength (MPa)	Weibull modulus
SN5	1.0	98.9	915.9 ± 52.0	18.7
SN4	1.5	99.2	877.2 ± 40.5	22.8
SN6	2.0	99.0	944.7 ± 29.5	33.9
SN7	2.5	98.0	804.2 ± 77.9	10.8

important role on sintering and densification of the green bodies as well as on the properties of the ceramic. Natansohn et al. (1991) investigated the role of surface oxygen and found that there was an optimal oxygen content at which the fracture strength of Si_3N_4 reached its maximum. They found that the approach was only effective when the oxygen content was adjusted by thermal treatment, i.e., oxidation of the Si_3N_4 powder. However, other methods of oxygen adjustment, such as silica additions or chemical and physical treatment did not result in Si_3N_4-based ceramics of equivalent properties. Nevertheless, whether this is the case for the material under investigation remains to be clarified in the future. Further investigations are also needed to ascertain the optimal surface oxygen content of the coated Si_3N_4, related to the maximum strength and reliability of the Si_3N_4 ceramics. However, the sintering scheme was optimized in this regard, as discussed in the following section.

5.2.3.4 Influence of GPS Scheme on the Properties of Gelcast Silicon Nitride Ceramics

To determine an optimal GPS scheme for Si_3N_4, three additional schemes with various second holding times were selected besides that of SN4 sample, which are as follows: 1750 °C × 0.3 MPa (N_2) × 1.5 h + 1900 °C × 6 MPa (N_2) × 1.0 h (SN5), 2.0 h (SN6), and 2.5 h (SN7). The resulting properties of Si_3N_4 sintered with those schemes are shown in Table 5.14 and Fig. 5.51.

As can be seen from Table 5.14, a maximal bending strength as high as 944.7 MPa (Batch SN6) could be achieved, and that the differences in the bending strength for SN4, SN5, and SN6 samples are relatively small. There is also a maximum of the Weibull modulus as high as 33.9 (Batch SN6), but the differences in the Weibull modulus are relatively large. This signifies that in the current investigation, the sintering scheme has much influence on the Weibull modulus, but little effect on the bending strength of the material. Micrographs of the fracture surfaces of the ceramics are shown in Fig. 5.52. It can be seen from (a) to (d) in Fig. 5.52 that there is a more obvious evidence of roughness and pullout on the fracture surface of the SN6 sample with the highest Weibull modulus. According to earlier studies, high Weibull modulus is theoretically and experimentally connected to a strong R-curve effect (Cook 1988; Tuan 1994) . In addition, previous studies also found that bridging and

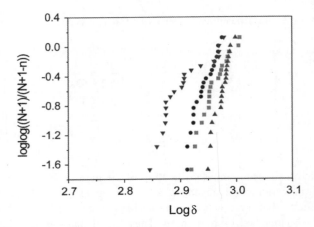

Fig. 5.51 Properties of silicon nitride ceramics with different holding times in the two-stage GPS. ■: SN4, ●: SN5, ▲: SN6, ▼: SN7

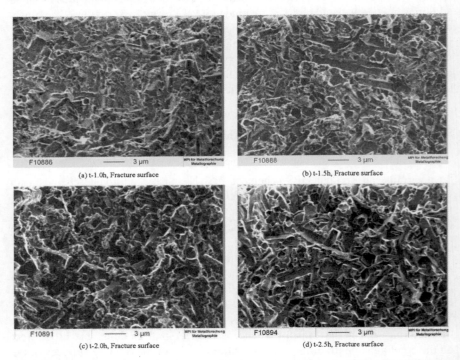

Fig. 5.52 Micrographs of the silicon nitride ceramics with various second holding times, t, in the two-stage GPS

pullout was one of major reasons for the R-curve behavior (Fang 2000; Yang 1999) Therefore, both the toughening mechanisms are proposed to be the dominant factors for the high reliability of Si_3N_4 ceramics observed in our investigation. However, further investigations have been carried out to explore the mechanism responsible for the high reliability of this material.

5.2.3.5 Summary

Surface modification of the Si_3N_4 powder, i.e., coating the powder with its Y–Al oxide sintering aids and oxidation of the coated powder, obviously increases the dispersibility of the powder in the aqueous media, because the surface of the treated powder is similar to that of the oxides, which exhibit good dispersion in the aqueous solution. Both the coating process and proper scheme of oxidation ultimately improve the bending strength and reliability of the Si_3N_4 ceramics fabricated by gelcasting and GPS. CIP of the gelcast green bodies obtained from the coated and oxidized Si_3N_4 powder as well as an optimized scheme of GPS are observed to further improve the properties of the material. Si_3N_4 ceramics with three-point bending strength and two-parameter Weibull modulus as high as 944.7 ± 29.5 MPa and 33.9 were prepared with the optimal processing strategy, by combining (i) coating of the starting Si_3N_4 powder with sintering aids, (ii) oxidation of the coated powder with a proper scheme, (iii) gelcasting the surface-modified powder, (iv) CIP of the gelcast green bodies after binder burnout, and (v) optimized GPS.

5.3 Gelcasting of Silicon Carbide Ceramic and Silicon Nitride-Bonded Silicon Carbide Ceramic

5.3.1 Gelcasting of Concentrated Aqueous Silicon Carbide Ceramic

The colloidal behavior of SiC powder in aqueous solution was investigated by preparing high concentrated aqueous SiC slurries and casting complicated shapes of uniform SiC green bodies.

5.3.1.1 Experimental Procedure

(1) Materials and Reagents

The SiC used is a commercially available powder (Rs07, Huamei abrasive and sharpener Corporation, P.R. China), which was manufactured by carbothermic reduction method, and its physical and chemical characteristics are shown in Table 5.15. The yttria and alumina powders (Jiangsu Wuxian Special Ceramics Plant, P.R.

Table 5.15 Physical and chemical characteristics of silicon carbide powder

d_{50} (μm)	S_{BET} (m^2 g^{-1})	SiC (wt%)	Free Si (wt%)	O (wt%)	Free C (wt%)	Fe$_2$O$_3$ (wt%)
1.48	4.19	>98	0.45	0.49	0.66	<0.5

China), serving as the sintering aids for SiC, had the average particle size of 1 μm with a purity of more than 99%. Furthermore, AM (C$_2$H$_3$CONH$_2$) and MBAM [(C$_2$H$_3$CONH)$_2$CH$_2$] were applied in the process of gelcasting as monomers for polymerization, which was initiated by an initiator, ammonium persulfate [(NH$_4$)$_2$S$_2$O$_8$] and accelerated by a catalyst, TEMED. All these reagents used were chemically pure.

(2) Procedure

The gelcasting process employed was similar to that of the previous studies (Young 1999; Omatete 1997). First, SiC powder (with or without sintering aids) was suspended in a premix solution prepared by dissolving AM and MBAM in demonized water. To improve the dispersion of the powder and fluidity of the suspension, an organic base, TMAH [(CH$_3$)$_4$NOH] was applied as a dispersant. The slurry was then degassed for 10 min after having been milled for 16 h. Two degassing methods, mechanical vibration and vacuum pumping, were employed to compare their effect. The suspension was degassed for another 3 min when the initiator and catalyst were added. All the abovementioned operations were carried out at room temperature. Subsequently, the slurry was cast into a nonporous mold, which was then kept at 70 °C. After polymerization of the monomers, the green bodies were demolded and dried at room temperature under controlled humidity, to avoid cracking and nonuniform shrinkage owing to rapid drying. Binder burnout was operated before sintering, which is not discussed in this paper.

(3) Measurements of Physical and Chemical Properties

The zeta potentials of the SiC powder were measured using Zetaplus (Brookhaven Instruments Corp., USA). Diluted HCl and TMAH solutions were used for pH adjustment. The apparent viscosity of the suspensions was examined by a rotary viscometer (Model NXS-11, Chendu Instrument Plant, P.R. China). The microscopic morphology was observed using SEM (CSM950, OPTON, Germany). During the process, pores were found in the green bodies, owing to the production of a gas during the reaction between the SiC powder, the dispersant, and the premix solution. Subsequently, a self-made gas-gathering device, shown in Fig. 5.28, was designed to ascertain the reaction. The gas gathered was analyzed using a gas chromatography (Gas Chromatograph 3400, Beijing Analytical Instrument Plant, P.R. China). The condition of the reaction was similar to that of polymerization, except that the experiments for collecting the gas were longer. The reagents (SiC powder, premix solution, or dispersant) were contained in a gas-tight reactor immersed in an 80 °C water bath. The reagents were agitated using a magnetic stirrer and the experiment lasted for 30 min. A certain volume of air was pressurized as a gas vehicle in the reactor, so that the volume of the resultant gas can be calculated by detecting the volume ratio of the resultant gas to N$_2$ in the collected gas.

5.3.1.2 Colloidal Behavior of the Silicon Carbide Powder

The pH-dependent zeta potentials of the SiC powder in the aqueous solutions are shown in Fig. 5.53. As shown in the figure, the IEP of SiC in the deionized water is acidic, below which the zeta potentials are positive, and above which the zeta potentials are negative; however, at a strongly basic pH value, the absolute values of the zeta potentials are relatively high.

During synthesis, transportation, and storage, the SiC powder is inevitably oxidized, forming an oxygen-rich layer on its surface. The oxidized layer is chemically similar to the surface of silica (Hackly 1997). When the powder is treated with water vapor or dispersed in deionized water, the layer is hydrated, causing silanols (Si–OH) on the surface of the SiC powder. Silanols are amphoteric, relatively strongly acidic, and weakly alkalescent (Crimp 1986). In all but low pH range (<pH 3.3 in Fig. 5.54), silanols are observed to react with OH^-, leaving $Si–O^-$ with negative charges on the surface of the powder, resulting in negative zeta potentials. In a low pH range,

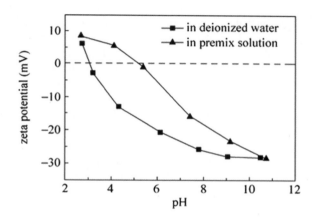

Fig. 5.53 The pH-dependent zeta potential of the silicon carbide powder

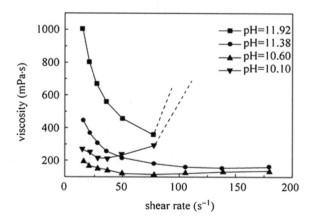

Fig. 5.54 Rheological behavior of the silicon carbide suspensions at different pH (at 50 vol% solids loading)

silanols react with H^+, yielding $Si-OH_2^+$ with positive charges on the surface of SiC powders, resulting in positive zeta potentials. The reactions can be expressed as follows:

$$Si - OH + OH^- \rightleftharpoons Si - O^- + H_2O$$

$$Si - OH + H^+ \rightleftharpoons Si - OH_2^+$$

As silanols are relatively strongly acidic, they can easily react with OH^-, and thus, the absolute values of the zeta potential of the SiC powder are higher at strongly basic pH than at strongly acidic pH.

The zeta potential curve for the SiC powder in the premix solution is similar to that of the SiC particles in the deionized water. However, the IEP of the former is higher than that of the latter, while the gradient in the zeta potential at the IEP of the former is lower than that of the latter. The adsorption of the monomers on the surface of the SiC particles, with a high probability, contributes to the shift of IEP.

Nevertheless, when the SiC powder is suspended in an aqueous solution, the absolute values of its zeta potentials are larger at high pH, inducing greater electrostatic interparticle repulsion, which is important to establish a colloidally stable suspension. Hence, SiC powder is easy to disperse in alkaline solution.

Figure 5.55 shows the curves for the rheological property of the slurries (at 50 vol% solids loading) with various pH values. The slurry with a pH of 10.60 or 11.38 is found to possess good rheological property. When the pH value is lower than 10.1, the zeta potential of the SiC particle is not sufficient to establish a stable suspension, and hence, the slurry is of shear-thickening and sharply increases in viscosity at high shear rate. When the pH value is higher than 11.92, the ionic strength in the suspension becomes too high, causing a viscous slurry. Therefore, the suspensions

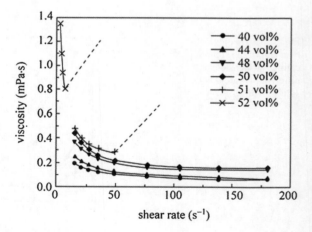

Fig. 5.55 Rheological behavior of the silicon carbide suspensions at different solids loading (at pH 10.5)

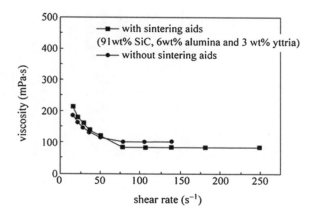

Fig. 5.56 Effect of sintering aids on the rheological behavior of the silicon carbide slip (at 50 vol% solids loading and pH 10.5)

need a proper pH value range (10.5–11.5), i.e., the dispersant, TMAH, should be added to the slurries in an appropriate quantity.

Figure 5.55 shows the curves for the rheological behavior of the SiC slurries (at pH = 10.5) at different solids loading values. As shown in the figure, when the solids loading values are less than 50 vol%, the viscosity of the SiC slurry increases gently with the increase in the solids loading value, but when they are more than 50 vol%, the fluidity decreases sharply with the increase in the solids loading. The operation in the gelcasting process is observed to respond to shear rate at 100 s^{-1}, and thus, the suspension at 51 or 52 vol% does not fit gelcasting.

The suspensions, with a formulation of 91 wt% SiC, 6 wt% Al_2O_3, and 3 wt% Y_2O_3, were developed for consolidation in this study. The viscosities of these slurries were not increased, because the alumina and yttria powders were also well dispersed in the basic solution, as demonstrated in Fig. 5.56.

5.3.1.3 Consolidation of Silicon Carbide Suspension

When casted into a mold at constant temperature of 70 °C, the SiC suspensions were consolidated within 10 min. Figure 5.57 shows the SEM micrographs of the fracture surfaces of the SiC green bodies, which were casted from the slurries at 50 vol% solids loading and a pH of 11.52, degassed by the two methods. Sample A was prepared by mechanical vibration, while sample B was prepared via vacuum pumping. As shown in Fig. 5.57, there are more and larger pores in sample A, and fewer and smaller pores in sample B.

When milled or when initiator and catalyst are added, the SiC slurries are inevitably puffed, leaving air bubbles in them. If not excluded, the air bubbles may get inflated and yield pores in the green bodies when the suspensions are consolidated at 70 °C. On the other hand, AM reacts with TMAH and discharges ammonia, and the reaction can be expressed as:

$$C_2H_3CONH_2 + OH^- = C_2H_3COO^- + NH_3(g)$$

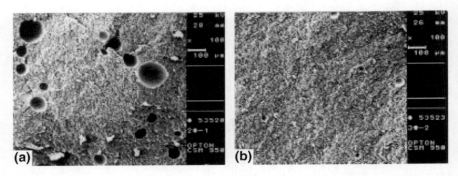

Fig. 5.57 The fracture surfaces of the silicon carbide green bodies (casted from the slurries at 50 vol% solids loading and a pH of 11.52): **a** by mechanical vibration and **b** by vacuum pumping

At room temperature, the degree of the reactivity is very low and the ammonia produced is dissolved in the suspensions owing to its high solubility (Ninghuai Zhou 1980). However, at 70–80 °C, the reaction is strong and the solubility of ammonia decreases greatly. If the rate at which the reaction gives out ammonia is higher than that at which the slurries discharge the gas, then pores will be formed in the green bodies. Because of these two reasons, degassing must be applied before casting the slurries.

Comparison of sample A with sample B revealed that vacuum pumping plays a better role in degassing, but it still cannot eliminate holes in the green bodies. The pores in sample B may probably be owing to the resultant gas discharged from an unexpected reaction due to the slurry's high pH of 11.52.

To confirm the reaction discharging the resultant gas in the suspensions, a series of experiments was designed for gathering the gas. As it was difficult to collect the gas during consolidation, owing to the gas being trapped in the green bodies, the experiments for gathering the gas produced by Si hydrolysis in the suspensions were operated in the suspensions with lower solids loading and for longer duration without initiator and catalyst. Table 5.16 shows the results of the experiments. The solution formulation in Experiment No. 1 is that of the slurry with a pH of 10.5 after milling. Its result suggests that the suspensions do discharge an extra gas, namely, H_2 gas at 70–80 °C. Ammonia could not be detected because of its high solubility in water. Experiment No. 2 and 3 clarify that the H_2 gas production is the result of SiC powder reacting with TMAH. According to the chemical compositions of the SiC powder, free Si is the only possible substance that is capable of reacting with TMAH, which is authenticated by Experiment No. 4.

Several early investigations (Hackly 1997; Popper 1960; Ezis 1974) reported that Si has a tendency to hydrolyze in water, which is an exothermic reaction:

$$Si(s) + 2H_2O = SiO_2(s) + 2H_2(g)$$

Furthermore, there were difficulties associated with Si processing in an aqueous environment. When mixing Si with aqueous solution, H_2 evolution was observed. The degree of reactivity increased with decreasing particle size and increasing alkalinity,

5.3 Gelcasting of Silicon Carbide Ceramic …

Table 5.16 Results of the gas chromatography experiments

Experiment No.	Reactants (reacting at 80 °C for 30 min)		H_2 (ml)
	Powder	Solution	
1	SiC/200 g	Premix solution/250 ml, TMAH[b]/12.5 ml	195.7
2	SiC/200 g	Premix solution/250 ml	none
3	SiC/200 g	H_2O/250 ml, TMAH/12.5 ml	291.3
4	Si/0.25 g	H_2O/250 ml, TMAH/12.5 ml	280.9
5	Oxidized SiC/200 g	H_2O/250 ml, TMAH/12.5 ml	4.08
6	SiC/200 g	Premix solution/250 ml, TMAH/25 ml	380.4

[a]Premix solution: Acrylamide (AM) ($C_2H_3CONH_2$), N, N'-methylenebisacrylamide (MBAM) [$(C_2H_3CONH)_2CH_2$], and deionized water
[b]TMAH: Tetramethyl ammonium hydroxide [$(CH_3)_4NOH$], 10 wt%

and severe gassing was reported for pH values greater than 10. According to the authors' knowledge, there has been no early study on the reaction between SiC powder and alkaline solution, partly because of the fact that free Si in SiC powder is commonly less than 1 wt%, and thus, the reaction is slow and inconspicuous, and partly because in early researches, SiC slurries were not heated to as high as 70–80 °C with a pH value beyond 10.

Experiment No. 5 illustrates that when free Si on the surface of the powder is reduced by oxidation in air at 800 °C for 2 h, there is less resultant H_2 gas. However, when compared with Experiment No. 2, Experiment No. 6 shows that the rate of the reaction, and thus, the volume of H_2 gas produced, are proportional to the concentration of TMAH. The two experiments indicate that there are at least two methods to decrease or even eliminate the resultant H_2 gas, leading to pores in the green bodies: one is to reduce the content of free Si on the surface of the SiC powder by oxidation or other methods, and the other is to decrease the concentration of TMAH, namely, pH of the SiC suspensions.

Figure 5.58 shows the SEM micrographs of the fracture surfaces of the SiC green bodies casted from slurries degassed by vacuum pumping at 50 vol% solids loading and pH of 10.5. It shows a uniform green body without pores. Figure 5.59 shows some complex shapes gelcasted from the suspensions. The density of these bodies was 54% of the theoretical value, while their strength was 31.6 MPa ($3 \times 4 \times 30$ mm), indicating that the green bodies are machinable. Furthermore, after binder burnout, there were no shrinkage, crack, and distortion in them.

5.3.1.4 Conclusion

(a) The surface of the SiC powder is similar to that of silica, and hence, its colloidal behavior is similar to that of silica. SiC can be suspended in basic solution, but a proper pH value should be controlled. Too low or too high pH value may result in viscous slurries.

Fig. 5.58 The fracture surface of the silicon carbide green body prepared by vacuum pumping (casted from slurries at 50 vol% solids loading and pH of 10.5)

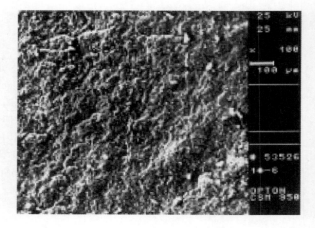

Fig. 5.59 Complex shapes of the silicon carbide green bodies obtained by gelcasting

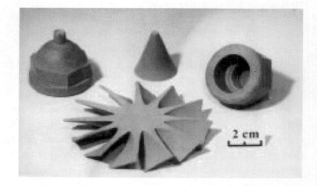

(b) The rheological property of SiC slurry decreases with the increase in the solids loading. Slurries prepared at solids loading as high as 50 vol% possess good fluidity and is suitable for gelcasting.

(c) When SiC slurry is casted at 70 °C, free Si on the surface of the SiC powder reacts with the dispersant, an organic base providing alkalescency, and gives out H_2 gas. If the reaction is violent, pores will be formed in the green bodies. By vacuum pumping and controlling the pH value of the suspensions, complicated shapes of uniform SiC green bodies are prepared.

5.3.2 Gelcasting of Aqueous Slurry with Silicon Nitride-Bonded Silicon Carbide

Gelcasting of aqueous slurry with coarse SiC (~1 mm in size) and fine Si particles was investigated to fabricate Si_3N_4-bonded SiC (SNBSC) materials.

5.3.2.1 Experimental

(1) Materials Preparation

Three grades of SiC powders, 18–34 mesh (0.75–1.4 mm), 34–70 mesh (0.36–0.75 mm), and >70 mesh (<0.36 mm), and two commercial metallurgical grades of Si powders, 200 mesh (127 μm) and 600 mesh (42 μm), were used. A typical chemical analysis for Si and SiC powder is given in Table 5.17. The main impurities in both the powders were Fe, Ca, and Al. Furthermore, a surface oxide layer represented another common impurity in the Si powders. The weight ratio of SiC:Si in the samples was 79:21, unless specified.

The essential components used for gelcasting were the reactive organic monomers: monofunctional AM ($C_2H_3CONH_2$) and difunctional MBAM [$(C_2H_3CONH_2)_2CH_2$]. These monomers were dissolved in distilled water to give a premix solution. The premix solution was subjected to free radical-initiated vinyl polymerization in the presence of an initiator. Ammonium persulfate, $(NH_4)_2S_2O_8$, was chosen as the initiator in this work. The green bodies formed by gelcasting were dried and fired at 1450 °C for 10 h in an alumina tube furnace in N_2 atmosphere. The heating and cooling rates used were 3 °C min^{-1}.

(2) Specimen Analysis

The apparent solid density measurements of the specimens at different stages, namely, after drying, debinding, and sintering, were determined by the Hg immersion method. Flexural strengths of both the green body and sintered body were examined with an Instron 4505 instrument. Three-point bending method was employed for the test, with a cross-head speed of 3.0 mm min^{-1} and a span of 90 mm. Bars of green body for the strength test were directly gelcasted into a mold of 11-mm width and 10-mm height. A sample sheet was gelcasted and sintered at 1450 °C, from which bars of 12 × 10 × 150-mm size were cut using a diamond saw for flexural-strength test. Three bars were measured for each specimen.

Crystalline-phase analysis was performed on a Rigaku Geigerflex X-ray diffractometer using Ni-filtered CuKα radiation. TGA was performed to study the binder burnout process and mass gain of Si oxidation. The samples were heated in a platinum crucible at a rate of 3 °C min^{-1}, which was the same as the heating rate used for nitridation of the green body in the furnace. All the analyses were carried out in a TG 92 SETARAM instrument. Specimens of green body and sintered materials were examined with a JEOL 840A SEM. Cross-section of the selected samples was cut using a diamond saw. Fracture surfaces of the materials were obtained from the three-point bending test.

Table 5.17 Chemical composition of the starting powders

	Si	SiC	Fe	Al	Ca	Other
Si (%)	98.2		0.8	0.5	0.3	2.1
SiC (%)	0.5	98.1	0.9	0.2	0.2	0.1

5.3.2.2 Gelcasting of Silicon Carbide/Silicon Mixtures

The gelcasting process is schematically shown in Fig. 5.60. A mixture of starting SiC and Si powders was dispersed in a premix solution that had been prepared by dissolving suitable amounts of AM and MBAM in water. Stable SiC/Si suspensions with 50–72 vol% solids loading could be obtained by adding a commercial dispersant (DURAMAX D-3005) which is mainly composed of acrylic polymer, and mixing the particles with the solution for around 20 min. Subsequently, an initiator solution was added. After homogeneous mixing, the suspension was cast into a nonporous metal mold. All the abovementioned operations were carried out at room temperature. The slurry-containing mold was then placed in an oven at 80 °C. Gelation or consolidation of the suspension occurred through polymerization of the monomers, and was completed in 10–30 min to form a wet green part that was strong enough to be demolded owing to the formation of a three-dimensional polymer network. After

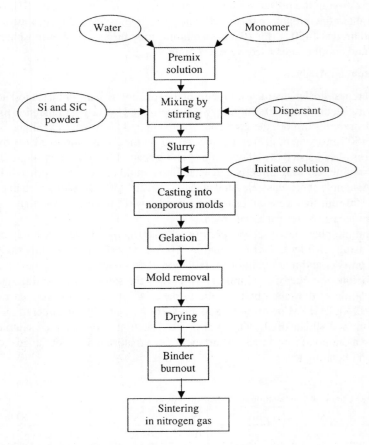

Fig. 5.60 Flowchart of silicon carbide/silicon gelcasting process

5.3 Gelcasting of Silicon Carbide Ceramic ...

drying at about 100 °C, the strength of the green body was further increased as a result of the strengthened interaction between the polymer network and particles (see Fig. 5.61). This forming mechanism produced complex-shaped parts with different sizes. Examples of the green parts of different shapes prepared by this process are shown in Fig. 5.62.

It was found that the concentration of initiator and the temperature used for gelation have great influence on the gelling rate and appearance of the green body. When the volume ratio between the initiator and monomer varied from 3 to 10, the gelling process accelerated and was completed in a short time. The gelling process was also accelerated by increasing the temperature. At room temperature (25 °C), the gelling process required more than 2 h to get completed, and it resulted in green bodies that could be easily deformed. On the other hand, it took less than 30 min for the completion of gelling when the samples were heated at 60–80 °C, producing good green bodies with little defects.

Fig. 5.61 SEM image of the fracture surface of the gelcast green body in which the weight ratio of silicon carbide:silicon is 70:30 and the monomer content in the suspension is 2.2 wt%

Fig. 5.62 Gelcast green parts with complex shapes and high green strength

5.3.2.3 Effects of Solids Loading and Monomer Content on Density

High density and low porosity are important for SNBSC refractory materials, because the properties, especially fracture strength, of the materials depend on the density after firing. For fine ceramics such as Al_2O_3, ZrO_2, and Si_3N_4, the green density of the gelcast samples may range from 50 to 65% of the theoretical density, but the final density of 90–99% of the theoretical value can be readily achieved after sintering owing to large shrinkage occurring through the liquid-phase or solid-state sintering mechanism (Prabhakaran 2000; Huang 2000; Omatate 1991) . However, for the reaction-bonded SNBSC materials, very little shrinkage occurs during firing and the voids are partially filled with the products of Si nitridation. Therefore, the density of the final product is critically dependent on that of the green body. Aqueous gelcasting is carried out in nonporous molds through which water cannot be removed during the gelling process. This is fundamentally different from slip-casting, where water in the slurry is absorbed by plaster molds through capillary action. Consequently, the green density produced by gelcasting strongly relies on the solids loading of the suspension. A high green density generally requires a high loading of the solid particles in the slurry; however, if the solids content is excessively high, then the slurry could have poor flow ability owing to high viscosity, resulting in inadequate particle compaction and low ultimate density.

The effect of solids loading on density was examined and the results are shown in Fig. 5.63, where various solids content were dispersed in the same monomer solution in which the weight ratio between the monomer and water was kept at 4:25. It was observed that both the green density and final density (after sintering) increased significantly with increasing solids loading in the slurries. We found that the density

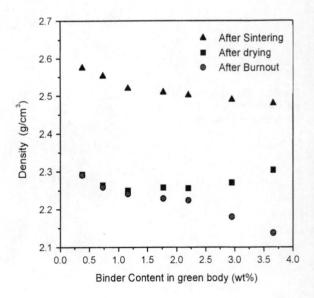

Fig. 5.63 Effect of solids loading on the density of the gelcast samples at different stages (after drying at 110 °C, polymer burnout at 600 °C, and after sintering at 1450 °C for 10 h in nitrogen atmosphere)

5.3 Gelcasting of Silicon Carbide Ceramic …

values of the dried samples at 110 °C were higher than those of the burnout samples at 600 °C. This is because the sample mass was reduced owing to the burnout of the organic binder and the sample volume did not change at the burnout temperature.

Slurry with higher solids loading, such as 70 vol%, comprising a mixture of coarse SiC and fine Si particles showed better stability when compared with those with lower solids loading of 50–60 vol%. It was observed that sedimentation of the coarse particles occurred in the slurry with lower solids loading of 50–60 vol%; however, the sedimentation of the coarse particles was almost absent in the slurry with higher solids loading of more than 70 vol%, owing to higher parking density and viscosity. In addition, lesser drying shrinkage was observed for the green body with higher solids loading because of the reduction in the water content in the suspension.

The sample density was also affected by the amount of organic monomer dissolved in the slurry. In the present work, the organic component was mainly introduced from the monomer in the initial solution, which polymerized and produced a polymer network to bind the solid particles after the addition of the initiator. Figure 5.64 shows the influence of various organic monomers on the densities of the samples at different stages, where the ratio between the ceramic powder volume and monomer solution volume was kept identical for all the samples (i.e., all the samples had the same solids loading of 72 vol% for their slurries), but the dissolved monomer content in the monomer solution was different, and the weight ratios of the monomer to water were 3:100, 6:100, 10:100, 16:100, 20:100, 30:100, and 40:100, respectively. The percentages of the organic binder, based on the ceramic powder weight, were 0.38, 0.72, 1.16, 1.77, 2.20, 2.95, and 3.66 wt%, respectively.

With regard to the samples containing less than 1.16 wt% organic binder, there was little difference in the density before and after organic burnout at 600 °C. However, as the organic content increased to 3.66 wt% (relative to the weight of the solid particles) or 40 wt% (based on the water weight), the density significantly decreased

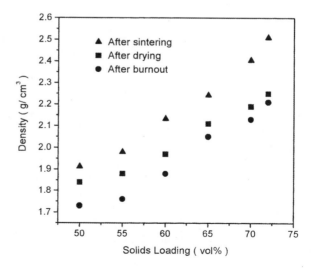

Fig. 5.64 Influence of the organic monomer content on the density of the gelcast samples with same solids loading of 72 vol% (silicon carbide:silicon=79:21 in weight ratio) at different stages (after drying at 110 °C, polymer burnout at 600 °C, and after sintering at 1450 °C for 10 h in nitrogen atmosphere)

from 2.304 g cm^{-3} in the dried sample to 2.138 g cm^{-3} after organic burnout. This was because of the fact that some space of the air pores were occupied by the organic binder mass, and higher density was reached with increased organic binder mass for the dried samples. However, more air pores were left after organic burnout, which resulted in lower density when compared with the drying stage of the respective samples, especially those with high organic monomer content of 3.66 wt%. Thus, the difference in the density before and after burnout increased with the organic monomer content in the samples, as shown in Fig. 5.64.

As discussed earlier, lower organic content in the samples with the same solids content is helpful for the density increase after either polymer burnout or nitridation sintering. For example, samples containing 0.38 and 3.66 wt% organic polymer showed a density of 2.575 and 2.480 g cm^{-3}, respectively, after nitridation at 1450 °C for 10 h. It can be observed from Fig. 5.64 that the final density of the samples after sintering strongly depends on the density after the organic burnout stage and not after the green density at the drying stage. This is because the magnitude of the increase in mass was almost the same during the sintering stage of these samples with same Si content.

The results suggest that smaller amount of monomer was selected for these ceramic systems consisting of coarse particles for gelcasting. This is in sharp contrast to that observed in the gelcasting of ultra-fine particle systems, where the organic monomer content was often around 2–4 wt% [10]. Furthermore, there is a major difference in the size of the starting powders between the SNBSC material and fine ceramics. The particle size of SiC, used for making SNBSC, ranges from a few hundred microns to above 1 mm, which is much greater than that used for the fine structural ceramics (usually $d < 1$ μm). The coarse powders have a smaller surface area than the fine ones, and consequently, a green body produced using coarse powders requires lesser amount of polymers for binding the particles when compared with those made using fine powders. The abovementioned results suggest that solids loading and binder content are the two major factors that have strong influence on the density of SNBSC materials formed by the gelcasting process. To achieve high final density of the material, the solids loading in the slurry should be high and the amount of monomer used should be low, to provide reasonable strength to the green body.

5.3.2.4 Binder Removal and Silicon Oxidation During the Burnout Process

Polymers in the gelcast green body must be carefully removed before nitridation, to suppress sample bloating at nitridation temperature and to avoid excessive amounts of carbon remaining in the system. For fabrication of fine ceramics, such as Al_2O_3 and ZrO_2, binder removal is carried out between 600 and 800 °C, at which there would be little change in the chemistry of the oxide powders. However, in the SNBSC system, Si oxidation could occur at the polymer burnout temperature, producing a silica layer on the surface of the Si powder, although SiC particles are less affected at this temperature range. It is expected that the oxidation of Si would have some effect on the subsequent Si nitridation. To study this effect, TGA was performed between 500 and 800 °C in air for a gel sample that was formed through polymerization of the organic monomer. Residue of the polymer gel was obtained when the sample was fired below 550 °C in air; however, above this temperature, the gel was thoroughly burned off in 2 h (Table 5.18). The weight changes of the Si powder samples were determined at the same temperature range using TGA. According to XRD analysis, silica was the sole product of Si oxidation between 600 and 800 °C, and the reaction can be represented as follows (Moulson 1979):

$$Si + O_2 \rightarrow SiO_2 \qquad (5.19)$$

This reaction is accompanied by a weight gain and thus, the quantity of silica produced by oxidation of Si could be calculated from this equation when the weight gain is measured from TGA. Assuming that the starting Si powders are spherical-shaped and homogeneously covered by a silica layer after oxidation, the thickness of the silica layer could be calculated based on the quantity of silica produced and the Si particle size. In this study, the average Si particle size was 43.5 µm. The mass gain measured by TGA at different temperatures and the calculated results of the thickness of the silica layer are summarized in Table 5.18. As shown in the table, when the temperature increased from 500 to 700 °C, the mass gain of Si increased from 0.058 to 0.498 wt%, corresponding to an increase in the silica layer thickness from 0.014 to 0.122 µm. At 800 °C, the mass gain was too high to be employed in the present study. From these results, a suitable burnout temperature of 600 °C was selected, at which the binder removal could be completed in 2 h with only a very small amount of Si oxidation.

Table 5.18 Results of gelcast samples fired at different temperatures in air for polymer burnout, and then fired at 1450 °C for 10 h in nitrogen atmosphere

Heat treatment conditions	500 °C for 2 h in air	550 °C for 2 h in air	600 °C for 2 h in air	650 °C for 2 h in air	700 °C for 2 h in air	800 °C for 2 h in air
Residue for gel (%)	8.080	0.003	0.000	0.000	0.000	0.000
Mass gain for Si (%)	0.058	0.114	0.202	0.363	0.498	1.828
Silica layer thickness (µm)	0.014	0.028	0.049	0.082	0.122	0.448
Density of SNBSC after firing at 1450 °C (g cm^{-3})	2.536	2.562	2.564	2.560	2.544	2.541

The sample contained 1.77 wt% polymer (relative to the weight of the solid particles).

The influence of Si oxidation on the final density of the material was studied on the gelcast SiC/Si samples containing 1.77 wt% of the polymer. The samples were first fired at various temperatures between 500 and 800 °C for 2 h in air for polymer burnout, and then at 1450 °C for 10 h in N_2 atmosphere. The results are shown in Table 5.18. Samples that had polymer burnout below 700 °C yielded a high density of 2.56 g cm^{-3} after nitridation at 1450 °C. Those that underwent initial firing at 700 °C or above in air showed a relatively lower density around 2.54 g cm^{-3}. The result indicates that a small amount of silica on the Si surface does not markedly hinder nitridation or decrease the final density. It is known that nitridation of the Si powder compact occurs in three stages: induction, growth, and termination (Sheldon 1988). At the induction stage, the first step is the decomposition of the native vitreous silica layer covering the Si particles (Sheldon 1988; Giridhar 1998), which exposes the underlying Si to the N_2 gas. This can take place when the partial pressure of oxygen is below a certain value. It was reported (Giridhar 1998) that when the partial pressure of air was below about 4 Torr in 1 atm of N_2 at 1370 °C, a small amount of Fe (as low as 55 ppm by weight) could drastically raise the removal rate of vitreous silica with respect to pure oxide. However, if the thickness of the surface oxide layer on Si powder is too high, then the exposure of surface silica would be more difficult, which could hinder Si nitridation and result in lower density of the SNBSC materials.

5.3.2.5 Effects of Firing Temperature and Starting Composition

Densification of the SNBSC materials relies on the formation of α- and β-Si_3N_4 as the bonding phases of SiC. Si_3N_4 is formed as a result of in situ reaction between Si and N at elevated temperatures, and its formation is dependent on the nitriding temperature and the starting composition. The Si nitridation process controls the density and strength of the SNBSC materials. Figure 5.65 shows the relationship between the density and nitriding temperature for SiC/Si mixtures containing different amounts of Si. The general trend observed is that the density of the materials increases with the increasing reaction temperature, but the degree of the increase is different in different systems. After firing at 1350 °C, the sample with a mass ratio of Si:SiC = 15:85 demonstrated higher density than that with a mass ratio of Si:SiC = 30:70. At this low firing temperature, Si powders in all the samples only partially reacted with N to form Si_3N_4 (see Fig. 5.66). As Si has a lower density (2.3 g cm^{-3}) than both SiC and Si_3N_4 (3.1–3.2 g cm^{-3}), partial nitridation will, in general, result in lower density of the samples that have higher Si contents. The melting temperature of Si is around 1400 °C, above which Si nitridation is accelerated. Thus, the density of the materials fired below 1400 °C is dependent on the SiC:Si ratio as well as the amount of Si_3N_4-bonding phases formed in the system. At 1450 °C, Si in all the samples was fully nitrided and this facilitated significant increases in the density, particularly for the sample with a high Si concentration.

Fig. 5.65 Relationship between density and nitriding temperature of SNBSC materials with various percentages of silicon in the starting compositions

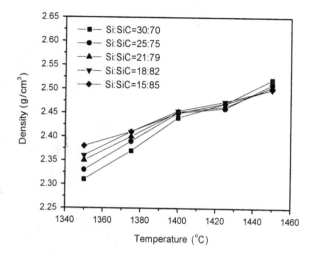

Fig. 5.66 Comparison of the XRD spectra of the sintered samples at 1450 °C × 10 h and 1350 °C × 10 h

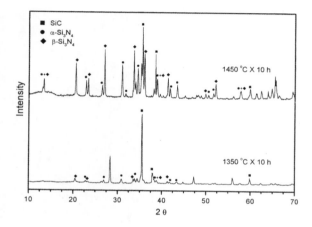

The XRD analysis revealed that the ratio of β-Si_3N_4 and α-Si_3N_4 increased from 1350 to 1450 °C. The formation mechanisms of Si_3N_4 include direct reactions of N with all the three phases of Si, namely, solid, liquid, and gas (Boyer 1978; Barsoum 1991). It is presumed that α-Si_3N_4 is mainly formed by a gas-phase reaction between the gaseous Si and N, and that the major growth of β-Si_3N_4 occurs in the liquid phase and, to a minor extent, as a result of the reaction between the solid Si and N (Ziegler 1987; Jennings 1983, 1988). The increase in β-Si_3N_4 is attributed to the melting of Si above 1400 °C. SEM images of the specimens fired at 1350 and 1450 °C, respectively, reflect their microstructure difference. As shown in Fig. 5.67a, thin α-Si_3N_4 whiskers were produced through the gaseous reaction at 1350 °C. However, at 1450 °C, relatively coarse β-Si_3N_4 grains were developed (see Fig. 5.67b).

Fig. 5.67 SEM images of silicon nitride bonding phase in the SNBSC material at **a** 1350 °C and **b** 1450 °C

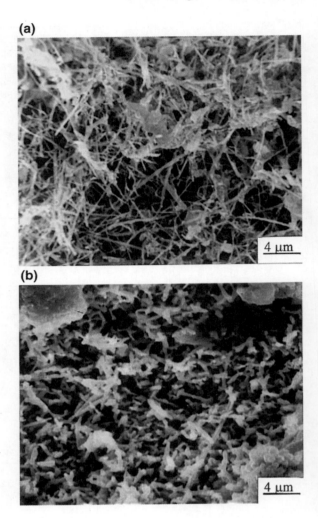

5.3.2.6 Strength of the Samples

As discussed earlier, gelcasting of a ceramic green body relies on the polymerization of the monomer in the suspension, which produces a strong three-dimensional network that holds the ceramic particles together after drying. Thus, the flexural strength of a gelcast green body would be proportional to the monomer content in the specimen. Table 5.19 shows the relationship between the green strength and the monomer content. The green strength increases from 1.32 to 2.11 MPa when the monomer content varies from 0.38 to 2.95 wt% (based on the amount of ceramic powder). The bonding of the polymer network to SiC particles can be clearly viewed from the fracture surface of the green sample containing 2.2 wt% monomer (see

5.3 Gelcasting of Silicon Carbide Ceramic ...

Table 5.19 Effect of monomer content on the fracture strength of the green body and fired materials

Polymer content in the green body (wt%)	0.38	0.73	1.77	2.95
Green strength before polymer burnout (MPa)	1.32	1.48	1.75	2.11
Flexural strength sintered at 1450 °C for 10 h (MPa)	38.59 ± 1.4	34.3 ± 2.2	25.8 ± 2.2	13.5 ± 1.9
Density after polymer burnout (g cm^{-3})	2.570	2.554	2.473	2.325

Fig. 5.61). Although higher monomer content is helpful for achieving greater green strength, densities for both the green body after polymer burnout and the final material after firing at 1450 °C are inversely proportional to the initial monomer content (Table 5.19). As discussed earlier, removal of excess of polymer in the samples could produce more pores after polymer burnout. The results indicate that a high density for the green body after polymer burnout is essential for achieving high strength of SNBSC materials after nitridation.

A comparison was made between two SNBSC samples formed by gelcasting and slip-casting, respectively. The results are shown in Table 5.20. Both the samples had the same amounts of SiC and Si powders, but the gelcast sample contained an additional 0.38 wt% of the monomer, whereas the slip-cast sample contained about 4–5 wt% of clay added to the SiC/Si slurry to improve its flowability for casting. It was observed that demolding of the gelcast sample required a much shorter time (<1 h) when compared with that for the slip-cast sample (>10 h). This can be a significant advantage with regard to the gelcasting technology for commercial production. After nitridation at 1450 °C for 10 h, the sample formed by gelcasting showed a higher density than that obtained by slip-casting. In addition, the XRD results unveiled different phase components in these two samples after nitridation. Although SiC and Si_3N_4 (in both α and β forms) were the major phases in the samples, additional silicon oxynitride and glass phases were observed in the slip-cast product (see Fig. 5.68). It is presumed that while clay facilitates slip-casting, it introduces oxides that are

Table 5.20 Comparison of the gelcast and slip-cast SNBSC materials

	Starting powders	Time from cast to demold	Mold materials	Density after sintering (g cm^{-3})	Phases after sintering
Gelcasting	SiC	20–60 min	Nonporous materials	2.587	Si_3N_4
	Si				SiC
Slip-casting	SiC	≥10 h	Plaster	2.576	Si_3N_4
	Si				SiC
	Clay				Si_2N_2O
					Glass

Fig. 5.68 Comparison of the XRD spectra of the gelcast and slip-cast SNBSC materials

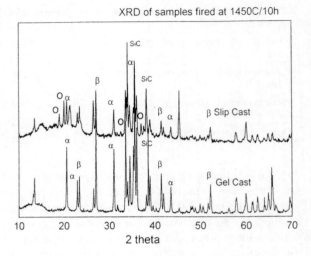

responsible for the formation of silicon oxynitride and glass after sintering. In future, it would be interesting to compare the high-temperature performance of these two samples to further evaluate the two green body-forming technologies.

5.3.2.7 Conclusions

Ceramic systems with coarse SiC and Si powders (up to 1 mm in size) have been successfully gelcasted to form complex-shaped parts. When compared with the more commonly used slip-casting technique, gelcasting of SNBSC materials has the advantages of fast demolding time, high green strength, and better control of the final-phase assemblage. Density of the SNBSC materials prepared by gelcasting is observed to be strongly dependent on the solids loading and the organic binder (monomer) content in the slurry. A high solids loading in the slurry is desirable for achieving high density of the materials after sintering. It has been found that slurry with a solids loading of up to 72 vol% can be casted and gelled to produce strong wet green body in 10–30 min at 60–80 °C. A high monomer content has been observed to render high green strength to the gelcast Si/SiC parts, but excessive amounts of monomer could lead to lower density after polymer burnout and sintering.

References

Baader, F. H., Graule, T. J., & Gauckler, L. J. (1996). *Indian Ceramic, 16,* 36.
Barsoum, M., Kangutkar, P., Koczak, M. J. (1991). Nitridation kinetics and thermodynamics of silicon powder compacts. *Journal of the American Ceramic Society* 74, 1248–1253.

References

Bergstrom, L., & Paugh, R. J. (1989). Interfacial characterization of silicon nitride powders. *Journal of the American Ceramic Society* 72(1), 103–109.
Bergstrom, L. (1994). US Patent No. 5340532, 23.
Bergstrom, L., & Pugh, R. J. (1989). *Journal of the American Ceramic Society, 72*, 103.
Bleier, A., & Omatete, O. O. (1993). *Materials Research Society Symposium Proceedings, 289*, 109.
Boyer, S. M., & Moulson, A. J. (1978). A mechanism for the nitridation of Fe-contaminated silicon. *Journal Materials Science, 13*, 511–516.
Brow, K., & Pantano, C. G. (1986). *Journal of the American Ceramic Society, 69*, 314.
Busca, G., Lorenzelli, V., Baratoon, M. I., et al. (1986a). FT-IR charaterization of silicon nitride Si_3N_4 and silicon oxynitride Si_2N_2O surfaces. *Journal of Molecular Structure, 143*, 525–528.
Busca, G., Lorenzelli, V., Porcile, G., Baraton, M. I., Quintard, P., & Marchand, R. (1986b). FT-IR Study of the Surface Properties of Silicon Nitride. *Materials Chemistry and Physics, 14*, 123–140.
Carisey, T., Laugier, W. A., & Brandon, D. G. (1995). *Journal of the European Ceramic Society, 15*, 1.
Castanho, M., Fierro, J. L. G., & Moreno, R. (1997a). *Journal of the European Ceramic Society, 17*, 383.
Castanho, M., Moreno, R., & Fierro, J. L. G. (1997b). *Journal of Materials Science, 32*, 157.
Coll. Interface Sci., 180, 413–20 (1996).
Cook, R. F., & Clarke, D. R. (1988). Fracture stability, R-curves and strength variability. *Acta Metallurgica, 36*(3), 555–562.
Crimp, M. J., Johnson, R. E., Halloran, J. W., & Feke, D. L. (1986). Colloidal behavior of silicon carbide and silicon nitride. In L. L. Hench (ed.) Science of Ceramic Chemical Process. Wiley, New York, pp. 539-549.
Dai, J.-Q. (2001). The Effects of Surface Characterization of Silicon Nitride Powders on Preparation of Concentrated Suspensions and Properties of Sintered Ceramics, Ph.D. Thesis, Tsinghua University, Beijing, China.
Dai, J. Q., Xie, Z. P., Huang, Y., Ma, L. G., & Zhou, L. J. (1999). *Journal of Material Science Letters, 18*, 1739.
Dai, J.-Q., Huang, Y., Xie, Z.-P., et al. (2001). FTIR study of Si_3N_4 powder. *Chinese Journal of Spectroscopy Laboratory, 18*(1), 78–83. (In Chinese).
Ezis, A. (1947). The fabrication and properties of slip-cast silicon nitride. In J. H. Burke, A. E. Gorum and R. N. Katz (eds.) *Ceramics for high performance applications*, Brook Hill, Chestnut Hill, MA, pp. 207–222.
Ezis, A. (1974). The fabrication and properties of slip-cast silicon nitride. In J. H. Burke, A. E. Gorum, & R. N. Katz (eds.) pp. 207–220 in *ceramics for high performance applications*. Brook Hill, Chestnut Hill, MA.
Fagerholm, H., Johansson, L., & Rosenholm, J. B. (1994). A Surface Study on Adsorption of Lignosulphonate on Mixed Si3N4–Y2O3 Powder Dispersions. *Journal of the European Ceramic Society, 14*, 403–409.
Fagerholm, H., Johansson, L. S., Graeffe, M., & Rosenholm, J. B. (1996). Surface charge and viscosity of mixed Si_3N_4–Y_2O_3 suspensions containing lignosulphonate. *Journal of the European Ceramic Society, 16*, 671–678.
Fang, Y., Yu, F., & White, K. W. (2000). Microstructural influence on the R-curve behavior of a 70%Si_3N_4–30% barium aluminum silicate self-reinforced composite. *Journal Materials Science, 35*(11), 2695–2699.
Garg, A. K., & De Jonghe, L. C. (1990). Microencapsulation of Silicon Nitride Particles with Yttria and Yttria-Alumina Precursors. *Journal of Materials Research, 5*(1), 136–142.
Giridhar, R. V., & Rose, K. (1998). Conditions for thermal nitridation of Si in $_{N2-O2}$ mixtures. *ECS Journal of Solid State Science and Technology, 135*(111), 2803–2807.
Graule, T. J., Gauckler, L. J., & Baader, F. H. (1996). *Indian Ceramic, 16*, 31.
Greil, P. (1989a). *Materials Science and Engineering A, 109*, 27.

Greil, P. (1989b). Processing of silicon nitride ceramics. *Materials Science and Engineering* 109, 27–35 (1989).

Greil, P., Nitzsche, R., Friedrich, H., & Hermel, W. (1991). *Journal of the European Ceramic Society, 7,* 353.

Hackley, V. A., & Malghan, S. G. (1994). The surface chemistry of silicon nitride powder in the presence of dissolved ions. *Journal Materials Science, 29,* 4420–4430.

Hackley, V. A., Wang, P. S., & Malghan, S. G. (1993). Effects of soxhlet extraction on the surface oxide layer of silicon nitride powders. *Materials Chemistry and Physics, 36,* 112–118.

Hackly, V. A., Paik, U., Kim, B. H., & Malghan, S. G. (1997). Aqueous processing of sintered reaction-bonded silicon carbide: 1, dispersion of silicon powder. *Journal of the American Ceramic Society, 80*(8), 1781–1788.

Harame, D. L. (1984). Integrated Circuit Chemical Sensors, Ph.D. Thesis. Stanford University, Stanford, CA.

Huang, Y., Ma, L. G., Yang, J. L., Xie, Z. P., & Xu, X. L. (2000). Surface oxidation to improve water-based gelcasting of silicon nirtride. *Journal Materials Science, 35,* 3519–3524.

Huang, Y., Zhou, L. J., Tang, Q., Xie, Z. P., & Yang, J. L. (2001). Water-based gelcasting of surface coated silicon nitride powder. *Journal of the American Ceramic Society, 83*(4), 701–707.

James, R. O., & Parks, G. A. (1980). Surface and colloid science. vol. 2, edited by E. Matijevic (Wiley Interscience, New York, 1980).

Janney, A. (1990-1-16). US Patent no. 4894194.

Janney, M. A., & Omatete, O. O., US Patent no. 5145908.

Janney, M. A., & Omatete, O. O. (1991-7-2). US Patent no. 528362.

Janney, M. A., et al. (1991). *Journal of the American Ceramic Society, 81,* 581.

Jennings, H. M. (1983). Review on reaction between silicon and nitrogen, Part 1 Mechanisms. *Journal Materials Science, 18,* 951–967.

Jennings, H.M., Dalgleish, B.J., & Pratt, P.L. (1988). Reaction between silicon and nitrogen. Part 2 Microstructure. *Journal of Materials Science, 23,* 2573–2583.

Joshi, R. N., & Ronald, R. A. (1994). metallorganic surfactants as sintering aids for silicon nitride in an aqueous medium. *Journal of the American Ceramic Society, 77*(11), 2926–2934.

Kim, J. S., Schubert, H., & Petzow, G. (1997). Sintering of Si_3N_4 with Y_2O_3 and Al_2O_3 added by coprecipitation. *Journal of the European Ceramic Society, 5,* 311–319.

Kitahara, A., & Watanab, A. (1992). *Electric phenomena of interface* (in Chin.). Peking University Press, Beijing, China.

Kitahara, A., & Watanabe, A. (1992). *Electric phenomena of interface* (Peking University Press, Beijing) p. 49.

Kitahara, A., & Watanabe, A. (1992). (Peking University Press, Beijing) p. 49.

Kulig, M., & Greil, P. (1991). Surface chemistry and suspension stability of oxide-nitride powder mixtures. *Journal Materials Science, 26,* 216–224.

Laarz, E., Lenninger, G., & Bergstrom, L. (1997). Aqueous silicon nitride suspensions: Effect of surface treatment on the rheological and electrokinetic properties. *Key Engineering Materials, 132–36,* 285–288.

Levine, S., & Smith, A. L. (1971). *Discussions of the Faraday Society, 52,* 290.

Levine, S., Neale, G., & Epstien, J. (1996). *Journal of Colloid and Interface Science, 57,* 424.

Malghan, S. G. (1992). Dispersion of Si3N4 powders: Surface chemistry interactions in aqueous media. *Colloids and Surfaces, 62,* 87–99.

Messing, G. L., Mazdiyasni, K. S., McCauley, J. W., & Haber, R. A. (eds.) (1987). Ceramic powder science. American Ceramic Society, Westerville, OH.

Mezzasalma, S., & Baldovino, D. (1996). *Journal of Colloid and Interface Science, 180,* 413.

Mezzasalma, S., & Baldovino, D., Characterization of silicon nitride surface in water and acid environment: a general approach to the colloidal suspensions. Journal.

Moulson, A. J. (1979). Review, reaction-bonded silicon-nitride: its formation and properties. *Journal Materials Science, 14,* 1017–1051.

References

Natansohn, S., & Pasto, A. E. (1991). Improved processing methods for silicon nitride ceramics. *American Society of Mechanical Engineers*, Paper No. 91-GT-316.

Natansohn, S., Pasto, A. E., & Rourke, W. J. (1993). Effect of powder surface modifications on the properties of silicon nitride ceramics. *Journal of the American Ceramic Society, 76*(9), 2273–2284.

Ninghuai Zhou, S. I. (1980). *Chemical data*. Beijing: Higher Education Press.

Nilsen, K., Danforth, S. C., & Wautier, H., Dispersion of laser-synthesized si_3n_4 powder in nonaqueous systems, pp. 537–47 in Advances in Ceramics.

Nilsen, K., & Danforth, S. C., Dispersion of laser-synthesized Si_3N_4 powder in nonaqueous systerms. In *Advances in ceramics*, vol. 21: Ceramic Powder Science, pp. 537–547.

Of Silicon Nitride Prepared for Extrusion. (1996). *Journal of the European Ceramic Society, 16*, 1127–32.

Omatate, O. O., Strehlow, R. A., Armstrong, B. L. (1991). Forming of silicon nitride by gelcasting. In *Proceedings of the Annual Automotive Technology Development Contractors' Meeting, Society of Automotive Engineers*, Warrendale, PA, pp. 245–251.

Omatete, O. O., Janney, M. A., & Streklow, R. A. (1991). *American Ceramic Bulletin, 70*, 1641.

Omatete, O. O., Janney, M. A., & Nunn, S. D. (1997). Gelcasting: From laboratory development toward industrial production. *Journal of the European Ceramic Society, 17*, 407–413.

Peuckert, M., & Greil, P. (1987). Oxygen distribution in silicon nitride powders. *Journal Materials Science, 22*, 3717.

Popper, P., & Ruddlesden, S. N. (1960). The Preparation, Properties, and Structure of Silicon Nitride. *Transactions and journal of the British Ceramic Society, 60*, 603–623.

Prabhakaran, K., & Pavithran, C. (2000). Gelcasting of alumina from acidic aqueous medium using acrylic acid. *Journal of the European Ceramic Society, 20*, 1115–1119.

Rahaman, M. N., Boiteux, Y., & De Jonghe, L. C. (1986). Surface characterization of silicon nitride and silicon carbide powders. *American Ceramic Society Bulletin, 65*(8), 1171.

Raider, S. I., Flitsch, R., Aboaf, J. A., & Pliskin, W. A. (1976). Surface oxidation of silicon nitride films. *Journal of the Electrochemical Society, 123*, 560–565.

Ramis, G., Busca, G., Lorenzelli, V., Baraton, M. I., Merle-Mejean, T., & Quintard, P. (1989). FT-IR characterization of high surface area silicon nitride and carbide, pp. 173–84 In L. C. Dufour, C. Monty, and G. Petot-Ervas (eds.) *Surfaces and interfaces of ceramic materials*. Kluwer, Dordrecht, The Netherlands.

Ramis, G., Busca, G., & Lorenzelli, V. et al. (1989). FT-IR characterization of high surface area silicon nitride and carbide. In L. C. Dufour, C. Monty and G. Petot-Ervas (eds.) *Surfaces and interfaces of ceramic materials*. Kluwer Academic, Dordrecht, The Netherlands, pp. 173–184.

Rolf, W., Gabriele, S. (1997). *Key Engineering Materials*, 132–136, 374, Part 1 22–26.

Sanchez, R. T., Garcia, A. B., & Cesio, A. M. (1996). *Ibid,16*, 1127.

Schwelm, M., Kaiser, G., Schulz, W., Schubert, H., & Petzow, G. (1993). The effect of alcohol treatment on the rheological of Si3N4. *Journal of the European Ceramic Society, 11*, 283–289.

Sheldon, B. W., & Haggerty, S. (1988). The nitridation of high purity, laser–synthesised silicon powder to form reaction bonded silicon nitride. *Ceramic Engineering and Science Proceedings, 9*(7–8), 1061–1072.

Shih, W. H., Kisailus, D., & Shih, W. Y. (1996). Rheology and consolidation of colloidal alumina-coated silicon nitride suspensions. *Journal of the American Ceramic Society, 79*(5), 1155–1162.

Stadelmann, H., Petzow, G., & Greil, P. (1989). Effects of surfaces purification on the properties of aqueous silicon nitride suspensions. *Journal of the European Ceramic Society, 5*, 155–163.

Sun, J., Gao, L., Guo, J. K., & Yan, D. S. (1996). *Acta Metallurgica Sinica, A9*, 489.

Sanchez, R. T., Garcia, A. B., & Cesio, A. M., Changes in surface characteristics.

Tang, Q., Xie, Z. P., Yang, J. L., & Huang, Y. (1998). Coating of silicon nitride and its colloidal behavior. *Journal of Material Science Letters, 17*(14), 1239–1241.

Tuan, W. H., Lai, M. J., Lin, M. C., Chan, C. C., & Chiu, S. C. (1994). Mechanical performance of alumina as a function of grain size. *Materials Chemistry and Physics, 36*(3–4), 246–251.

Wang, H. T., Liu, X. Q., & Meng, G. Y. (1997). *Materials Research Bulletin, 32*, 1705.

Yang, J. L., Xie, Z. P., & Huang, Y. (1997). *Journal of the Chinese Ceramic Society, 25,* 679.

Yang, J., Oliveira, F. J., Silva, R. F., & Ferrera, J. M. F. (1999a). Pressureless sinterability of slip cast silicon nitride bodies prepared from coprecipitation-coated powders. *Journal of the European Ceramic Society, 19,* 433–439.

Yang, J. F., Sekino, T., & Niihara, K. (1999b). Effect of grain growth and measurement on fracture toughness of silicon nitride ceramics. *Journal Materials Science, 349*(22), 5543–5548.

Young, A. C., Omatete, O. O., Janney, M. A., & Menchhofer, P. A. (1991a). Gelcasting of Alumina. *Journal of the American Ceramic Society, 74*(3), 612–618.

Young, A. C., Omatete, O. O., Janney, M. A., et al. (1991b). *Journal of the American Ceramic Society, 74,* 612.

Zhou, N. H. (1980a). *SI chemical data* (p. 123). Beijing, China: Higher Education Press.

Zhou, N. H. (1980). *SI chemical data.* Higher Education Press, Beijing, p. 123.

Zhou, L. J., Huang, Y., Xie, Z. P., & Cheng, Y. B. (2000). Gas-discharging reactions and their effect on the microstructures of green bodies in gelcasting of non-oxide materials. *Materials Letters, 45*(8), 51–57.

Ziegler, G., Heinrich, J., & Wotting, G. (1987). Review: Relationships between processing, microstrature and properties of dense and reaction-bonded silicon nitride. *Journal Materials Science, 22,* 3041–3086.

Chapter 6
Applications of New Colloidal-Forming Processes

Abstract The colloidal-forming process based on gelcasting was applied for preparing microbeads of ceramics, improving the breakdown strength of the rutile capacitor, developing the thin-wall rutile tube for ozone generator, and producing the refractory nozzle of zirconia (ZrO_2) and lead zirconate titanate (PZT). The microbeads of ceramics were used as the grinding media, gel-pen ball, and healthcare product with far-infrared function for the human body. Owing to the low-cost associated with the production of the gel beads as well as because of its superior properties, they have been widely used and marketed. When compared with the extrusion formation, the breakdown strength of rutile capacitor prepared by gelcasting was significantly improved. Rutile ceramics with high dielectric constant and high breakdown strength have been used in the thin-wall tube of the ozone generator. The ZrO_2 refractory nozzle prepared by gelcasting has a high melting point above 2680 °C and a good thermal shock resistance to withstand the temperature rising from the room temperature to 2200 °C in several seconds. Hence, it is considered as a suitable material for making nozzles for precision casting of Cu–Cr molten alloy. When compared with die pressing, gelcasting can provide a more homogenous microstructure as well as more homogenous piezoelectric property of the PZT samples. By using the same sintering procedure, gelcast samples were observed to exhibit slightly stronger piezoelectric effects than the die-pressed ones, and this can be attributed to both density difference and pore size.

Keywords Gelcasting · Gel-bead forming · Microbeads · Gel-pen ball · Rutile capacitor · Ozone generator · Lead zirconate titanate · Refractory nozzle

6.1 Ceramic Microbeads

6.1.1 The Forming Principle of Ceramic Microbeads Based on Gelcasting

According to the interfacial tension principle, formation of gel beads is a process that employs the rapidly solidifying slurry with water-soluble monomer. By adjusting the ball-forming media as well as the job parameters of the slurry characteristics, the key processes of molding, dressing shape, and shape maintenance can be accomplished. In this work, a new preparation process using colloidal-forming technique for the production of ceramics microbeads is proposed, and the preparation equipments are designed and manufactured.

The forming principle of ceramics microbeads:

The basic principle of this process is based on the interface tension between two different surfaces. When the slurry is dropped into the oil medium, solidification occurs and subsequently, beads are formed. The interface tension can be described by the difference in the surface tensions when two kinds of liquids have been mutually saturated, i.e., $\gamma_{12} = \gamma_1 - \gamma_2$. According to this formula, without the influence of the external force, the beads can always be formed when γ_{12} is large and the time is sufficiently long. The time required for the formation of the beads becomes shorter when γ_{12} is larger.

In the preparation process, two kinds of mediums are employed. One is the water-based ceramic slurry ($\gamma_1 \approx 50$ mN m^{-1}), and the other is the oil-based medium ($\gamma_2 \approx 20$ mN m^{-1}). When the interfacial tension γ_{12} is about 30 mN m^{-1}, the slurry droplet can rapidly become a bead in the oil medium.

The design of the process parameters is as follows:

$mg \sim f + f'$

When falling into the medium, the bead is mainly influenced by three forces, namely, the downward gravitational force, mg, the upward buoyancy force, f', and the viscous force, f (Fig. 6.1).

$$mg \sim f + f' \qquad (6.1)$$

According to the Stokes' formula, when the bead moves in the liquid, the viscous force can be expressed as:

$$f = 6\pi r \eta \upsilon \qquad (6.2)$$

Fig. 6.1 The applied forces on the beads

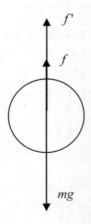

6.1 Ceramic Microbeads

where η is the liquid's viscous coefficient, υ is the bead's velocity, and r is the radium of the bead.

The buoyancy can be given as:

$$f' = \rho_0 v g \qquad (6.3)$$

where v is the volume of the bead and ρ_0 is the density of the liquid.

It can be concluded from the abovementioned formula that once the size of the bead is determined, the factors that affect the falling time of the bead are the density and viscosity of the medium, i.e., the buoyancy and viscosity of the liquid medium. Therefore, the medium chosen must be in line with the following two conditions (or at least one of them): (1) the density of the slurry should be close to that of the medium (to increase the buoyancy), and (2) the medium should have some viscosity (to increase the adhesion force).

The vertical downward gravitational force is larger than the sum of the buoyancy and viscous forces, which are vertical upward when the beads fall into the liquid; hence, the beads exist in accelerated motion state. The viscous force is increased with the increase in the bead velocity. When the velocity of the bead reaches a certain value, υ_0, with the sum of their values being zero, the bead will exist in a uniform speed state, which can be calculated using the following formula:

$$mg - \rho_0 v g - 6\pi r \eta \upsilon_0 = 0 \qquad (6.4)$$

where υ_0 is the end speed.

From the abovementioned formula, it can be deduced that:

$$\upsilon_0 = \frac{2}{9} \cdot \frac{(\rho - \rho_0)g r^2}{\eta} \qquad (6.5)$$

There exists a whereabouts distance, S, in which the bead moves from the accelerated to the uniform speed state. When the bead travels the distance, S, from the two formulas mentioned earlier, it can be deduced that

$$S = 1.11 \times 10^4 \frac{r^4 \rho^2}{\eta^2} \left(1 - \frac{\rho_0}{\rho}\right) \qquad (6.6)$$

Thus, the sphericity and diameter of the microbead can be determined by the balance status distance, S.

The following are the advantages with regard to the advantages of gel-beads-forming process:

Gel-beads–forming process is suitable for various ceramic materials. For the production of ceramic microbeads of Al_2O_3, zirconia (ZrO_2), Si_3N_4, SiC, etc., only the process parameters need to be adjusted, and the equipments need not be replaced. When compared with other forming methods of ceramic microbeads, this method has shorter production cycle and higher forming efficiency. Furthermore, the microstructure of the products formed is uniform and dense. In addition, this method does not have the limitations of decortications and fragmentations which occur in traditional rolling method of beads production. Moreover, this method does not generate hollow microbeads that were produced by the melting method. Thus, the gel-beads-forming method is a simple process, has short production cycle, its process parameters can be easily adjusted (the diameter of the microbeads is between 0.1 and 3 mm), and automatic and continuous production are easy to realize.

6.1.2 The Process of Preparing Microbeads

The technical route of colloidal-forming process for microbeads production is presented in Fig. 6.2.

The process can be more specifically described as follows:

Step 1: The first step comprises the preparation of high concentrated ceramic suspensions with low viscosity and high solids loading. The ceramic powders, organic additives, and dispersants are mixed in the aqueous solution uniformly. Generally, the solids loading of ceramic powders in the aqueous solution is greater than 50 vol%.

Step 2: By adding the curing agents, the suspension is dropped into the oil-based medium by dedicated equipment, and then is solidified into a microbead through the synthetic effect of the curing agents and temperature.

Step 3: After forming, cleaning, and drying, the microbeads are sintered.

Step 4: By using the centrifugal rotation-type polishing machine, the surface of the sintered ceramic ball can be polished.

Step 5: After cleaning, the abovementioned balls are separated into the final products with different diameters.

6.1.3 The Properties of Ceramic Microbeads

6.1.3.1 Use of Ceramic Microbeads as a Grinding Media

The following is the application overview of ceramic microbeads:

(1) Ultrafine Grinding in Non-metallic Minerals Industry

Ultrafine grinding is a key technology in the deep processing of non-metallic minerals. With its rapid development, ceramic microbeads, as the indispensable grinding media for ultrafine grinding process, show a more vast market potential and application fields. Many overseas and domestic ultrafine materials, such as paint, printing

6.1 Ceramic Microbeads

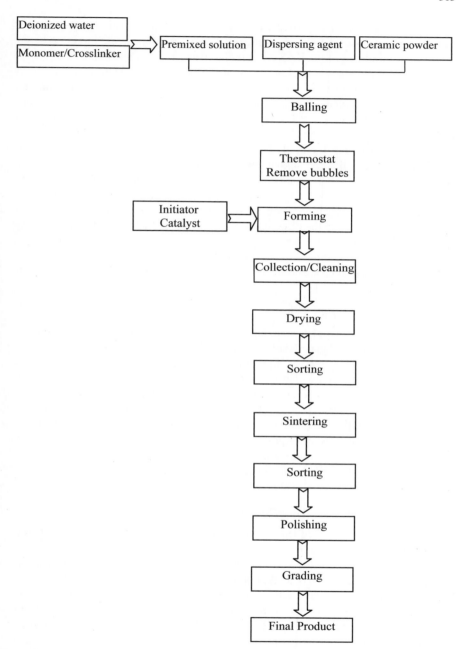

Fig. 6.2 The technical route of gel bead-forming process

ink, dye-based color, magnetic materials, electronic materials, coatings, optical materials, cosmetics, food and pharmaceutical products, etc., need ceramic microbeads to carry out ultrafine grinding, open the reunion, and for homogenizing.

China's non-metallic minerals industry earns foreign exchange through the export of 2.6 billion U.S. dollars in 2001 and 2.8 billion U.S. dollars in 2002. Kaolin and heavy calcium carbonate as paper coatings play a vital role in the inorganic non-metallic minerals processing. The demand of ceramic microbeads that can be used as the grinding media for ultrafine powders is growing. From the foreign markets, it can be seen that the annual output of heavy calcium carbonate and kaolin has reached 57 million tons, and that the annual amount of ceramic microbeads as the grinding media has approximately reached 8000 tons. According to the statistics provided by the China Paper Association and China Inorganic Materials Association, at present, the market capacity of China's non-metallic mineral ceramic microbeads is about 8000 tons every year. On the other hand, for the paper industry, the market capacity is 3000–5000 tons every year and it is growing rapidly with 15% growth rate.

Irrespective of being domestic or foreign, the market of paper coating is in a high-growth period, and the growth of its demand is fast. It also brings the high growth of the grinding media.

(2) Inks, Paints, and Dope Industry

Grinding media in inks, paints, and dope industry can be divided into two categories: ceramic microbeads and glass beads. Ceramic microbeads have some excellent characteristics such as high density, high efficiency, low wear-abrasion, pollution-free, etc. The products can be divided into three categories: alumina, zirconium silicate, and ZrO_2, which can be used for the grinding and dispersion of high viscosity materials. Currently, the annual output of inks, paints, and dope in China has reached 3 million tons; yet, the international market has reached 50 million tons. Owing to the different technologies and equipments, the grinding media are different and the amount of it is less than the non-metallic minerals. At present, there are 50–80 tons of zirconium silicate microbeads that are used in the domestic market, and the amount in the international market is 2000–3000 tons.

(3) Machinery equipment industry

Ceramic materials have excellent properties of resistance to heat, acid, alkali, and wear, which are not observed in the metal products. Ceramic bearing ball, which has a wide application in the field of aerospace, military, electronics, metallurgy, energy, etc., is the most advanced alternative to the traditional steel bearing ball. With the development of the machine tool technology, the spindle speed has become increasingly high. For example, the maximum spindle speed had been improved from 5000 r min^{-1} in the early 1980s to the current 30000 r min^{-1}. As the ordinary steel bearings cannot meet the requirements, it is necessary to develop new ultra-high-speed bearing, which are used under high temperature, vacuum, non-magnetic, or strong corrosive medium.

(4) Deburring and Polishing of Workpiece

Deburring is an indispensable aspect in workpiece processing. In this step, the workpiece is put into the polishing equipment, and then in the course of operation, the cutting force generated during vibration and rotation of the workpiece, is used to achieve effects, such as deburring, derusting, and high brightness and smoothness of its surface. The workpieces are machined into different products such as micromotor shafts, metal parts, die castings, copper, etc., using different polishing equipments. The main equipment has a centrifugal polishing machine, vibration polishing machine, water swirl polishing machine, and sandblasting equipment. According to statistics, the demand for cylindrical grinding media in the domestic market is 300–500 tons per year and the international market sales are greater.

In addition, ceramic microbeads have been widely used as a part of advanced ball-point pen and other wear-resistant pieces.

The application fields of high-performance ceramic beads can be divided according to the difference in their diameter, as shown in Table 6.1.

The following are the performance requirements of the grinding media:

The widespread applications of ceramic microbeads are based on its excellent performance. With the increase in the application, there is a need for higher quality ceramic microbeads.

The main application of ceramic microbeads in industrial production is the grinding media in the ultrafine grinding device. The average diameter of these microbeads is less than 6 mm in the mixing mill, and less than 1 mm when used for ultrafine grinding. The particle size of the grinding media has a direct impact on the grinding efficiency and product size. It is presumed that if the size of the media is greater, then the product size becomes larger and the production yield becomes higher; however, the opposite of this presumption is true. Generally, the particle size of the grinding media is determined in accordance with the raw material and requirement size. To improve the grinding efficiency, the size of the grinding media must be 10 times larger than the average feeding size.

The density of the grinding media also plays an important role in grinding efficiency. If the density of the media is greater, then the grinding time is shorter. Hence, it is necessary to consider the density when selecting the material system. To increase

Table 6.1 The application fields of ceramic beads with high performance

No.	Diameters (mm)	Applications
1	ϕ 0.4–0.8	Sandblasting, derusting, surface treatment
2	ϕ 0.4–3.0	Ultrafine grinding of non-metallic minerals, paints, dope, printing inks, food, pharmacy
3	ϕ 0.38, ϕ 0.5, ϕ 0.7, ϕ 1.0	Ball-pen beads
4	>ϕ 3.0	Bearing (hybrid and complete), decorations, ball screw, bike ball
5	>ϕ10	Grinding media, ball valve with acid and alkali resistance, and high-temperature resistance

the grinding effect, the hardness of the grinding media must be higher than the grinding materials. Based on our experience, we believe that it is better to have the Mohs hardness of greater than 3 for the grinding media than the grinded materials. The density and diameter of the grinding media that are commonly used are shown in Table 6.2.

The grinding media is one of the key factors which decide the grinding efficiency. Figure 6.3 shows the performance requirements of the ceramic microbeads as the grinding media. The main factors that impact the performance of the microbeads include the following:

(1) Wear resistance. It is directly related to the property of the microbeads. In the grinding process, the wearing of the microbeads should be decreased, because the grinded material will be polluted by the wear, and at the same time, the grinding equipment may be damaged by the broken microbeads.
(2) Sphericity. If the sphericity of the microbeads is better, then the distribution of roundness is narrower and the diameter distribution of the grinded materials is more uniform.

Table 6.2 The density and diameter of the grinding media

Grinding media	Glass (Leaded)	Glass (Unleaded)	Alumina series	$ZrSiO_4$ series	ZrO_2	Steel ball	Chrome ball
Density (g cm^{-3})	2.5	2.9	3.6 ± 0.3	3.7 ± 0.1	5.9 ± 0.1	7.6	7.8
Diameter (mm)	0.3–3.5	0.3–3.5	0.3–3.5	0.3–1.5	0.3–3.5	0.2–1.5	1.0–2.0

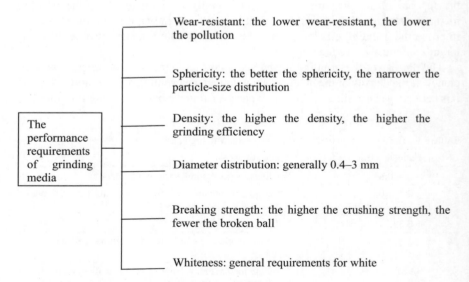

Fig. 6.3 The performance requirements of the grinding media

(3) Density. Under the same grinding conditions, the microbeads with higher density have larger impact force as well as produce excellent grinding effect.
(4) Breaking strength. It is one of the most important performance indicators. It shows the greatest impact force and can be used to judge the occasion in which the microbeads are employed.
(5) Color. The main application areas of the ceramic microbeads are paper and paint industry, and hence, many white microbeads are manufactured.
(6) Different applications. For different situations, microbeads with different diameters are required, as shown in Table 6.1.

6.1.3.2 Use of Ceramic Microbeads in Ball Pen

Over the past two decades, the development of ball-pen manufacturing industry in our country has attracted worldwide attention. The production scale and export volume has been ranked first in the world. While developing pen products, the evolution and improvement of the related parts and components are also desired by the pens industry. The performance of the ball pen mainly depends on the quality of the pen head and ink. The quality of the pen head includes its inherent structure and its coordination with the bead holder, bead, and ink. The beads, as an important component of the ball-point pen, influence the writing performance based on its material and machining accuracy.

In the 1970s, the materials selection of the beads was examined by the technical staff of our domestic pens industry. Owing to its characteristics of high hardness, wear-resistance, and corrosion-resistance, ceramics were considered to be the best material. However, at that time, because of the poor preparation process and tooling technology of high-performance ceramics, the obtained Al_2O_3 ceramic beads, with poor surface finish and low yield, were unable to satisfy the demand of commercialization. Currently, ceramic beads have been equipped in the top-quality ball-point pens in Japan, Switzerland, and Germany, whose pens industries are well developed. However, the molding technology and processing have always been kept as a secret. Nowadays, the ceramic body of the beads molded by the new process of gel-beads formation has obtained a breakthrough in the process aspect of molding and tooling. The obtained ceramic microbeads sized 1.2, 1.0, 0.9, 0.8, 0.7, 0.6, 0.5, and 0.38 mm have passed the test carried out by the related technology department of China, satisfying the technical requirement of the pens industry.

Tungsten carbide alloy is now widely used as the material for the ball-pen beads. These beads are produced with powder metallurgy technology, based on tungsten carbide sintered by the hard refractory metal compounds and binder metal. The metallurgical structure consists of the tungsten carbide phase and the binder phase. ZrO_2, with three crystalline transformations, does not have much practical engineering value because of the volume change during the crystalline transformation at different temperature regions. On the other hand, tetragonal zirconia polycrystals (TZP) comprise single-phase tetragonal ZrO_2. The 3Y–TZP is a kind of TZP with yttria as a stabilizer. The contents of yttria range from 1.75 to 3.5% (mol). Its strength and fracture toughness are up to 1.5 GPa and 15 MPa m$^{-1/2}$, respectively. Currently, 3Y–TZP has the highest room temperature mechanical properties in the field of ceramic materials, and hence, it is known as ceramic steel. This material has been widely used in harsh loading conditions and harsh industrial environments, and is therefore considered to be the best replacement material for tungsten carbide pen beads.

Figure 6.4 shows the micrographs of the cross-section and surface of 3Y–TZP ceramic beads produced by Heibei Yonglong Bangda New Materials Company.

Figure 6.5 presents the micrograph of the cross-section of the tungsten carbide-Tungsten carbide (WC) alloy beads produced by a company in Shanghai.

It can be seen from Fig. 6.4 that the microstructure of 3Y–TZP is uniformly dense; all the grains are small equiaxed-shaped and all the grain sizes are between 0.3 and 0.5 μm. It can be seen from Fig. 6.5 that the tungsten carbide grains are bonded

(a) cross-section (fracture surface)   (b) surface (polished and etched)

Fig. 6.4 3Y-TZP ceramic beads

Fig. 6.5 The cross-section of tungsten carbide alloy beads

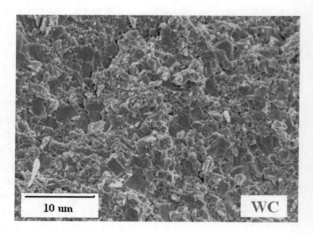

6.1 Ceramic Microbeads

and surrounded by the binder phase, with uneven grain size and larger grain flakes reaching 3.0–5.0 μm.

As the ZrO_2 ceramic grain size is smaller and more homogeneous than the tungsten carbide-ceramic materials, and as the surface of the ZrO_2 beads can reach a very high surface finish after processing, the abrasion of the bead holder will be very small. In addition, ZrO_2 ceramic beads also have good insulation properties and non-magnetic characteristics; hence, they can avoid the electrochemical corrosion engendered when the bead rolls in the bead holder, and effectively reduce the damage to the bead holder. Therefore, ZrO_2 ceramic beads can easily maintain the stability gap between the bead and the bead holder. Thus, the writing time is longer, the lineation is uniform, and the feeling is fine.

The following is the affinity test for the bead materials and ink:

When writing with a ball-point pen, because of the tangential friction force, the bead rolls in the bead holder, adheres and carries ink to the surface of the paper, and finishes the writing process. The good coordination between the bead and the parts that come in contact with it is the premise of continuous writing and results in uniform ink-out. According to the description about the transport condition in fluid mechanics, a good affinity between solid and liquid is needed to make them achieve a balance.

At the interface of the bead, air, and ink, a tangent plane is made along the contact surface of the solid wall and liquid. The angle between the tangent plane and the solid wall in the liquid is the contact angle of the two phases, as shown in Fig. 6.6. The liquid infiltrates the solid when θ is an acute angle, otherwise, the opposite occurs. The occurrence of infiltration depends on the characteristics of the liquid and solid. If the infiltration is better, then the affinity is better, and the liquid can easily adhere to the surface of the solid. The contact angles of different inks and bead materials have been measured at room temperature by the CS400 contact angle apparatus. Two other kinds of ceramic materials that have good mechanical properties and chemical stability have been measured at the same time, as a reference for comparison.

Use of surface modification treatment to improve the capacity of affinity with ink:

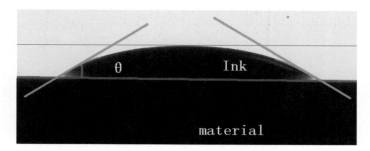

Fig. 6.6 The ink contact angle

The ink, bead, and bead holder are the three major factors that decide the writing performance of the ball-point pen. A ball-point pen with excellent writing performance is a perfect outcome of the combination of these three aspects. When using a ball-point pen, owing to the tangential friction force, the bead rolls in the bead holder, adheres and carries ink to the surface of the paper, and finishes the writing process. Hence, it is essential that the bead and ink have perfect affinity so that the ink adheres to the bead. An earlier test showed that the contact angle between the ink and 3Y–TZP ceramic beads is smaller than that between the ink and the traditional tungsten carbide beads. Therefore, ceramic beads satisfy the basic requirements of a ball-point pen. ZrO_2 ceramic bead is wear-resistant, corrosion-resistant, non-conductive, non-magnetic, and has a dense microstructure. The ZrO_2 ceramic beads produce little abrasion with the bead holder, and the gap between the two is capable of maintaining long-term stability. The writing length of the line is 2–4 times greater than that of the traditional beads. Moreover, ZrO_2 ceramic bead with small and uniform grains can be machined with high precision. In addition, the comfortable feeling is another outstanding feature.

However, ZrO_2 ceramic beads prepared only by grinding and polishing process have some drawbacks, such as poor ink capacity, hydroplaning, and disconnections when writing. In this paper, a surface modification technology for ceramic is analyzed, intending to improve the affinity of the beads with the ink, increase the adaptive capacity of the beads to different inks and bead holders, and expand its application scope. The result shows that the ceramic beads after surface modification treatments have better affinity with ink, and maintain a long-term writing characteristic and comfortable writing feeling.

The 3Y–TZP beads produced by Heibei Yonglong Bangda New Materials Company were treated using surface modification process. Before the treatment, the diameter of the bead was $0.7 + 0.001$ mm, the roundness tolerance was less than or equal to 0.20 μm, and the surface roughness (Ra) was less than or equal to 7 nm.

Figure 6.7 illustrates the process of milling and surface treatment.

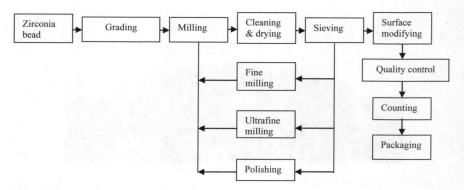

Fig. 6.7 The process of milling and surface treatment

6.1 Ceramic Microbeads

Table 6.3 The test result of contact angle

Materials of the bead		WC	Al_2O_3	Si_3N_4	ZrO_2 (3Y-TZP)
Contact angle (θ)	Neutral ink	28.6°	27.2°	24.9°	22.5°
	Oil-based ink	33.7°	31.0°	30.1°	27.0°
Affinity		Increasing $\longrightarrow$			

To improve the comparability of the test results, all the measured beads were assembled with the bead holder of the same size and materials in a same assembly machine. In addition, a kind of commercial neutral ink was used. The line length and the volume of ink referred to the national standard: GB/T2625-2003 Neutral pen and ink cartridge. Furthermore, Tianjin RH-01C zoned circular writing instrument and analytical balance were employed. The S-450 scanning electron microscope (SEM) was used to observe the changes in the surface morphology of the samples before and after treatment. AJ-III atomic force microscope (AFM) was used to test the sample surface morphology and roughness.

The contact angle of different bead material and ink was measured by the CS-400 contact angle apparatus at room temperature. The results are shown in Table 6.3. It can be seen that the contact angles of the ink and the ceramic materials are all smaller than those of the ink and tungsten carbide alloy, based on the inherent characteristic of the material. The contact angle of ZrO_2 ceramics and ink is the smallest, and thus, the affinity between the ceramic bead and the ink is the best, and it is very much easier to bring out the ink for smooth and even writing.

The writing performance testing of ZrO_2 ceramic beads:

Lead brass, alpaka, and stainless steel are the most commonly used bead-holder materials. In this study, the 0.5-mm ceramic beads produced by Heibei Yonglong Bangda New Materials Company were assembled with the bead holder made from different materials, and the beads were tested at the National Pens Testing Center. Table 6.4 shows the actual length of the lines wrote using different pen points.

It can be seen from Table 6.4 that the length of the lines gets significantly improved after using ZrO_2 ceramic microbeads assembled with the bead holder made from different materials. Lead brass has a good machining performance and low price, but is susceptible to stress corrosion. The average length of the lines wrote using ceramic microbead pen was 600 m, which meets the international standard. On the other hand, the length of the lines wrote using ceramic microbeads assembled with alpaka was 2200 m, which makes alpaka as the best alternative to the much more expensive stainless steel. ZrO_2 ceramic microbeads pen presents universality, because it is applicable to bead holders made using different materials as well as to a variety of inks.

The influence of surface treatment on writing performance:

Table 6.5 shows the average of the 10 test results of the 0.7-mm tungsten carbide and ZrO_2 ceramic microbeads pen assembled with a bead holder made up of lead brass.

Table 6.4 The test results of the length of the lines wrote using ceramic beads pen

Pen holder	Ink	Test results (m)										Average
		1	2	3	4	5	6	7	8	9	10	
Lead brass	Neutral	564	593	575	569	671	627	577	568	589	666	600
	Neutral	2211	2183	2175	2166	2219	2194	2237	2208	2216	2235	2204
Alpaka	Waterborne	1313	1274	1181	1243	1195	1301	1186	1207	1215	1217	1233
Stainless steel	Neutral	2031	2069	2087	2117	2102	2048	2093	2101	2027	2111	2079

6.1 Ceramic Microbeads

Table 6.5 The average of the 10 test results

0.7-mm bead		Ra (nm)	Ink amount (mg m^{-1})	Line length (m)	Break situation	Remark
			Average of 10 times	Average of 10 times		
WC		12–15	1.53	208	Not present	Bead holder is made up of lead brass
ZrO$_2$	Before modification treatment	<7	1.45	605	Exists	
	After modification treatment	8–10	1.66	517	Not present	

It can be seen from Table 6.5 that the average ink amount of the ceramic bead before modification treatment is only 1.45 mg m^{-1}, which is less than that of the tungsten carbide bead (1.53 mg m^{-1}). Although the surface roughness of the ceramic bead pen after modification treatment is still smaller than that of the tungsten carbide one, the ink amount is observed to increase to 1.66 mg m^{-1}. The tungsten carbide bead produces a lot of wear and tear with the bead holder, which causes the line to become uneven when the length reaches about 150 m. However, when the length reaches 200 m, the pen cannot write normally, because the bead nearly breaks away from the bead holder. With regard to the ceramic bead without modification treatment, slipping and breaking occurs at the beginning, but the line length achieved is about 3 times of that of the tungsten carbide bead. Furthermore, the ink streams evenly, especially after the length reaches 200 m. The modification treatment can prevent slipping and breaking at the beginning, and the line length of the ceramic bead is about 2.5 times of that of the tungsten carbide bead. Thus, it can be concluded that after modification treatment, the ceramic bead pen has smaller surface roughness, higher ink amount, and is better than the tungsten carbide bead of the same specifications, owing to the more even line and prevention of slipping and breaking.

The influence of surface treatment on surface microstructure:

Figure 6.8 shows the surface morphology of the ceramic bead after modification treatment, where a is the SEM photo and b is the AFM photo, and whose surface Ra is about 8–10 nm. Figure 6.9 illustrates the surface morphology of the tungsten carbide bead observed through AFM, whose surface Ra is about 15 nm. Figure 6.10 presents the surface morphology of the ceramic bead after mechanical grinding, whose surface Ra is no more than 7 nm.

It can be observed that the ZrO$_2$ ceramic beads have dense microstructure and smooth surface. From Fig. 6.8, it can be seen that after modification treatment, the ceramic bead could form microgrooves and stripes whose width is less than 0.2 μm on the smooth surface. The stripes look like rivers on the plain. Figure 6.9 shows the surface morphology of the tungsten carbide beads. As the crystal grains of tungsten carbide are about 10 times larger than those of ZrO$_2$, and as the hardness of its alloy-coupling phase among the crystal grains is lesser than that of the tungsten carbide

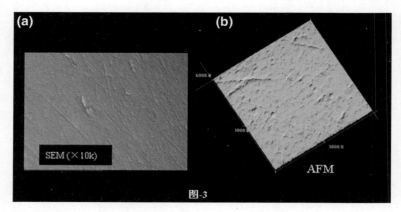

Fig. 6.8 Ceramic beads after modification treatment

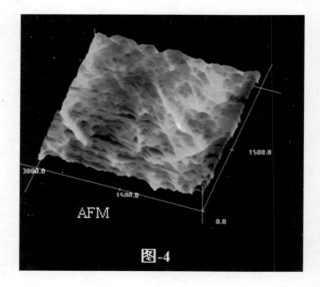

Fig. 6.9 The surface morphology of tungsten carbide beads

crystal grains, the crystal grains easily form large pits when abrasive machining occurs during the coupling phase. Although these pits can improve the affinity of the ink toward the tungsten carbide beads, the microstructure, like peaks and valleys, takes a lot of wear and tear with the bead holder and does not fit well with the bead holder whose hardness is significantly different from that of the bead. Figure 6.10 shows the ceramic beads after mechanical grinding. The surface of the beads endures mechanical damage, and fracture occurs among the crystal grains; thus, the smooth base plane gets damaged and the crystal grains bump down significantly, easily taking up a lot of wear and tear with the bead holder.

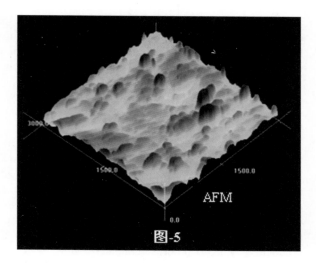

Fig. 6.10 Ceramic beads after mechanical grinding

The surface of the ceramic beads after modification treatment has both smooth base plane and microgrooves, which can improve its affinity toward the ink. It is neither different from the pits abraded between the large tungsten carbide crystal grains and coupling phase nor the inequalities of the crystal grains after mechanical grinding. The microstructures, just like plains and rivers, can ensure less wear between the smooth base plane and the bead holder, while the affinity toward the ink is improved through the rivers; hence, it is considered as an ideal method of surface treatment.

6.1.3.3 Use of Ceramic Microbeads with Far-Infrared Function in Healthcare

Many international research institutes have made the following conclusions regarding far-infrared radiation: the far-infrared radiation that has the same wavelength as the human body's radiation can be totally absorbed by the human body. The far-infrared radiation can penetrate through the skin to stimulate the inner organs of the body. It can accelerate blood circulation, promote metabolism, dredge the meridians, remove humidity and algidity, and increase blood and nutrition supply. When the feet are in the far-infrared energy field, the powerful energy can reach everywhere in the human body through the meridians in the feet. Thus, it can bring out the healthcare effects described earlier.

The far-infrared ceramic beads are obtained from high-purity inorganic ceramic raw material by employing an advanced manufacturing process known as "Colloidal Injection Molding (CiM)". They have many specific electrical, magnetic, and thermal functions, and can radiate far-infrared rays with specific wavelength and high emissivity. The far-infrared radiation can be totally absorbed by the human body. The ceramic beads are very peculiar far-infrared radiation materials. After heating, the ceramic microbeads with far-infrared function may bring out healthcare effects such as accelerating the blood circulation, promoting metabolism, dredging the meridians, removing humidity and algidity, and increasing blood and nutrition supply.

6.1.4 Summary

Gel-beads-forming process is suitable for various ceramic materials, such as Al_2O_3, ZrO_2, Si_3N_4, SiC, etc. This method can be realized by adjusting the process parameters, and does not require the replacement of equipments. When compared with other forming methods of ceramic microbeads, this method has a shorter production cycle and higher forming efficiency. The microstructure of the products is uniform and dense. Thus, the gel-beads-forming method is a simple process, has a short production cycle, its process parameters can be easily adjusted (the diameter of the microbeads is between 0.1 and 3 mm), and automatic and continuous production can be easily realized.

The ceramic microbeads have very wide applications, such as in ultrafine grinding of non-metallic minerals, dispersing all kinds of inks, paints and dopes, as well as deburring and polishing of metal workpieces. The ceramic microbeads prepared by gel-beads-forming process have very good comprehensive properties such as wear resistance, whiteness, crushing strength, and density.

ZrO_2 pen-beads prepared by gel-beads-forming process have the following advantages:

(1) ZrO_2 ceramic beads have dense microstructure, small and uniform crystal grains, high surface-machining accuracy, produce little wear, have a stable gap with the bead holder, produces even lines, as well as longtime writing.
(2) ZrO_2 ceramic beads have the best affinity toward ink, and while writing, the ink streams easily, smoothly, and evenly.
(3) After the test of assembling with the bead holder made from different materials, the line length of the ZrO_2 ceramic beads pen is observed to be significantly better than that of the traditional materials, and presents universality to bead holders made from different materials as well as to a variety of inks.
(4) Surface modification treatment could improve the surface micromorphology of the ZrO_2 ceramic beads forming structures like plains and rivers, which can significantly improve the affinity toward the ink and prevent slipping and breaking at the beginning. At the same time, the surface modification treatment avoids formation of structures like peaks and valleys, and brings down the wear with the bead holder significantly, ensuring that the ceramic beads pen has a long line and fine hand-feeling.

6.2 Improving the Breakdown Strength of the Rutile Capacitor

Gelcasting is a near-net-shaped method combining the physical chemistry of the polymer, colloidal chemistry, and ceramic technology (Omatete et al. 1991; Young et al. 1991). The main steps (Janney et al. 1991; Janney et al. 1992; Janney, 1990) of gelcasting are as follows: first, the ceramic powder (e.g., alumina or silicon nitride), along with small quantities of gel initiators, catalyst, monomers, crosslinker, and sintering aids is thoroughly mixed to form the homogeneous suspension with high solids loading and low viscosity. Second, gelation is initiated after the suspension is casted into a non-pore mold and when the temperature is elevated. Thus, the three-dimensional network structure resulting from this in situ polymerization holds the ceramic particles together. Finally, after unmolding and drying, the green bodies possess high homogeneity and high strength, and after sintering, such bodies exhibit good mechanical property. This method is a promising colloidal in situ forming technique, and has been widely utilized in the production of various ceramic material systems (Nunn and Kirby 1996; Sun et al. 1996; Waesche and Steinborn 1997).

Rutile capacitor exhibits an excellent mechanical and electrical performance, possesses high dielectric constant, low loss angle, and large breakdown strength. It is an important material particularly applied as a high-power capacitor. It has been widely used in high-frequency appliances, such as broadcast emitters, radar, high-frequency welding machines, smelting furnaces, etc. For these applications, properties such as dielectric constant, loss angle, and breakdown strength are considered to be very important (Li et al. 1995). Usually, the dielectric constant of rutile ceramic is 60–80, ranging from 0.5 Hz to 5 MHz at room temperature, and the tangent of loss angle under 1 MHz is about $4 \sim 5 \times 10^{-4}$ (testing temperature is 20 ± 5); such a low loss angle is beneficial in decreasing the energy loss of the rutile ceramic under high frequency. Another important property of the dielectric materials is the ability to withstand large field strengths without electrical breakdown. According to the national standard of the People's Republic of China, the breakdown strength of the rutile ceramic applied under DC field should be above 10 kV mm^{-1}.

In the past decades, rutile capacitors with complex shapes and large sizes were mainly produced through extrusion-forming process and subsequent machining (Yang et al. 1999). However, the whole procedure was time-consuming (about 2½ months), and machining of dried green bodies resulted in extensive pollution and wastage of raw materials. Furthermore, it was difficult to produce rutile capacitors with a large size, complex shape, or containing numerous additives in the raw materials, using the traditional process. To overcome these disadvantages, the present authors developed a new route based on gelcasting (Ma et al. 2001), by which the production cycle can be reduced to 3 weeks, and the dust caused by machining can be eliminated. Thus, both production costs and pollution could be significantly reduced, while the functional property could be considerably improved, especially the breakdown strength.

Table 6.6 Composition of the rutile mixture (wt%)

Main starting materials	Additives				
TiO_2	CaF_2	H_2WO_4	$BaCO_3$	ZnO	Bentonite
100	2	2	1	1	5

The composition of the rutile mixture applied to extrusion-forming process is shown in Table 6.6. The commercial rutile powder and additives (Coke-oven Plant of Shanghai, China) are of industrial purity, and the particle sizes of the additives are at the micrometer level. The reagents that were used in this study are as follows.

Solvent: Deionized water with a conductivity of 1.02 $\mu S\ cm^{-1}$.

Monomer: Acrylamide (AM) obtained from Mitsui Toatsu Chemical Inc., Japan.

Crosslinker: Methylenebisacrylamide (MBAM) obtained from Hongxing Biological and Chemical Factory of Beijing, China.

Initiator: $(NH_4)_2S_2O_8$ obtained from Beijing Third Reagent Works, China.

Catalyst: N, N, N', N'-tetraethylmethylenediamine (TEMED) obtained from Xingfu Fine Chemical Institute.

Dispersants: A poly(methacrylic acid)-ammonium (PMAA-NH_4) solution obtained from Chemical and Engineering University of Nanjing, China.

The rutile mixture, according to Table 6.6, was calcined for 2 h at 1170 °C and then ball-milled for 50 h, to obtain the calcined rutile mixture with d_{50} of 2 μm. The gelcasting process of the rutile mixture is presented in Fig. 6.11. First, the calcined rutile mixture was dispersed in a premix solution, prepared by dissolving AM and MBAM in deionized water. With the aid of a dispersant, the concentrated slurry with low viscosity was obtained after ball-milling for 24 h. The slurry was then degassed in vacuum for 7 min after the addition of the initiator and catalyst. All the abovementioned steps were operated at room temperature. Subsequently, the slurry was casted into a steel mold heated in an oven at 70 °C. After coagulation, the demolded green bodies were dried, debindered, and sintered.

The particle-size distributions were measured using BI–XDC particle-size analyzer (Brookhaven Instrument Corp., USA). The zeta potentials were measured using Zeta Plus analyzer (Brookhaven Instrument Corp., USA). With 0.5% solids volume loading, the tested powder was dispersed in demonized water, and the pH was adjusted by the addition of HCl or NaOH. The apparent viscosity of the suspension was measured using a viscometer (MCR-300 mode, Physica Corp., Germany). During the measurement, concentric cylinder (gap size of 1.13 mm) was used, and the shear rate was increased from 0.1 to 100 L s^{-1} at 25 °C. The calcined rutile mixtures were examined using an X-ray diffraction (XRD) instrument (D/MAX IIIB mode, Rigaku Corp., Japan). The green body and sintered body were observed using S-450 SEM (Hitachi, Japan). Disk-shaped green bodies of $\Phi\ 30 \times 2$ mm were prepared for the measurement of the electrical properties. Silver slurry was pasted on the two main sides of the sintered bodies to form electrodes. The dielectric properties of these samples were measured at 1 MHz under room temperature, and the breakdown strength was measured in Si oil, by applying a DC field between the two opposite electrodes.

6.2 Improving the Breakdown Strength of the Rutile Capacitor

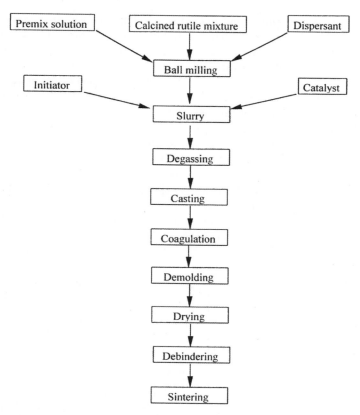

Fig. 6.11 The flowchart of gelcasting of calcined rutile mixture

6.2.1 The Influence of Sintering Additives on the Flow Behavior

Generally, various additives are added to the rutile powder, such as $BaCO_3$, ZnO, CaF_2, H_2WO_4, and bentonite, to improve the sintering behaviors and the electroproperties. Bentonite, usually employed as an extrusion aid, can also lower the sintering temperature, and hence, was considered as a sintering aid in our experiment. At a relatively lower temperature, fusible phase resulted from $BaCO_3$, ZnO, and CaF_2, causing a decrease in the sintering temperature. Furthermore, H_2WO_4 was found to lead to fine crystal grain of the rutile powder, and thus, may play a role in increasing the dielectric properties of the rutile powder.

As the zeta potentials of all these components are significantly different, it is difficult to appropriately disperse all these powders for gelcasting. Rutile powders containing different aids (Table 6.6) were dispersed in deionized water. To maintain definite solids volume loading, the quantity of the deionized water was decided by means of the content of different sintered aids and their densities. H_2WO_4 particles were difficult to disperse in the basic medium owing to their acidity; however, the concentrated rutile suspension must be prepared in a basic condition. According to Fig. 6.12, $BaCO_3$, ZnO, and CaF_2 have a slight influence on the viscosity of the rutile suspension, while bentonite has a relatively large influence. From the viscosity measurement, it can be observed that the concentrated pure rutile suspension exhibits an obvious shear-thinning behavior (Mikulasek et al. 1998; Mikula´sU ek et al. 1997), which is similar to that of the pure alumina gelcasting slip. However, according to Fig. 6.13, the rheological characteristics of the rutile mixture slurries prove to be shear thickening owing to the existence of additives, and if the quantity of bentonite is more, then the viscosity changes are more obvious. When absorbing water, bentonite particles dilate up to 30 times (Du and Tang 1986), significantly changing the volume fraction, which results in an increase in the slurry viscosity and shear-thickening phenomenon. As the viscosity of the slurry with 5 wt% of bentonite is too large (about 6 Pa s), the slurry is not suitable for gelcasting.

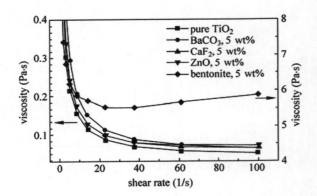

Fig. 6.12 Influence of the sintering additives on rutile flow behavior

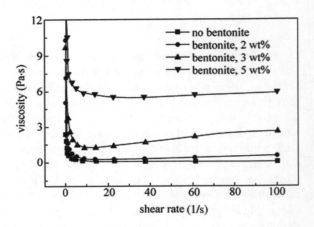

Fig. 6.13 Influence of the addition of bentonite on rutile flow behavior

6.2.2 Calcining of the Rutile Mixture

In general, calcining can remove volatile impurities, chemicals-bonded water, physical-absorbed water, and can improve the purity of the raw materials (Liu, 1990). Furthermore, it can also help to form stable crystal phases and densify the particles, which results in improved properties of the ceramic parts.

In our experiment, the rutile mixture was calcined at high temperature and subsequently ball-milled for 50 h. The XRD pattern (Fig. 6.14) revealed some additional peaks besides the rutile peaks for the mixture calcined at 900 °C[1] ([1] To observe the diffraction peak evidently, the ratio of the rutile to sum of additives was changed to 1:1). When compared with the XRD pattern of the unmilled mixture, more complex peaks that do not represent the rutile were found in the XRD pattern of the ball-milled mixture. This indicates that new phases appear during the ball-milling procedure. However, almost no other peaks besides the rutile were found in the XRD pattern of the rutile mixture calcined at 1170 °C (Fig. 6.15); even after ball-milling, other peaks failed to appear. Therefore, it can be concluded that when calcined at 1170 °C, phases in the mixture will be unified even after ball-milling. In summary, when calcined at 1170 °C, the additives can form fusible phase unobservable by X-rays; thus, the harmful influence of H_2WO_4 and bentonite mentioned earlier could be removed, and the interaction between the rutile mixture and water can also be avoided. Thus, suitable suspension for gelcasting can be obtained easily.

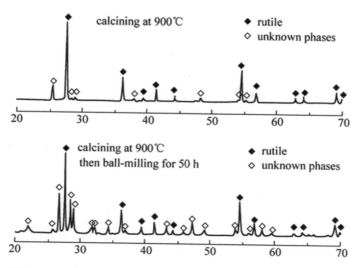

Fig. 6.14 XRD pattern of the rutile mixture calcined at 900 °C

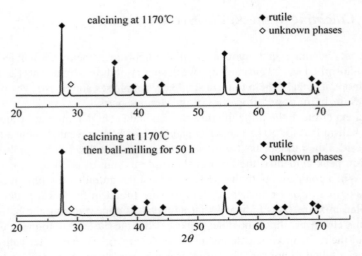

Fig. 6.15 XRD pattern of the rutile mixture calcined at 1170 °C

6.2.3 The Rheological Behavior of the Calcined Rutile Mixture

Figure 6.16 shows the relationship between the zeta potential and pH value of the rutile mixture calcined at 1170 °C. When a dispersant (PMAA-NH$_4$ solution) is added, the zeta potential of the rutile mixture is greatly improved. The dispersant is found to result in the change in the particle surface charge as well as the electrochemical double layer at the solid/liquid phase, because of its adsorption at the particle interface. The negative groups on the polymer chain make the zeta potential more negative, imparting stability to the particles against flocculation. The adsorption

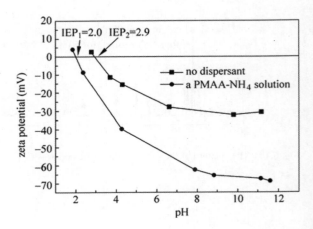

Fig. 6.16 Relationship between the zeta potential and pH value

6.2 Improving the Breakdown Strength of the Rutile Capacitor

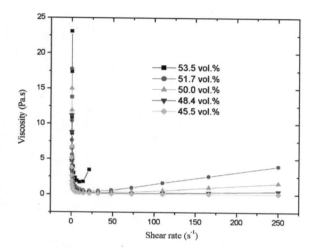

Fig. 6.17 Rheological behavior of the rutile suspension at different solids loading

obviously decreases the electrophoretic mobility, shifts the isoelectric point (IEP) of the system to low pH values, and improves the absolute value of the zeta potential (Strauss et al. 1993; Chen et al. 1998).

Figure 6.17 shows the curves for the rheological behavior of the rutile mixture calcined at 1170 °C at different solids loading. The figure indicates that the viscosity of the slurry increases gently with the increase in the solids loading, and that the slurries are shear-thinned and subsequently shear-thickened. When the solids loading of the slurry is more than 52 vol%, the slurry quickly changes its behavior from shear-thinning to shear-thickening, and its lowest viscosity becomes higher than 1 Pa s; thus, the slurry cannot fit gelcasting. However, when the solids loading of the slurry is less than 52 vol%, it can be used for gelcasting, owing to its relatively low viscosity (shear rate <80 s^{-1}).

6.2.4 Gelcasting of the Calcined Rutile Mixture

Figure 6.18 presents a micrograph of the green bodies gelcasted from the slurry of the rutile mixture calcined at 1170 °C with 51.7 vol% solids loading. It can be observed that the monomer has been polymerized efficiently, and that the homogeneous microstructure can be achieved. Figure 6.19 indicates that all close pores in the rutile parts are removed, which results in a good microstructure after sintering. The functional properties were measured, and are listed in Table 6.7. Among these parameters, the breakdown strength was greatly increased from the original 12.6 kV mm^{-1} to more than 23.6 kV mm^{-1}.

Flaw minimization and processing optimization of the ceramic materials can be successfully accomplished by colloidal processing (Lange, 1989; Pugh and Bergströ¨m 1994). During the gelcasting process, aggregates of the powders could

Fig. 6.18 Micrograph of the green bodies gelcasted from the slurry of the rutile mixture

be removed in a well-dispersed suspension and homogeneous green bodies can be obtained from the suspension through adequate degassing (Zhou et al. 2000); thus, the microstructure of the gelcast parts after sintering is homogeneous. It is well known that the dielectric breakdown phenomenon is structure-sensitive, and that the breakdown strength values vary widely according to the characteristics of the microstructure of the dielectric materials (Yamashita et al. 1980). By gelcasting, the microstructure of the sintered bodies of the rutile mixture can be improved, and the breakdown can be significantly enhanced.

6.2.5 Summary

The high-power rutile capacitor can be prepared by extrusion-forming process and mixing various sintering aids and processing aids. Among these aids, $BaCO_3$, ZnO, and CaF_2 have a slight influence on the viscosity of the rutile suspension, while bentonite and H_2WO_4 have a notable influence. Particularly, if the bentonite content is more, then the mixture exhibits an increase in viscosity. The rutile mixture applied to extrusion-forming process is calcined at 1170 °C, which results in the fusible phase, and thus, the harmful influence of bentonite and H_2WO_4 mentioned earlier can be removed; in addition, the interaction between the rutile mixture and water can also

6.2 Improving the Breakdown Strength of the Rutile Capacitor

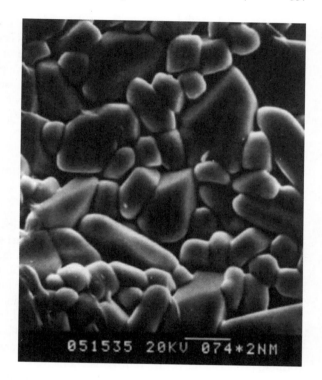

Fig. 6.19 The microstructure of the parts produced by gelcasting

Table 6.7 Properties of the rutile sintered body formed by gelcasting

Tested sample	ε_1	$tg\delta_1$			EnP (kV mm^{-1})	EnP (kV mm^{-1})[a]
		$20 \pm 2\,°C$	$85 \pm 2\,°C$	H_2O		
1	81.9	0.5	0.3	0.3	25.0	13.2
2	80.9	0.4	0.2	0.2	22.0	12.7
3	81.1	0.4	0.2	0.2	22.5	13.2
4	80.3	0.3	0.1	0.3	22.6	12.9
5	80.2	0.4	0.2	0.4	22.7	11.3
6	80.9	0.6	0.4	0.5	25.0	12.2
7	81.0	0.4	0.2	0.5	24.6	14.1
8	80.8	0.4	0.2	0.4	23.6	11.7
9	81.7	0.4	0.2	0.5	22.7	12.3
10	80.6	0.5	0.3	0.5	25.2	12.2
Average value	80.9	0.4	0.2	0.4	23.6	12.6

[a]The breakdown strength of the rutile sintered body by extrusion-forming process

be avoided. Hence, a suitable suspension for gelcasting can be obtained easily. With the aid of the dispersant, PMAA-NH$_4$ solution, a concentrated rutile suspension of 51.7 vol% with low viscosity (about 0.40 Pa s) can be successfully prepared. Thus, by employing calcining, the complex rutile mixture applied to extrusion-forming process becomes suitable for gelcasting and good microstructure as well as superior properties can be achieved. For example, the breakdown strength is improved from the original 12.6 kV mm^{-1} to more than 23.6 kV mm^{-1}.

6.3 Thin-Wall Rutile Tube for Ozone Generator with High Dielectric Constant

Ozone is one of the strongest oxidants extensively applied in water treatment, chemical oxidation, food protection, medical treatment, etc. In recent years, the requirement for high-capacity ozone generators is rapidly increasing. Earlier studies have indicated that the yield of ozone is directly proportional to the dielectric constant and wall thickness of the dielectric body in the ozone generator (Chu and Wu 2002; Wang 1990). The dielectric constant of the rutile ceramic is about 80–100, which is 15–30 times greater than that of glass and enamel. Furthermore, the rutile ceramic possesses high breakdown voltage. These two characteristics indicate that the rutile ceramic is the optimal material for the dielectric body of the ozone generators. However, the forming process of large-size thin-wall ceramic tube is very complicated. Technical difficulty and high cost are the main reasons that restrict the application of rutile ceramic in ozone generators.

The essential components of the colloidal injection molding process are the reactive organic monomers: monofunctional AM, $C_2H_3CONH_2$, and difunctional MBAM, $(C_2H_3CONH)_2CH_2$. The weight proportion of AM and MBAM is about 100:1–25:1. The monomers are dissolved in deionized water with 10–30% weight ratio, to obtain a premix solution. Furthermore, ammonium persulfate, $(NH_4)_2S_2O_8$, is used as an initiator (0.1–0.5 vol% of the slurry), TEMED is employed as a catalyst (0.05–0.25 vol% of the slurry), and PMAA-NH$_4$ solution is used as a dispersant (0.2–0.5 wt% of the ceramic powder). A concentrated 52 vol% of TiO$_2$ suspension with low viscosity is prepared by adding the TiO$_2$ powder and dispersant to the premix solution, followed by ball-milling for 24 h.

A stainless steel mold was used to obtain the thin-wall ceramic tubes. An antisticking agent was pre-coated on the in-wall of the mold, which was preheated at about 80–100 °C. The large-size thin-wall rutile tubes were prepared by employing the machine used for the colloidal injection molding of the ceramics. The processing parameters include injection speed of about 15–20 mm s^{-1}, injection pressure of 0.3 MPa, and pressure holding time of 1–2 min.

6.3.1 Results and Discussions

Appropriate sintering schedule is important for obtaining high-performance ceramics. Thermal analysis revealed that there is a strong exothermic effect at 357 °C; thus, the corresponding decomposition of the organic substances in the green body require a slow temperature rising speed from 300 to 400 °C. Within the temperature ranges from 950 to 1050 °C and from 1100 to 1200 °C, the shrinkage rates of the TiO_2 body are more rapid than ever, and hence, the heating-up process should be mild at these two stages to avoid formation of closed gas porosity. At a temperature from 1200 to 1300 °C, the liquid phase will appear in the body and ceramic-forming reaction occurs; therefore, control of holding time is very important at this stage. Figure 6.20 shows the sintering schedule curve of the thin-wall rutile tube. From Fig. 6.21 that presents the SEM fractograph of the sintered body, it can be seen that the tube possesses fine crystal grains and high degree of densification.

The ozone generator requires a dielectric body with high dimensional precision, and the thin-wall rutile tubes prepared through CIMC can satisfy this criterion. The wall thickness of the sintered tube is found to be 1.5–2 mm and its inner diameter is observed to be about 26 mm, with its length reaching about 300 mm. Furthermore, the largest eccentricity of the tubes is found to be 0.15. Figure 6.22 shows the configuration of the large-size thin-wall rutile tubes prepared by CIMC for the ozone generator.

The dielectric constant of the dielectric body is the important factor that determines its electric performance. In this research, a frequency spectrometer was used to measure the dielectric constant of the obtained rutile tube. Figure 6.23 shows the variation of the tubes' dielectric constants with frequencies at different temperatures. This figure indicates that the tubes' dielectric constants are about 85 at low frequency, and the values rise constantly with the increasing frequency. Figure 6.24 shows the variation of the tubes' dielectric constants with temperatures at different frequencies. The figure indicates that the variation of the tubes' dielectric constant is inconspicuous when the temperature is increasing. Thus, the tubes' dielectric performance will not be influenced when the temperature in the discharge chamber varies owing to the

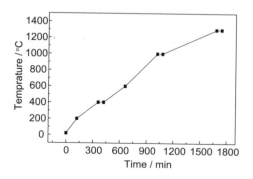

Fig. 6.20 Sintering schedule curve of the thin-wall rutile tube

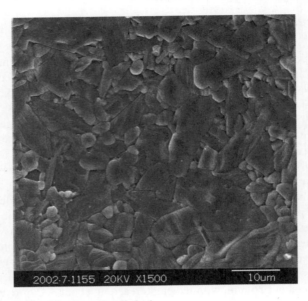

Fig. 6.21 Fractograph of the sintered rutile body

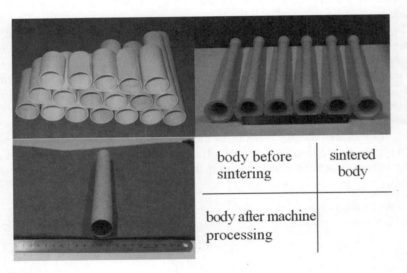

Fig. 6.22 Configuration of the rutile tubes for ozone generator

corona discharge. This characteristic can make the ozone generator work stably and safely.

The breakdown voltage of the tubes was also measured in this study. The average spot breakdown voltage of 20 samples was 16.8 kV mm^{-1}, and the value of the

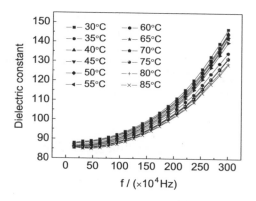

Fig. 6.23 Variation of the dielectric constants with temperatures at different frequencies

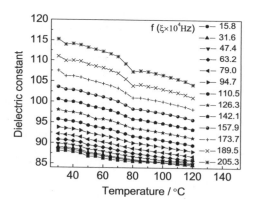

Fig. 6.24 Variation of the dielectric constants with frequencies at different temperatures

surface was about 10.1 kV mm^{-1}. This performance perfectly satisfied the demand of the ozone generator.

The obtained tubes were applied in the ozone generator. The yield of ozone was about 50 mg L^{-1} under the following experimental conditions: power supply of 300 W; frequency of 7000 Hz; input voltage of 220 V; and current of 0.3 A. At the same time, the generator's Corona Voltage sharply decreased from 4000–10000 V to 800–1000 V.

6.3.2 Summary

The large-size thin-wall rutile ceramic tubes with eminent properties were prepared by colloidal injection molding of ceramics, and the technological parameters of forming and sintering were determined in this research work. The wall thickness of the obtained tubes was 1.5–2 mm and its inner diameter was about 26 mm, with a length of about 300 mm. The electric properties of the tube were also measured under different conditions. The abovementioned tubes were applied in the ozone generator, and the yield of ozone was about 55 mg L^{-1} under the following experimental conditions: power supply of 300 W; frequency of 7000 Hz; input voltage of 1000 V; and current of 0.3 A. Furthermore, the generator's Corona Voltage was greatly reduced from 4000–10000 V to 800–1000 V.

6.4 Refractory Nozzle of Zirconia

The refractory nozzle for precision casting of a 50%Cu–50%Cr molten alloy is required with a high melting point above 2200 °C and a good thermal shock resistance to withstand the temperature rising from the room temperature to 2200 °C in several seconds. To accurately control the molten alloy flux, the nozzle should also possess a precise dimension that remains unchanged during casting.

ZrO_2, with a melting point of 2680 °C, is usually used as a refractory material at an extremely high temperature. Owing to its microcracks induced by martensitic phase transformation, ZrO_2 materials have high toughness and good resistance to thermal shock and flux scour. Hence, it is considered as a suitable material for making the nozzle for precision casting of Cu–Cr molten alloy. To avoid deformation caused by the shrinkage of green body during sintering, some coarse particles in the ingredient are found to be useful.

Gelcasting is a near-net-shaped technique used for the manufacturing of homogeneous ceramic green bodies, owing to its characteristic in situ consolidation of the suspension. However, few researches on the gelcasting of the suspensions with coarse particles have been reported. For these suspensions with coarse particles, the settlement stability as well as aggregation stability should be considered. In this study, the effects of various kinds of dispersants on the rheological behavior and the settlement stability of ZrO_2 suspension with coarse particles were studied, and such suspension can be employed to produce a refractory nozzle for precision casting of Cu–Cr molten alloy using gelcasting.

Three kinds of ZrO_2 were used in this study: ZrO_2 A with a size range of 0.2–0.074 mm, ZrO_2 B with a size range of 0.044–0 mm, and ZrO_2 C with a size of d_{50} = 1 μm. ZrO_2 A and B are electric-fused CaO-stabilized cubic structures, and ZrO_2 C is a TZP containing 3 mol% Y_2O_3. By choosing cubic ZrO_2 as the major raw material, we were able to avoid the volume change caused by transformation during sintering and service. Furthermore, some Y-TZP containing complex phase was used as one of the ingredients for the manufacture of the nozzle, which is believed to improve thermal shock resistance of the ceramic parts.

Three kinds of dispersants, namely TAC (tri-ammonium citrate), arabic gum, and APA (ammonium polyacrylate), were used to prepare ZrO_2 suspensions. They were found to separate the particles in the suspensions by electrostatic, steric, and electrosteric repulsive forces, respectively. AM and MBAM, in the mass ratio of 50:1, were dissolved in the deionized water to obtain 15 wt% premix solution. ZrO_2

was added to the premix solution in the proportion of A:B:C = 4:8:1. The mixture was ball-milled for 48 h to obtain a suspension.

The rheological curve of the suspension was measured with MCR300 Rhometer (Physica Corp., Germany). To measure the sediment stability, the suspension was cast into a Φ20-mm glass tube with the height of 30 cm, and the height of the supernatant relative to the limpid liquid layer was measured at a certain interval to indicate the sediment stability of the suspension.

The density of the nozzle was measured according to the Archimedes' law; the phase composition of the nozzle was analyzed with a D/max-RB X-Ray Diffractometer (Rigaku, Japan); and the microstructure was analyzed with a S-450 SEM (Hitachi, Japan).

6.4.1 Rheological Behaviors of Zirconia Suspensions with Different Dispersants

In ZrO_2 suspensions, the trend of particle flocculation owing to van der Waals attraction is inhibited because of the repulsion caused by the dispersants. However, owing to the Brownian motion of the solvent molecules, some particles may overcome the repulsion to move closer to each other and form stable or metastable particle clusters. Solvent enclosed in the clusters loses its fluidity to some extent and becomes non-free solvent. The viscosity of the suspensions indicates the volume of the free solvent, and thus, the presence of more clusters in the suspension results in higher viscosity of the suspension.

In the suspension stabilized by TAC, the particles are separated by electrostatic repulsion, which is governed by ζ potential on the particle surface. Appropriate TAC ratio (0.2 wt% to the ZrO_2 mass) yields a higher ζ potential and few particle clusters, and subsequently, low viscosity of the suspension. Exceeding TAC amount may suppress the double-layer thickness near the particle surface and reduce the ζ potential because of high ion strength, and increase the viscosity of the suspension. ZrO_2 suspension stabilized with TAC shows a shear-thinning behavior. When sheared, the particle clusters are destroyed by the velocity gradient in the suspension, and water enclosed in these clusters is released and becomes free solvent, and thus, the viscosity of the suspension decreases. As the shear rate rises, more and more clusters are destroyed. When the shear rate reaches more than 300 s^{-1}, almost all the clusters are destroyed, and the viscosity of the suspension remains, by and large, unchanged and independent of the TAC content (Fig. 6.25a).

On the other hand, the viscosity of the suspension stabilized with arabic gum changes with the shear rate in an another way (Fig. 6.25b). If the amount of the arabic gum is not sufficient to cover the surface of the particles in the suspension, as in the case of 1 wt% to ZrO_2 mass, there may not be adequate repulsion between the particles to separate them, and the clusters may be formed by the bridging effect. These clusters will be broken when sheared, and the suspension will show a shear-thinning behavior.

Fig. 6.25 Rheological curves of 55 vol% zirconia suspensions with different dispersants

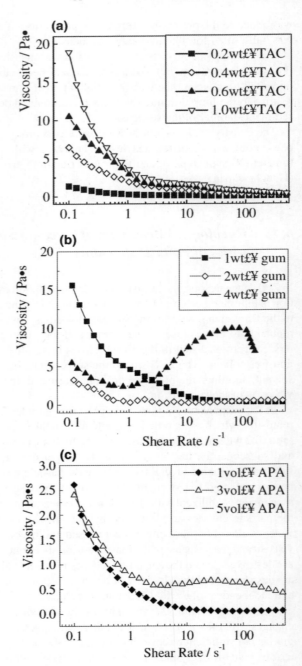

6.4 Refractory Nozzle of Zirconia

If the amount of the arabic gum is sufficient to cover the surface of the particles (2 and 4 wt% to ZrO_2 mass), then few clusters remain in the suspension, and subsequently, the suspension will show an inconspicuous shear-thinning behavior. In this case, at a higher shear rate, the osmotic pressure in the zone among the particles, which is caused by the concentration difference of the macromolecular chains and which avoids the particles to move closer in the static state, will prevent the particles from moving apart. As a result, the viscosity of the suspension increases with the rise in the shear rate, and the behavior of shear-thickening is observed.

The ZrO_2 suspension stabilized with APA (Fig. 6.25c) shows an average rheological behavior when compared with those stabilized with arabic gum and TAC. The shear-thinning behavior of this suspension is not as severe as that of those stabilized with TAC at low shear rates, owing to the steric repulsion. Furthermore, the shear-thickening behavior is also not as severe as that of those stabilized with arabic gum at high shear rates, because of the existence of electrostatic repulsion between the APA chains.

6.4.2 Sediment Stability of Zirconia Suspension with Different Dispersants

The sediment volumes of 55 vol% ZrO_2 suspensions stabilized with different dispersants are compared in Fig. 6.26. During the incipient stage of settlement, ZrO_2 particles undergo an accelerated course, and the increasing velocity of the particles improves the velocity gradient near them. Hence, in the shear-thinning suspensions, the settlement of the particles decreases the viscosity of the suspensions. On the other hand, decreased viscosity accelerates the particles according to Stokes' equation. As a result, settlement equilibrium is achieved in a short time period for suspensions with a low viscosity or a severe shear-thinning behavior. This can be illustrated by ZrO_2 suspensions stabilized with TAC, which exhibit the worst settlement stability (Fig. 6.26). Because of the impediment of the macromolecular chains to the

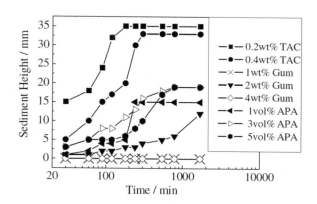

Fig. 6.26 Sediment height of zirconia suspensions with various dispersants

motion of the particles, the settlement of the suspensions with APA and arabic gum is lesser than that with TAC. With regard to the suspensions stabilized with APA, the one containing 1 vol% APA exhibits the highest viscosity at low shear rate but most severe shear-thinning behavior (see Fig. 6.25c), and subsequently, its settlement height changes slowly at the beginning but increases more quickly after 180 min. The suspension stabilized with arabic gum shows the best settlement stability owing to its high viscosity and obvious shear-thickening behavior.

6.4.3 Preparation of Zirconia Refractory Nozzles

Although ZrO_2 suspension with arabic gum has the best settlement stability, it is not the best choice for the preparation of the refractory nozzle, as the solids loading of the ZrO_2 suspension, which is important to avoid deformation during the sintering of the green bodies, could not be increased to more than 60 vol% by the steric stabilization of the arabic gum. With a compromise to fluidity and settlement stability, ZrO_2 suspension with 69 vol% solid and 1 vol% APA was prepared to shape the refractory nozzle for precision casting of Cu–Cr alloy. A 2-μl initiator (8 wt% $NH_4S_2O_8$ solution) and a catalyst (tetramethylethylenediamine) were added to 1 ml suspension, and then stirred for 3 min to make it disperse uniformly. The stirred suspension was cast into a mold, and kept at about 40 °C for about 30 min to ensure solidification of the suspension. The demolded green body was dried at 110 °C for 24 h, and then was sintered at 1580 °C for 2 h. The prepared nozzle was found to satisfy the requirement of precision casting of Cu–Cr molten alloy, and its microstructure, phase composition, and some final properties are shown in Fig. 6.27, Fig. 6.28, and Table 6.8, respectively. As shown in Fig. 6.27, there are some pores in the nozzle, which may be responsible for the good thermal shock resistance of the nozzle. On the other hand, the underfired nozzle may shrink to counteract the dilation caused by the elevated temperature when it is used, which also improves the thermal stability. The XRD pattern (Fig. 6.28) shows that besides some $CaZr_4O_9$, which is a metastable phase synthesized during the production of electric-fused CaO-stabilized ZrO_2, the nozzle mainly consists of cubic and tetragonal ZrO_2, and the complex phases also contribute to the thermal stability. The good resistance to flux scouring may be owing to the coarse fused Ca–ZrO_2 particles.

6.4.4 Summary

ZrO_2 suspensions containing coarse particles and stabilized with different dispersants show different rheological behavior and settlement stability. Generally, ZrO_2 suspensions with obvious shear-thinning and shear-thickening behavior have the worst and best settlement stability, respectively. With respect to both fluidity and settlement stability, APA was chosen as the dispersant to prepare 69 vol% ZrO_2 suspensions

6.4 Refractory Nozzle of Zirconia

Fig. 6.27 SEM micrograph of the zirconia nozzle

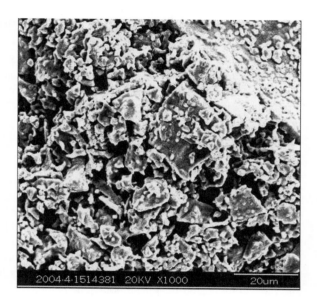

Fig. 6.28 XRD pattern of the zirconia nozzles

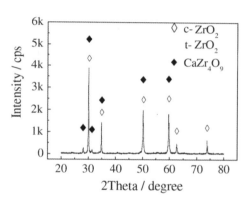

Table 6.8 Some properties of zirconia refractory nozzles

Density (g cm^{-1})		Three-point flexural strength (MPa)			Sintering linear shrinkage 1580 °C × 2 h
Top of the nozzle	Bottom of the nozzle	Green body	Sintered	Sintered, 1100 °C water-cooled	
4.85	4.93	20.6	35.9	23.6	4.17%

to make a refractory nozzle for precision casting of Cu–Cr alloy by gelcasting. The nozzle had little density difference between the different parts, and exhibited good resistance to thermal shock and flux scouring.

6.5 Water-Based Gelcasting of Lead Zirconate Titanate

In the last several decades, piezoelectric materials such as lead zirconate titanate (PZT) have found an increased demand in a wide field of applications including actuators, sensors, and transducers. The hardness and fragility of the material makes it difficult for mechanical processing, and fabrication is often a time-consuming task. In some methods used for producing the material, such as tape casting or injection molding, the ceramic phase is dispersed in a nonaqueous solvent, and a significant volume fraction of the organic binders are required to support the structure. As the organic substances bring about a series of material problems, such as debinding issues during sintering and in homogeneity of the sintered parts, an aqueous solvent with low organic content is highly desirable in the processing of the materials.

The gelcasting was first developed by Janney and Omatete at the Oak Ridge National Laboratory (ORNL), USA (Young et al. 1991). It is an attractive ceramic-forming process for making high-quality complex-shaped ceramic parts by means of in situ polymerization, through which a macromolecular network is created to hold the ceramic particles together. The gelcast parts have low polymer content, and thus, complicated binder burnout schedule can be avoided. The gelcast green bodies also have high homogeneity and high mechanical strength, which is of great advantage for handling the parts before firing and for producing large castings. Since its invention in 1991, a wealth of literature describing the application of gelcasting for the production of structural ceramics, such as Al_2O_3, ZrO_2 (Bengisu and Yilmaz 2002), SiC (Vlajic and Krstic 2002; Yi et al. 2002), Si_3N_4 (Huang et al. 2000), mullite (Liu et al. 2001), etc., has been published. However, little has been reported about its application in the production of piezoelectric ceramics.

The colloidal chemistry of the PZT powders in an aqueous solvent and the rheological behavior of the PZT suspension were determined in this study. Furthermore, microstructures and piezoelectric properties of the gelcast samples from suspensions with different solids loading were also investigated and compared with those of the die-pressed samples. Using the same sintering method, the gelcast samples were found to exhibit stronger piezoelectric effects than the die-pressed ones, which may be owing to their microstructural difference.

The PZT [$Pb(Zr_{0.52}Ti_{0.48})O_3$] near the morphotropic phase boundary (MPB) composition, prepared by the mixed-oxide processing technique, was chosen for this study. Stoichiometric amounts of Pb_3O_4, ZrO_2, and TiO_2 were mixed together in ethanol and thoroughly ball-milled for 24 h. After drying, the mixture was calcined at 850 °C for 2 h, then ground and passed through 200-mesh sieves. The ultimate powder was characterized using XRD before being used. The particle-size distribution of the powder was measured by X-ray sedimentation method using a Brookhaven BI-XDC-type analyzer (Fig. 6.29).

In the gelcasting system, deionized water was used as the medium. AM ($C_2H_3CONH_2$) and MBAM [$(C_2H_3CONH)_2CH_2$] were used as the monomer and crosslinker, respectively. In addition, TEMED and $(NH_4)_2S_2O_8$ were used as the catalyst and initiator, respectively. AM and MBAM were dissolved in deionized water

6.5 Water-Based Gelcasting of Lead Zirconate Titanate

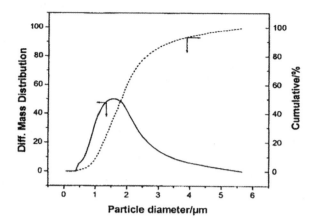

Fig. 6.29 Particle-size distribution of the PZT powder

to obtain the premix solution. The gelcasting process was similar to that of the previous studies. First, the PZT powder was added to the premix solution (containing 20 wt% of AM) and thoroughly ball-milled for 24 h. The slurry was then de-aired by a vacuum pump for 10 min. After adding the initiator and catalyst, the slurry was cast into a nonporous mold without additional pressure. Subsequently, the mold was kept at 80 °C for 24 h without controlling the humidity, for polymerization and drying of the green body. TAC was selected as the dispersant to improve the flowability of the ceramic slurry. The zeta potentials of the powders were measured using a Zeta Plus analyzer (Brookhaven Instrument Corp., USA). The PZT suspensions at a solids loading of 0.05 vol% were used for the analysis. Viscosities of the ceramic slurries were measured using an advanced MCR300 Rheometer (Physica Corp., Germany) with a concentric-cylinder setting. The microstructures of the samples were observed using SEM. The density of the sintered samples was measured by employing the Archimedes' method. To compare the homogeneity of the piezoelectric property of the samples prepared by gelcasting with those by prepared by die pressing, two 60 × 10 × 1-mm PZT strips derived from both the methods under the same sintering procedure, i.e., a sintering temperature of 1250 °C and a hold time of 2 h, were tested. The samples were divided into six identical 10 × 10 × 1-mm square segments and then polarized in a Si oil bath by applying an electric field of 3 kV mm^{-1} at 120 °C for 30 min. The piezoelectric coefficient d_{33} of the samples was determined by using a ZJ-3A piezometer (Institute of Acoastics, Academia Sinica). The pore-size distribution of the green ceramics was measured using the mercury porosimetry (Autopore 9220).

6.5.1 Colloidal Chemistry and Rheological Behavior

The solids loading required for gelcasting is generally higher than that required either for spray drying or slip casting. Solids loading of at least 50 vol% is preferred for the gelcasting slurries. Low solids loading and hard agglomerates in the ceramic suspensions can result in deleterious surface and bulk defects in the finished ceramics. Therefore, a key factor for the successful production of ceramics by gelcasting technique is to produce a flowable and stable slurry with possible high solids loading. Homogenous dispersion of the powder in the premix solution and stability of the suspension are mainly determined by the attractive and repulsive forces between the particles in the system. The former generally arises from the van der Waals forces, while the latter can arise from the electrostatic repulsion or steric repulsion of the polymeric or surfactant material present on the particle surfaces (Israelachvili, 1992). The magnitude of the van der Waals forces is mainly determined by the nature of the particle surface and the solvent, and hence, it is essentially a constant. On the other hand, the repulsive forces can be modified over a wide range by the addition of the dispersants. Electrostatic repulsion is directly dependent on the zeta potential of the powders. If this value is higher with the same polarity, then the electrostatic repulsion between the particles is more significant. On the other hand, when this value is close to the IEP, the particles tend to flocculate.

The zeta potential values of different PZT aqueous suspensions at different pH values are shown in Fig. 6.30. The zeta potential value of pure PZT suspension is observed to change from 33.4 mV at a pH of 1.7 to −35.1 mV at a pH of 11.9, with an IEP at a pH of about 7.2, suggesting that neutral environment may be disadvantageous to acquire good dispersion. Addition of 0.5 wt% TAC is found to result in much more negative potential values. Furthermore, a value of 2.2 mV for pure PZT suspension at a pH of 7 is decreased to about −28.7 mV, and the IEP is also shifted to a pH of about 3.6. Addition of monomer shows little effect on the IEP position of the two systems, but it slightly decreases the absolute value of the zeta potential. This might

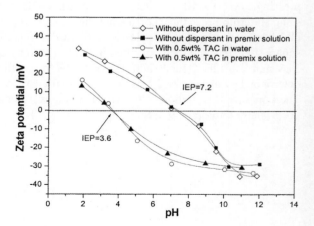

Fig. 6.30 Influence of the dispersant and monomer on the zeta potential of the PZT suspension

Fig. 6.31 Influence of solids loading on the viscosities of the PZT suspensions with and without the dispersant at a steady shear rate of 0.1 s^{-1}

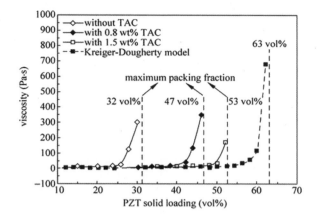

be ascribed to the screening of the charged PZT particles caused by the uncharged AM molecules (Barnes, 1989). With regard to the two systems containing TAC, the zeta potentials decrease slightly with the increase in the pH value from the neutral environment to about 11.7, suggesting that a basic environment might have little effect on the decrease in the viscosities of the two systems.

The influence of solids loading on the apparent viscosity of several PZT slurries is shown in Fig. 6.31. The viscosity data were obtained at the shear rate of 0.1 s^{-1}. Generally, there exists a complex nonlinear relationship between the viscosity and solids volume fraction, which is closely related to many factors, such as the continuous phase viscosity, particle-size distribution, particle shape, etc. The curves in Fig. 6.31 have the typical shape of a plot of solids loading versus viscosity: low viscosity at low solids loading and sharp increase at high solids loading. At one particular time point, more and more particles would be added making the suspensions to "jam up", causing continuous three-dimensional contact throughout the suspension, and thus, making the flow impossible, i.e., the value of viscosity tends to infinity. The particular phase volume at which this happens is called the maximum packing fraction, Φ_m (Barnes, 1989). Before the addition of the dispersant, the system has a measurable viscosity of about 31 vol% solids loading. However, addition of a little more PZT powder to approximately 32 vol% may lead to a "solidified" slurry; thus, the viscosity will increase sharply and make it practically impossible to measure the viscosity using a rheometer with a concentric-cylinder setting. Thus, the Φ_m value has been determined to be 32 vol%. Additions of 0.8 and 1.5 wt% of TAC are observed to increase the Φ_m value to about 47 and 53 vol%, respectively, suggesting that TAC is very effective in increasing the solids loading of the PZT powder. However, increasing the dispersant amount further to about 2.2 wt% is found to decrease the maximum packing fraction again, indicating that 2 wt% of TAC in the suspension has achieved the maximum coverage of the PZT powder and that excess dispersant may be harmful in decreasing the viscosity. The relationship between viscosity and solids volume percentage for the monodispersed suspension can be modeled by the Krieger–Dougherty (K–D) model as follows:

$$\eta = \eta_0 \left(1 - \frac{\varphi}{\varphi_m}\right)^{-[\eta]\varphi_m}$$

where η is the viscosity of the suspension and η_0 is the viscosity of the solvent. The true volume fraction of the powder dispersed in the suspension is represented by the variable Φ. The intrinsic viscosity $[\eta]$ of the suspension is a function of the particle geometry, and a value of 2.5 is appropriate for the spherical particles. A maximum packing fraction of 0.63 ± 0.002 is considered to be appropriate for random close packing at a low shear rate. The K–D model curve with $[\eta] = 2.5$ and $\Phi_m = 0.63$ is also presented in Fig. 6.31. Although there are obvious discrepancies between the experimental curves and the K–D model curves in Fig. 6.31, the experimental curves have the typical shape of a plot of solids loading versus viscosity as that of K–D model: low viscosity at low solids loading and sharp increase at high solids loading. The discrepancy may partly be ascribed to the inhomogeneity of the particle size. On the other hand, particle flocculation will lead to a lower value of Φ_m, because the flocs themselves are not close-packed (Barnes, 1989; Starov et al. 2002) and they can trap part of the continuous phase, thus, leading to a bigger "effective phase volume" and a greater viscosity than that of the primary particles. Therefore, the discrepancy also suggests that the PZT suspension without TAC is composed of a large amount of agglomerates. The results also reveal that TAC can effectively break up the agglomerates in the suspension.

Figure 6.32 shows the apparent viscosity as a function of shear rate for different PZT slurries. Here, 30 vol% solids loading was used for systems having no dispersant in the case of "jamming up". Adjustment of the pH value to basic condition is observed to have little effect on the decrease in the viscosity, while addition of TAC is found to greatly decrease the viscosity. This indicates that electrostatic stabilization mechanism alone is not sufficient for good dispersion of the PZT powders. Effect of

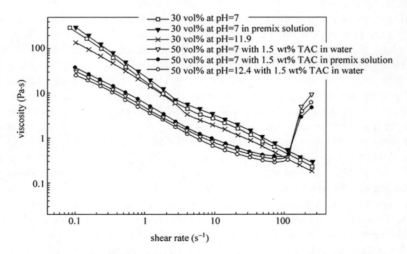

Fig. 6.32 Influence of the dispersant and monomer on the viscosity of the PZT suspension

6.5 Water-Based Gelcasting of Lead Zirconate Titanate

the dispersant can be attributed to both electrostatic and steric stabilization mechanisms. Because of the relative short chain of the TAC macromolecule than that of the polymeric dispersant, the latter stabilization mechanism can be more precisely denoted as a semi-steric mechanism. Generally, viscosity below 100 Pa s is indicative of a flowable slurry; thus, a TAC content of 1.5 wt% is suitable for casting and excess of TAC has little effect on the viscosity. Consistent with the zeta potential results, increasing the pH of the system containing TAC to about 12.4 is observed to show little effect on the viscosity. However, all the systems show shear-thinning behaviors at low shear rates. Shear-thinning behavior indicates that the flow can bring about a more favorable particle spatial rearrangement. For the concentrated suspensions, it is generally accepted and has been verified that the flow-induced structure is a layered structure, which can result in a low resistance of the particle movement between different layers, and thus, a low viscosity is obtained (Barnes, 1989). Owing to the damage of the layered structure at very high shear rate, the critical shear rate (γ_c), the viscosity increases again and a shear-thickening behavior appears in the systems containing TAC. This shear-thickening behavior can also be ascribed to the increase in the phase volume fraction at high solids loading. Furthermore, addition of AM is observed to exhibit little (increase in the value of η from 30.5 to 38.6 Pa s at the shear rate of 0.1 s^{-1}) effect on the viscosity. These results are consistent with the zeta potential results, indicating that the dispersants work by increasing the stability and resisting particle agglomerates by both electrostatic and semi-steric stabilization mechanisms. After developing concentrated PZT suspension with low viscosity, complex-shaped PZT objects were successfully fabricated (Fig. 6.33).

Fig. 6.33 Illustration of the ultimate PZT ceramic parts fabricated by the gelcasting approach

6.5.2 Microstructure and Properties

The SEM micrographs of the fracture surfaces of the gelcast PZT green body and those of the sintered PZT samples are shown in Fig. 6.34. From Fig. 6.34a, it can be observed that there exists a little linear-shaped organic substance between the PZT particles, which is responsible for the high strength of the green body. Figure 6.34b and c shows the green body obtained at a solids loading of 30 (without dispersant) and 50 vol%, respectively. As the former is formed from a low solids loading suspension with many agglomerates, a porous and cracked morphology is obtained. On the other hand, Fig. 6.34c indicates that after increasing the solids loading to 50 vol% and breaking up the agglomerates by adding TAC, a homogeneous microstructure can be obtained. Moreover, Fig. 6.34c also shows a more homogeneous microstructure than the die-pressed sample shown in Fig. 6.34d. The density and shrinkage value of the PZT samples prepared under different conditions are consistent with this finding (Table 6.9). As expected, increase in the solids loading results in less shrinkage, higher density, and green strength. Although gelcasting results in a slightly greater

Fig. 6.34 SEM photos of the gelcast green body and sintered PZT parts: **a** Green body, **b** 30 vol% solids loading and without dispersant, **c** 50 vol% solids loading with 1.5 wt% of TAC as the dispersant, and **d** die-pressed samples

6.5 Water-Based Gelcasting of Lead Zirconate Titanate

Table 6.9 Comparison of some parameters of PZT samples prepared under different conditions

Sample	Green strength (MPa)	Sintered density (g cm^{-3})	Total shrinkage after sintering (%)
Gelcasting at 30 vol% solids loading	17	7.228	18.6 with cracks
Gelcasting at 50 vol% solids loading	20	7.762	15.9 without cracks
Die pressing[a]	3.6	7.661	14.8 without cracks

[a]Containing 1.5 wt% of PVA as the binder and pressed at 80 MPa

shrinkage than die pressing, a higher density value was also obtained, thus, a more uniform structure can be observed in Fig. 6.35c than in Fig. 6.34d. In addition, as the 50 vol% slurry was relatively concentrated and could solidify instantly after casting at 80 °C, no sedimentation was observed. Consequently, no obvious density difference between the different sections of the ceramic parts was observed. Moreover, based on our observation, no cracking occurred in the gelcast body after drying, even though no special drying conditions, such as temperature and humidity, were maintained.

The d_{33} coefficient results of the samples prepared by gelcasting (50 vol% solids loading) and die pressing are illustrated in Fig. 6.35. A more homogenous piezoelectric property of the gelcast samples than that of the die-pressed ones was observed by contrasting the S.D. of the two series of data (Table 6.10). This also suggests that a higher homogeneity can be obtained by gelcasting than by die pressing. We also

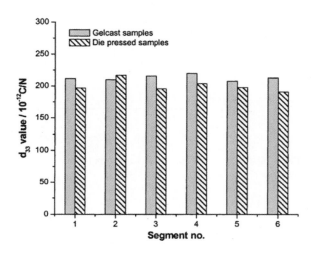

Fig. 6.35 Comparison of the piezoelectric properties of the PZT samples derived from different methods

Table 6.10 Comparison of the piezoelectric property of the PZT samples prepared by gelcasting with a 50 vol% solids loading and those prepared by die pressing with a pressure of 100 MPa

Sample	Average d_{33} 10^{-12} C/N	S.D. of d_{33} 10^{-12} C/N
Gelcasting	213.2	3.933
Die pressing	200.2	8.513

found that gelcast samples have a slightly higher average d_{33} value than die-pressed ones. This is consistent with a previous study and may be attributed to the density difference (Kiggans et al. 1999). Furthermore, study of their microstructures by examining the pore-size distributions of the two kinds of green bodies (see Fig. 6.36) revealed that the gelcast sample has a smaller median pore size than the die-pressed sample, which is indicative of a more homogeneous microstructure and may partly account for the piezoelectricity difference. The results in Fig. 6.35 and Table 6.10 are in good agreement with those obtained from SEM examination and density measurement, suggesting that the homogeneously dispersed PZT particles in the suspension obtained by the addition of dispersant and thorough milling were effectively conserved in the green body during the in situ polymerization, and subsequently in the sintered samples.

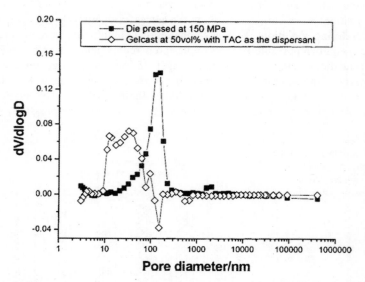

Fig. 6.36 Comparison of the pore-size distributions of the PZT green bodies obtained from different methods

6.5.3 Summary

The experimental results suggest that gelcasting can be successfully used for the fabrication of piezoelectric ceramics. When compared with die pressing, gelcasting can provide a more homogenous microstructure as well as more homogenous piezoelectric property of the PZT samples. By using the same sintering procedure, the gelcast samples were observed to exhibit slightly stronger piezoelectric effects than the die-pressed ones, and this may be attributed to both density difference and pore size.

References

Barnes, H. A. (1989). *Journal of Rheology, 33*, 329.
Barnes, H. A., Hutton, J. F., & Walers, K. (1989). *An introduction to rheology*. Oxford: Elsevier.
Bengisu, M., & Yilmaz, E. (2002). *Ceramic International, 28*, 431.
Chen, X. B., Cheng, H. M., & Ma, J. M. (1998). A study on the stability and rheological behavior of concentrated TiO_2 dispersions. *Powder Technology, 99*(2), 171–176.
Chu, J., Wu, C. (2002). *Ozone technology and its application*. Beijing: Chemistry Industry Publishing House.
Du, H. Q. & Tang, S. Q. (1986). *Ceramic material and recipe*. Light Industry Press.
Huang, Y., Ma, L. G., Tang, Q., Yang, J. L., & Xie, Z. P. (2000a). Surface oxidation to improve water-based gelcasting of silicon nitride. *Journal Materials Science, 35*(14), 3519–3524.
Huang, Y., Ma, L. G., Tang, Q., et al. (2000b). *Journal Materials Science, 35*, 3519.
Israelachvili, J. (1992). *Intermolecular & Surface Forces* (2nd ed.). London, UK: Academic Press.
Janney, M. A. (1990). Method for forming ceramic powders into com-plex shapes. US Patent 4894194, 1–16.
Janney, M. A., & Omatete, O. O. (1992-97). Method for molding ceramic powders using a water-based gelcatsing. US Patent 5145908.
Kiggans, J. O., Tiegs, T. N., Montgomery, F. C., Maxey, L. C., & Lin, H. T. (1999). Ceramic Engineering and Science Proceedings (Vol. 20, pp. 19–26). American Ceramic Soc.
Lange, F. F. (1989). Powder processing science and technology for increased reliability. *Journal of the American Ceramic Society, 72*(1), 3–15.
Li, B. R., Wang, X. Z. & Zhang, X. L. (1995). *Inorganic dielectric* (50–56). China: Science and Technology University of Huazhong Press.
Liu, K. S. (1990). *Principle of ceramic technology*. Science and Technology University of Huanan Press.
Liu, Y. F., Liu, X. Q., Wei, H., et al. (2001). *Ceramic International, 27*, 1.
Ma, L. G., Huang, Y., Yang, J. L., et al. (2001). Improving the break down strength of rutile capacitor by gelcasting. *Journal of Material Science Letters, 20*(14), 1285–1288.
Mikula´ sU ek, P., Wakeman, R. J. & Marchant, J. Q. (1997). The influence of pH and temperature on the rheology and stability of aqueous titanium dioxide dispersions. *Chemical Engineering Journal, 67*(2), 97–102.
Mikulasek, P., Wakeman, R. J., & Marchant, J. Q. (1998). Crossflow microfiltration of shear-thinning aqueous titanium dioxide dispersions. *Chemical Engineering Journal, 69*(1), 53–61.
Nunn, S. D., & Kirby, G. H. (1996). Green machining of gelcast ceramic materials. *Ceramic Engineering and Science Proceedings, 17*(3), 209–213.
Omatete, O. O., Janney, M. A., & Sterklow, R. A. (1991). Gelcasting-a new ceramic forming process. *Ceramic Bulletin, 70*(10), 1641–1649.

Pugh, R. J. & Bergström, L. (Eds.). (1994). *Surface and colloid chemistry in advanced ceramics processing*. New York: Marcel Dekker.

Starov, V., Zhdanov, V., Meireles, M., et al. (2002). *Advances in Colloid and Interface Science, 96*, 279.

Strauss, H., Heegh, H., & Sreienitz, I. (1993). Effect of PAA adsorption on stability and rheology of TiO_2 dispersion. *Chemical Engineering Science, 48*(2), 323–332.

Sun, L., Gao, L., Guo, J. K., & Yan, D. S. (1996). Gelcasting of nano-size Y-TZP. *Acta Materialia, 9*(6), 489–492.

Vlajic, M. D., & Krstic, V. D. (2002). *Journal Materials Science, 37*, 2943.

Waesche, B., & Steinborn, G. (1997). Influence of slip viscosity on the mechanical properties of high purity alumina made by gelcasting. *Key Engineering Materials, 132–136*(1), 374–377.

Wang, J. (1990). Measurement of Ozone Concentration, Yield and Power Consumption for Ozone Generator (Trade standard of town building in PRC: CJ/T3028, Beijing 1990).

Yamashita, K., Koumoto, K., Takata, M., & Yanagida, H. (1980). *Japanese Journal of Applied Physics, 19*, 867.

Yang, J. L., Xie, Z. P., Ma, J. C., Huang, Y., Zhao, J. S., & Fan, Q. S. (1999). Study on gelcasting processing of high power rutile capacitor. *Key Engineering Materials, 161–163*, 517–520.

Yi, Z. Z., Xie, Z. P., Huang, Y., et al. (2002). *Ceramic International, 28*, 369.

Young, A. C., Omatete, O. O., Janney, M. A., & Strehlow, A. (1991a). Gelcasting of alumina. *Journal of the American Ceramic Society, 74*(31), 612–618.

Young, A. C., Omatete, O. O., Janney, M. A., & Menchhofer, P. A. (1991b). *Journal of the American Ceramic Society, 74*, 612.

Zhou, L. J., Huang, Y., & Xie, Z. P. (2000). Gelcasting of concentrated silicon carbide suspension. *Journal of the European Ceramic Society, 20*, 85–90.

Chapter 7
New Methods and Techniques Based on Gelation

Abstract The reliability and cost of the products, which are closely related to the preparation process, are the two major concerns in the industrial production of high-performance ceramics. The preparing, forming, and sintering processes of the ceramic powders are thus becoming increasing important in the recent decades in the ceramic research field. As a precondition of the materials design and realization of the high-performance ceramic products of the desired shapes, the ceramic-forming process usually plays a crucial role in determining the reliability and cost of the products. The traditional methods of producing ceramic products have been widely used in the field of fine ceramics, but they do not meet the requirements of high-performance ceramics. Therefore, numerous research efforts have been directed to devise new methods for the ceramic-forming process. In this chapter, some new methods and techniques based on gelation are introduced, such as solid freeform fabrication (SFF) of alumina ceramic parts; freeform fabrication of aqueous alumina–acrylamide gelcasting suspensions; ceramics with special porous structures fabricated by freeze-gelcasting; solidifying concentrated Si_3N_4 suspensions for gelcasting by ultrasonic effects; laser machining technique of alumina green ceramic; etc. Furthermore, the forming mechanism is analyzed and the characteristics of the green and sintered bodies of the ceramic parts using the new forming methods or new techniques are introduced.

Keywords Solid freeform fabrication · Freeze-gelcasting · Ultrasonic effects · Laser machining technique

Currently, with the development of gelation, several new forming techniques for preparing high-quality green bodies have been developed, such as gelcasting, colloidal vibration casting, direct coagulation casting, temperature-induced flocculation, selective laser sintering, solid freeform fabrication (SFF), freeze-gelcasting., etc. Many new techniques such as laser machining technology have also been developed.

In this chapter, the investigation status and new achievements with regard to the novel forming techniques based on gelation, including SFF, freeze-gelcasting, as well as new preparing methods such as ultrasonic effects to prepare concentrated Si_3N_4 suspension and novel laser drilling technique for ceramics are reviewed and summarized in detail.

SFF is the near-net-shaped approach to produce parts or prototypes directly from a computer-aided design (CAD) file without hard tooling, dyes, or molds. It has attracted increasing interest because of its advantages over conventional fabrication methods. To date, several SFF techniques have been developed. Laminated object manufacturing (LOM) process uses tape-cast ceramic sheets to fabricate ceramic parts (Griffin et al. 1994). Ceramic-binder extruded filaments are being used as feedstock in fused deposition of ceramics (FDC) processing (Agarwals and Weeren 1996; Danforth 1995). The stereo lithographic (SLA) techniques are being developed to produce ceramic parts via ultraviolet curing of a highly concentrated suspension of ceramic particles in a photo-polymerizable liquid (Griffin and Halloran 1996; Hinczewski et al. 1998). However, many of these approaches directly rely on the ceramic feedstocks that contain a large number of organic species, which result in debinding issues. Selective laser sintering (SLS) is a form of SFF. In this process, a thin layer of powder is spread across the build platform by a roller mechanism. The first slice or cross-section of the object is selectively drawn on the layer of the powder using a CO_2 laser. The laser energy heats the powder to a temperature just below the melting point of the material and bonds the powder into a solid mass. The laser power is modulated so that the powder only in the areas described by the object's geometry is fused. The build platform is lowered and the roller deposits another layer of powder across the build platform. The steps are repeated until the complete object is formed. The primary advantage of SLS process is the flexibility in selecting material systems when compared with other SFF techniques (Pham and Gault 1998). Freeze-casting technique is a good processing technique for the fabrication of porous ceramics. It can produce porous materials with controllable porosity and open interconnected pores. Particularly, the pore morphology and pore size can be tailored in a certain range by easily altering the parameters in the processing technique, such as slurry concentration, freezing temperature, and cooling rate. Moreover, freeze-casting is an environment-friendly method. In the preparation process, ceramic slurries with different solids loading values are first frozen, then the frozen bodies are dried in a lyophilizer to sublimate the frozen medium, and finally, the porous samples are obtained by sintering the green bodies at different temperatures. Till date, aqueous and non-aqueous slurries have been successfully used to fabricate the porous materials by freeze-casting. Gelcasting is a novel near-net-shaped forming method for fabricating complex-shaped ceramic bodies. It was invented by Young et al. (1991), Omateta et al. (1991), and is gaining increasing attention. In this process, the monomer, crosslinker, initiator, catalyst, and ceramic powder are thoroughly mixed in water to form a homogeneous suspension with high solids loading and low viscosity; subsequently, by means of in situ polymerization, a macromolecular network is created to bind the ceramic particles together. When compared with other casting methods, gelcasting results in much more homogeneous materials with little density difference over the parts and produces green body with higher flexural strength.

With these new methods and techniques, high-performance ceramic parts with high reliability and complex shapes can be mass-prepared by automation. This may pave the way for the industrial production of advanced ceramics in the near future.

7.1 Development Overview and Application of Solid Freeform Fabrication

7.1.1 Development Overview of Solid Freeform Fabrication

SFF, which was developed in the 1980s, has triggered a huge change in the way of thinking about the manufacturing technology and production efficiency. Many countries in the world have given great importance to the new forming technology, and it was also considered to be a major breakthrough in the filed of manufacturing.

In today's society, with the continuing diversification demand for high-performance materials and low-cost manufacturing technology, product competition has become more and more intensive, with the update cycle becoming shorter. Thus, a designer is needed to design the new products rapidly and create a product sample in the shortest time. The use of traditional method to produce product samples requires various machine tools, molds, mechanical appliances, as well as high-level skilled workers. Furthermore, it is time-consuming, expensive, has a long cycle, and cannot adapt to the rapid change in ceramic industrialization. As an advanced preparation technology, SFF was developed based on discrete analysis and accumulation.

SFF is a comprehensive embodiment of multi-discipline and technology, the emergence of which accompanied the high-speed development of computer, laser, digital control and material technology, as well as the combination of various disciplines. The concept of SFF appeared in the late 1970s, but the thinking prototype of using the layered manufacturing principle to deposit three-dimensional objects could be traced back to the nineteenth century. In 1892, the idea of using layered manufacturing to form topographic map was proposed by J. E. Blanther in his patent. In 1902, the principle of using photosensitive polymer to manufacture plastic was mentioned by Carol Bease in his patent, and it was also the preliminary envisage of the first SFF—Sterolithography. Currently, with the development of material science, SFF has been developed rapidly. It has been used in the fields of electrical engineering, toys, light industry, aviation, space, medicine, film, modeling of artificial organs, archaeology, industrial design, etc. SFF not only provides convenience for manufacturing, but also brings new ways and techniques for materials preparation. It is an advanced manufacturing technique that generates accurate geometrical objects directly from a three-dimensional computer image without part-specific tooling or human intervention. This signifies that designers have the freedom to produce physical models of their drawings more frequently, allowing them to check the assembly and function of the design, as well as to discuss the downstream manufacturing issues with an easy-to-interpret, unambiguous prototype. Consequently, errors are minimized and product development costs and lead time are substantially reduced. With the emergence of SFF technology, industrialization of advanced ceramics is predicted to have a bright future.

However, as the green bodies prepared by this technique will contain a large amount of organic binder, long and complicated burnout heat treatments are required to eliminate it. Gelcasting is also a near-net-shaped fabrication technique employing polymerization of the monomer. The gelcasting system contains only a small amount of binder. In this work, a new freeform fabrication approach combining SFF and gelcasting is described, which uses selective gelation of concentrated aqueous gelcasting suspensions. The forming mechanism of this method is similar to the widely studied acrylamide (AM) gelcasting system. Furthermore, the process of the SFF of alumina ceramic parts is also introduced.

7.1.2 Application of Solid Freeform Fabrication

7.1.2.1 Solid Freeform Fabrication of Alumina Ceramic Parts Through a Novel Lost Mold Method

(1) The Combination of Selective Laser Sintering and Gelcasting

To put forward a more efficient SFF approach, a novel lost mold method has been successfully designed by combining the SLS and gelcasting technique. Figure 7.1 shows a graphic depiction of this process. First, a composite polymer powder is developed for SLS to fabricate sacrificial molds having a negative of the desired structure for gelcasting. Then, homogeneous alumina slurry with high solids loading and low viscosity is poured into the molds. During the sintering of the ceramic parts, the polymer mold is entirely removed. As the solidified green bodies have high strength even at elevated temperature, the desired geometry of the ceramic parts is successfully retained after sintering. In this way, many complex alumina parts have been produced successfully.

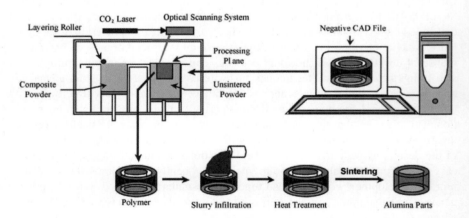

Fig. 7.1 Schematic chart of the steps followed in the lost mold approach for fabricating alumina ceramics

7.1 Development Overview and Application of Solid Freeform … 363

In this study, AM and N,N′-methylenebisacrylamide (MBAM) were dissolved in deionized water to obtain a premix solution. Triammonium citrate (TAC) was selected as the dispersant to produce well-dispersed homogeneous ceramic suspension. The slurry containing alumina powder and dispersant in the premix solution were thoroughly ball-milled for 24 h. After de-gassing for 15 min, polymerization was initiated by the addition of initiator and catalyst. The slurry was casted into the sacrificial plastic molds, which was fabricated using an AFS SLS machine. The zeta potential of the ceramic particles was determined by a Zeta Plus analyzer (Brookheaven Instrument, USA) and the viscosity of the alumina slurries was determined by an advanced MCR300 Rheometer (Physica Corp., Germany). The details about the sintered mold structure and pyrolysis of the polymers were obtained via thermogravimetric analysis (TGA) in air using a Dupont Thermal Analyzer 2000. The microstructure of the laser-sintered polymer mold was observed by a HITACHI S-450 scanning electron microscopy (SEM).

(2) **Rheological Property of the Alumina Suspension**

It is of prime importance to obtain a stable and homogeneously dispersed ceramic suspension to acquire high-quality ceramic parts by gelcasting. The rheological behavior or viscosity is related to the suspension property, which is determined by the attractive and repulsive forces of the particles in the system. Attractive van der Waals forces can be counteracted by repulsive forces resulting from either overlapping of the electrical double layers (electrostatic stabilization) and/or layering of the materials adsorbed onto the surface (steric stabilization). The magnitude of the van der Waals forces is mainly determined by the nature of the particles and the solvent, while the repulsive forces can be modified over a wide range by the addition of dispersants (Ferreira and Diz 1992). The repulsive forces can be characterized by the zeta potential, and the potential of the particles as a function of pH value is shown in Fig. 7.2.

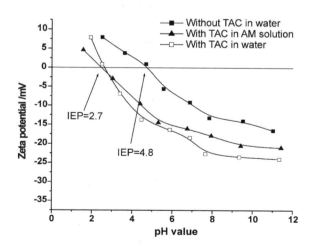

Fig. 7.2 Zeta potential versus pH value of different alumina suspensions

The 0.1 vol% alumina suspensions were regulated by HCl and NaOH to fixed pH value, and ultrasonicated for 2 min prior to analysis. The electrostatic repulsion between the particles was found to be more significant if this potential with the same polarity was higher. Contrarily, when close to the isoelectric point (IEP), the particles tended to flocculate. From Fig. 7.2, it can be seen that the IEP of the powder is at about a pH of 4.8 and has a zeta potential value of -10 mV at pH 7. With the addition of TAC (1 wt% of the alumina), the IEP shifted to about 2.7 mV and the zeta potential at pH 7 doubled, suggesting that TAC has an electrostatic stabilization mechanism to bring about good dispersion. However, addition of the monomer did not shift the IEP, but slightly decreased the relative value of the zeta potential. This indicates that the uncharged AM molecule either screens the charge that is developed by the powders in the solution or preferentially gets absorbed onto the surface of the alumina particles. The viscosity of the alumina slurries as a function of the shear rate is illustrated in Fig. 7.3. It can be seen that TAC can greatly decrease the viscosity, which is in good agreement with the zeta potential results. The shear thickening behavior of the slurries with TAC at higher shear rate is owing to the increase in the "phase volume fraction" at high solids loading (Barnes et al. 1989). For a dispersant, there exists an optimum concentration at which just the adequate dispersant exists to provide maximized coverage of the ceramic powder, and any excess dispersant may be harmful in decreasing the viscosity (Mcnulty et al. 1999). The optimum concentration of TAC has been observed to be about 0.3 wt% of the alumina powder.

(3) **Selection of Mold Materials and Heating Procedure**

A large range of materials can be used in SLS, such as nylon, ABS, sand, wax, metals, polycarbonate, etc. However, some problems must be considered when selecting a material. During heat treatment, the sacrificial mold should burn out clearly without leaving any residue that may be detrimental to the properties of the ceramic parts. Therefore, all organic constituent materials system should be adopted. In this study, a polystyrene-based composite powder was developed to make the sacrificial mold.

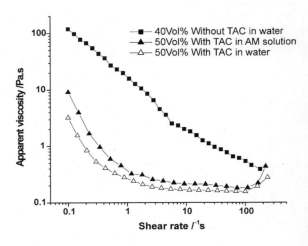

Fig. 7.3 Effect of the dispersant and monomer on the viscosity of different ceramic slurries

7.1 Development Overview and Application of Solid Freeform ...

In fact, polystyrene is frequently used in lost foam-casting owing to its thermal degradation property (Chang and Shih 1999). As a general polystyrene product may result in large geometry distortion upon heat treatment, the composite powder is specially processed by reactive blending of PS, rubber, antioxidant, impact modifier, and other additives through twin-crew extruder. It is also necessary to cool the part when it becomes too hot to prevent distortions in the final piece. Therefore, wax is not a suitable material because it needs a long cooling cycle on the machine, though it can be removed easily. From the SEM photograph of the cross-section of a sintered mold shown in Fig. 7.4, a porous network structure can be observed, which is helpful in preventing distortion.

The details of the TGA curves (4 °C min^{-1} in air) of the sintered molds and that of a green cast alumina ceramic sample (50 vol% solids loading) are shown in Fig. 7.5.

The mold material starts evaporating at about 85 °C, and loses 94% of its weight at 85–300 °C. At a higher temperature, a slow burnout occurs which ends at about

Fig. 7.4 SEM photograph of the microstructure of the sintered polymeric mold materials

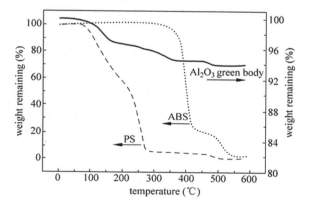

Fig. 7.5 TGA curves of the polymeric mold, ABS resin, and a cast alumina sample

510 °C without leaving any residue. The gelcasting sample starts evaporating at about 120 °C, and before reaching 500 °C, only a 6% mass loss occurs owing to the trapped water and burnout of the cross-linked polymerized AM network. Because of the lower mass loss temperature and porous structure of the polymeric mold, the green ceramic structures have sufficient strength to hold the structure during evaporation. As a result, the geometry of the ceramic part is well preserved. The mold material should evaporate first before the evaporation of the binder, to prevent collapse of the ceramics caused by the deformation or flow of the plastic mold during heating. Indeed, collapsing occurs when other polymer materials such as ABS are used as the mold materials (see Fig. 7.5). The binder removal and detailed sintering procedure of dry ceramic powder-infiltrated molds were determined based on the TGA results. The heating procedure consists of three steps. During the first part of the procedure (room temperature to 600 °C), the mold material and crosslinked polymer binder evaporate. In the second step, 1 h of constant temperature at 600 °C is maintained to remove the remnants of the decomposed polymers. At higher temperature, densification of alumina occurs. A final sintering temperature of 1550 °C and a hold time of 2 h are used for the alumina samples. The ultimate sintered ceramic parts are shown in Fig. 7.6.

Generally, the mass ratio of the polymer mold to alumina slurry is about 5:100 before sintering. The polymeric mold material can be sintered with a relatively low-powered laser in a short time period. A polymer mold with a dimension of 2 × 2 × 2 cm needs just 30 min to be fabricated. Using adaptive slicing that generates different slice thicknesses between 0.08 and 0.3 mm based on the local slope of the part, the build rate can also be increased. Thus, the cost of the mold is acceptable and this process can be considered as an economical SFF method for fabricating ceramic parts.

Fig. 7.6 Illustration of the ultimate alumina ceramic parts fabricated by the lost mold approach

7.1.2.2 Freeform Fabrication of Aqueous Alumina–Acrylamide Gelcasting Suspensions

(1) The Process of Fabricating Ceramic Parts

The ceramic powder used in this work was α-Al_2O_3. Furthermore, AM, MBAM, $(NH_4)_2S_2O_8$, N,N,N′,N′-tetraethylmethylenediamine (TEMED), and dibasic ammonium citrate were applied as monomer, crosslinker, initiator, catalyst, and dispersant, respectively. All these reagents were chemically pure.

In the process, the suspension was first prepared. The dispersant with 0.5 wt% dry powder weight was completely dissolved in the premix solution (5–20 wt%), prepared by dissolving AM and MBAM in deionized water, and subsequently, the alumina powder was added. The suspension was degassed for about 10 min under vacuum after milling for 24 h. The solids loading value of the final suspension was ~50 vol%.

Second, similar to the tape-casting process, a blade was used to spread the viscous ceramic suspensions to produce a layer with the desired thickness before polymerization over a build platform. A heating system was used to adjust the gelation rate. A nitrogen system was also employed to avoid oxygen that inhibits the polymerization process. According to the computer information, localized gelation of the selected areas of the suspension layer was subsequently induced by a jet of droplets of deionized water. When a layer of the part was formed, the build platform was lowered and a new layer of the suspension was deposited onto the already-polymerized part to continue the manufacturing process. This process was repeated until the part was built. Figure 7.7 shows the schematic representation of the selective gelation process. After formation, the green parts were washed in water, and then dried at room temperature for 24 h. According to the TGA result, debinding was performed with a heating rate of 1 °C min^{-1} up to 550 °C for 2 h, and then at 3 °C min^{-1} up to 1580 °C for 2 h.

The rheological properties were examined using a Rheometer US200 (Physica Corp., Germany). The density of the sintered bars was determined by the Archimedes' method. The shrinkage was determined from the physical measurements of the sintered parts. The room temperature flexure strength was measured using the three-point bending method with a specimen of 4 × 3 × 36 mm (with a span of 30 mm and

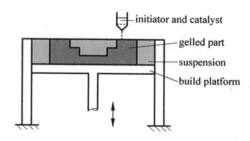

Fig. 7.7 Schematic representation of selective gelation process

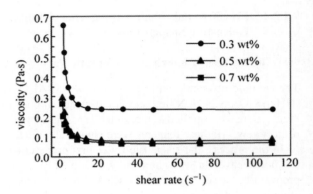

Fig. 7.8 Rheological behavior of the alumina suspensions at different concentrations of the dispersant

a crosshead speed of 0.5 mm min^{-1}). The microscopic morphology was observed by SEM (OPTON, CSM950).

(2) **Rheological Properties of the Alumina Suspensions**

Homogeneous and high-loaded suspensions are important for this method. To obtain ceramic parts with superior properties, the volume fraction of the ceramic powder must be at least 50%. Furthermore, the viscosity of the suspension must be as low as possible to allow a good recoat of the suspension on the gelled layers. Usually, to obtain good homogeneity and low viscosity of the suspension, which are in contradiction with a high ceramic loading, the use of a dispersant is necessary. In this study, dibasic ammonium citrate was used as the dispersant in the alumina suspension system. To determine the optimal concentration of the dispersant, dispersion of the alumina powder in the remix solution was studied. Suspensions containing 50 vol% of alumina was prepared with various amounts of dispersant varying from 0.3 to 0.7 wt% with respect to the powder. Figure 7.8 shows the rheological behavior of the alumina suspensions at different concentrations of the dispersant.

From Fig. 7.8, it can be seen that the minimum viscosity, corresponding to the best state of dispersion of the alumina particles, was obtained with an addition of 0.5 wt% of the dispersant. Furthermore, the suspensions exhibited a shear-thinning behavior that is favorable for tape-casting of the green layers in the range of the shear rate tested. It could be observed that the viscosities of the suspensions were adequately low to be utilized in the gelation process.

(3) **Properties of Alumina Ceramic Parts**

In this selective gelation method, a heating system is important for the fabrication of the ceramic green bodies. At room temperature, the reactive degree of this gelcasting system is very low. However, at 60–70 °C, the reaction is strong. Thus, by adjusting the system temperature, the gelation rate, i.e., the process time, could be adjusted at a wide range.

Several nozzles with different diameters were used to fabricate the ceramic parts. Earlier experiments showed that the nozzle diameter has a significant effect on the thickness of the gelation layer. When the diameter was large (e.g., ~0.2 mm), the build layers up to 1-mm thickness gelled completely. However, when the diameter of the nozzle was ~0.08 mm, the edge of the gelled object was ~0.02 mm. When the layers were thicker (~0.5 mm), the interlayer defects and bad interlayer adhesion were observed using a nozzle with a smaller diameter. In fact, a decrease in the nozzle diameter was observed to cause a reduction in the flow rate, leading to a decrease in the diffusion distance of the initiator and catalyst in the suspension; thus, the gelation layer thickness simultaneously decreased. Therefore, the build layer thickness should correspond to the diameter of the nozzle.

Various ceramic green parts were prepared using this forming process (Fig. 7.9). The thickness of the individual green layers varied from 200 μm to 1 mm. Some rectangular alumina bars approximately $6 \times 5 \times 50$ mm were also prepared by this method, using a 200-μm thick layer suspension. The density of the green rectangular bars fabricated by this forming method was ~53.2%. The anisotropic shrinkage behavior was observed in the bars where the shrinkage was greater in the build direction than in the build plane. The shrinkage for the rectangular alumina bars was ~14.3% in the $X - Y$ plane, and ~16.7% in the Z direction.

The density of the sintered bars was ~91.2% of the theoretical density, which was still insufficient for making the ceramic parts. However, the microstructure of the sintered bars was rather homogeneous, as shown in Fig. 7.10.

In the future, the final density may be increased by improving the process parameters. The sintered bars were observed using SEM to evaluate the microstructures for build-related defects or to determine the evidence of layer-wise build strategies. There were no discernible signs of layer-wise fabrication, and it was difficult to distinguish individual laminated layers inside the sample (Fig. 7.11). Furthermore, the microstructure perpendicular and parallel to the build direction were similar.

The rectangular alumina bars were machined into test bars of $4 \times 3 \times 36$ mm and polished with a diamond paste. Subsequently, the flexure strength tests were carried out either perpendicular or parallel to the build direction. The strength of

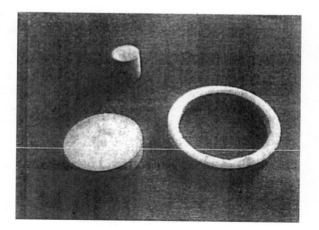

Fig. 7.9 Green parts fabricated by selective gelation

Fig. 7.10 SEM micrograph of the fracture surface of the sintered samples

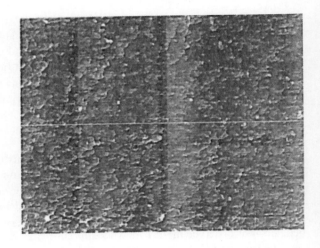

Fig. 7.11 SEM micrograph of a polished sintered sample surface (the arrow indicates the build direction)

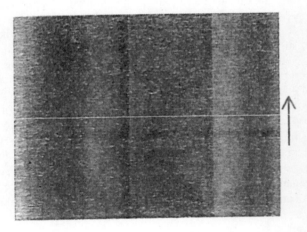

the specimen bars was 284 MPa in the perpendicular direction and 281 MPa in the parallel direction. The mechanical properties were similar regardless of the test direction.

A new technique was developed for freeform fabrication of ceramics that contain low amount of organic binders. This kind of localized suspension gelation method can be used to produce ceramic parts. This study proved that it was possible to obtain rather dense ceramic parts with a homogeneous microstructure.

7.2 Development Overview and Application of Freeze-Gelcasting

7.2.1 The Combination of Gelcasting and Freeze-Casting Technique

An increasing number of applications that require porous ceramics have appeared in the last decades, e.g., the filtration of molten metals and gases, high-temperature thermal insulation, support for catalytic reactions, etc. The advantages of using porous ceramics are usually controllable permeability, large surface area, low density, and high specific strength, which can be tailored for each specific application by controlling the microstructure of the pores. According to the microstructure, porous ceramics can be classified into several general categories (Young et al. 1991) based on pore distribution: disordered, long-range ordered, and gradient, among which, the last two categories have recently begun to attract much attention, because they are expected to have both higher fluid permeability and higher mechanical strength. Furthermore, processing methods used for the production of porous ceramics were recently reviewed by Gauckler et al. (Omateta et al. 1991), including replica, sacrificial template, and direct foaming methods, while many novel methods are being developed.

As a wet-shaping technique, gelcasting is widely used in advanced ceramics manufacturing, because high-strength green bodies can be obtained and complex component geometries can be achieved. Besides the application of fabricating dense ceramics, the gelcasting technique has also been used for fabricating porous ceramics, among which direct foaming and pore-forming agent additives (Tari 2003; Xie and Huang 2001) are the two main methods. In the former method, pores are introduced by incorporating air into a suspension, that subsequently solidifies and retains the structure of the air bubbles. Pores in the ceramics acquired by this method usually have relatively large sizes in the order of several hundreds of micrometers and are spherical-shaped. In the latter method, a suspension containing ceramic powders and natural or synthetic pore-forming agents is prepared and subsequently processed by gelcasting. The pore-forming agent is then burned out at temperatures between 200 and 600 °C, leaving pores with the same shape. Strong green bodies can be acquired by this gelcasting method, while complex pore shape or special structures are difficult to achieve.

Another special wet-shaping technique to fabricate porous ceramics is freeze-casting. Particular features of this technique are the steps of freezing a slip in a mold, demolding in the frozen state, followed by freeze-drying, and sintering. The frozen suspension liquid not only acts as a binder between the powder particles, but also acts as the template of the pore channels. By controlling the freezing direction and temperature gradient, aligned pore channels and a porous gradient can be achieved (Studart et al. 2006; Araki and Halloran 2005; Koh et al. 2006). Besides water (Fukasawa et al. 2001; Zhang et al. 2006), camphene (Meng et al. 2000; Menon et al.

2006) and naphthalene–camphor eutectic (Guha et al. 2001), which can be frozen and sublimated at relatively high temperatures, were developed as the alternative systems. A major problem of freeze-casting is the low strength of the green bodies; when the frozen suspension is volatilized, the green bodies become very frangible and difficult to handle, and further efforts are underway to improve the green bodies' strength (Scheffler and Colombo 2005).

Both gelcasting and freeze-casting techniques have their own merits and drawbacks. In this work, the advantages of these two methods have been combined. To accomplish this purpose, t-butyl alcohol (TBA) was chosen as a candidate for the vehicle of the freezing process capable of near-room-temperature manufacturing. At the following step, polymerization of AM was realized in the TBA slurry, which significantly improved the strength of the ceramic green bodies. We also investigated the effect of temperature on the microstructure and mechanical properties of the green bodies and final sintered ceramics, and obtained porous ceramics with unique long-range ordered and gradient porous structures. The features and mechanisms of this freeze-gelcasting method were analyzed and presented in the following section.

7.2.2 Fabrication of Ceramics with Special Porous Structures

7.2.2.1 Procedure of Freeze-Gelcasting Technique

Alumina with a mean particle size of 0.2 μm was used as the ceramic powder. TBA (Chemical Purity, China) was used as the vehicle, and the main properties are shown in Table 7.1. A premix solution of monomers in TBA was prepared with a concentration of 14.5 wt% of AM ($C_2H_3CONH_2$) and 0.5 wt% of MBAM [$C_2H_3CONH)_2CH_2$]. As the liquid phase in the slurry is a multi-component system, i.e., a mixture of TBA with dispersant and AM monomer, the freezing temperature of this solution was lower than pure TBA, decreasing to about 8 °C. A 40 wt% water solution containing the initiator [$(NH_4)_2S_2O_8$] was prepared for the process of gelation. Citric acid and acetic acid were added to produce stable alumina suspensions in liquid TBA (Menon et al. 2006).

Table 7.1 Some physical properties of TBA (compared with water)

	Liquid density (g ml^{-1})	Boiling temperature (°C)	Saturated vapor pressure at 40 °C (KPa)	Freezing temperature (°C)	Volume increase during freezing (%)	Morphological characteristics after freezing
Water	1	100	3.4	0	10	Dendritic crystal
TAB	0.79	82.5	6.4	25.3	2	Straight prisms

7.2 Development Overview and Application …

Figure 7.12 shows the schematic illustrations of the TBA-based freeze-gelcasting method for the fabrication of porous ceramics with unidirectional pore channels.

First, slurries with 25 vol% solids loading were prepared by ball-milling The complete process for the fabrication of ceramics the alumina powder in warm premixed

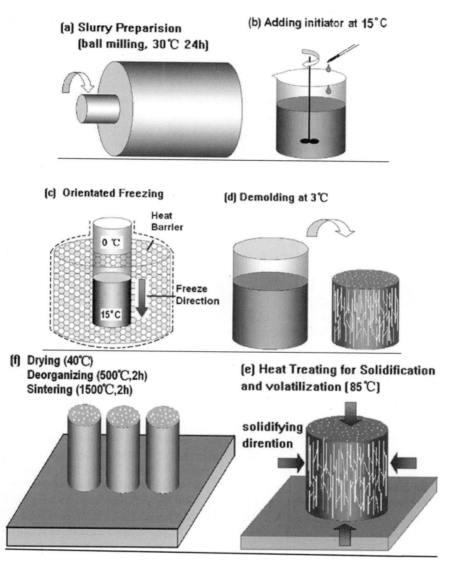

Fig. 7.12 The complete process for the fabrication of ceramics with long-range ordered porous structure by freeze-gelcasting technique

TBA/AM solution at 25 °C for 24 h, containing 1 wt% (to 100 wt% alumina powder) of citric acid as the dispersant, and with a pH value of 3 adjusted using acetic acid (Fig. 7.12a). The prepared slurries were found to be stable and possessed good flowability for casting. The initiator solution (40 wt% in water) was then added at appropriate proportions to the alumina suspension while stirring, and the temperature was maintained at 15 °C, so that polymerization of AM could not be promoted (Fig. 7.12b). The slurries prepared were then poured into polyethylene molds with a diameter of 12 mm, packed by a heat barrier layer at the bottom and surrounding sides. The whole specimen was then placed in a freezing atmosphere at around 0 °C for over 1 h. The TBA froze from top to bottom, forming complete unidirectional crystalline prisms penetrating the whole body as the template of the pore channels (Fig. 7.12c). The frozen bulks were easily removed from the molds at a low temperature (3 °C, Fig. 7.12d), and then heat-treated at 85 °C (Fig. 7.12e). In this thermal treatment step, melting of TBA, gelation of AM, and evaporation of TBA rapidly occurred continuously. Frozen TBA melted first, while AM rapidly polymerized simultaneously under the influence of the initiator. Subsequently, TBA evaporated and formed the pore channels. This procedure occurred with a rapid speed from the exterior to the interior of the samples. Thus, porous green bodies with high strength could be obtained by these freeze and gelation steps, and the final ceramic parts were fabricated by the processes of drying, de-organization, and sintering (Fig. 7.12f).

The process of fabricating gradient porous ceramics is similar to that described earlier, with the main difference in the freezing step. Molds containing slurry and packed by heat barrier layers were placed under conditions with a temperature gradient. One end was set below the freezing point of TBA and the other end was exposed to air at a temperature in its melting range. Thus, the slurry froze from the cold bottom with crystals of TBA. On the other end, solidification and volatilization occurred near the hot surface. These two procedures proceeded simultaneously in opposite directions and formed a novel pore-gradient structure. The step to fabricate gradient frozen parts is illustrated in Fig. 7.13, in which two different temperature conditions, about 12 °C mm^{-1} and 5 °C mm^{-1} in (a) and (b), respectively, were put into effect.

The porosities of the sintered samples were measured using the water displacement method based on the Archimedes' principle. The microstructures were observed using SEM (JEOL, JSM-6400, Japan) and the pore structures were directly observed from the fractured samples. The pattern of frozen TBA solution was examined by optical microscopy (BX50, Olympus).

To determine the unidirectional compression strength of the ordered porous alumina (green bodies and sintered bulks), cylinders with a diameter of 12 mm and height of 25–30 mm were fabricated and tested on a CSS110 materials testing machine (Changchun Testing Machine Institute, China), with a crosshead speed of 1 mm min^{-1}. By setting the dense side under tension, the maximum bending strength of the gradient porous specimens was determined by three-point bending tests (test bars of $4 \times 3 \times 36$ mm^3) using an AG-2000G Shimadzu universal materials testing machine, with a crosshead speed of 0.5 mm min^{-1}.

7.2 Development Overview and Application ...

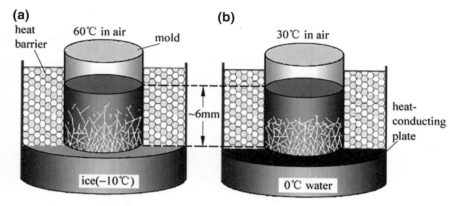

Fig. 7.13 Freezing step under different temperature conditions to achieve novel gradient porous structures **a** with higher temperature gradient of ~12 °C mm^{-1} and **b** with lower temperature gradient of ~5 °C mm^{-1}

7.2.2.2 Characteristics of Freeze-Gelcasting Process

To date, water and camphene have been successfully used as freezing vehicles. Each vehicle has its own benefits and drawbacks. In this study, TBA was used as the freezing template to achieve pore channels, because of its advantages over other freeze-casting vehicles. First, TBA can be removed via rapid volatilization (80 °C, 30 min), without any collapse or obvious shrinkage of the green body. However, in the case of water- or camphene-based freeze-casting, they have to be removed by the sublimation drying method, which takes a long time (over 48 h). Second, unlike the dendrite structure of frozen water or camphene, TBA normally presents a form of long straight ice prisms without any branches at its freezing temperature, which is important for specific applications, such as fabricating unidirectional porous ceramics used as infiltrators. Third, the freezing process can be carried out at near room temperature (below 8 °C); thus, complex treatment can be handled.

To understand the mechanisms of pore formation using frozen TBA as a template, its freezing behavior was observed by optical microscopy. TBA containing 15 wt% AM was dropped onto a pre-cooled (~0 °C) slide glass, and then frozen immediately. Figure 7.14a shows the typical microstructure on the front edge of the frozen AM/TBA solution. It can be seen that TBA prisms grew straight in one direction, and did not intersect each other. It should be noted that AM had no obvious influence on the pattern of frozen TBA, and precipitation of AM under low temperature was not observed.

Using TBA as a template in its frozen state, green bodies with pore channels were fabricated. Unlike the circular cross-sections in aqueous- or camphene-based freeze-casting, these pore channels had straight polygonal cross-sections, which can be clearly seen in Fig. 7.14b. This channeled structure existed throughout the whole green body and was inherited in the subsequent processes.

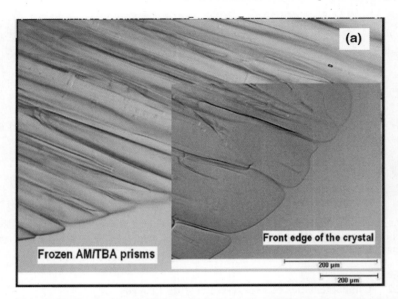

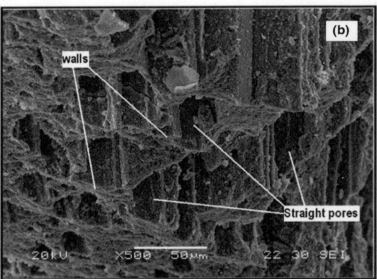

Fig. 7.14 Mechanism of pore formation using TBA as the template **a** Growth of TBA ice prisms and **b** Pore channels in dry green body (obtained from slurry with solids loading of 25 vol%)

Fig. 7.15 Effect of temperature on the polymerization of AM in TBA-based slurry with 4 wt% of initiator

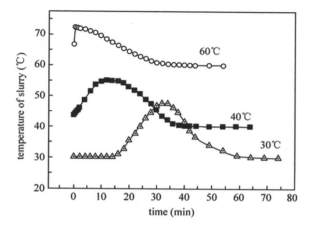

As a critical step in this freeze-gelcasting technique, the following process of gelation was induced by heat treatment at a high temperature (85 °C). Two processes occurred at this higher temperature: first, the liquid TBA crystals melted, and the AM polymerized simultaneously, converting the liquid medium into an alcogel. Subsequently, the liquid TBA from the alcogel evaporated, rather than subliming as in the case of typical freeze-casting. An important aspect is the speed of polymerization of AM in TBA, for which the heating temperature is a critical factor. Figure 7.15 shows the gelation process in TBA/AM/alumina slurry, where the sharp rise in the temperature indicates that the polymerization of AM occurs in the slurry, owing to the exothermic nature of the reaction. Slurries containing 4 wt% $(NH_4)_2S_2O_8$ initiator could be gelated at temperatures above 30 °C, and the polymerization time became shorter when more initiator was added. However, when the temperature increased to 60 °C, polymerization occurred with a very rapid speed and no obvious induction time was required.

After gelation, the compression strength of the alumina green body reached 5–10 MPa, which was much higher than that prepared by the common freeze-casting method without gelation, and made the material more convenient to handle in fabrication. The mechanical properties were affected by the initiator's proportion. Figure 7.16 illustrates the dependence of the compression strength of the green bodies with unidirectional pores on the amount of initiator in the ceramic suspensions. Under the temperature of 60 °C, if the amount of the initiator, $(NH_4)_2S_2O_8$, in the slurry was less than 1 wt%, then there was no significant solidification. When 3 wt% of initiator was added, the green bodies solidified rapidly by heat treatment, and reached the maximum compression strength of 10 MPa. On the other hand, the strength decreased when excessive initiator was added to the slurry, because the modular chain length of the polyacrylamide was restricted when too much chain growth occurred in the meantime.

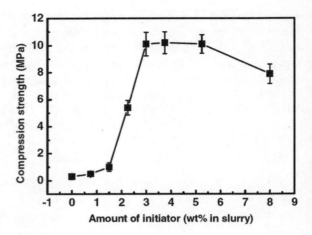

Fig. 7.16 Influence of the amount of initiator, $(NH_4)_2S_2O_8$, on the strength of the porous green bodies

7.2.3 Microstructure and Properties of Porous Alumina Ceramics

7.2.3.1 Long-Range Unidirectional Porous Structure of Alumina

As it was feasible to form pore channels in ceramic bodies as replicas of TBA ice prisms, unidirectional freezing was employed to control the growth of the solvent. The whole process is shown in Fig. 7.12, in which the steps of freezing and heat treatment can be completed rapidly (in this case, in about 5 min).

Figure 7.17 shows the typical microstructures of the unidirectional pore channels in an alumina body after sintering. Penetrating channels were aligned unidirectionally over a long range, and were observed at a vertical, horizontal, and 45° angle orientation to the TBA freezing direction, which are shown in Fig. 7.17a, b, and c, respectively. From the figure, it can be noted that the walls surrounding the channels were almost fully dense (Fig. 7.17d) after sintering.

By comparing the microstructure shown in Fig. 7.14b with that shown in Fig. 7.17c, it can be observed that the pore channels in the green bodies did not obviously shrink after sintering at 1,500 °C. The unidirectional channeled pore structure existed throughout the sintered body, except for a thin dense outer layer of about 0.5-mm thickness on the side wall that touched the mold. The channel sizes remained almost the same from top to bottom. The porosity and mechanical properties of these sintered samples acquired from the slurry of 25 vol% solids loading are summarized in Table 7.2.

These physical properties can be explained by the formation of open pore channels and fully dense alumina walls. From the measurements of total porosity and open porosity, it can be concluded that most (>90%) of the pores in the sintered bodies were open pores, indicating that the pore channels were almost fully open to the surface. It also can be seen that the ceramic walls had about 94% density, which is in good

7.2 Development Overview and Application …

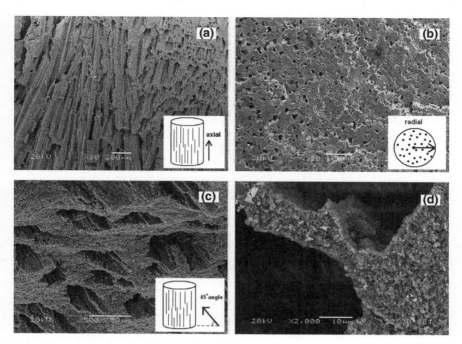

Fig. 7.17 Microstructures of the porous ceramics with long-range unidirectional pore channels fabricated by the freeze-gelcasting method observed at different orientations: **a** Vertical to freezing direction, **b** parallel to freezing direction, **c** at 45° angle direction, and **d** the dense walls formed between pores

Table 7.2 Properties of sintered alumina with long-range ordered pore channels acquired by freeze-gelcasting

Total porosity (%)	Open porosity (%)	Closed porosity (%)	Density of the walls (% TD)	Axial compression strength (MPa)	Radial compression strength (MPa)
58 ± 3	52.3 ± 2	5.7 ± 2	94 ± 2	153 ± 30	54 ± 14

agreement with the SEM images. These dense walls were formed because during the freezing of TBA, most of the ceramic particles in the slurry were discarded by the growing ice prisms and became concentrated between the crystals (Shanti et al. 2006). Because of the high density of these walls, the ceramics' compression strength in an axial direction reached 150 MPa, after being sintered at 1500 °C for 2 h.

7.2.3.2 Gradient Porous Structure of Alumina

Freeze-gelcasting of alumina/TBA/AM slurry has also been used to fabricate gradient porous ceramics. In the freezing step shown in Fig. 7.13, the TBA near the top surface at the melting temperature range can easily be volatilized with an increasing solid content. When the temperature is sufficiently high (>50 °C), polymerization of AM will occur simultaneously. On the other hand, the slurry freezes in the inner region from the cold bottom side. In this way, a pronounced pore gradient can be fabricated, with more and larger pores at the bottom, and lesser and smaller pores at the top.

As illustrated in Fig. 7.13a, gradient porous alumina with a thickness of over 5 mm was successfully fabricated with a freezing temperature gradient of 12 °C mm^{-1} and sintered at 1500 °C. SEM observations revealed that the samples retained their gradient porous structures with a long range (>3 mm, Fig. 7.18). Figure 7.18b and c shows the dense and porous regions at a higher magnification. It is obvious that the pores become smaller and fewer with larger distance from the bottom plate. Sintered parts showed almost complete densification without any noticeable defects, either in the dense region near the top surface or in the alumina walls in the porous region near the bottom (Fig. 7.18d).

As a low solids content of 25 vol% was used to prepare the alumina/TBA/AM slurry, in a normal sense, the sintered sample without any special treatment should have a high degree of porosity. However, the top surface was exposed to air to evaporate the molten TBA, where the solid content was higher. After sintering, this layer showed a density of over 90%, with some small pores, as shown in Fig. 7.18b. At the freezing side, a high porosity of over 65% was achieved, and the concentrated alumina walls were sintered without destroying the porous structure. In the middle position, the evaporation of the molten TBA and the freezing of the slurry were synchronized, and hence, the porosity was on a medium scale.

To understand the function of the temperature on the formation of the gradient porous structure, a freezing process with a smaller temperature gradient (5 °C mm^{-1}) was applied (Fig. 7.13b). On comparing the porous structures achieved by the two different temperature conditions, obvious differences in the pore gradients were observed. In addition, the microstructure of the sample acquired under 5 °C mm^{-1} was also revealed (Fig. 7.19).

By setting the bottom of the sample as the starting position, the porosity in each relative position was measured, and the results are shown in Fig. 7.20. It can be seen that in contrast with method (a), which had a larger temperature gradient, the pore gradient acquired by method (b) was less uniform over the thickness of the sample, and there was a great divide in the middle of the height. These results suggest that the gradient of the pore structure can be determined by controlling the temperature gradient. Furthermore, method (b) produced weaker green bodies than method (a); no obvious polymerization occurred at the dense end, because in the freezing step under this gradient temperature, the top end was neither "hot" enough to initiate the polymerization of AM, nor "cold" enough to avoid evaporation of TBA solvent. Hence, the gelation of AM/TBA solution on the top side was restricted. This resulted in low green body strength before firing. By setting the dense side as the tension

7.2 Development Overview and Application ...

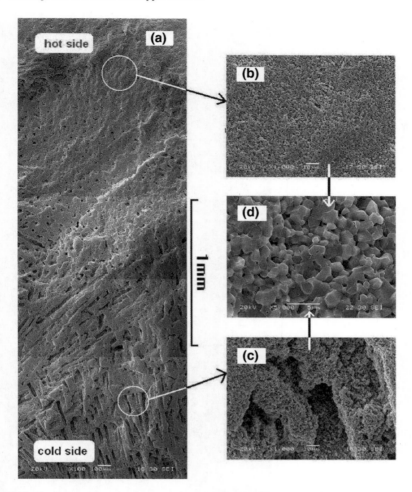

Fig. 7.18 Gradient porous alumina ceramics fabricated by freeze-gelcasting technique under the freezing temperature gradient of 12 °C mm^{-1}

face, the bending strength of the gradient porous alumina fabricated under these two different temperature gradients was measured, and the results are shown in Table 7.3.

The effect of AM polymerization on the gradient porous structure was also investigated. The experiment was carried out under the temperature condition given in Fig. 7.13a, but without AM in TBA, and the microstructure obtained is shown in Fig. 7.21. By comparing the microstructure given in Fig. 7.18 with that presented in Fig. 7.21, it can be observed that the pore distribution in Fig. 7.18 is more uniform and homogeneous than that in Fig. 7.21. The main reason for this difference is that when the polymerization of AM takes place, a rapid rise in the temperature on the hot side occurs, indicating that the temperature gradient through the body is enhanced.

Fig. 7.19 Gradient porous alumina ceramics fabricated by the freeze-gelcasting technique under the freezing temperature gradient of 5 °C mm^{-1}

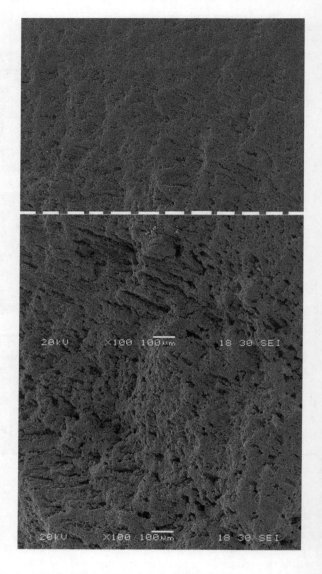

7.2.4 Mechanical Properties and Applications of Alumina Ceramics with Ultra-Low Density

The preparation of ceramics with ultra-low density is similar to that of porous ceramics by freeze-gelcasting. From the paper "Ceramics with Ultra Low Density Fabricated by Gelcasting: An Unconventional View" by Chen Ruifeng, many meaningful conclusions could be obtained (Chen et al. 2007).

7.2 Development Overview and Application ...

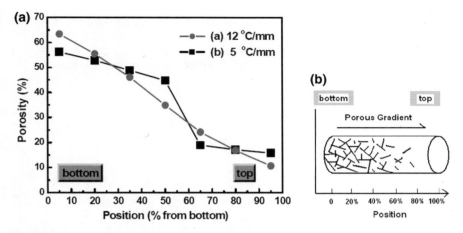

Fig. 7.20 Distribution of porosity over thickness in the sintered gradient porous alumina, with the bottom as the starting position **a** Top temperature = 60 °C, bottom temperature = −10 °C and **b** Top temperature = 30 °C, bottom temperature = 0 °C

Table 7.3 Bending strength of porous alumina with different gradients

Fabrication method	Green strength (MPa)	Strength after sintering at 1500 °C (MPa)
Method (a)	11 ± 1	260 ± 20
Method (b)	2 ± 0.5	226 ± 15

The main properties of dry green bodies with low relative density (from 5 to 15% of initial solids loading) are shown in Table 7.4. As a result of polymerization, the strength of the green body samples was significantly improved; they were much stronger than the ceramic green bodies with ultra-high porosity fabricated by the other methods. Even when the relative green body density was as low as 5%, the compression strength was over 1 MPa, which was strong enough to allow the materials to be handled in the subsequent procedures.

The density and mechanical properties of the sintered samples are summarized in Table 7.5. The results show that after being sintered at a high temperature, the samples acquired from 5 and 10 vol% slurries were so light that the bulk densities were even lower than that of water (1 g cm^{-3}). As the initial solids loading increased, the compression strength increased from 31 to 84 MPa.

The effect of sintering temperature on the mechanical properties was also investigated, and the results are shown in Table 7.6.

Table 7.7 lists some of the other important properties of ultra-low-density alumina ceramics fabricated using this gelcasting method, in which the open porosity varied from 60 to 92%, the pore size varied from 0.1 to 2.2 μm, and the specific surface area varied from 4 to 15 m^2 g^{-1}. It can be concluded that in these light-weight alumina

Fig. 7.21 Pores in alumina ceramics fabricated by TBA freezing without AM addition

7.2 Development Overview and Application …

Table 7.4 Properties of the green bodies with ultra-low density (from 5 to 15% of initial solids loading)

Solids loading of the slurry (vol%)	Compression strength (MPa)	Linear shrinkage after drying (%)	Relative density (%)
5	1.1 ± 0.2	0.42–0.48	5.05–5.1
10	2.1 ± 0.2	0.25–0.35	10.1–10.2
15	2.9 ± 0.2	0.14–0.22	15.1–15.2

Table 7.5 Properties of alumina with different initial densities after being sintered at 1500 °C for 2 h

Solids loading of the slurry (vol%)	Bulk density after sintering (g cm^{-3})	Compression strength (MPa)
5	0.48–0.52	31.5 ± 3.1
10	0.99–1.02	56.4 ± 3.9
15	1.55–1.60	83.9 ± 4.3

Table 7.6 Properties of alumina (made from slurry containing 5 vol% of solids loading) after being sintered at different temperatures

Sintering temperature (°C)	Bulk density after sintering (g cm^{-3})	Compression strength (MPa)
1400	0.31–0.33	11.1 ± 1.1
1500*	0.49–0.51	31.5 ± 1.5
1600	0.79–0.82	50.6 ± 2.1

Note Symbol "*" refers to the examples mentioned in the "Application" section

Table 7.7 Open porosity, pore size, and specific surface area of low-density alumina acquired by TBA-based gelcasting

Solids loading of the slurry (vol%)	Sintering temperature (°C)	Open porosity (%)	Mean pore size (μm)	Specific surface area (m^2 g^{-1})
5	1400	92	0.1, 1.0	14.82
	1500*	88	1.2	10.69
	1600	80	2.2	6.52
10	1400	85	0.1, 0.8	8.75
	1500	75	1.0	4.97
15	1400	77	0.1, 0.6	5.2
	1500	64	0.8	4.00

Table 7.8 Applications of alumina with low density obtained using TBA-based gelcasting (Scheffler and Colombo 2005)

Application	Required properties
Filter	Open porosity of >80%, pore size of -1 μm
Supports for catalysts	Specific surface area of >10 m^2 g^{-1}, compression strength of >10 MPa
Artificial bones	Bulk density of <1 g cm^{-3}, compression strength of >50 MPa
Lightweight parts for high-temperature manufacturing	Bulk density of <0.5 g cm^{-3}, compression strength of >30 MPa

ceramics, the increase in porosity is often related to larger pore sizes, decreased strength, and greater specific areas.

With controllable microstructures and properties, the gelcasting technique for the fabrication of ceramics with ultra-low density can be expanded from the laboratory to actual industrial applications, as illustrated in Table 7.8. For each application, ceramics with the required properties can be produced, as shown in Tables 7.5, 7.6, and 7.7. For example, alumina acquired from 5 vol% slurry and sintered at 1500 °C (designated with an asterisk in Tables 7.6 and 7.7) had an open porosity of 88%, a pore size of 1.2 μm, and a compression strength over 30 MPa, which makes it suitable for infiltrating bacteria in biomedicine.

In this section, a new freeze-gelcasting technique that combines the merits of freeze-casting and gelcasting has been introduced. For this method, alumina/TBA/AM slurry was prepared, with TBA as the freezing vehicle and AM as the gelation agent. Freezing occurred rapidly at a relatively high temperature (0–5 °C), and generated frozen TBA prisms with diameters of several tens of micrometers as templates of pore channels. By controlling the freezing direction, unidirectional porous ceramics with a long range were successfully fabricated, with almost all the pore channels being open pores. The temperature gradient was found to have a greater influence on the porosity and pore-size distribution in the ceramic body, over the thickness of the sample. By choosing appropriate temperature gradient in the freezing step, a uniform linear pore gradient could be achieved, with high porosity (>65%) at the bottom and low porosity (<10%) at the top after sintering. Polymerization of AM in TBA was realized by subsequent heat treatment, providing a way to acquire green porous ceramic bodies with high strength, i.e., compression strength over 10 MPa of unidirectional porous alumina; thus, more complex manufacturing can be handled. The porous ceramics demonstrated good mechanical properties, i.e., the unidirectional porous alumina reached a compression strength of 150 MPa after being sintered at 1500 °C, because the pore channels formed by the TBA template were surrounded by almost completely dense walls without any noticeable defects. These results show the prospect of the freeze-gelcasting technique for fabricating porous ceramics with possible applications, such as infiltration and catalytic support. In addition, this technique can be applied in many other ceramic systems.

7.3 Solidification of Concentrated Silicon Nitride Suspensions for Gelcasting by Ultrasonic Effects

7.3.1 The Forming Method of Gelcasting Using Ultrasonic Effects

Gelcasting is a near-net-shaped forming method for the production of advanced ceramic materials. The green body prepared by gelcasting has a similar homogeneous microstructure like the precursor suspensions and shrinks more uniformly during drying and sintering, than those prepared using conventional methods. As a result, the structure homogeneity and reliability of the ceramics are improved. Hence, ceramics scientists have paid great attention to the gelcasting technique since its initial development. However, the controllability of the gelcasting process is not good, and the amounts of initiator and catalyst added to the ceramic suspensions are restricted within a narrow range. Ceramic suspensions are observed to solidify very quickly when casted with slightly more initiator and catalyst, while the addition of slightly less initiator and catalyst is found to result in an incomplete solidification of the suspensions. Therefore, in most cases, initiator and catalyst are added to the cooled ceramic suspensions to prolong the processing time, and then the polymerization of the monomers are initiated to solidify the suspensions by increasing the temperature of the mold. The temperature-induced polymerization of the monomers results in another problem, namely, the asynchronous solidification of the suspension, which can cause cracks or potential cracks between the parts solidifying at different times, owing to the thermal transmission from the outside to the center of the suspension. Hence, it is necessary to develop another mechanism to solidify the suspensions homogeneously. Yang et al. (2002) reported that a pressure-induced solidification of the ceramic suspensions could avoid the effect of temperature gradient on gelcasting process. However, this method was found to increase the complexity of the equipment—microwaves can heat the suspension uniformly, but it is not very easy to control and it usually overheats the suspension and forms pores in it; also, it is not applicable for metal mold, which is necessary for the formation of a good surface of the green body. On the other hand, to reduce the cracks caused by excessive shrinkage during drying of the green body, the content of the solid powders in the suspensions for gelcasting must not be less than 45 vol%. Unfortunately, most suspensions prepared directly with raw Si_3N_4 powders could not reach this lower limit. In this study, acid-washing and preoxidation were used to increase the solid volume fraction of the Si_3N_4 suspensions from 40 to 50 vol%, and ultrasonic waves were used to accelerate the decomposition of the initiator, $(NH_4)_2S_2O_8,$ as well as for the homogeneous solidification of Si_3N_4 suspensions for gelcasting. The obtained results showed that the flexibility of gelcasting could be improved with the ultrasonic effect.

The Si_3N_4 powder, M11 (Stark Corp., Germany), was used as the raw material. Furthermore, 15 wt% of AM premix solution, containing some MBAM (2 wt% to

AM), was used as the disperse media, and 10 wt% TMAH (tetramethyl ammonia hydroxide) solution was used as the dispersant. The groups on the surface of Si_3N_4 powders were analyzed with infrared Fourier transformation spectroscopy (Model FTIR-750, Nicolet Corp., USA), and the chemical state of Si on the surface of Si_3N_4 powders was observed using X-ray photoelectron spectroscopy (XPS) (Model PHI-5300ESCA, USA). The absorption spectra of the solutions were recorded using UNICAM500 spectral photometer (Unicam Corp., USA). The rheological curves of the Si_3N_4 suspension were measured with a MCR300 Rheometer (Physica Corp., Germany). After the addition of $(NH_4)_2S_2O_8$ (8 wt% solution) and TEMED in the volume ratio of 2:1 to the de-aired suspension, the temperature of the suspension during solidification was recorded using a computer linked with the temperature detector. The strength of the samples was measured by three-point flexural tests. The frequency and capacity of the ultrasonic wave were 28 kHz and 0.3 w cm^{-2}, respectively.

7.3.2 Preparation of Concentrated Silicon Nitride Suspensions

It is well known that it is difficult to prepare concentrated suspensions of more than 40 vol% solids loading for gelcasting with raw M11 Si_3N_4 powder, because of the presence of hydrophobic groups such as amine structures (Si_2NH and $SiNH_2$) on the powder's surface. Generally, various surface treatments are used to modify the surface groups of Si_3N_4 powders to improve its dispersibility in the aqueous suspensions. In this work, preoxidation after acid-washing was used to eliminate the hydrophobic groups and to introduce a hydrophilic group such as silanol (Si–OH) on the surface of the Si_3N_4 powders. First, Si_3N_4 powders were mixed with 5% HCl solution in the ratio of 100 g: 200 ml, and the mixture was ball-milled for 24 h to ensure effective acid-washing. Subsequently, the slurry was filtered and washed thrice with deionized water during filtration, and the obtained mud was dried at 110 °C for 24 h. Finally, the dried mud was smashed and calcined at 600 °C for 6 h for the preoxidation of Si_3N_4 powders. The preoxided Si_3N_4 powders were analyzed with diffusion reflection infrared Fourier transformation (DRIFT) spectrum (Fig. 7.22a) and XPS spectrum (Fig. 7.22b). Figure 7.22a shows that after peroxidation, the hydrophobic amine groups (corresponding to the absorption peak at about 3400 cm^{-1}) in the untreated Si_3N_4 powders were eliminated, and the hydrophilic Si–OH groups (corresponding to the absorption peak at about 3750 cm^{-1}) were introduced on the powder's surface. In addition, the peak of the binding energy of Si (2p) obviously shifted to larger values after preoxidation (Fig. 7.22b). The shifted peak was confirmed to consist of Si (2p) peak of Si_3N_4 (101.8 eV) and Si_2N_2O (102.3 eV), after data-processing. Thus, it can be presumed that Si_2N_2O appeared on the surface of the Si_3N_4 powder. Because of the formation of Si_2N_2O, the amount of oxygen on the Si_3N_4 surface was increased, and the raised oxygen amount increased the zeta potential in the Si_3N_4

7.3 Solidification of Concentrated Silicon Nitride Suspensions ...

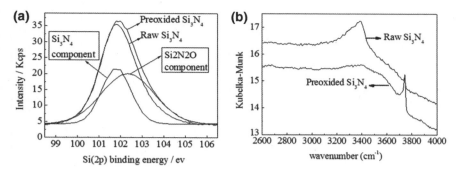

Fig. 7.22 a DRIFT spectrum and b XPS spectrum of raw and preoxided silicon nitride powders

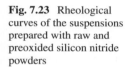

Fig. 7.23 Rheological curves of the suspensions prepared with raw and preoxided silicon nitride powders

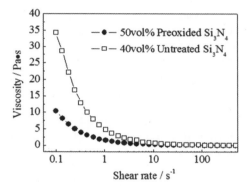

surface (Nagel 1988). As a result of the combination of the abovementioned effects, the solid volume fraction of the Si_3N_4 suspensions was increased from 40 (untreated powders) to 50 vol% (preoxided powders) with the preoxidation treatment, and the rheological behavior was also improved (Fig. 7.23).

7.3.3 Ultrasonic Accelerated Solidification

The temperature changes in the Si_3N_4 suspensions without and with $(NH_4)_2S_2O_8$ (3 μl ml^{-1} suspension) during ultrasonic treatment are shown in Fig. 7.24. During the process of ultrasonic solidification, the temperature increase in the suspension may be owing to two reasons: (1) ultrasonic heating effect and (2) polymerization of the monomers.

In Fig. 7.24, curve A shows the ultrasonic heating effect and curve B shows the overall temperature rise caused by monomer polymerization and ultrasonic heating effect. If the ultrasonic heating effect (curve A) is deducted from the overall temperature rise in the suspension (curve B) during ultrasonic treatment, the temperature

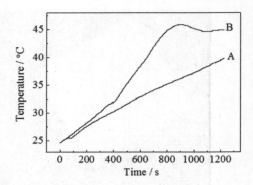

Fig. 7.24 Temperature rise in 50 vol% silicon nitride suspensions under ultrasonic condition (A) without the initiator and (B) with the initiator

increment owing to the polymerization of the monomers in the Si_3N_4 suspension can be obtained, which is shown as curve C in Fig. 7.25. The temperature increment owing to polymerization induced by ultrasonic waves (curve C) and heating at 30 °C in a water bath without ultrasonic waves (curve D), are compared in Fig. 7.25. The acceleration of the polymerization of the monomer by ultrasonic waves can be predicated by comparison, because the Si_3N_4 suspension had a much shorter idle time under ultrasonic waves.

The polymerization of AM monomers was initiated by free radicals. The free radicals from $(NH_4)_2S_2O_8$–TEMED system were found to be the superoxide anion (Xiao et al. 1991), which could react with the hydroxylamine solution according to the formula:

$$2O_2^{-\bullet} + NH_2OH + H^+ = NO_2^- + H_2O_2 + H_2O \tag{7.1}$$

The resultant nitrite ions, after coloration reaction with sulfanilic acid and α-naphthylamine, could be quantified by the specific absorption peak at a wavelength of 530 nm (Wang and Luo 1990; Xiao et al. 1999). The linear relationship between the concentration of the nitrite ions and the adsorption coefficient of the solution at 530 nm was calibrated with a $NaNO_2$ solution, as shown in Fig. 7.26. The original

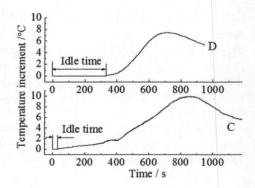

Fig. 7.25 Temperature increment caused by the solidification of 50 vol% silicon nitride suspensions (C) under ultrasonic treatment and (D) at 30 °C water bath

Fig. 7.26 The dependence of the absorption coefficient at a wavelength of 530 nm on the concentration of the nitrite ions in the solution

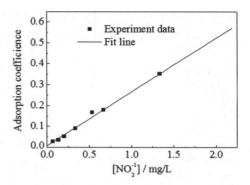

mole concentration of the superoxide anions could be determined by doubling that of the nitrite ions.

According to the abovementioned principle, the concentration of the superoxide anions in $(NH_4)_2S_2O_8$–TEMED aqueous solution (6 μl of 8 wt% $(NH_4)_2S_2O_8$ and 3 μl of TEMED per 1 ml of deionized water) under various conditions was determined (Fig. 7.27). From this figure, it can be seen that the concentration of the superoxide anions at 40 °C is much greater than that at 25 °C, but less than half of that treated by ultrasonic waves, although the solution temperature during ultrasonic treatment never reached 40 °C. Figure 7.27 shows that the concentration of the superoxide anions changed little with time, indicating that the decomposition of $(NH_4)_2S_2O_8$ reached equilibrium in a shorter time under a specified condition. Hence, the solidification of the suspension is difficult to control at room temperature. However, the cage effect, which restricts the diffusion and promotes the recombination of the free radicals in the solution, was weakened under ultrasonic treatment or at raised temperature, and this effect resulted in higher concentrations of the superoxide free radicals (Fig. 7.27). Figure 7.27 shows that the ultrasonic treatment had a much greater effect on the decomposition of $(NH_4)_2S_2O_8$ in the solution than temperature. Therefore, the accelerated solidification of Si_3N_4 suspensions by the ultrasonic waves could be attributed to the increased decomposition degree of the initiator, $(NH_4)_2S_2O_8$.

Fig. 7.27 The effects of ultrasonic waves and temperature on the concentration of the superoxide anions in $(NH_4)_2S_2O_8$–TEMED solutions

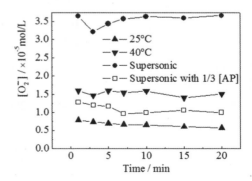

7.3.4 Comparison of Thermal- and Ultrasonic-Activated Solidifications

Thermal-activated solidification was accomplished by dipping the glass beaker containing the Si_3N_4 suspension at 25 °C (Fig. 7.28) into a 40 °C water bath in a thermostat, while ultrasonic-activated solidification was accomplished by dipping the beaker containing a suspension at 25 °C into a 25 °C water bath in the ultrasonic generator. In both the cases, the beaker was fixed in a polyurethane shoe to avoid the influence of axial thermal transfer from the bottom of the beaker. The water surface outside the beaker was about 10 mm higher than the suspension surface in the beaker. The temperature changes at points A, B, and C (Fig. 7.28) were recorded after the beaker was dipped into the thermostat or after the ultrasonic generator was turned on.

The temperature increase owing to the polymerization at different points in the Si_3N_4 suspensions during solidification activated by ultrasonic waves and temperature is compared in Fig. 7.29. It should be noted that the effects of ultrasonic heating and heating at 40 °C in a water bath were deducted according to the same method mentioned earlier. In the case of solidification by heating at 40 °C in a water bath, an annular temperature gradient appeared in the suspension after the beaker was dipped into the water, and hence, the solidification proceeded radically from the beaker wall to the center of the suspension (Fig. 7.29b). This made the solidification of the suspension near the center constrained by the outer solidified ring. As a result, the center part of the green body endured radial and circumferential tensile stresses, which may result in cracks in the green body during drying and demolding. The asynchronous solidification caused by the temperature gradient could be suppressed using ultrasonic waves. Figure 7.29a shows that solidification of the Si_3N_4 suspension began from the edge to the center of the beaker almost simultaneously. The temperature difference at the points A, B, and C during solidification may be owing to the thermal transfer from the solidifying suspension to the water out of the beaker.

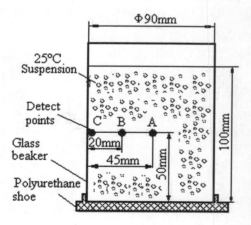

Fig. 7.28 Schematic representation of the apparatus to study the asynchronous solidification of silicon nitride suspensions

7.3 Solidification of Concentrated Silicon Nitride Suspensions ...

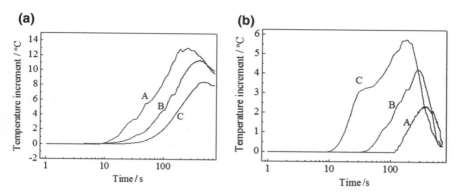

Fig. 7.29 The temperature increments at different positions in the silicon nitride suspensions during solidification activated by **a** ultrasonic waves and **b** temperature (heating at 40 °C in a water bath)

The strength of the Si_3N_4 green bodies prepared by the gelcasting of 50 vol% suspension is shown in Table 7.9. In cases I and II, 3 and 1 μl of $(NH_4)_2S_2O_8$ solution was added to 1 ml of suspension, respectively. Table 7.9 indicates that the thermal-solidified green body has a higher strength than that solidified with ultrasonic waves, with respect to case I. This may be owing to the fact that ultrasonic treatment yielded too many superoxide anions in the suspension (Fig. 7.27), thus reducing the length and crosslinkage of the polyacrylamide, which bound the Si_3N_4 powders and determined the strength of the green body. Similarly, the suspension ultrasonically solidified with less amount of $(NH_4)_2S_2O_8$ solution (case II) resulted in a higher strength of the green body. However, the suspension with less amount of $(NH_4)_2S_2O_8$ solution could not be solidified by thermal activation (40 °C) alone. This shows that in the case of suspension with less amount of initiator, thermal activation could not yield sufficient amount of free radicals to initiate the solidification of the suspension, while the ultrasonic waves could. Hence, ultrasonic treatment can be used to improve the controllability of the solidification of Si_3N_4 suspensions, thus improving the flexibility of the gelcasting process.

In addition, Si_3N_4 suspensions with less amount of initiator were solidified, yielding high-strength green body (43.6 MPa). This suggests that Si_3N_4 suspensions with less amount of initiator could be solidified using ultrasonic waves, yielding adequate

Table 7.9 Strength of the silicon nitride green bodies prepared by the gelcasting of 50 vol% suspension

	I (MPa) 3 μl of $(NH_4)_2S_2O_8$ per 1 ml of suspension	II (MPa) 1 μl of $(NH_4)_2S_2O_8$ per 1 ml of suspension
Ultrasonic-activated	23.6	43.6
Thermal-activated (40 °C water bath)	30.5	Incomplete solidification

strength to maintain the configuration of the green body, which is in favor of debonding of the green body. However, this suggestion should be confirmed in the future studies.

The hydrophobic amine groups were removed from the surface of the Si_3N_4 powders by preoxidation, and thus, the solid volume fraction of Si_3N_4 suspensions was raised from 40 to 50 vol% and the rheological behavior of Si_3N_4 suspensions was improved.

Both temperature and ultrasonic waves promoted the decomposition of $(NH_4)_2S_2O_8$, thus accelerating the solidification of concentrated Si_3N_4 suspensions for gelcasting. However, ultrasonic treatment resulted in apparently more superoxide anionic free radicals than temperature increase. As a result, the Si_3N_4 suspensions could be solidified with less amount of initiator, and hence, the controllability of suspension solidification and flexibility of gelcasting process could be improved. At the same time, asynchronous solidification of the suspension caused by the temperature gradient, which led to crack potential during the drying and debonding of the green bodies, could be suppressed by ultrasonic-activation solidification. Additionally, Si_3N_4 suspensions with less amount of initiator yielded a green body with high strength after ultrasonic-activated solidification. This demonstrates that Si_3N_4 suspensions with less amount of initiator could be solidified by ultrasonic waves.

7.4 Novel Laser Machining Technology for Alumina Green Ceramic

7.4.1 Laser Machining Technology

The machining of ceramics to their final dimensions by conventional methods is extremely laborious and time-consuming. Laser machining is a non-contacting, abrasion-less technique, which eliminates tool wear, machine-tool deflections, vibrations, and cutting forces, reduces limitations with respect to shape formation, and inflicts less sub-surface damage. Hence, as an advanced manufacturing method, it is used in different areas such as electronics, aerospace, material processing, etc. (Chryssolouris et al. 1997; Chryssolouris 1991). Structural ceramics demonstrate good high-temperature performances, abrasion resistance, anti-corrosion, and other unique properties, and have been widely used in the fields of machinery, chemicals, petroleum, textiles, aviation, aerospace, etc. However, the current marketing characteristic of structural ceramics is multi-purpose and limited. Hence, if the manufacturing costs are high, the promotion and application of structural ceramics are limited. A new laser machining method has been proposed to produce ceramics that satisfy the requirements mentioned earlier. This technology combines laser machining and gelcasting technique, to machine complex-shaped ceramic parts. This method builds parts by selectively removing the materials, layer by layer, as specified by a computer program, to generate geometrical objects directly from a three-dimensional computer image without part-specific tooling or human intervention.

Gelcasting is a near-net-shaped forming method combining polymer physical chemistry, colloidal chemistry, and ceramic technology. This process utilizes the polymerization of a monomer in the suspension solution of the ceramic powder via a free radical initiator to produce strong green bodies, which can be removed from the mould rather quickly. When compared with other casting methods, gelcasting can obtain much more homogeneous materials with little density difference over the parts and produce green bodies with higher flexural strength (Chen et al. 1999; Huzzard and Blackburn 1997). The laser machining process of ceramics in the sintered state has some shortcomings owing to the presence of spatter and microcracks. Hence, many researches have been employing ceramics in their green states. At present, this study has attempted to discuss the laser machining mechanism by means of laser-material interaction parameters and to realize the laser machinability of alumina green bodies by changing the laser parameters. Laser machining of ceramics is also used extensively in the microelectronics industry for scribing and via-hole drilling (Lumpp and Allen 1997).

7.4.2 Practical Application of Laser Machining Technology

7.4.2.1 Spatter-Free Laser Drilling Process

Laser drilled holes are inherently associated with spatter deposition owing to the incomplete expulsion of ejected material from the drilling site, which subsequently resolidifies and adheres onto the material surface around the hole periphery. In addition, high hardness and brittleness may lead to fracture (microcracks) of the ceramic material during laser machining. Hence, to prevent spatter deposition and microcracks during laser machining, many techniques based on either chemical or physical mechanisms have been developed (Orita 1988; Sharp et al. 1997; Low et al. 2001; Murray and Tyrer 2000); however, the results are not encouraging.

Gelcasting is an attractive ceramic-forming process for making high-quality, complex-shaped ceramic parts by means of in situ polymerization, through which a macromolecular network is created to hold the ceramic particles together. In the process, ceramic slurry with high solids loading, obtained by dispersing ceramic powders in a premix monomer solution, is casted in a mold of the desired shape. After the addition of initiator and catalyst, the entire system is solidified in situ and a green body with excellent mechanical properties, but with only a few percents of polymer, is obtained. The dried green body can be machined easily and no extra binder burnout program is required (Young et al. 1991). In this study, an anti-spatter and anti-crack laser drilling method that uses Nd:YAG laser to drill holes directly on a gelcast green ceramic body is reported. Owing to the difference between the microstructure of the green body and that of the sintered ceramics, formation of spatter and microcracks can be successfully prevented in the green body.

The ceramic powder used in this study was an alumina powder with a mean particle size of 2 μm. AM and MBAM were used as the monomer and crosslinker, respectively. AM and MBAM were dissolved in demonized water to obtain the premix solution. TEMED and $(NH_4)_2S_2O_8$ were used as the catalyst and initiator, respectively. The gelcasting process was similar to that described in the previous studies. First, the ceramic powder was added to the premix solution (containing 20 wt% of AM) and thoroughly ball-milled. Then, the slurry was de-aired by a vacuum pump. After adding initiator and catalyst, the slurry was casted into a nonporous mold without additional pressure. After polymerization and drying, a green body with high mechanical strength was obtained, and subsequently, laser drilling was carried out. The detailed flowchart of the gelcasting and drilling process is shown in Fig. 7.30. The microstructure of the sample was observed by SEM, and the hole size was measured through optical microscopy. The mechanical strength of the samples was determined by the three-point flexure test. A pulsed Nd:YAG laser, typical of many

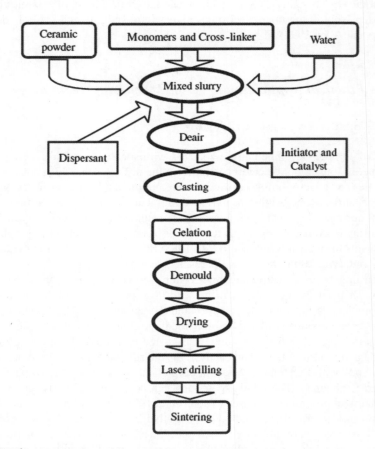

Fig. 7.30 Detailed flowchart of the gelcasting and laser drilling process

7.4 Novel Laser Machining Technology for Alumina Green Ceramic

lasers used in the industry for drilling and fine-cutting applications, was selected for processing. The laser beam had a focal spot size of about 100 μm and a wavelength of 1.064 μm. Drilling was operated at a pulse duration of 0.2 ms and pulse repetition rate of 2 Hz, without any assist gases.

7.4.2.2 The Basic Mechanism of Laser Drilling and the Comparison of Hole Morphologies

Figure 7.31 shows the SEM photos of the holes drilled on the green alumina samples obtained from the slurry with 50 vol% solids loading. In addition, holes on sintered ceramics drilled under the same conditions are also illustrated for comparison. The basic mechanism of laser drilling is based on the thermal action of light. When high-intensity light strikes the material surface, the photons of the beam are absorbed and converted into thermal energy. The temperature rises locally and a combination of melting and vaporization occurs. Under the continual effect of the laser beam, the molten material is ejected by the material vapor and a hole is thus formed. When the laser pulse comes to an end, the decrease in the flux density, the blurring of the light-spot edges, as well as the power decrease at the pulse trailing edge, contribute to an increase in the resolidified liquid residue in the hole, resulting in recast layers and a spatter-deposited surface (Metev and Veiko 1996). As can be seen in Fig. 7.31a, the holes drilled on the sintered ceramics are encircled with large areas of irregular spatter deposits. In contrast, it is clearly evident from Fig. 7.31b that very little spatter is present around the holes drilled on the gelcast green body. Figure 7.31b also exhibits better surface quality and a smooth hole periphery. Figure 7.31c–f shows further details of the surface quality of the holes presented in Fig. 7.31a and b, respectively. From Fig. 7.31c and e, it can be seen that the spatter deposits at the periphery of the hole on the sintered alumina show characteristics of a resolidified molten phase, and some microcracks can be seen, which may be owing to the laser-induced thermal shock and stress. On the other hand, the periphery of the hole on the green body shown in Fig. 7.31d and f exhibits a structure composed of particles, without any cracks.

7.4.2.3 The Different Characters of Holes on Different Green and Sintered Bodies

Figure 7.32 compares the cross-section features of the two samples, which also indicates that the spatter on the sintered ceramics does not appear on the green body.

From the fracture microstructures of the two samples shown in Fig. 7.33, it can be seen that the sintered alumina has a polycrystalline character, i.e., it is composed of grains of 7–8 μm in size. On the other hand, the green body is composed of particles of about 2 μm in size.

As a result of the irregular spatter deposition, the consistency/repeatability of the geometry of the holes on the sintered body is inferior to that of the holes on the green

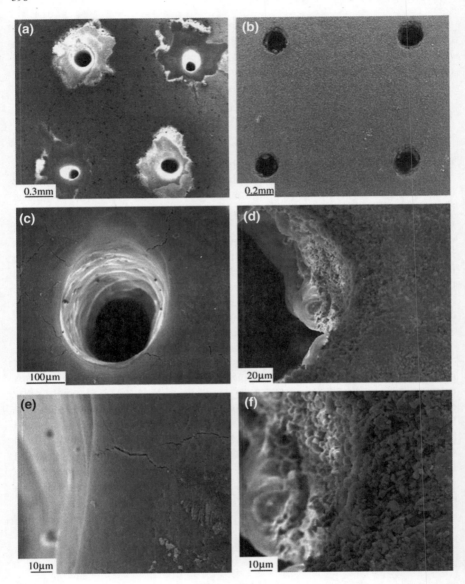

Fig. 7.31 Comparison of hole morphologies on different samples. **a**, **c**, and **e** are the holes on the sintered ceramics. **b**, **d**, and **f** are the holes on the gelcast green body

7.4 Novel Laser Machining Technology for Alumina Green Ceramic

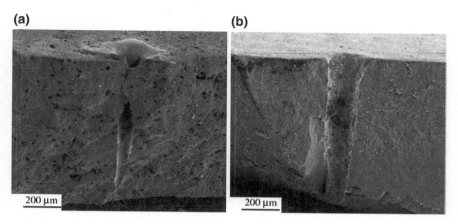

Fig. 7.32 Illustration of the cross-section features of the holes on different samples. **a** and **b** are the holes on the sintered ceramics and gelcast green body, respectively

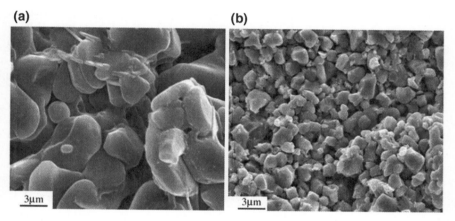

Fig. 7.33 SEM micrographs showing the fracture microstructures of the two samples. **a** and **b** are the fracture graphs of the sintered ceramics and gelcast green body, respectively

body. This can be clearly seen by comparing the standard deviation of the entrance hole diameter and hole depth values obtained by the two methods (see Table 7.10).

The relatively large standard deviation of the hole depth values are owing to the difficulty in precisely obtaining the fractured hole sections for measurement. More regular and uniform hole shapes can be obtained on the gelcast green samples than on the sintered ceramics. In addition, the holes drilled on the green body show larger diameter and depth values than those on the ceramics. Furthermore, lower total heat of evaporation and melting may result in larger hole depth and hole diameter values under the same pulse energy (Metev and Veiko 1996). This indicates a lower heat of evaporation for the green bodies. Thus, from the thermal nature of the radiation

Table 7.10 Comparison of the entrance hole diameter and hole depth values by two different drilling methods (6 pulses)

Sample	Average hole diameter[a] (mm)	Standard deviation (mm)	Average hole depth (mm)	Standard deviation (mm)
Gelcast green body	0.21	0.016	1.1	0.18
Sintered ceramic body	0.19	0.033	0.8	0.25

[a]Calculated from 10 holes

and from the examination of the microstructure, it is reasonable to conclude that the absence of spatter and microcracks in the gelcast green body (which contains about 4 wt% of polymer material) may be attributed to the lower heat of evaporation and the relatively loose microstructure of the green body, when compared with the sintered ceramics.

The conventional laser drilling of green alumina bodies containing PVA or PVB as binders is not a suitable method, because the propagation of shock waves and thermal stresses induced by the laser may cause damage to the alumina body (Metev and Veiko 1996). The influence of AM content in the premix solution on the green body strength derived from 50 vol% alumina slurries is shown in Fig. 7.34.

For the PVA system, the samples were die-pressed at a pressure of 75 MPa.

It can be seen that an AM concentration of 20 wt% can provide a strength of about 20 MPa. In comparison, die-pressed samples prepared using even a 10 wt% of PVA solution as the binder show a much lower strength of about 7.3 MPa. Consequently, the holes are damaged under the radiation of the laser beams. This can be clearly seen from the SEM micrographs shown in Fig. 7.35. A conical shape in the upper part of the hole was formed during laser drilling. The hole also exhibits a conical shape in

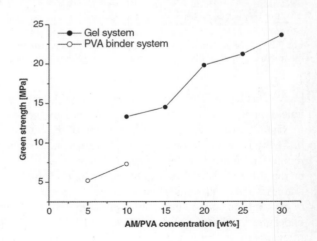

Fig. 7.34 Influence of AM/PVA concentrations on the green body strength

7.4 Novel Laser Machining Technology for Alumina Green Ceramic

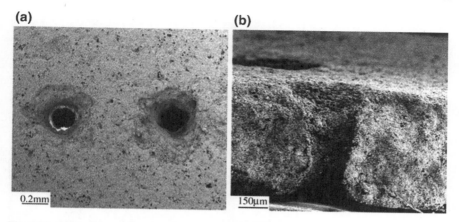

Fig. 7.35 SEM micrographs of the holes drilled on the conventional die-pressed green alumina body

the lower part near the bottom of the material. This implies that the relatively high-quality holes obtained on the gelcast green body can be attributed to its excellent mechanical properties.

In addition, the absence of spatter and microcracks in the holes produced by the method described earlier is independent of the Nd:YAG laser and special drilling parameters, although these conditions can affect the hole quality to a large extent. Figure 7.36 shows the compact hole arrays obtained after varying the pulse energy and pulse repetition rate, which exhibits reasonably encouraging results. Furthermore, when this laser drilling method was tested on Si_3N_4 samples, similar results were obtained (see Fig. 7.37). It should be pointed out that gelcasting is a highly versatile

Fig. 7.36 SEM photos of compact hole arrays drilled on the gelcast alumina green body

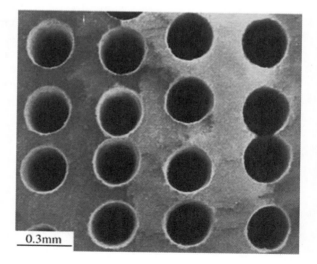

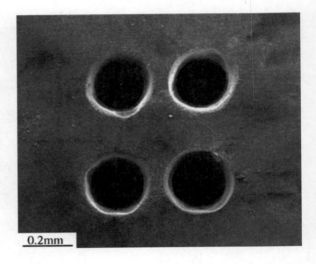

Fig. 7.37 SEM photos of compact hole arrays drilled on the gelcast silicon nitride green body

fabrication process. It is not limited to the usage with any particular ceramic powder. It can be quickly adapted for use with new materials and new applications, and is observed to be as effective with the metal powders, as with the ceramic powders. Thus, laser processing based on the gelcasting technique can be applied to a variety of materials.

Thus, in this study, a novel anti-spatter and anti-crack laser drilling method based on the gelcasting technique was developed. Laser drilling of a gelcast alumina green body proved to be highly effective in preventing spatter and microcracks during laser percussion drilling of the material. Furthermore, microholes with regular and uniform shapes, but without cracks, were successfully obtained by this technique. From the thermal nature of the drilling processing and from the examination of the microstructure, it was deduced that the absence of spatter and cracks in the gelcast green body might be attributed to the lower heats of evaporation and melting, as well as the relatively loose microstructure of the green body, when compared with the sintered ceramics. In addition, it should be pointed out that gelcasting is a highly versatile fabrication process. Its use is not limited to any particular ceramic powder. It can be quickly adapted for use with new materials and new applications, and it is as effective with metal powders, as with ceramic powders. Hence, the laser processing method used in this study can be applied to the laser machining of a variety of materials.

References

Agarwals, M. K., & Weeren, R. (1996). *American Ceramic Society Bulletin, 75,* 60.
Araki, K., & Halloran, J. W. (2005). Porous ceramic bodies with interconnected pore channels by a novel freeze casting technique. *Journal of the American Ceramic Society, 88*(5), 1108–1114.

References

Barnes, H. A., Hutton, J. F., & Walers, K. (1989). *An introduction to rheology.* Oxford: Elsevier Press.

Chang, A. S., & Shih, T. S. (1999). Permeability of coating in the lost foam casting process. *International Journal of Cast Metals Research, 12,* 263–275.

Chen, Y. L., Xie, Z. P., Yang, J. L., & Huang, Y. (1999). Alumina casting based on gelation of gelatine. *Journal of the European Ceramic Society, 19,* 271–275.

Chen, R., Huang, Y., Wang, C. A., & Qi, J. (2007) Ceramics with ultra low density fabricated by gelcasting: an unconventional view. *Journal of the American Ceramic Society,* 1–31.

Chryssolouris, G. (1991). *Laser Machining: Theory and Practice.* New York: Springer-Verlag.

Chryssolouris, G., Anifantis, N., & Karagiannis, S. (1997). Laser assisted machining: an overview. *Journal of Manufacturing Engineering—Transaction of ASME, 119,* 4B, 766–769.

Danforth, S. C. (1995). *Mat Tech. 144,* 10.

Ferreira, J. M. F., & Diz, H. M. M. (1992). Effect of slurry structure on the slip casting of silicon carbide powders. *Journal of the European Ceramic Society, 10,* 59–64.

Fukasawa, T., Ando, M., Ohji, T., & Kanzaki, S. (2001). Synthesis of porous ceramics with complex pore structure by freeze-dry processing. *Journal of the American Ceramic Society, 84*(1), 230–232.

Griffin, M. L., & Halloran, J. W. (1996). *Journal of the American Ceramic Society 79,* 2601.

Griffin, C., Daufenbach, J., & Mcmillim, S. (1994). *American Ceramic Society Bulletin 73,* 109.

Guha, S., Ray, B., & Mandal, B. M. (2001). Anomalous solubility of poly acrylamide prepared by dispersion (Precipitation) polymerization in aqueous tert-butyl alcohol. *Journal of Polymer Science Part A: Polymer Chemistry, 39,* 3434–3442.

Hinczewski, C., Corbel, S., & Chartier, T. (1998). *Journal of the European Ceramic Society, 583,* 18.

Huzzard, R. J., & Blackburn, S. (1997). A water-based system for ceramic injection moulding. *Journal of the European Ceramic Society, 17,* 211–216.

Koh, Y. H., Song, J. H., Lee, E. J., & Kim, H. E. (2006). Freezing dilute ceramic/camphene slurry for ultra-high porosity ceramics with completely interconnected pore networks. *Journal of the American Ceramic Society, 89*(10), 3089–3093.

Low, D. K. Y., Li, L., Corfe, A. G., et al. (2001). Spatter-free laser percussion drilling of closely spaced arrayholes. *International Journal of Machine Tools and Manufacture, 41,* 361–377.

Lumpp, J. K., & Allen, S. D. (1997). Excimer laser machining and metallization of vias in aluminum nitride. *IEEE Transactions on Components, Packaging B, 20,* 241–246.

Mcnulty, T. F., Shanefield, D. J., Danforth, S. C., & Safari, A. (1999). Dispersion of PZT for fused deposition of ceramics. *Journal of the American Ceramic Society, 82,* 1757–1760.

Meng, G. Y., Wang, H. T., Zheng, W. J., & Liu, X. Q. (2000). Preparation of porous ceramics by gelcasting approach. *Materials Letters, 45,* 224–227.

Menon, M., Decourcelle, S., Ramousse, S., & Larsen, P. H. (2006). Stabilization of ethanol-based alumina suspensions. *Journal of the American Ceramic Society, 89*(2), 457–464.

Metev, S. M., & Veiko, V. P. (1996). *Laser assisted microtechnology.* Berlin: Springer-Verlag.

Murray, A. J., & Tyrer, J. R. (2000). New numerical evaluation tech-niques for laser processed ceramic substrates. *Journal of Laser Applications, 12,* 251–260.

Nagel, A. (1988). Slip casting of sinterable silicon nitride, Ph.D. Thesis, University of Stuttgart, Germany.

Omateta, O. O., Janney, M. A., & Strehlow, R. A. (1991). Gelcasting—a new ceramic forming process. *Ceramic Bulletin, 70*(10), 1641–1649.

Orita, N. (1988). Laser cutting method for high chromium steel and a device to carry out that method, US Patent 4,774,392.

Pham, D. T., & Gault, R. S. (1998). A comparison of rapid prototyp-ing technologies. *International Journal of Machine Tools and Manufacture 38,* 1257–1287.

Scheffler, M., & Colombo, P. (2005). *Cellular ceramics: structure, manufacturing, properties and applications.* Weinheim: Wiley-VCH, Chichester: John Wiley [distributor].

Shanti, N., Araki, K., & Halloran, J. W. (2006). Particle redistribution during dendritic solidification of particle suspensions. *Journal of the American Ceramic Society 89*(8), 2444–2447.

Sharp, C. M., Mueller, M. E., Murthy, J., et al. (1997). A novel anti–spatter technique for laser drilling: Applications to surface texturing. In *Proceedings of the 6th International Congress on Applications of Lasers and Electro-Optics, San Diego* (pp. 41–50).

Studart, A. R., Gonzenbach, U. T., Tervoort, E., & Gauckler, L. J. (2006). Processing routes to macroporous ceramics: A review. *Journal of the American Ceramic Society, 89*(6), 1771–1789.

Tari, G. (2003). Gelcasting ceramics: a review. *American Ceramic Society Bulletin, 82*(4), 42–46.

Wang, A., & Luo, G. (1990). *Plant Physiology Communication, 6,* 55.

Xiao, H., Fu, W., Zhao, B., Yang, F., & Xin, W. (1991). *Acta Biophysica Sinica 8,* 334.

Xiao, H., He, W., Fu, W., Cao, H., & Fan, Z (1999). *Progress in Biochemistry and Biophysics 26,* 180.

Xie, Z. P., & Huang, Y. (2001). New progress of gelcasting technology application in ceramic processes. *Journal of Ceramics, 22*(3), 142–146.

Yang, Jinlong, Liang, Su, Ma, Liguo, & Huang, Yong. (2002). *Key Engineering Materials, 224–226,* 667.

Young, A. C., Omatete, O. O., Janney, M. A., et al. (1991). Gelcasting of alumina. *Journal of the American Ceramic Society, 74*(3), 612–618.

Zhang, Fa-Zhi, Kato, Takeaki, Fuji, Masayoshi, & Takahashi, Minoru. (2006). Gelcasting fabrication of porous ceramics using a continuous process. *Journal of the European Ceramic Society, 26,* 667–671.

Chapter 8
Novel In-situ Coagulation Casting of Ceramic Suspensions

Abstract In this chapter, direct coagulation casting of ceramics via controlled release of high valence counter-ions and novel in-situ coagulation casting of ceramic suspensions via dispersion removal are discussed systematically. Direct coagulation casting of ceramics via high valance counter-ions (DCC–HVCI), as a new direct coagulation casting method, is proposed based on the strong coagulation ability of high valance counter-ions (HVCI) on ceramic suspension. The influence of types and addition content of dispersant, pH value, and solid loading on zeta potential and viscosity of ceramic suspensions has been systematically investigated. Several new methods for controlled release of HVCI have been proposed, such as using temperature as the leading controlling factor, chemical reaction and pH value as assisting controlling means, the increase of solubility of calcium iodate, calcium phosphate reacting with hydrochloric acid and decomposition of citric acid salt, and so on. It has been proved from both experimental and theoretical calculation results that the suspension of DCC–HVCI is coagulated at the primary minimum. This avoids the problems such as cracks formation and inner stress resulted from monovalence ions coagulation method which has been proved to coagulate at the second minimum. The results of this chapter are the basis for the deep research and applications of DCC. Based on the stabilization mechanism of ceramic suspension, we systemically investigate another new ceramic colloidal forming method, namely, dispersion removal coagulation casting (DRCC). Dispersant reaction, hydrolysis, crosslink, and separation have been used to control the dispersion removed processing. In the second part, the coagulation mechanisms of different ceramic suspension systems with different stabilization mechanisms are systematically investigated and the idea of the dispersion removal has been proposed. Different kinds of oxide and non-oxide ceramics and green bodies with excellent performance have been successfully prepared by dispersion removal method. The results are the basis for the further research and applications of ceramic colloidal forming. It provides a new idea and new method for the further study of the colloidal processing of advanced ceramics and also provides a new way for the industrialization of advanced ceramics with complex shapes.

Keywords Ceramic suspension · Direct coagulation casting · High valence counter-ions · In-situ coagulation · Dispersion removal · Stabilization mechanism

Colloidal processing is a ceramic forming method to produce near-net-shape complex ceramic parts with high reliability and low cost (Lewis 2000). It includes slip casting, tape casting, injection molding, gel casting, direct coagulation casting (DCC), and so on (Tiller and Tsai 1986; Hotza and Greil 1995; Fanelli et al. 1989; Young et al. 1991; Gauckler et al. 1999a). Usually, a large amount of organic binders is used in injection molding process which is difficult for de-bindering of large parts with thick cross section. Gel casting is a well-established colloidal processing method for preparing high-quality, complex-shaped ceramic parts by means of in-situ solidifying through a macromolecular network which is created to hold the ceramic particles together. However, the main component of the commonly used monomer acrylamide system is a neurotoxin, which limits the application of gel casting (Yang et al. 2011a).

DCC is a novel near-net-shape method to prepare homogenous green ceramic bodies using high solid loading suspensions. Its principle lies in the in-situ coagulation of a powder suspension via a reaction-rate-controlled internal-enzyme-catalyzed reaction that produces the desired pH or increasing ionic strength in-situ after casting (Gaucker et al. 1999b; Graule et al. 1995). In the DCC process, coagulation of the ceramic suspensions is carried out by shifting the force between the particles in the suspension from the repulsive to the attractive regime. The force between the particles is shifted from the repulsive to the attractive regime either by shifting the pH of the suspension toward its isoelectric point (IEP) or by increasing ionic strength (Baader et al. 1996a, b; Balzer et al. 2001, 1999). DCC of pH shift to IEP is a reliable processing to prepare ceramic parts; however, the wet strength of green body is too low. The DCC via increasing ionic strength with one valence ion suitable for all suspensions, but the mechanical properties of sintered samples are poor for vertical cracks in microstructure. Also, the coagulation time is too long, usually about 1–3 days (Yang et al. 2002). A hydrolysis-assisted solidification (HAS) concept for net shaping of ceramic green parts from aqueous suspensions was introduced by Kosmac et al. (1997). The process is based on thermally activated hydrolysis of aluminum nitride powder added to ceramic suspensions with high solid loading. During hydrolysis of AlN, water is consumed and ammonia is formed, which in turn may increase the pH of the suspension. Both mechanisms can be used to increase the viscosity and ultimately to set a cast or injection-molded ceramic green body. However, this method is not suitable for silicon carbide or silicon nitride systems.

The high valence counter-ions (HVCI) are high valence ions with opposite charge to that of the ceramic particles in the suspensions. The minimum concentration of electrolyte required for the coagulation of a stable colloidal suspension is defined as the critical coagulation concentration (CCC) (Teot and Daniels 1969). The CCC of counter-ions is found to be inversely proportional to the sixth power of its valence, according to so-called Schulze–Hardy rule. For counter-ions of valences 3, 2, and 1, this rule suggests that the CCC ratio is $3^{-6}:2^{-6}:1^{-6}$, or roughly, 1:11:729. The Schulze–Hardy rule can be interpreted by DLVO theory, which considers the electrostatic repulsion force and the van der Waals attraction force between two interacting particles (Metcalfe and Healy 1990; Hsu and Kuo 1997). A high solid loading suspension with a low viscosity can hardly be obtained in the presence of HVCI. Previous results of our group showed that with ion exchange resins, the removal of

SO_4^{2-} in Al_2O_3 suspension improved the solids loading from 55 to 65 vol%, while the removal of Ca^{2+} and Mg^{2+} in Si_3N_4 suspension improved the solids loading to 58 vol%. Viscosity of both samples without HVCI at a shear rate of 100 s^{-1} is still lower than 1 Pa s (Yang et al. 1997a; Yang and Huang 2010). Recently, Wen et al. investigated the influence of HVCI on the rheological properties of alumina suspensions systematically. As expected, the addition of HVCI to the solution induced a complete coagulation of the suspension, and it could be prospective to develop a novel ceramic forming technique with a controllable release of HVCI method (Wen et al. 2011). Prabhakaran et al. reported a novel near-net-shape DCC method using ammonium poly(acrylate) as dispersant, and the generation of Mg^{2+} ions from sparingly soluble MgO. The Mg^{2+} ion reacts with the un-adsorbed ammonium poly(acrylate) in the dispersion medium to form precipitate of Mg–poly(acrylate), and it induces the desorption of dispersant from the alumina particle surfaces, which leads to insufficient dispersant surface coverage and the coagulation of suspension (Prabhakaran et al. 2008a, b, c). We also developed a new direct coagulation method by high valence counter-ions (DCC–HVCI) using calcium iodate, calcium phosphate, and thermo-sensitive liposomes as coagulating agent (Yang et al. 2011b).

A novel DCC method is first proposed. The ceramic suspensions are in-situ coagulated by directly shifting the isoelectric point of the electrostatic stability suspension via dispersant reaction or hydrolysis without adjusting pH and increasing the ions strength. The coagulation mechanisms of electrostatic stability ceramic suspensions via the isoelectric point shift have been expounded. The green compressive strength is significantly increased within shortened coagulation time. The technology of DCC is enriched and perfected. Three new approaches for controlling removal of dispersions have been proposed: In-situ coagulation of electrostatic stability ceramic suspensions by dispersant reaction or hydrolysis, in-situ coagulation of semisteric stability ceramic suspensions by dispersant separation via low temperature induced, and in-situ coagulation of semisteric stability ceramic suspensions via dispersion removal have been achieved by dispersant separating out from solvent. The coagulation processing is simplified without degassing and coagulation agent. The homogeneity of green body is increased, and the line shrinkage is decreased significantly. It provides the basis for the preparation of high-reliability ceramic parts.

8.1 Direct Coagulation Casting of Ceramic Suspension by High Valence Counter-Ions

8.1.1 Direct Coagulation Casting by Using Calcium Iodate as Coagulating Agent

In this section, an alumina suspension with high solid loading and low viscosity was prepared using tri-ammonium citrate (TAC) as dispersant. Different amounts of calcium iodate powder were added to the suspension at 10 °C, and the suspension was cast into a steel mold after. The influence of calcium iodate on the rheological properties of alumina suspension was investigated. A novel direct coagulation method (DCC–HVCI) for alumina suspension was produced by controlled release of high valence counter-ions from calcium iodate with the increase in the temperature from 55 to 70 °C. Cracks appeared in increasing one valence ionic strength method were not observed during formation in this method. Complex parts are prepared by this method with smooth surface and no deformation during demolding.

A CT3000SG alumina powder (Almatis, Ludwigshafen, Germany) of average particle size 0.33 μm (measured using Malvern Master Size Analyzer 2000, Malvern, Worcestershire, UK) and surface area 8.08 m^2/g (Automated Surface Area and Pore Size Analyzer, Nova 4000; Quantachrome, Boynton, FL) was used for suspension preparation. Suspensions with negatively charged ions on the surface of the alumina particles were prepared by adding 0.1–0.8 wt% TAC based on alumina powder as the dispersant. Analytical reagent grade calcium iodate $(Ca(IO_3)_2 \cdot 6H_2O)$ was used as source of high valence counter-ions Ca^{2+}. Deionized water was used for preparation of suspensions.

Alumina suspension was prepared by tumbling the alumina powder, water, and TAC in polyethylene containers along with zirconia grinding media for 4 h. The suspensions were cooled down at 10 °C. Various amounts of calcium iodate were added to the suspensions and mixed thoroughly by continuing the tumbling process for another 30 min, when a homogenous suspension was obtained. The viscosity of the suspension was measured using NXS-11A (Chengdu, China) rotational viscometer. To measure viscosity at different temperatures, a water bath was used to keep temperature at desired value. The cylinder used for suspension viscosity measurement was dipped in water bath and the temperature of water bath was controlled by thermocouple. In all the viscosity measurements, the dial readings were taken immediately after 30 s of rotation of the spindle, and viscosity was calculated by multiplying the readings with factors provided in the viscometer manual. All viscosity measurements were taken at a shear rate (D_s) of 100 s^{-1}, unless noted otherwise. The conductivity of calcium iodate solution was tested using a DDS-2A ion conductivity meter (Shanghai, China). 1.5 g calcium iodate was added to 100 mL water. The zeta potential of alumina suspension with and without calcium iodate addition was measured using a Nano-ZS90 Zetasizer (Malvern). The concentration of suspension was 0.5 wt%.

For wet compressive strength measurements, cylindrical bodies with 25.5 mm in diameter and a height between 25 and 30 mm were cast. After coagulating at different temperatures, the samples were demolded, and their wet green strength was measured immediately by an AG-IC 20KN (Shimadzu, Japan) testing machine with a crosshead speed of 0.5 mm/min. The wet green strength was determined as the yield point, extrapolated from the tangents to the curve. The pictured data represent mean values from 3 to 10 samples. The fractured surface of the green body was dried at 80 °C for 24 h. The dried samples were sintered at 1500–1550 °C for 2 h at a heating rate of 5 °C/min. Microstructure was observed by a SSX-550 (Shimadzu, Japan) scanning electron microscope (SEM).

8.1 Direct Coagulation Casting of Ceramic Suspension

(1) Effect of tri-ammonium citrate (TAC) on the viscosity of the suspension

The TAC is a well-known dispersant for alumina in aqueous suspensions (Luther et al. 1995; Jia et al. 2002). The TAC molecules adsorb onto the particle surface and form a unimolecular layer at the optimum dispersant concentration. The less viscous powder suspension indicates well dispersion in the liquid medium. The TAC also shifts the IEP of alumina from 9.2 to about 4 (Cai et al. 2005). Stable alumina suspension can be prepared at pH 7–8. Figure 8.1 shows the effect of various amounts of TAC on the viscosity of 50 vol% aqueous alumina powder suspensions. The suspension possessed a minimum viscosity of 0.2 Pa s with 0.3 wt% of TAC. The viscosity of the suspension increased marginally with further addition of TAC and reached a value of 0.6 Pa s with 0.8 wt% of TAC. When dispersant content is 0.1–0.2 wt%, citrate cannot completely cover the surface of the particles which causes agglomeration and flocculation. When the amount of dispersant is 0.3 wt%, citrate covers the surface of the particles sufficiently, and the suspension viscosity reaches the lowest value. However, excessive dispersant will lead to high ion concentration in suspension which compresses the electrical double layer between two particles, causing instability of the suspension.

(2) The solubility of calcium iodate and its conductivity

It is well known that the solubility of most of inorganic salts increases at an elevated temperature. The salt will be dissolved easily when the solvent is at a higher temperature rather than a low temperature. Calcium iodate is a salt that follows this rule with small amount dissolved in water. There are two forms of calcium iodate at room temperature, $Ca(IO_3)_2·H_2O$ and $Ca(IO_3)_2·6H_2O$. The solubility of $Ca(IO_3)_2·6H_2O$ is greatly affected by the temperature (Li 2003). As shown in Fig. 8.2, at 10 °C the solubility of calcium iodate is 0.17 g in 100 g water; however, nearly eightfold of solubility is achieved by increasing the temperature to 60 °C and more Ca^{2+} ions

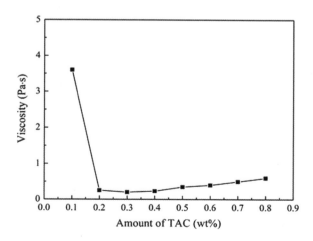

Fig. 8.1 Effect of concentration of tri-ammonium citrate (TAC) on viscosity of 50 vol% aqueous alumina suspension

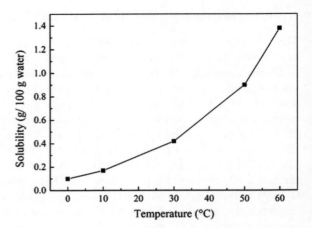

Fig. 8.2 The solubility curve of calcium iodate [Ca(IO$_3$)$_2$·6H$_2$O]

exist in water. Figure 8.3 shows the conductivity of calcium iodate solution at different temperatures and times. From the data, we can see that calcium iodate dissolves quickly in water and reaches equilibrium about 5–7 min after adding water at certain temperature. The solubility increases with the increase of temperature. The dissolved equation can be expressed as follows:

$$\text{Ca(IO}_3)_2 = \text{Ca}^{2+} + 2\text{IO}_3^{-} \tag{8.1}$$

(3) **Influence of calcium iodate concentration on the viscosity and Zeta potential of suspension**

As described in previous section, CCC is the minimum concentration of ions to coagulate the suspension, which can be reflected by an abrupt increase in the viscosity of the suspension. Figure 8.4 shows the influence of calcium iodate concentration

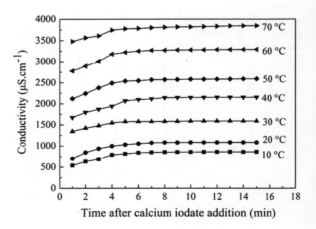

Fig. 8.3 Conductivity of calcium iodate solution as a function of temperature and time

8.1 Direct Coagulation Casting of Ceramic Suspension

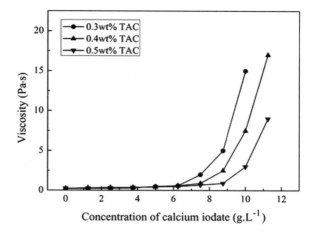

Fig. 8.4 Influence of calcium iodate concentration on the viscosity of alumina suspension with different TAC contents

on the viscosity of alumina suspension with different TAC contents. The suspension used for measurement is 10 °C, and the experiment was conducted at 10 °C. In the suspension with 0.3 wt% TAC, when the concentration of $Ca(IO_3)_2$ is less than 6.25 g/l, the viscosity is less than 0.5 Pa s. When the $Ca(IO_3)_2$ concentration exceeds 6.25 g/l, suspension viscosity increases sharply as a function of concentration of calcium iodate. More than 5 Pa s is achieved within 30 min after addition of $Ca(IO_3)_2$ for concentration exceeds 6.25 g/l. Therefore, the CCC of $Ca(IO_3)_2$ in 50 vol% alumina suspension with 0.3 wt% TAC is 6.25 g/l. When dispersant content is 0.4 and 0.5 wt%, similar trend is observed. When the dispersant content is 0.4 or 0.5 wt%, the CCC of $Ca(IO_3)_2$ in the suspension is 7.5 or 8.75 g/l, respectively. It can be seen that the CCC of calcium iodate in the present suspension increases with the increase in TAC content. In direct coagulation casting process, suspension with low viscosity (<1 Pa s at $Ds = 100$ s^{-1}) is required before casting. The suspension viscosity should remain low enough for sufficient period of time after addition of coagulating agent for proper mixing and degassing before casting. For suspension with 0.3–0.5 wt% TAC, the $Ca(IO_3)_2$ concentration below 6.25–8.75 g/l is suitable for casting, respectively.

The high valence Ca^{2+} ions diffuse into the electrical double layer of alumina particles surface through electrostatic attraction, reduces the zeta potential, and decreases the repulsive force between particles. As a result, the rheological properties of suspension become poor. According to the DLVO theory, with the increase in Ca^{2+} concentration, the gravitational potential energy between particles hardly changes, but the repulsion potential energy decreases rapidly. The dispersed particles collide and agglomerate, leading to the abrupt increase in viscosity of the suspension.

It is noted that the amount of TAC has profound influence on the CCC of $Ca(IO_3)_2$ in the suspension. The mechanism can be proposed as follows. Citrate and Ca^{2+} occurs in the following reaction:

$$2C_6H_5O_7^{3+} 3Ca^{2+} = (C_6H_5O_7)_2Ca_3 \downarrow \tag{8.2}$$

Fig. 8.5 Viscosity of alumina suspension versus increasing concentrations of calcium iodate treated at 70 °C for 30 min

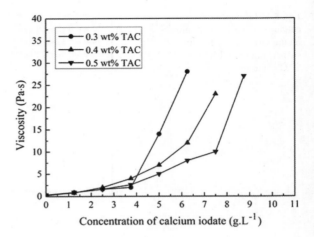

When Ca^{2+} ions enter into electrical double layer, Ca^{2+} ions react with citrate and form calcium citrate. Redundant citrate will be added into electrical double layer, make repulsion potential energy between particles not affected. Therefore, more content of citrate in suspension resists the impact of the high valence ions. More Ca^{2+} ions are needed to reach a sharp change in the viscosity of the suspension when excessive citrate exists in the suspension.

Figure 8.5 shows the viscosity of alumina suspension with different calcium iodate concentrations treated at 70 °C for 30 min. The suspension viscosity increases with the increase in the calcium iodate concentration. In the suspension containing 0.3 wt% TAC, the viscosity of the suspension is not affected by the concentration less than 3.75 g/l, but the viscosity changes sharply with concentrations larger than 3.75 g/l. The change in viscosity of suspension containing 0.3 wt% of dispersant is the fastest among the samples we used here, and this value increases with the amount of dispersant which is also observed at lower temperature.

The solubility of $Ca(IO_3)_2$ increases with the rise of temperature. At the temperature of 10 °C, the solubility is 0.17 g. However, it reaches 1.38 g at 60 °C, which is 8 times higher than that at 10 °C. Therefore, with the increase of temperature, the solubility of calcium iodate increases, and more Ca^{2+} ions dissociated in suspension. Figure 8.6 shows the zeta potential with different concentrations of calcium iodate at different temperatures. The pH value of the suspension is 8.5 during the zeta potential measurement. The zeta potential decreases gradually with the increase of calcium iodate. The zeta potential at higher temperature is lower than that of lower temperature. The zeta potential decreases to near zero as calcium iodate exceeds 6.25 and 3.75 g/l at the temperature of 10 and 70 °C, respectively. The shift of zeta potential to zero means the great change in viscosity occurs in this condition. When Ca^{2+} concentration reaches a certain value, the electrical double layer on the surface of alumina particles is strongly compressed, and zeta potential of the particle surfaces decreases. The interparticle repulsion potential energy reduces and the dispersed particles agglomerate, leading to increase in viscosity. It is reported that iodate ions

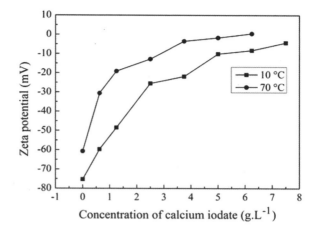

Fig. 8.6 Zeta potential of suspension with different calcium iodate concentrations at different temperatures

also have strong coagulation ability even at low concentration of 0.01 M. The iodate ions decrease the zeta potential and shift the IEP to low pH. The phenomenon is attributed to strong absorption of iodate on alumina particles (Franks et al. 1999).

Figure 8.7 illustrates the relationship between viscosity of alumina suspension and 6.25 g/l calcium iodate and 0.3 wt% TAC treated at different temperatures for 30 min. The viscosity measurements were taken 30 min after the addition of calcium iodate. Alumina slurry without calcium iodate showed only a marginal increase in the viscosity with time at 70 °C during a period of 30 min. This slight increase in the viscosity could be due to the evaporation of water during the measurement. However, in the case of the suspension containing 6.25 g/l, the viscosity increased rapidly on heating. The slurry that was heated at 50–70 °C showed a viscosity of 40–50 Pa s, which is high enough for coagulating the slurry. The coagulated body is sufficiently stiff to be removed from the mold immediately after gelation at 70 °C

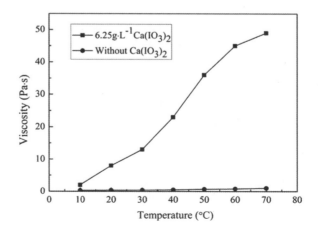

Fig. 8.7 Viscosity of alumina suspensions with and without calcium iodate concentration treated at different temperatures for 30 min ($D_S = 50$ s^{-1})

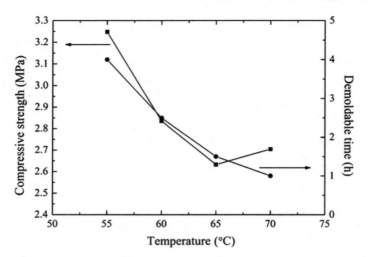

Fig. 8.8 The relationship between compressive strength and demoldable time of wet green bodies coagulated at different temperatures

for 1 h. The rapid coagulation of the slurries on heating could be due to an increase in the solubility of the calcium iodate particles at high temperature.

(4) **The properties and microstructure of green bodies and sintered sample**

Strength of wet-coagulated body is one of the parameters that determines whether the process is suitable for fabricating complex parts. The higher the strength of green body is, the easier the wet green body could be demolded without any deformation. Figure 8.8 shows the change in compressive strength with temperature increasing from 50 to 70 °C and demoldable time of wet green bodies coagulated at different temperatures. The demoldable time decreases with the increase in temperature. At the temperature of 70 °C, the coagulation time reduces to 1 h which production rate could be increased. The compressive strength of the wet green bodies decreases slightly with the increase of temperature in the range of 2.6–3.2 MPa which is much larger than increasing one valence ion's strength in previous reports (Blazer et al. 1999; Yang et al. 2002).

Figure 8.9 illustrates stress–strain behavior of the wet-coagulated body prepared from DCC–HVCI coagulated at 70 °C for 1 h with addition of 6.25 g/l calcium iodate and 0.3 wt% TAC. The wet-coagulated body showed elastic deformation up to the yield point and showed plastic deformation with considerable strain hardening effect after the yield point. This suggests that the samples have high strength. The sample breaks down at the stress of 2.7 MPa, with a small deformation of 2% which contributes to the high compressive strength.

A SEM photograph of fractured surface of the green alumina body via DCC–HVCI coagulated at 70 °C for 1 h with addition of 6.25 g/l calcium iodate and 0.3 wt% TAC is presented in Fig. 8.10. The SEM photograph of fractured surface of the alumina green body prepared by the coagulation casting process showed uniform microstructure,

8.1 Direct Coagulation Casting of Ceramic Suspension 415

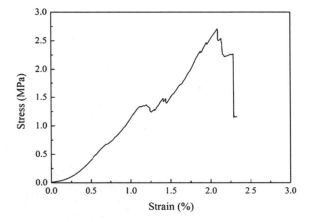

Fig. 8.9 Stress–strain behavior of the wet green body coagulated at 70 °C for 1 h with addition of 6.25 g/l calcium iodate

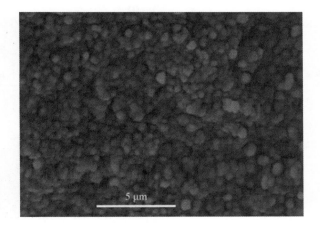

Fig. 8.10 SEM photograph of fractured surface of alumina green body prepared by DCC–HVCI method coagulated at 70 °C for 1 h with addition of 6.25 g/l calcium iodate

and no agglomeration or pores were observed. The uniform microstructure indicated less cracks in the green body which leads to high mechanical properties, compared with one valence ionic strength method. The new direct coagulation procedure could fabricate near-net-shape ceramic parts. Figure 8.11 shows a photograph of coagulated bodies and sintered samples with different shapes prepared by DCC–HVCI method coagulated at 70 °C for 1 h with addition of 6.25 g/l calcium iodate. The suspension coagulated in the mold could be removed without any deformation by keeping in an oven at 70 °C for 1 h immediately after cooling the mold to room temperature. Figure 8.12 shows the relationship between relative density and sintered temperature of ceramics prepared by DCC–HVCI method. The density increases with increasing the sintered temperature. The relative density reaches about 99.4% when the sintered temperature is 1550 °C. The microstructure of sintered sample (etching by 2 wt% HF for 10 min) with 6.25 g/l calcium iodate coagulated at 70 °C for 1 h and sintered

Fig. 8.11 Complex parts prepared by DCC–HVCI method coagulated at 70 °C for 1 h with addition of 6.25 g/l calcium iodate; **a** green bodies, **b** sintered sample

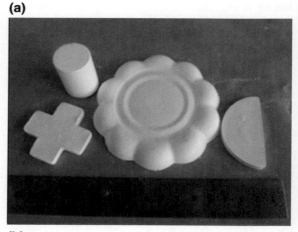

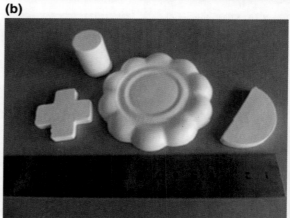

Fig. 8.12 The relationship between relative density and sintered temperature with 6.25 g/l calcium iodate coagulated at 70 °C for 1 h

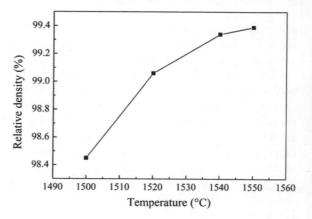

at 1550 °C for 2 h is presented in Fig. 8.13. It can be observed that the microstructure is uniform and there is no crack in the sample.

(5) **Other properties of alumina coagulated bodies**

The schematic illustration of the coagulation process via DCC–HVCI is shown in Fig. 8.14. Ca^{2+} ions react with citrate which leads to the instability of the suspension. Both factors contribute to the rapid coagulation at high temperature.

The compressive strength of wet-coagulated samples is an important parameter in colloidal forming process. Green body prepared by injection molding showed high strength with the presence of binder in the sample (Fanelli et al. 1989). The compressive strength of gel-casting green body was also very high (~40 MPa) because of three-dimensional network produced by polymer (Young et al. 1991). Early reports on DCC methods showed that low strength of green body was obtained as the particles are only attracted by van der Waals force (Hesselbarth et al. 2001; Wyss et al. 2005).

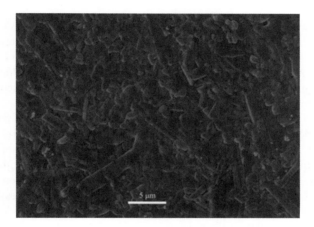

Fig. 8.13 The microstructure of sintered sample (etching by 2 wt% HF for 10 min) with 6.25 g/l calcium iodate coagulated at 70 °C for 1 h and sintered at 1550 °C for 2 h

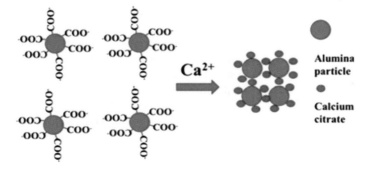

Fig. 8.14 Schematic illustration of the coagulation process via DCC–HVCI

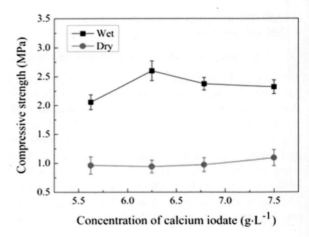

Fig. 8.15 Compressive strength of coagulated bodies with different amounts of calcium iodate

However, the strength of wet-coagulated samples prepared by DCC–HVCI was found to be higher than that prepared by increasing monovalent electrolytes (urea) (Xu et al. 2012a, b). It is crucial to investigate the effect of the process parameters, such as calcium iodate content, coagulation temperature, and solid loading on the properties of green body. The wet and dried strength were also compared to illustrate the mechanism of coagulating process and fracture behavior.

Figure 8.15 shows the compressive strength of coagulated bodies with different amounts of calcium iodate. The samples were coagulated at 65 °C for 1.5 h with solid loading of 50 vol%. The strength of wet samples is higher than that of dried ones. For wet-coagulated bodies, the strength increases in the beginning with increase in calcium iodate with optimum strength at the concentration of 6.25. Then the strength decreases slightly with further addition of calcium iodate. The compressive strength of wet samples varies from 2.0 to 2.7 MPa which is 1 order of magnitude higher than that of coagulated by monovalent electrolytes (Gauckler et al. 1999a; Baader et al. 1996a; Balzer et al. 1999). For dried green bodies, the strength does not affect much by the concentration of calcium iodate and the strength is in the range of 1.0 MPa.

Figure 8.16 displays the compressive strength of coagulated bodies versus temperature. The concentration of calcium iodate is 6.25 g/l, and the solid loading is fixed at 50 vol%. The strength of wet samples is also higher than that of dried ones which is observed in samples with different amounts of calcium iodate previously. The strength of wet-coagulated bodies decreases slightly from 3.2 to 2.7 MPa with temperature increasing from 55 to 70 °C. However, the strength of dried samples shows an opposite trend compared with wet ones. The strength increases gradually in the range 0.5–1.8 MPa with increase in temperature. The coagulation mechanism of DCC–HVCI is the compression of electrical double layer by the high valence counter-ions that leads to the close contact of suspended particles. During the process, the aggregation of particles results in the shrinkage of the suspension and the loss of water. In previous work, a decrease in coagulation time with increase in coagulation temperature was observed (Xu et al. 2012a). Table 8.1 presents the shrinkage

Fig. 8.16 Compressive strength of coagulated bodies versus temperature

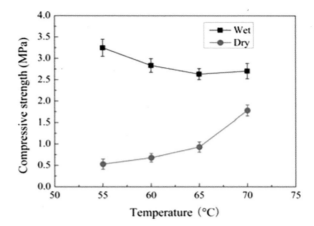

Table 8.1 Shrinkage rate and water remaining of wet samples after coagulating at different temperatures

Temperature (°C)	Shrinkage rate (%)	Water remaining (%)
55	2.2	87
60	1.8	89
65	1.5	95
70	1.7	92

and water remaining of wet samples after coagulating at different temperatures. The shrinkage rate decreases with the increase of coagulation temperature which is similar with the observed compressive strength. Large shrinkage rate means the strong aggregation of particles, and water is squeezed out of the sample which leads to high green density and contributes to high compressive strength. Also, cracks are easy to generate at high temperature which are detrimental to the mechanical properties of green body (Xu et al. 2012b). The increased strength of dry samples with calcium iodate concentration may be the slow decomposition of excessive calcium iodate during drying process, which enhances the mechanical properties of samples. The compressive strength of coagulated bodies as a function of solid loading is plotted in Fig. 8.17. The concentration of calcium iodate is 6.25 g/l with coagulated temperature of 65 °C for 1.5 h. Again, the strength of wet samples is higher than that of dried ones. The strength increases in the beginning with increase in solid loading with optimum strength at 54 vol%. Then, the strength decreases slightly at the solid loading of 56 vol%. The strength increases gradually in the range 1.0–1.5 MPa with increase in the solid loading.

Figure 8.18 displays the fracture behavior of samples after compressive strength testing. The wet green body shows elastic deformation after strength testing, and only small cracks appear in the sample. However, the dried coagulated body cracks into small pieces after the testing. The phenomenon agrees with the strength difference

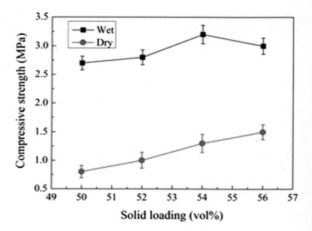

Fig. 8.17 Compressive strength of coagulated bodies as a function of solid loading

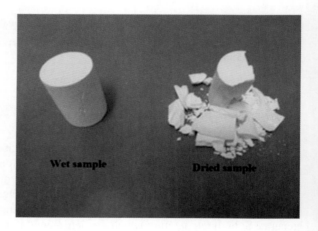

Fig. 8.18 Fracture behavior of samples after compressive strength testing

between the wet body and dried one. Figure 8.19 shows the schematic description of the fracture behavior of wet and dried green bodies. As shown in Fig. 8.19a, the particles in the wet-coagulated body are held together by van der Waals force in close contact with sufficient amount of Ca^{2+} release from calcium iodate. The coagulated particles are assumed to fall into the primary minimum by addition of 0.01 mol/l divalent electrolytes according to the theoretical calculation from DLVO theory (Yang and Sigmund 2003). In this sense, the wet-coagulated body shows high compressive strength. Also, the strengthening effect of capillary forces contributes to the high strength of wet-coagulated samples (Isrealachvili 1991). Wet particles also attract each other with an adhesive force associated with the liquid bridges between particles. Therefore, the presence of water changes the system from one with a hard-core repulsive interaction to one with both a repulsive and an attractive interparticle interaction (Barabási et al. 1999; Jaeger and Nagel 1992; Hornbaker et al. 1997). However, when water is removed from the wet-coagulated body after

8.1 Direct Coagulation Casting of Ceramic Suspension

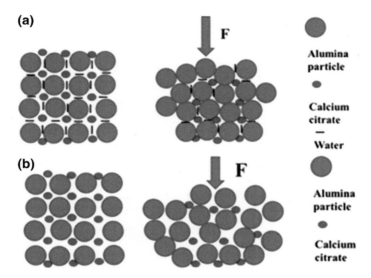

Fig. 8.19 Schematic description of the fracture behavior of wet (**a**) and dried (**b**) green bodies

sufficient drying, dry powers interact only through repulsive contact forces and some weak attraction and are easy to break the bridge by low stress, Fig. 8.19b.

It is also noted that the compressive strength of wet samples prepared by DCC–HVCI is about one order higher than that prepared by DCC. Results of Hans et al. showed that the strength of wet green bodies is drastically influenced by the local arrangements of particles. The local arrangement of pH destabilized suspensions is homogeneous microstructures, whereas the local arrangement of increasing ionic strength destabilized suspensions is heterogeneous microstructures (Wyss et al. 2002, 2004, 2005).

Heterogeneous microstructures exhibit up to one order of magnitude higher elastic properties and yield strengths than their homogeneous counterparts (Schenker et al. 2008; Krall and Weitz 1998). The calculations conform to our measured strength results. The observed low strength of increasing monovalent ionic strength is due to the reversible coagulation at the second minimum, and the coagulated sample is sensitive to external forces (Balzer et al. 1999). Vertical cracks are observed during the coagulation process which is detrimental to the wet strength (Yang et al. 2002). The present system is coagulated by divalent counter-ions at the primary minimum, and the particles are assumed to get closer with strong network.

In addition to compressive strength, the drying behavior of green body is also of great importance in the coagulation process. Drying characteristics of the wet-coagulated body at 80 °C is plotted in Fig. 8.20. The shrinkage rate of coagulated bodies is 1.5% after demolding. The drying process can be divided into three stages. The first stage occurs in the first 2 h after drying. The shrinkage rate changes rapidly during this period and 30% of water loses. The second stage is from 2 to 8 h after

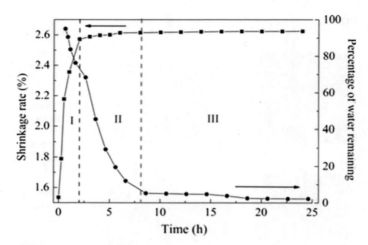

Fig. 8.20 Drying characteristics of the wet-coagulated body prepared by 50 vol% alumina suspension with 6.25 g/l calcium iodate coagulated at 65 °C for 1.5 h and dried at 80 °C for 24 h

drying. A small shrinkage rate changes, but the majority of water loses in this period. Less than 5% of water stays in the sample after 8 h. Slight change in both shrinkage rate and water loss is occurred in the third stage. The drying rate stops after 20 h with 2% of water remaining in the sample. The linear drying shrinkage observed for the coagulated bodies prepared from slurries of solids loading 50 vol% is 2.6%.

The pore size distribution can illustrate the pore structure in the compacted body. Early studies showed that the green body prepared by dry powder pressing displayed double peaks or multiple peaks in pore size distribution measurement for agglomeration or solid bridge in the sample which causes the nonhomogeneous microstructure (Yang et al. 1997b). Colloidal forming methods can avoid the drawbacks if direct coagulation is carried out without rearranging the particles in a low viscosity and homogeneous suspension (Lewis 2000). Figure 8.21 shows the pore size distribution and mean pore diameter of the green bodies. The sample with the powder content of 50 vol% displays a narrow average pore channel size, showing a mean pore diameter of 40 nm in spite of a wide particle size distribution. Figure 8.22 shows the microstructure of green body prepared from 50 vol% suspensions with 6.25 g/l calcium iodate coagulated at 65 °C. Homogeneous microstructure is observed in the sample without large pores. These results suggest that well-dispersed suspension, sufficient time for degassing and casting could be performed by the DCC–HCVI process.

Fig. 8.21 Pore size and distribution of green body prepared from 50 vol% suspension with 6.25 g/l calcium iodate coagulated at 65 °C

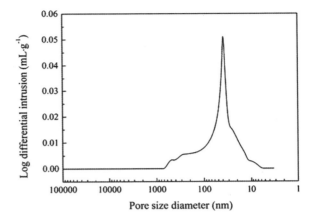

Fig. 8.22 Microstructure of green body prepared from 50 vol% suspension with 6.25 g/l calcium iodate coagulated at 65 °C

8.1.2 Direct Coagulation Casting by Using Calcium Phosphate as Coagulating Agent

The purpose of this section is to introduce a DCC method based on the idea of controlled release of high valence counter-ions to coagulate positively charged alumina suspension. Calcium phosphate was used as the high valence counter-ions source. Influence of the content of calcium phosphate on the viscosity of the suspension with positive charge on particles and the properties of green body were investigated. A novel direct coagulation method for alumina suspension was produced by controlled release of high valence ions from calcium phosphate with increase the temperature to 55–70 °C.

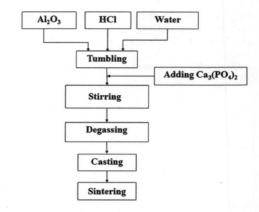

Fig. 8.23 Flowchart of the direct coagulation casting process

(1) **Preparation Procedure**

The flowchart of the preparation process is shown in Fig. 8.23. Alumina suspensions were prepared by tumbling the alumina powder, water, and hydrochloric acid in polyethylene containers along with zirconia grinding media for 4 h. The suspensions were cooled down at 15 °C. Various amounts of calcium phosphate were added to the suspensions and mixed thoroughly by continuing the tumbling process for another 30 min, when a homogenous suspension was obtained. The suspension was degassed in vacuum condition for 5 min. The suspension was cast in a steel mold. To prevent the evaporation of water, a piston was used. The samples were placed in a water bath with a heating rate of 5 °C/min and heated at different temperatures for a period of time. The coagulation time is defined as the time from the sample is heated to the sample is removed from the mold without any deformation.

Zeta potential was measured by a Zeta Potential Analyzer (Brookhaven Instrument Corp., Holtsville, NY) via the electrophoresis light scattering method. Suspension containing 0.5 wt% of alumina powder was prepared for the zeta potential measurements. HCl and NaOH solutions were employed to adjust the pH values. The pH value of the suspension was measured by using a LE438 pH meter (Mettler, Toledo, Switzerland).

(2) **Zeta potential and viscosity of alumina suspension with different pH values**

In direct coagulation casting process, a stable suspension is needed to reduce the number of bubbles during the molding process and to improve the reliability of ceramics for the absence of crack in sintered samples. Zeta potential of particle surface can be used to characterize the nature of surface electric charge of powder particles in suspension. Larger absolute value of zeta potential suggests better dispersion of the suspension. Zeta potential of alumina suspension at different pH values is shown in Fig. 8.24. With the pH increases, the zeta potential of alumina in the water decreases from 43 to −40 mV. The IEP of the alumina suspension used in this experiment is about 7, slightly different from previous reports with IEP of 8–9 (Hidber et al. 1996). The IEP of alumina depends strongly on the crystallographic form and history of the

Fig. 8.24 Zeta potential of alumina suspension with different pH values

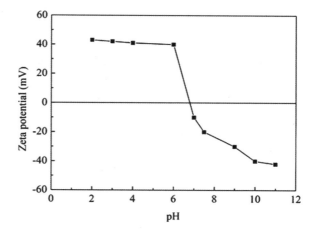

powder (i.e., prior surface treatment) and any impurities. One possible explanation for the different IEPs measured for the different forms of alumina is that the relative surface densities of aluminum and oxygen groups will vary depending on the structure of the surface. Another possibility is the instrument error (Karaman et al. 1997; Cai et al. 2005). When the suspension pH is above 7, the alumina particle surface is negatively charged. When the suspension pH is below 7, the surface of alumina particles is positively charged. Stable suspension can be prepared either at pH 3–5 in acidic regime with positive charge on particles or at pH 10–12 in basic regime with negative charge on particles, with absolute zeta potential value of above 40 mV (Franks and Meagher 2003; Sprycha 1989).

Positively charged suspension was prepared at acidic regime with the addition of hydrochloric acid. Figure 8.25 shows the viscosity of alumina suspension with 50 vol% solid loading at different pH values. At first, the viscosity of suspension decreases slightly with increase of pH. The lowest viscosity suspension can be

Fig. 8.25 Viscosity of alumina suspension with 50 vol% solid loading at different pHs

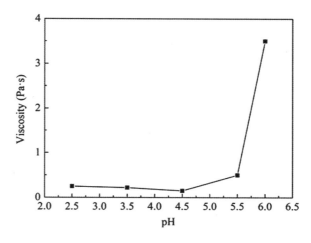

prepared at pH = 4.5. Then the viscosity increases rapidly above pH of 4.5. The alumina surfaces are charged positively at pH 4.5 by reaction of the surface hydroxyl groups with added acid. Generally, the decrease of pH is conducive to the accumulation of positively charged alumina surface. According to DLVO theory, however, excess hydrochloric acid increases the ionic strength of the suspension. Electric double layer of alumina particles is compressed by high ionic strength, hence reducing the electrostatic repulsion between the particles. In this sense, the viscosity increases slightly as the reduction of pH. Meanwhile, the hydrochloric acid decreases the surface tension of water. At low pH, it is difficult for concentrated suspension to bubble out which affect the microstructure of the green body, and hence will impact the mechanical properties of sintered samples. Suspension with pH of 4.5 is used in this section followed.

(3) **Effect of Calcium Phosphate on Suspension Properties**

Figure 8.26 shows the viscosity of suspension with different amounts of $Ca_3(PO_4)_2$ stirring for 30 min at 15 °C. It can be seen from Fig. 8.26 that the viscosity increases gradually with increasing the content of $Ca_3(PO_4)_2$. The viscosity increases slightly in the range of 0.2–0.35 Pa s with addition of 0.05–0.25 wt% $Ca_3(PO_4)_2$. Great change in viscosity occurs with addition of above 0.3 wt% $Ca_3(PO_4)_2$. However, the viscosity of suspension with less than 0.06 wt% $Ca_3(PO_4)_2$ is not high enough to coagulate at high temperature, whereas the viscosity of suspension with more than 0.25 wt% $Ca_3(PO_4)_2$ has been close to 1 Pa s and will affect the casting effect. Suspensions with 0.15 and 0.2 wt% $Ca_3(PO_4)_2$ were chosen for coagulation process.

In direct coagulation casting process, the viscosity of suspension is required to be low enough for casting process to avoid bubbles in the suspension. Then rapid increase in viscosity should be achieved in certain conditions to coagulate the suspension. Usually, temperature is a key factor in this process. Here, $Ca_3(PO_4)_2$ is added to the suspension at different temperatures. Viscosity of suspension with and without 0.2 wt% $Ca_3(PO_4)_2$ at different times at temperature of 15 and 60 °C is presented in

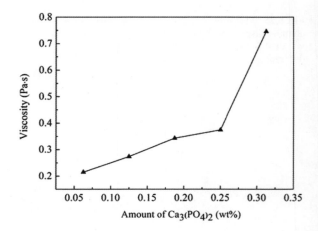

Fig. 8.26 Viscosity of suspension with different amounts of $Ca_3(PO_4)_2$ at 15 °C

8.1 Direct Coagulation Casting of Ceramic Suspension

Fig. 8.27. It can be seen from Fig. 8.27, at 15 °C, that the viscosity of suspension increases slightly with time, less than the viscosity 1 Pa s within 100 min. At 60 °C, the viscosity increases slightly without addition of calcium phosphate and increases slowly within 60 min after adding calcium phosphate; however, after 60 min the viscosity increases rapidly. The great change in viscosity results in the instability of the suspension and turns into a wet-coagulated body.

Figure 8.28 shows the pH of suspension after addition of 0.2 wt% $Ca_3(PO_4)_2$ at different temperatures. The pH values at both 15 and 60 °C increase gradually after adding $Ca_3(PO_4)_2$. The pH at 15 °C is slightly higher than that at 60 °C. The pH changes from acidic regime to basic regime. After addition of $Ca_3(PO_4)_2$, the following reactions occur:

$$Ca_3(PO_4)_2 + 2HCl \rightleftharpoons 2CaHPO_4 + CaCl_2 \tag{8.3}$$

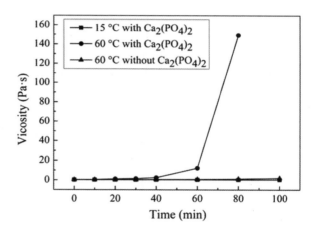

Fig. 8.27 Viscosity of suspension with and without 0.2 wt% $Ca_3(PO_4)_2$ at different times at temperature of 15 and 60 °C

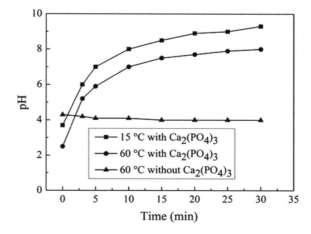

Fig. 8.28 pH of suspension with and without addition of 0.2 wt% $Ca_3(PO_4)_2$ at different temperatures

$$CaHPO_4 + 2HCl \rightleftharpoons Ca(H_2PO_4)_2 + CaCl_2 \quad (8.4)$$

$$Ca(H_2PO_4)_2 + 2HCl \rightleftharpoons 2H_3PO_4 + CaCl_2 \quad (8.5)$$

$$HPO_4^{2-} + H_2O \rightleftharpoons H_3PO_4 + OH^- \quad (8.6)$$

$$H_2PO_4^- + H_2O \rightleftharpoons H_3PO_4 + OH^- \quad (8.7)$$

In these reactions, hydrochloric acid is consumed to generate $CaHPO_4$, $Ca(H_2PO_4)_2$, and $CaCl_2$. The solubility of $Ca_3(PO_4)_2$, $CaHPO_4$, and $Ca(H_2PO_4)_2$ is 0.002, 0.043, and 1.8 g at 25 °C, respectively. $CaCl_2$ is a high solubility salt with solubility of 74.5 g at 20 °C. The reactions lead to more Ca^{2+} ions existing in the suspension. The hydrolysis of HPO_4 and H_2PO_4 generates OH which contributes to the pH shift to basic regime. The surface charge of particles turns into negative. When the Ca^{2+} concentration reaches a certain value, the electrical double layer on the surface of alumina particles is strongly compressed, and zeta potential of the particle surfaces decreased. The interparticle repulsion potential energy reduces and the dispersed particles agglomerate, leading to the increase of viscosity. It should be noted that the pH value of the suspension is lower at higher temperature. Alumina is an amphoteric oxide that can become charged due to surface ionization of acid or basic groups in aqueous solution. The reactions occurring at the oxide/water interface can be represented as follows (Tarì et al. 2000; Valdivieso et al. 2006; Mustafa et al. 1998):

$$Al-OH_2^+ (surface) \rightleftharpoons Al-OH(surface) + H^+ \quad (8.8)$$

$$Al-OH (surface) + OH^- \rightleftharpoons Al-O^- (surface) + H_2O \quad (8.9)$$

Halter et al. reported that an increase in temperature decreases the zeta potential, also indicating proton desorption from the surface for a given pH value (Halter 1999). The presence of proton in the suspension leads to the decrease of pH value at higher temperature. Figure 8.29 shows the zeta potential of suspension after calcium phosphate addition at 60 °C. The IEP of alumina suspension shifts to acidic region after the addition of calcium phosphate. This phenomenon may be due to the phosphate ions absorption on the alumina surface. The pH at high temperature is closer to IEP of alumina, to some extent, contributing to the instability of suspension.

(4) **Properties and microstructure of the dried and sintered body**

Curve of coagulated time at different temperatures is displayed in Fig. 8.30. The coagulated time decreases as the increase of temperature and addition of $Ca_3(PO_4)_2$ which is attributed to the rapid release of high valence counter-ions at higher temperature. At 70 °C, the coagulation time reduces to 100 min. Figure 8.31 shows the density of wet green bodies coagulated at different temperatures. The density of the

8.1 Direct Coagulation Casting of Ceramic Suspension

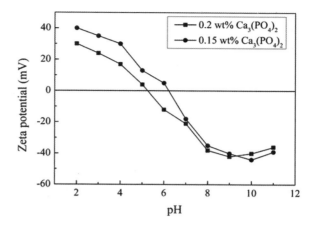

Fig. 8.29 Zeta potential of alumina suspension with different pH values after adding calcium phosphate at 60 °C

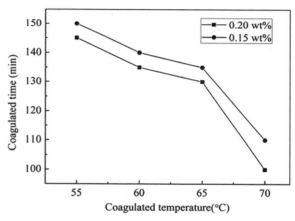

Fig. 8.30 Curve of coagulated time at different temperatures

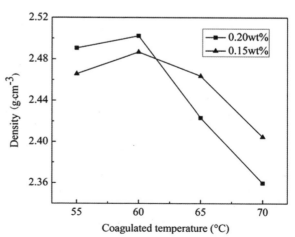

Fig. 8.31 Density of wet green bodies coagulated at different temperatures

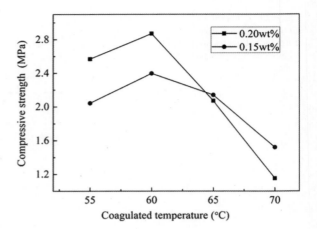

Fig. 8.32 Compressive strength of wet green bodies coagulated at different temperatures

wet green bodies increases first and then decreases slightly with the increase of temperature. Strength of wet-coagulated body is one of the parameters that determine whether or not the process is suitable for fabricating complex parts. High strength suggests that the samples can be demolded without any deformation. Figure 8.32 shows the change of compressive strength with temperature increasing from 55 to 70 °C. Similar phenomenon is observed in compressive strength at different coagulated temperatures as that of density. The compressive strength of the wet green bodies increases first and then decreases slightly with the increase of temperature. The strength increases with increase of $Ca_3(PO_4)_2$ addition. The optimum coagulated temperature is 60 °C with compressive strength of 2.8 MPa. Below 60 °C, high valence counter-ion is not completely released, thus affecting the coagulated effect. Above 60 °C, the demolding time becomes shorter with the increase of temperature for great amount of high valence counter-ions release at high temperature. Defects and cracks, however, are easy to generate at high temperature which impact the mechanical properties of the green bodies.

Microstructure of fractured surface of the green alumina body is presented in Fig. 8.33. The SEM photograph of fractured surface of the alumina green body prepared by DCC–HVCI method coagulated at 60 °C shows uniform microstructure and no agglomeration or pores is observed. However, pores appear in the sample coagulated at 70 °C. The microstructure can interpret the decrease in strength at a temperature higher than 60 °C. The uniform microstructure indicates fewer cracks in the green body which leads to high mechanical properties, compared with increasing monovalent ionic strength method. 8 Suspension coagulated in the mold could be removed without any deformation by keeping the mold in an oven at 60 °C for 135 min immediately after cooling the mold to room temperature. The properties of dried and sintered samples prepared by DCC–HCVI are summarized in Table 8.2. The samples used for measurement were coagulated at 60 °C for 135 min and sintered at 1550 °C for 2 h. It can be seen that the samples with 0.2 wt% calcium phosphate show higher linear shrinkage rate and relative density than that with 0.15

8.1 Direct Coagulation Casting of Ceramic Suspension

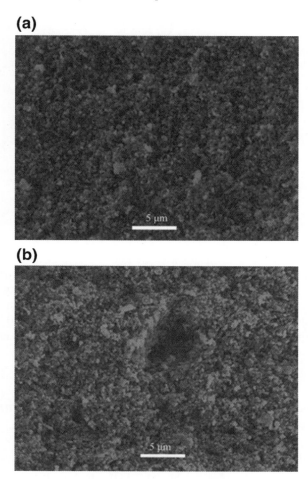

Fig. 8.33 SEM photograph of fractured surface of alumina green body prepared by DCC–HVCI method coagulated at 60 °C for 135 min (**a**) and at 70 °C for 100 min (**b**) with 0.2 wt% $Ca_3(PO_4)_2$

Table 8.2 Properties of the dried and sintered samples prepared by DCC–HVCI (coagulated at 60 °C for 135 min and sintered at 1550 °C for 2 h)

Calcium phosphate concentration (wt%)	Dried samples		Sintered sample	
	Relative density (%)	Linear shrinkage (%)	Relative density (%)	Linear shrinkage (%)
0.15	52.3	2.6	98.8	17.4
0.20	54.7	2.8	99.5	18.5

Fig. 8.34 The microstructure of sintered ceramics prepared by DCC–HVCI with 0.2 wt% calcium phosphate coagulated at 60 °C for 135 min and sintered at 1550 °C for 2 h

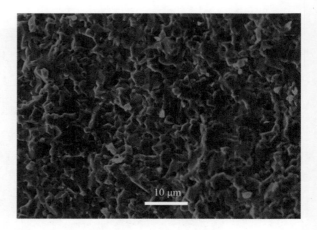

wt% calcium phosphate. The linear shrinkage rate of coagulated body and sintered sample is 2–3% and 17–20%, respectively. The relative densities of dried and sintered samples are 52.3–54.7% and 98.8–99.5% of theoretical density of alumina. The microstructure of sintered ceramics prepared by DCC–HVCI with 0.2 wt% calcium phosphate coagulated at 60 °C for 135 min and sintered at 1550 °C for 2 h is displayed in Fig. 8.34. The good microstructure suggests dense ceramic parts and high mechanical properties prepared by DCC–HVCI.

8.1.3 Direct Coagulation Casting by Using Thermo-Sensitive Liposomes as Coagulating Agent

Thermo-sensitive liposomes were prepared using reverse-phase evaporation method using natural lipid egg phosphatidylcholine (EPC) and cholesterol (CH). Inorganic salts containing high valence counter-ions (HVCI) are encapsulated by the liposomes. The phase transition temperature of the liposome is at 38 °C with 50 wt% addition of cholesterol. The encapsulation rate of liposomes reaches 85% for high valence anion (SO_4^{2-}) and 55% for high valence cation (Ca^{2+}). The liposomes are introduced into ceramic colloidal forming and dispersed in the suspension for identical charge with alumina particles at room temperature. The release of HCVI from the liposomes can coagulate the alumina suspension after heating at 38 °C for 3 h, but the demoldable time is 6–7 h. Dense ceramic products with relative density of above 98% and uniform microstructure can be prepared by this method without burnout process.

Generally, there are three methods to realize DCC–HVCI: (1) increasing the solubility of insoluble salts containing HVCI by increasing temperature; (2) chemical reactions to release HVCI; (3) soluble salts containing HVCI encapsulated by microcapsules, liposomes, etc. and controlled release by environmental trigger. The first and second methods have been reported in previous research. HVCI (Ca^{2+})

were controlled release from calcium iodate or reactions between calcium phosphate and hydrochloric acid by increasing temperature, which increased the viscosity of the concentrated alumina suspension. Complex ceramic parts have been successfully produced with high relative density and uniform microstructure. However, the present methods need to coagulate the suspension at temperature above 60 °C and the calcium is not suitable for some ceramic systems (Yang et al. 2011b; Xu et al. 2012a, b, c). Liposomes are artificially prepared vesicles composed of a lipid bilayer. A number of stimuli-sensitive liposomes have been designed as drug delivery systems (DDS) with special usage. These liposomes release the encapsulated drugs in response to pH, light, temperature, etc. (Yatvin et al. 1980; Frankel et al. 1989; Kono et al. 1994). Thermo-sensitive liposomes are known to release encapsulated water-soluble contents more quickly near their gel-to-liquid crystalline phase transition temperature in response to environmental temperature (Yatvin et al. 1978). Dipalmitoylphosphatidylcholine (DPPC) is usually used as the primary lipid in liposome preparation for its excellent temperature-sensitive behavior (Needham and Dewhirst 2001). However, as a synthetic lipid, DPPC is too expensive for daily use. Natural lipids have been developed to prepare thermo-sensitive liposomes with combination of cholesterol (Chelvi et al. 1995; Sharma et al. 1998).

In this section, thermo-sensitive liposomes were prepared using natural lipid and cholesterol by reverse-phase evaporation method and the properties were characterized. The release of high valence counter-ions from thermo-sensitive liposomes was examined. The liposomes are introduced into ceramic colloidal forming. Alumina suspension was coagulated by DCC–HVCI with controlled release of HVCI from thermo-sensitive liposomes.

(1) **Preparation Procedure**

Egg phosphatidylcholine (EPC) and cholesterol (CH) were purchased from Guoyao Group (Beijing, China). Chloroform and methanol were used in liposome preparation. Analytical purity $(NH_4)_2SO_4$ and $CaCl_2$ were used as HVCI sources.

Preparation of Thermo-Sensitive Liposomes: The liposomes were prepared by a reverse-phase evaporation method. Total amount of 0.3 g of egg phosphatidylcholine and cholesterol were dissolved in 20 ml organic solvent mixed by chloroform: methanol (2:1 v/v). Quantities of 4 ml salt solutions with $(NH_4)_2SO_4$ or $CaCl_2$ were added to the organic solution, and then ultrasonic emulsification was carried out for 15 min to obtain a homogeneous water/oil emulsion. The organic solvent was removed via a reverse-phase evaporation process at 45 °C and the liposome film was developed. Nitrogen was used to dry the film to eliminate the remained organic solvent. A volume of 8 ml distilled water was added to liposome film. The liposome was then dialyzed in a 10000 MW maximum permeability dialysis bag for 48 h with change of water every 8 h to remove the free ions in the suspension. The concentration of prepared liposome emulsion was 0.025 g lipid per 1 ml emulsion.

Preparation of Alumina Suspension: In DCC process, high solid loading and low viscosity suspension is needed. Alumina suspensions with solid loading of 50 vol% were prepared by tumbling the alumina powder, water, and the dispersant in polyethylene containers for 12 h. To prepare negatively charged suspensions, TAC was used as the dispersant with 0.3 wt% based on alumina powder. Suspension with positively charged ions on the surface of the alumina particles was prepared by adding 3 wt% hydrochloric acids (2 mol/l) based on alumina powder. Zirconia balls with diameter of 5–10 mm were used as grinding media. The mass ratio between grinding media and alumina powder is 1:2. The pH values of negatively and positively charged suspensions are adjusted to 9 and 4.5, respectively.

Coagulation Process: To coagulate the positively charged alumina suspension, liposome encapsulated with $(NH_4)_2SO_4$ solution was used while liposome encapsulated with $CaCl_2$ solution was used to coagulate the negatively charged suspension. Different amounts of liposome emulsion were added to the alumina suspension and stirred for 30 min to obtain a homogeneous suspension. The suspension was then degassed for 10 min under vacuum condition. The suspension was cast into a plastic mold with a piston as cover to prevent the evaporation of water, and the sample was heated in a water bath at 38 °C for 3–5 h. The coagulated bodies were rested at room temperature for another 4 h and demolded.

(2) **Properties of Thermo-Sensitive Liposomes**

To investigate the thermo-behavior of the liposome, a differential scanning calorimeter was used. The thermotropic behavior of liposome prepared with different amounts of cholesterol is shown in Fig. 8.35. It can be seen from Fig. 8.35 that the liposome prepared by EPC displays a wide range of phase transition temperature, from 15 to 25 °C. This indicates that the liposome releases the encapsulated contents in this wide temperature range, which is difficult to control the drug release process. However, with the addition of cholesterol, the liposome shows a sharp peak at 38 °C in the DSC measurement. Also, the peak is sharper with higher cholesterol content as the addition of cholesterol enhances the stability of the liposome (Lindner et al. 2004; Sabin et al. 2005). Liposomes prepared by EPC and CH with 1:1 in mass were used in the following experiments.

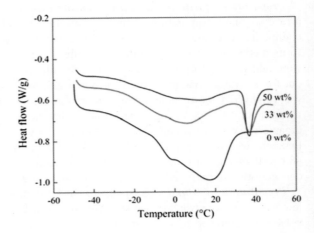

Fig. 8.35 DSC curves of liposomes prepared with different amounts of cholesterol

8.1 Direct Coagulation Casting of Ceramic Suspension

Generally, the particle size of liposome is an important factor. Large size indicates that more drugs can be encapsulated in the liposome, and the effect of drug release can be enhanced. However, the size of liposome in the present system should be controlled to avoid large pores presented in the sample after sintering of ceramics. Figure 8.36a shows the microstructure of the liposome under the optical microscope. Smooth surface and uniform particle size can be observed in the picture. Figure 8.36b shows the particle size distribution of liposome prepared by reverse-phase evaporation method with EPC:CH = 1:1 in mass ratio. The average size d_{50} is 10 μm. The size distribution is in single peak, and most of the particles are in the range from submicrometers to several tens of micrometers.

The zeta potential of liposome influences its dispersion state in the alumina suspension. The same charge of liposome and alumina is preferred in the suspension to prevent agglomeration for van der Waals attraction. Figure 8.37 shows the zeta

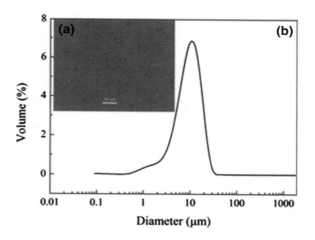

Fig. 8.36 Morphology (**a**) and particles (**b**) size distribution of liposomes prepared by reverse-phase evaporation method

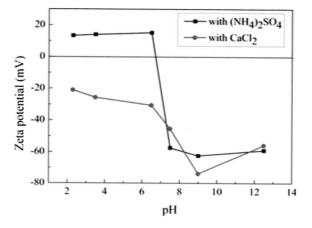

Fig. 8.37 Zeta potential of liposomes at different pHs encapsulated with different salts

potential curves of liposome at different pHs encapsulated with different salts. For the liposome encapsulated with $(NH_4)_2SO_4$ solution, the liposome is positively charged in the acidic condition, whereas negatively charged particles can be obtained in the basic region. The zeta potential shows high absolute value in the alkaline region, which indicates the well-dispersed liposome. The liposome encapsulated with $CaCl_2$ solution is negatively charged in all ranges of pH value with a maximum zeta potential at pH of 9. In our previous reports, stable alumina suspension can be obtained at pH 4.5 with hydrochloric acid with positively charged on particles or at pH of 8–9 with TAC as dispersant with negatively charged on particles (Xu et al. 2012a, b). The pH values of the prepared suspensions are 9 and 4.5, respectively. Therefore, the liposome encapsulated with $(NH_4)_2SO_4$ solution can be well dispersed in the acidic alumina suspension, while the liposome encapsulated with $CaCl_2$ solution is well dispersed in the basic alumina suspension.

The encapsulation rate of liposomes with different salts is displayed in Fig. 8.38. The liposome encapsulated with $(NH_4)_2SO_4$ shows higher encapsulation rate than that with $CaCl_2$. The encapsulation rate increases first with increasing the amount of salts, and then decreases gradually with an excessive addition of salts. The liposome encapsulated with 0.6 mol/l $(NH_4)_2SO_4$ has an optimum encapsulation rate of 85% while the liposome encapsulated with 0.4 mol/l $CaCl_2$ has an optimum encapsulation rate of 55%. Previous results showed that the presence of high valence ions (Ca^{2+}, La^{3+}) influenced the dispersion state of the liposome. Aggregation of the liposome occurred at a critical concentration of 0.7 mol/l for Ca^{2+} which can be interpreted using DVLO theory. The high ionic strength decreases the potential of the liposome suspension and leads to the instability of the system. Also, the packing order and structure of the PC monolayer films were influenced by high concentration divalent cations (Gaber et al. 1995). Liposomes encapsulated with 0.6 mol/l $(NH_4)_2SO_4$ and 0.4 mol/l $CaCl_2$ were used in the following experiments.

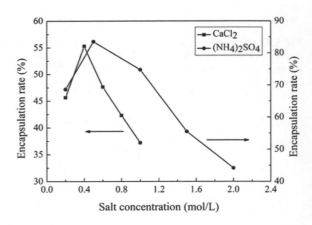

Fig. 8.38 Encapsulation rate of liposomes with different concentrations of salts

8.1 Direct Coagulation Casting of Ceramic Suspension

(3) Controlled Release and Coagulation

To examine the release properties of liposomes, solutions with 4 vol% liposomes encapsulated with $(NH_4)_2SO_4$ were used. Figure 8.39 shows the concentration and conductivity of liposome encapsulated with $(NH_4)_2SO_4$ solution at 38 °C. The liposome suspension was heated for a period of time and then cooled down to room temperature. The measurement was taken at room temperature. The concentration of $(NH_4)_2SO_4$ release from liposome at certain time was calculated from the relationship between concentration and conductivity of $(NH_4)_2SO_4$, Fig. 8.39a. It can be seen from Fig. 8.39b that the salt releases from the liposome quickly during the first 4 h after treatment. The concentration of $(NH_4)_2SO_4$ is ~18 mmol/l. However, the release rate slows down in subsequent time with maximum concentration of 20 mmol/l. The release of salt from liposome is a process of ion diffusion from high concentration region to low concentration region. The diffusion rate is lower when the concentration difference inside and outside of the liposome is smaller.

It can be seen from Fig. 8.39 that the HVCI release completely in liposome solution after a period of time at temperature above phase transition temperature. It seems to be a good method for controlled release of high valence counter-ions to coagulate ceramic suspension. Figure 8.40 shows the viscosity of alumina suspension after addition of 4 vol% liposomes encapsulated with $(NH_4)_2SO_4$ at different temperatures for 2 h. The viscosity of the suspension at room temperature is ~0.3 Pa s, which is low enough for casting process with sufficient time to degas. The viscosity increases gradually with increase in temperature below 38 °C. A rapid increase in viscosity is observed at 38 °C, which is in response to the phase transition temperature of liposome. The viscosity then increases slightly at elevated temperature. Figure 8.41 shows the viscosity of alumina suspension after addition of liposome for different times at 38 °C. The viscosity increases gradually after the addition of liposome.

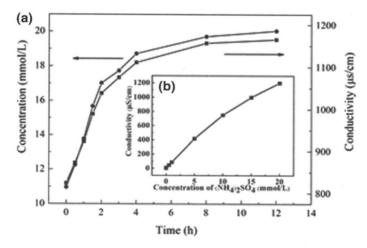

Fig. 8.39 Controlled release of $(NH_4)_2SO_4$ from liposome at 38 °C, **a** the relationship between $(NH_4)_2SO_4$ concentration and conductivity, **b** the concentration of $(NH_4)_2SO_4$ and conductivity at different times

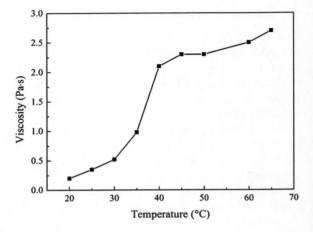

Fig. 8.40 Viscosity of alumina suspension after addition of 4 vol% liposome encapsulated with $(NH_4)_2SO_4$ at different temperatures for 2 h

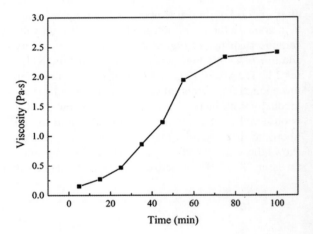

Fig. 8.41 Viscosity of alumina suspension after addition of 4 vol% liposome encapsulated with $(NH_4)_2SO_4$ for different times at 38 °C

The viscosity of the suspension reaches 2.3 Pa s after 80 min and only slightly increases on heating. The suspension coagulates at 38 °C for 3 h; the sample at this state is deformable and difficult to handle. However, the sample can be demolded without any deformation after resting at room temperature for another 3–4 h. The coagulation mechanism of the present system is proposed as follows. For a positively charged suspension, the coagulation process via controlled release from liposome encapsulated with SO^{2-} is displayed in Fig. 8.42. The liposome encapsulated with salt is well dispersed in the system for the same charge as alumina particles at 20 °C. When the temperature of the suspension rises to the phase transition temperature of liposome, the lipid film opens and the high valence counter-ions release from the liposome. The counter-ions SO_4^{2-} compress the electrical double layer and cause the instability of the suspension. The release of Ca^{2+} from liposome is similar to that of

8.1 Direct Coagulation Casting of Ceramic Suspension

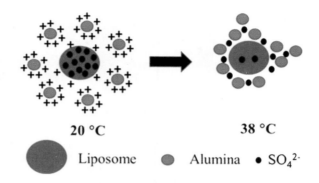

Fig. 8.42 Schematic representation of coagulation process induced by controlled release of high valence counter-ions from liposome at phase transition temperature

SO_4^{2-}. The coagulation mechanism of negatively charged suspension by increasing Ca^{2+} ions has been reported in the previous papers (Xu et al. 2012a, b, c).

(4) Properties of Coagulated Bodies

Figure 8.43 shows the photographs of coagulated and sintered samples with addition of liposome. The samples coagulated by 3 vol% liposome encapsulated with 0.4 mol/l $CaCl_2$ solution show insufficient coagulation phenomenon due to low concentration of high valence counter-ions in the suspension. When the concentration of liposomes increases to 5 vol%, the sample coagulates completely without any cracks. The samples coagulated by 2 vol% liposome encapsulated with 0.6 mol/l $(NH_4)_2SO_4$ solution show insufficient coagulation phenomenon due to low concentration of high valence counter-ions in the suspension. When the concentration of liposomes increases to 4 vol%, the sample coagulates well and no large pore is observed in the samples.

The properties of coagulated and sintered samples prepared by DCC–HVCI are summarized in Table 8.3. The coagulation time of the samples coagulated by liposome encapsulated with $CaCl_2$ is shorter than that of samples coagulated by liposome encapsulated with $(NH_4)_2SO_4$, even though the encapsulation rate of liposome encapsulated with $CaCl_2$ is lower. The shrinkage rate and relative density of samples coagulated by liposome encapsulated with $CaCl_2$ are lower. The shrinkage rate and relative density of green bodies are 3–4% and 54–56%, respectively. The shrinkage rate and relative density of sintered samples are 19–21% and above 98%, respectively. A previous paper reported that the relative density of sample coagulated by calcium iodate and calcium phosphate sintered at 1550 °C reaches ~99.4 and 99.5%, respectively (Xu et al. 2012a, b). This indicates that DCC–HVCI method can prepare dense ceramics. Figure 8.44 shows the relative density of samples prepared by DCC–HVCI from 5 vol% liposomes encapsulated with 0.4 mol/l $CaCl_2$ and 4 vol% liposomes encapsulated with 0.6 mol/l $(NH_4)_2SO_4$ sintered at different temperatures. The samples coagulated using $(NH_4)_2SO_4$ show higher relative density than samples coagulated using $CaCl_2$. However, both samples cannot reach relative density above 99%. With the same coagulation method (DCC–HVCI), the low relative density of samples coagulated by liposome may be the large size of liposomes that causes defects in the microstructure. The SEM photographs of coagulated and

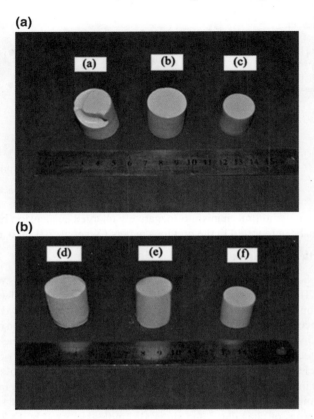

Fig. 8.43 Photographs of coagulated and sintered samples with addition of liposome; **a** encapsulated with CaCl$_2$, (**a**) 3 vol% (coagulated), **b** 5 vol% (coagulated), **c** 5 vol% (sintered); (**b**) encapsulated with (NH$_4$)$_2$SO$_4$, **d** 2 vol% (coagulated), **e** 4 vol% (coagulated), **f** 4 vol% (sintered)

Table 8.3 Properties of coagulated and sintered samples green body and sintered sample

Salt	ER (%)	CT (h)	Green body		Sintered sample	
			SR (%)	RD (%)	SR (%)	RD (%)
CaCl$_2$	55	6	3.98	54.5	19.3	98.2
(NH$_4$)$_2$SO$_4$	85	7	3.10	55.3	20.7	98.5

ER: encapsulation rate; CT: coagulation time; SR: shrinkage rate; RD: relative density

sintered samples are displayed in Fig. 8.45. The SEM photograph of fractured surface of the alumina green body and sintered sample prepared by DCC–HVCI showed uniform microstructure, but defects are also observed in the samples.

The present method provides a new way to coagulate ceramic suspension via controlled release of high valence counter-ions from liposomes. The liposomes can encapsulate different salts for different ceramic systems. The method provides wider applications range than the previous methods which are only suitable for alumina

8.1 Direct Coagulation Casting of Ceramic Suspension

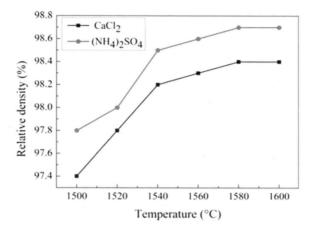

Fig. 8.44 Relative density of samples prepared by DCC–HVCI from 5 vol% liposomes encapsulated with 0.4 mol/l $CaCl_2$ and 4 vol% liposomes encapsulated with 0.6 mol/l $(NH_4)_2SO_4$ sintered at different temperatures

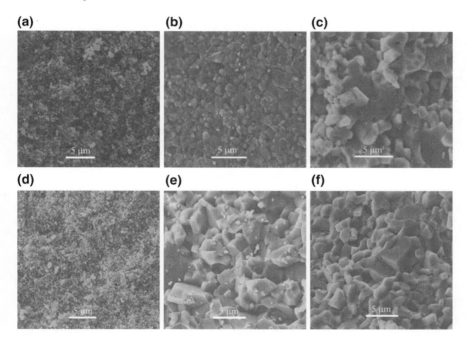

Fig. 8.45 Microstructure of coagulated and sintered samples coagulated by liposomes; **a** 5 vol% liposome encapsulated with $CaCl_2$ (coagulated), **b** 5 vol% (sintered), **c** 5 vol% (sintered and etched), **d** 4 vol% liposome encapsulated with $(NH_4)_2SO_4$ (coagulated), **e** 4 vol% (sintered), **f** 4 vol% (sintered and etched)

system (Yang et al. 2011b; Xu et al. 2012a, b, c). However, the size of liposomes is 10 times larger than that of ceramic particles which leads to defects in the microstructure that may deteriorate the mechanical properties of the sintered samples.

8.1.4 Direct Coagulation Casting from Citrate Assisted by pH Shift

(1) **Direct Coagulation Casting of Alumina Suspension from Calcium Citrate Assisted by pH Shift**

In this section, DCC of alumina suspension via controlled release of high valence counter-ions (DCC–HVCI) using calcium citrate as coagulating agent was reported. Hydrolysis of glycerol diacetate (GDA) shifts the pH of suspension to weakly acidic region which helps to decompose calcium citrate and release of calcium ions. The effect of concentration of glycerol diacetate and calcium citrate on the pH and viscosity of alumina suspension was investigated at 25 and 60 °C, respectively. Green body was prepared by DCC–HVCI with controlled release of calcium ions from calcium citrate at 60 °C 1 h. The properties of the green bodies and sintered samples were characterized.

The pH value of the alumina suspension and glycerol diacetate solution was measured by a LE438 pH meter (Mettler, Toledo, Switzerland) in the temperature range of 25–60 °C. The zeta potential of the 10 vol% alumina suspension with 0.3 wt% TAC was measured by a zeta potential analyzer (CD-7020; Colloidal Dynamics, Warwick, RI). 2 M HCl and NaOH were used to adjust the pH of the suspension. The measurement was carried out at room temperature. The rheological properties of the suspension were measured using a KINEXUS rheometer (Malvern Instruments, Worcestershire, UK) attached with C25 R0634 SS spindle and PC25 C0138 AL cylinder. In viscosity measurement, the shear rate was set at 100 s^{-1}. For the measurements of the concentration of calcium ions, nine 100 mL beakers are filled with 50 ml deionized water, and the pH is adjusted by ammonia. Then, 0.571 g of calcium citrate is dispersed into each beaker and stirred for 24 h. The above suspensions are centrifuged in plastic tubes (Centrifuge 800-1; Ronghua Instruments, Jintan, China) at 1790 g for 1 h. The clear supernatants are drawn out carefully and diluted using deionized water. The calcium concentration was measured using inductively coupled plasma–optical emission spectrometry (ICP-OES; Varian Vista MPX, Varian, Palo Alto, CA).

Tri-ammonium citrate is a common dispersant that has been widely used to prepare electrostatically stabilized suspension. The pH of as-prepared suspension is 9.2 with TAC as dispersant. Figure 8.46 shows the pH of alumina suspension at different temperatures. The pH value of the suspension decreases when the temperature is raised. The pH of suspension treated at 60 °C decreases to 8.0. The decrease in pH of the alumina suspension may be attributed to two factors. On the one hand, the pH decrease may be in part due to the increasing dissolution of alumina with temperature

Fig. 8.46 The pH of alumina suspension at different temperatures

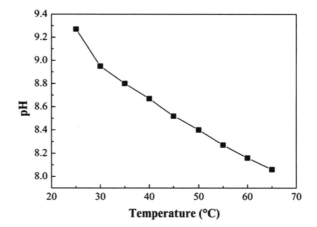

(Halter 2003). On the other hand, the decrease in pH at higher temperatures may be due to the evaporation of ammonia as gas.

$$NH_4^+ + OH^- = NH_3 \cdot H_2O \uparrow \tag{8.10}$$

The viscosity and zeta potential of alumina suspension at different pHs are presented in Fig. 8.47. It is indicated that both the viscosity and zeta potential increase with decreasing the pH of alumina suspension due to the approaching to the IEP of alumina which is at pH 4 with TAC as dispersant (Cai et al. 2005). A rapid increase in viscosity is observed when the pH below 8 as the zeta potential value approaches zero and the particles get closer, with van der Waals forces predominant in the system.

As one kind of esters, glycerol diacetate is stable in water, in acidic, and neutral solution. However, irreversible hydrolysis to the salt of two acetic acids and a glycerol occurs in alkaline region. The pH of the solution shifts to acidic region as the generation of acetic acid (Flory 1953).

Fig. 8.47 Viscosity and zeta potential of alumina suspension at different pHs

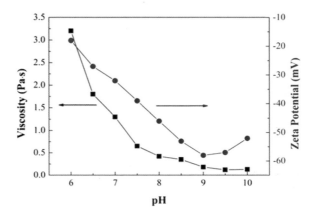

Fig. 8.48 Relationship between pH and concentration of glycerol diacetate solution heating at 60 °C for different times

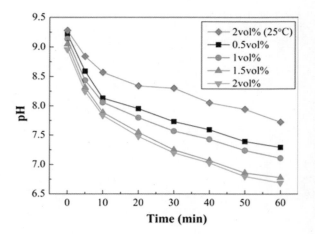

$$C_7H_{12}O_5 + 2H_2O \rightarrow 2C_2H_4O_2 + C_3H_8O_3 \quad (8.11)$$

Glycerol diacetate has been commonly used in practical applications for tailoring the pH internally in systems with sodium silicate binders (Haas 1989). To tailor the pH of alumina suspension, glycerol diacetate is used in the experiment. Figure 8.48 shows the relationship between pH and concentrations of glycerol diacetate solution heating at 60 °C for different times. The initial pH of the solution before glycerol diacetate addition is set at 9.4 adjusted by ammonia. It is observed that the pH of solution shifts to lower pH with increasing the concentration of glycerol diacetate. The hydrolysis is also promoted by increasing the temperature. The pHs of 2 vol% glycerol diacetate solution after treating at 25 and 60 °C for 1 h are 7.9 and 6.6, respectively.

Glycerol diacetate was added to 50 vol% alumina suspension, and the influence of glycerol diacetate on the pH of suspension was investigated. Figure 8.49a shows the pH of suspension with 2 vol% glycerol diacetate at different temperatures. The pH decreases from 9.0 to 7.9 when the temperature increases from 25 to 60 °C. Figure 8.49b shows the pH of suspension with 2 vol% glycerol diacetate treated at 60 °C for different times. The pH decreases to 6.7 heating at 60 °C for 1 h which is similar to that with glycerol diacetate solution only.

Figure 8.50 shows the viscosity of alumina suspension with addition of 2 vol% glycerol diacetate at 25 and 60 °C. The viscosity of suspension at 25 °C is stable with viscosity of around 0.15 Pa.s which is a little higher than suspension without glycerol diacetate addition. When the suspension is treated at 60 °C, the viscosity is higher than that treated at 25 °C and increases slightly with the holding time. The increase in viscosity is mainly attributed to the hydrolysis of glycerol diacetate which lowers the pH to near the IEP of alumina with absorbed TAC. However, the suspension is still in flow state despite the increase of temperature.

Figure 8.51 shows the viscosity of alumina suspension with addition of 0.5 wt% calcium citrate at 25 and 60 °C. It is observed that the viscosity of suspension at

Fig. 8.49 The pH of suspension with 2 vol% glycerol diacetate, **a** at different temperatures and **b** at 60 °C

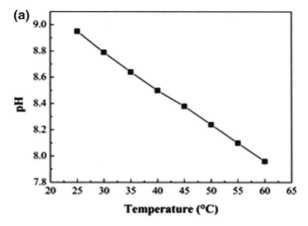

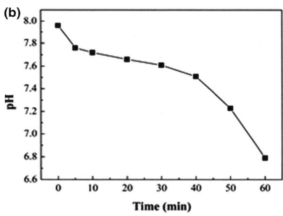

Fig. 8.50 Viscosity of alumina suspension with addition of 2 vol% glycerol diacetate at 25 and 60 °C

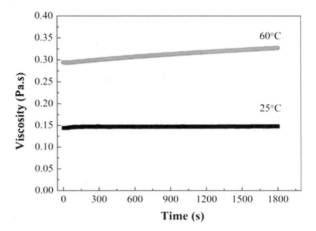

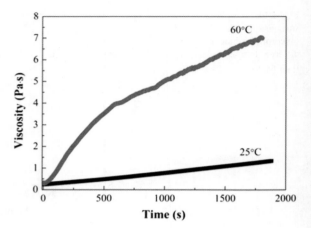

Fig. 8.51 Viscosity of alumina suspension with addition of 0.5 wt% calcium citrate at 25 and 60 °C

25 °C increases to about 1 Pa s after holding for 30 min. The suspension is still in a fluid state. The case at 60 °C is rather different. The viscosity increases remarkably and the viscosity reaches 7 Pa s which makes the suspension transfer to a semisolid state.

Calcium citrate is insoluble in water with minor solubility of 1 g/l at room temperature (Li 2003). The pH of saturated solution of calcium citrate is 9.6–9.8. It is indicated that the concentration of calcium ion is 5 mM in the pH range.

However, calcium citrate is a strong base–weak acid salt which has the ability of hydrolysis in water. The hydrolysis equations are listed as follows:

$$Ca_3(C_6H_5O_7)_2 + 2H_2O \rightleftharpoons 2CaHC_6H_5O_7 + Ca(OH)_2 \qquad (8.12)$$

$$2CaHC_6H_5O_7 + 2H_2O \rightleftharpoons Ca(H_2C_6H_5O_7)_2 + Ca(OH)_2 \qquad (8.13)$$

$$Ca(H_2C_6H_5O_7)_2 + 2H_2O \rightleftharpoons 2H_3C_6H_5O_7 + Ca(OH)_2 \qquad (8.14)$$

In weak alkaline solution, calcium hydroxide cannot be formed in a stable state and is likely to decompose into divalent calcium ions and hydroxyl.

$$Ca(OH)_2 \rightleftharpoons Ca^{2+} + 2OH^- \qquad (8.15)$$

Figure 8.52 shows the concentration of calcium ions in calcium citrate saturated suspension at different pHs. It is observed that the concentration of calcium ions increases gradually with decrease in pH. The concentration of calcium ions reaches 30 mM when the pH is lower than 7.5. It has been shown in Fig. 8.46 that the increase in temperature to 60 °C can lower the pH of suspension to around 8. The pH values of alumina suspension containing 0.5% calcium citrate after treating at 25 and 60 °C for 30 min are 9.2 and 8.3, respectively. Calcium citrate salts and calcium hydroxide

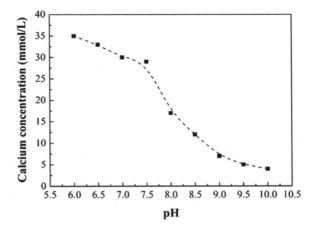

Fig. 8.52 Concentration of calcium ions in calcium citrate saturated suspension at different pHs

may be more easy to hydrolyze and decompose in lower pH (pH = 8.3) which can release higher concentration of calcium ions (ca. 12 mM). The critical coagulation concentration of calcium ion in 50 vol% alumina suspension with 0.3 wt% TAC is 10 mM (Wen et al. 2011). The compress of double electrical layer by calcium ions contributes to the increase in viscosity of the suspension.

Although the viscosity of suspension with calcium citrate treated at 60 °C is 7 Pa s, the viscosity is not high enough to coagulate the suspension. To further tailor the pH of suspension to a lower value, glycerol diacetate is added into the suspension. Figure 8.53 shows the viscosity and pH of alumina suspensions with addition of 2 vol% glycerol diacetate and 0.5 wt% calcium citrate at 25 and 60 °C, respectively. The

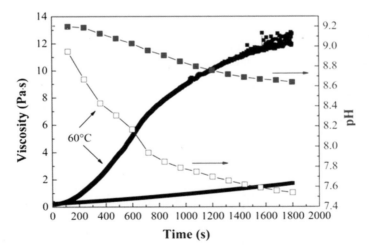

Fig. 8.53 Viscosity and pH of alumina suspensions with addition of 2 vol% glycerol diacetate and 0.5 wt% calcium citrate at 25 and 60 °C

Table 8.4 Coagulation of alumina suspension with different concentrations glycerol diacetate (GDA) and calcium citrate (CC)

Sample no.	GDA (%)	CC (wt%)	Phenomenon
1	1.5	0	NC
2	2	0	NC
3	0	0.5	IC
4	0	0.8	IC
5	1.5	0.8	C
6	2	0.5	C

NC, not coagulated; IC incomplete coagulated; C, Coagulated

pH of suspensions with glycerol diacetate and calcium citrate decreases to 8.6 and 7.5 treated at 25 and 60 °C, respectively. The viscosity of suspension treated at 25 °C is similar to that with calcium citrate addition alone, and the value is at around 1–2 Pa s. However, the viscosity of suspension increases rapidly after heating at 60 °C for 30 min with a viscosity of 13 Pa s, which is high enough to coagulate the suspension. The high viscosity is mainly attributed to the high concentration of calcium ions (>30 mM) above the critical coagulation concentration of calcium (10 mM) that the coagulation of suspension accords. Also, the shift of pH to approach to IEP of TAC-dispersed alumina suspension contributes to the increase in the viscosity of suspension. The increase in calcium ions can be explained in two ways. First, the hydrolysis of calcium citrate and decomposition of calcium hydroxide produce part of calcium ions. Also, as acetate acid is stronger than citric acid, the reaction between calcium citrate and acetate acid can produce calcium acetate. Calcium acetate is soluble in water which increases the concentration of calcium ions.

$$Ca_3(C_6H_5O_7)_2 + 3C_2H_4O_2 \rightarrow 3Ca(C_2H_3O_2)_2 + 2C_6H_6O_7 \tag{8.16}$$

Results above suggest that the addition of glycerol diacetate and calcium citrate into the suspension can be used to coagulate the mixtures at 60 °C for a period of time. Table 8.4 shows the coagulation experimental data of alumina suspension with different concentrations of glycerol diacetate and calcium citrate. The suspensions were heated at 60 °C for 1 h. It can be seen that the suspension cannot be coagulated by only addition of glycerol diacetate. Although the viscosity increases slightly as the pH of suspension decreases by glycerol diacetate hydrolysis, the viscosity is still very low (0.35 Pa s) and is in a fluid state. Samples with calcium citrate addition show incomplete coagulation. The samples are soft (pulpy) and easy to deform by handling. However, samples with both glycerol diacetate and calcium citrate addition can coagulate completely and the samples can be demolded without any damage after treated at 60 °C for 1 h. Also, the amount of calcium citrate used to coagulate the suspension decreases with increasing the concentration of glycerol diacetate. Figure 8.54a displays the picture of coagulated parts with different shapes. The coagulated samples are dried at 80 °C for 24 h. As no polymer has been used in the present system, a burnout process can be avoided. Ceramics can be prepared

8.1 Direct Coagulation Casting of Ceramic Suspension

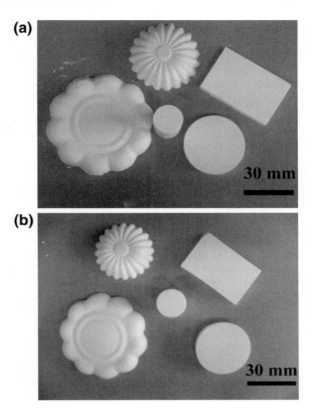

Fig. 8.54 Photographs of **a** green body and **b** sintered ceramics prepared by glycerol diacetate and calcium citrate coagulation

by directly sintering at 1550 °C for 2 h with a heating rate of 5 °C/min. Figure 8.54b shows the photographs sintered ceramics prepared by glycerol diacetate and calcium citrate coagulation.

Table 8.5 shows the properties of the green bodies and sintered samples prepared by citrate calcium and glycerol diacetate coagulation. The concentration of glycerol diacetate is 2 vol% for preparation process. For the green bodies, the linear shrinkage and relative density correspond to samples dried at 80 °C, whereas the compressive strength was measured for the wet green body. The compressive strength of green

Table 8.5 Properties of the green bodies and sintered ceramics prepared by glycerol diacetate and calcium citrate (cc) (coagulated at 60 °C for 1 h and sintered at 1550 °C for 2 h)

CC (wt%)	Green body			Sintered samples	
	Relative density (%)	Shrinkage (%)	Compressive strength (MPa)	Relative density (%)	Shrinkage (%)
0.5	56.2	2.19	1.06 ± 0.23	99.4	19.5
0.8	55.7	2.25	0.15 ± 0.35	99.0	18.9

body is in the range of 1.0 MPa which is high enough for demolding, and the strength is higher than DCC via increasing ionic strength monovalent salt (Balzer et al. 1999). The linear shrinkage and relative density of coagulated body are 2.1–2.3 and 55.7–56.2%, respectively. The samples are coagulated at 60 °C for 1 h and sintered at 1550 °C for 2 h. The linear shrinkage and relative density of sintered samples are 18–20 and 90.0–99.4% of theoretical density of alumina which is similar to DCC by pH shift (Gauckler et al. 1999a). Figure 8.55 shows the microstructure of (a) green body and (b) sintered ceramics prepared from suspension with 2 vol% glycerol diacetate and 0.5 wt% calcium citrate. The surface of sintered sample has been polished with a subsequent heat etching at 1500 °C for 30 min. It is observed that homogeneous microstructure has been obtained, and no agglomeration is presented in the sample which suggests the new processing method as a promising method to prepare near-net-shape ceramics with good mechanical properties.

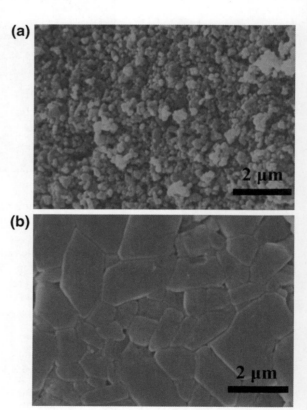

Fig. 8.55 Microstructure of **a** fractured surface of green body and **b** polished and heat etched surface of sintered ceramics prepared by DCC–HVCI method

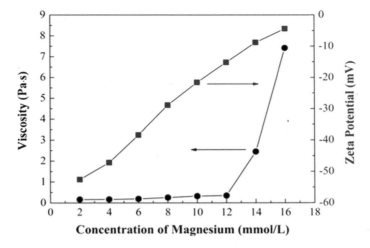

Fig. 8.56 Influence of magnesium ions on viscosity and zeta potential of alumina suspension

(2) **Direct coagulation casting of alumina using magnesium citrate as coagulating agent with glycerol diacetate as pH regulator**

In this section, DCC–HVCI using magnesium citrate as coagulating agent and glycerol diacetate as pH regulator at room temperature was investigated. Hydrolysis of glycerol diacetate shifts the suspension to lower pH, which increases the solubility of magnesium citrate and release of magnesium ions. Influence of concentrations of glycerol diacetate and magnesium citrate on the pH and rheological properties of alumina suspension was evaluated. The properties of green bodies and sintered samples were measured.

According to previous reports, high valence counter-ions may affect the electrical double layer of concentrated alumina particles which is illustrated via measuring the rheological behavior of the suspension (Wen et al. 2011). Figure 8.56 shows the influence of magnesium ions on viscosity and zeta potential of alumina suspension. 1 mol/l magnesium chloride solution was used for the experiment. The solid loadings of alumina suspensions are 55 and 10 vol% for viscosity and zeta potential measurements, respectively. The pH of the suspension is adjusted to 9.0 before measurement. The viscosity of suspension changes rarely when the concentration of magnesium ions is lower than 12 mmol/l. However, a rapid increase in viscosity is observed when the concentration of magnesium ions is higher than 12 mmol/l. The initial zeta potential of as-prepared suspension is ~62 mV which indicates that the alumina particles are negatively charged. The zeta potential increases with increasing the concentration of magnesium and approaches to zero when the concentration of magnesium is higher than 16 mmol/l. The above results can be explained by DLVO theory. When the concentration of magnesium reaches a certain value, the electrical double layer between alumina particles is strongly compressed and zeta potential of

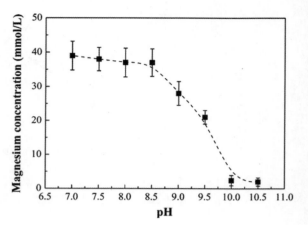

Fig. 8.57 Concentration of magnesium ions in magnesium citrate saturated suspension at different pHs

the particles decreases, leading to the increase in viscosity of alumina suspension (Johnson et al. 2000; Hunter 1987).

Magnesium citrate is an organic strong base and weak acid salt with low solubility in water, but it has the ability of hydrolysis in water. It was reported that magnesium citrate had minor dissolution in water at room temperature when pH > 10, but dissolved completely in water when the pH decreased to 8.0–9.0 (Yang and Sigmund 2003). In order to examine the concentration of soluble magnesium ions in magnesium citrate solution at different pHs, ICP analysis was conducted. 0.703 g magnesium citrate was mixed with 50 ml water at different pHs and stirred for 24 h. Figure 8.57 shows the concentration of magnesium ions in magnesium citrate saturated suspension at different pHs. It can be seen that the concentration of magnesium ions is very low (2–3 mmol/l) in the suspension at pH above 10.0. However, the concentration of magnesium ions increases with decreasing the pH of magnesium citrate suspension due to the increase of solubility of magnesium citrate. The concentration of magnesium ions reaches ca. 40 mmol/l when the pH is lower than 8.5. It has been indicated in Fig. 8.56 that concentrated alumina suspension can be coagulated by adding about 20 mmol/l of magnesium ions. So, it is feasible to control release of magnesium ions from magnesium citrate to coagulate alumina suspension via regulating the pH of the suspension.

Figure 8.58 shows the effect of concentration of magnesium citrate on the viscosity of 55 vol% alumina suspension. The pH of alumina suspension is set at 8.5. The viscosity was measured 30 min upon the addition of magnesium citrate. It is observed that the viscosity of suspension increases slightly with the concentration of magnesium citrate lower than 1.4 wt%. However, an evident increase in viscosity of the suspension is observed when the concentration of magnesium citrate is higher than 0.4 wt%. The viscosity reaches ca. 50 Pa s with addition of 0.8 wt% magnesium citrate which is high to transfer the suspension from liquid state to solid state. However, direct addition of magnesium citrate into concentrated alumina suspension at pH 8.5 may cause heterogeneous coagulation due to insufficient dispersion of magnesium citrate before coagulation starts. Hence, a suitable method to control

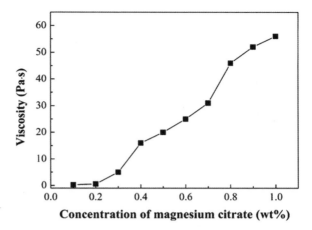

Fig. 8.58 Effect of concentration of magnesium citrate on the viscosity of 55 vol% alumina suspension at pH of 8.5

release of magnesium ions from magnesium citrate should be used when preparing homogeneous microstructure ceramics.

In order to investigate the influence of pH on the rheological properties of concentrated alumina suspension with magnesium citrate, 0.8 wt% magnesium citrate was added into 55 vol% alumina suspension at different pHs. Figure 8.59 shows the effect of magnesium citrate on viscosity of 55 vol% alumina suspension at different pHs. The viscosity of suspension changes slightly at pH 10.0 after 30 min. However, a rapid increase in viscosity is observed when the pH decreases to 8.0–9.0. Viscosity at pH 8.0 reaches 50 Pa s after resting at room temperature for 30 min which is larger than that at pH 9.0 (40 Pa s). The great difference of viscosity at pH of 8.0–9.0 and 10.0 is attributed to the solubility of magnesium citrate at different pHs. The increase in concentration of magnesium ions promotes the compression process of electrical double layer by magnesium ions which increases the viscosity of the suspension.

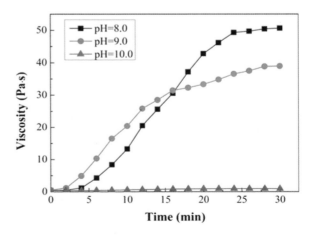

Fig. 8.59 Effect of 0.8 wt% magnesium citrate on viscosity of alumina suspension at different pHs

Fig. 8.60 Relationship between pH and concentration of glycerol diacetate solutions

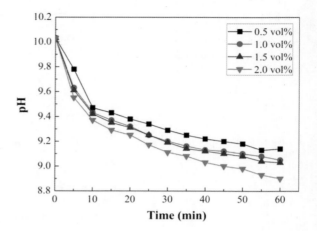

Glycerol diacetate is easy to hydrolyze and produce two acetic acids and a glycerol in alkaline region. The pH of the solution shifts to acidic region as the generation of acetic acid (Flory 1953). Our previous report showed that the hydrolysis was promoted by increasing the temperature. The hydrolysis rate increases with increasing the temperature from 25 to 60 °C (Xu et al. 2014a). However, the aim of this section is to coagulate the alumina suspension at room temperature. The temperature is set at 25 °C. Figure 8.60 shows the relationship between pH and concentration of glycerol diacetate solutions. The initial pH of the solutions is set at 10.0. It is observed that the pH of solutions shifts to lower pH by treating for different times. Lower pH is observed in solution with higher concentration of glycerol diacetate. The pH of solution with 2 vol% glycerol diacetate is 8.8 after resting at room temperature for 1 h.

In order to tailor the pH of alumina suspension, 2 vol% glycerol diacetate is used in the experiment. Figure 8.61 shows the effect of glycerol diacetate on pH and viscosity of 55 vol% alumina suspension. The initial pH of suspension before glycerol diacetate addition is set at 10.0 adjusted by ammonia. The hydrolysis rate of glycerol diacetate is lower in concentrated alumina suspension than in pure glycerol diacetate solution. The reason is that there is less water in concentrated suspension with equal volume of suspension. The pH decreases gradually after the addition of glycerol diacetate, and the pH decreases to 8.8 after 2 h which is the same value with treating the pure glycerol diacetate solution for 1 h. The viscosity of suspension changes slightly, and the viscosity is in the range of 0.24–0.32 Pa s with the minimum viscosity observed at pH 9.0–9.2.

As indicated in the results above, magnesium citrate can increase the viscosity of concentrated alumina suspension in pH range of 8.0–9.0 and the hydrolysis of glycerol diacetate can shift pH from 10.0 to 8.8. It is feasible to combine the two compounds to coagulate concentrated alumina suspension with a controllable method. Figure 8.62 shows the viscosity and pH of alumina suspension with addition of 0.8 wt% magnesium citrate and 2 vol% glycerol diacetate. The initial pH of the

8.1 Direct Coagulation Casting of Ceramic Suspension

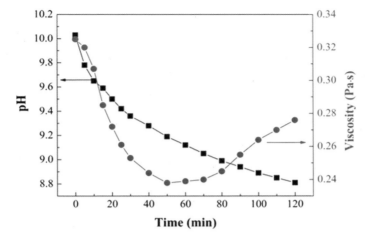

Fig. 8.61 Effect of glycerol diacetate on pH and viscosity of alumina suspension

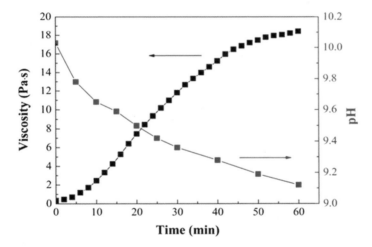

Fig. 8.62 Viscosity of alumina suspension with addition of 0.8 wt% magnesium citrate and 2 vol% glycerol diacetate

mixture is set at 10. The pH of the mixture decreases gradually and reaches 9.1 after resting at room temperature for 1 h. The viscosity of suspension increases gradually with magnesium citrate and glycerol diacetate addition. The viscosity reaches 17 Pa s after resting at room temperature for 1 h. Figure 8.63 presents the schematic illustration of coagulation process using glycerol diacetate as coagulating agent and glycerol diacetate as pH regulator. The increase in viscosity is due to the decrease of pH by glycerol diacetate hydrolysis, and hence promotes the decomposition of magnesium citrate. Magnesium ions diffuse into the electrical double layer of alumina

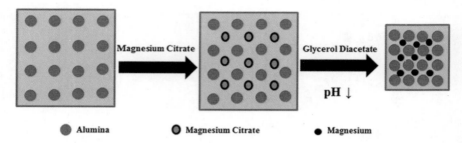

Fig. 8.63 Schematic illustration of coagulation process using magnesium citrate as coagulating agent and glycerol diacetate as pH regulator

particles through electrostatic attraction, reduces the zeta potential, and decreases the repulsive force between particles. The dispersed particles collide and agglomerate which leads to the abrupt increase in viscosity of suspension.

The controlled release of magnesium ions from magnesium citrate assisted by pH shift via hydrolysis of glycerol diacetate has been proved and can be used to coagulate concentrated alumina suspension at room temperature. Figure 8.64 shows the coagulation time and compressive strength of green bodies with different amounts of magnesium citrate. The concentration of glycerol diacetate is 2 vol%. It is observed that the coagulation time decreases with increase of concentration of magnesium citrate, from 6 h (0.5 wt% magnesium citrate) to 2.5 h (1.2 wt% magnesium citrate). The compressive strength increases with increasing the amount of magnesium citrate, from 0.7 to 1.8 MPa. The increase of magnesium citrate content provides more magnesium ions during the coagulation process which can shorten the coagulation time and improve the wet strength of green bodies.

The high green strength of new direct coagulation casting method ensures the fabrication of near-net-shape green bodies. Complex shape green bodies have been prepared by DCC–HVCI using magnesium citrate as coagulating agent with glycerol diacetate as pH regulator (Fig. 8.65a). The suspension coagulated in the mold could

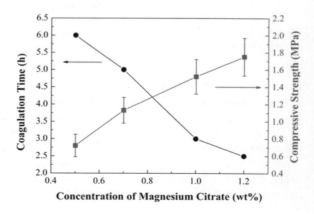

Fig. 8.64 Coagulation time and compressive strength of wet green bodies with different amounts of magnesium citrate

8.1 Direct Coagulation Casting of Ceramic Suspension

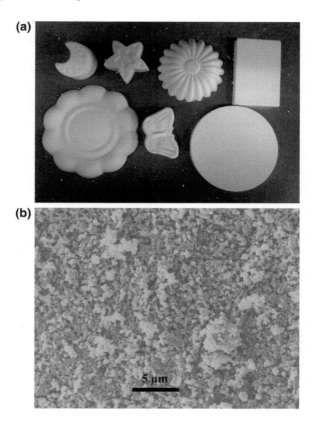

Fig. 8.65 Photographs (**a**) and microstructure (**b**) of green body prepared by magnesium citrate and glycerol diacetate coagulation

be removed without any deformation by treating at room temperature for 2.5–6 h. The SEM image of fractured surface of the alumina green body prepared by the coagulation casting process shows uniform microstructure and no agglomeration or big pores is observed (Fig. 8.65b). The uniform microstructure indicated less cracks in the green body to achieve high mechanical properties and reliability. The relative density of green body is 62.4%. The green samples can be directly sintered at high temperature without additional burnout process compared with gel casting and injection molding which contains large amount of binder in the green body. The density of ceramics sintered at 1550 °C is 3.96 g/cm^3, 99.5% of theoretical density of alumina.

(3) **Direct coagulation casting of yttria-stabilized zirconia using magnesium citrate and glycerol diacetate**

Yttria-stabilized zirconia (YSZ) ceramics have been widely used as an advanced structural material for its high strength and toughness and good wear resistance (Garvie et al. 1975; Lange 1982; Hannink et al. 2000; Evans and Cannon 1986). It is desirable to obtain a homogeneous microstructure and to fabricate components with complex shapes for various applications. Colloidal processing is recognized as an ideal route to prepare near-net-shape ceramics with tailoring rheological properties, minimal organic binder content, high wet strength, and excellent microstructural homogeneity (Alford et al. 1987; Lewis 2000). New colloidal processing methods such as gel casting and direct coagulation casting (DCC) have been applied to form green bodies from well-dispersed zirconia suspension (Xie et al. 2004; Adolfsson 2008; Prabhakaran et al. 2009; Zhang et al. 2005; Sun and Gao 2003).

Preparation of high solid loading and low viscosity YSZ suspension: High solid loading and low viscosity suspension are desirable for colloidal processing to avoid defects generated from agglomerates and to decrease the possibility of cracking and deformation during sintering due to high shrinkage. As a polyelectrolyte, ammonium polyacrylate has been used to disperse aqueous ceramic powder suspension in the pH range of 9–9.5 (Wang and Gao 1999; Ewais and Safari 2010; Shen et al. 2004). In order to prepare a well-dispersed zirconia suspension, suitable amount of ammonium polyacrylate should be considered. Figure 8.66 shows the effect of concentration of ammonium polyacrylate on viscosity and zeta potential of 30 vol% YSZ suspension. The pH of the suspension is set at 9.2. It is observed that the zeta potential of the suspension decreases to −50 mV for the amount of ammonium polyacrylate higher than 0.3 wt%. High absolute value of the zeta potential indicates high charge density on the surface of powders and large repulsion among particles. The viscosity decreases from 0.27 to 0.04 Pa s with the amount of ammonium polyacrylate rising from 0.1 to 0.4 wt%. Further addition of ammonium polyacrylate increases the viscosity of the suspension as the existence of excess amount of PAA–NH_4 which does not adsorb on the surface. The high concentration of the electrolyte causes the compression of the double layer and results in weaker repulsion (Tadros 1996).

Figure 8.67 shows the zeta potential of YSZ suspensions with and without ammonium polyacrylate at different pHs. The IEP of YSZ suspension without dispersant is

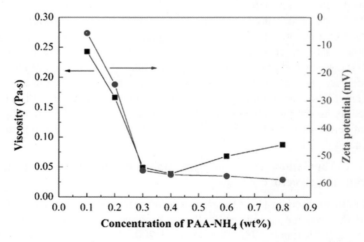

Fig. 8.66 Effect of concentration of ammonium polyacrylate on the viscosity and zeta potential of 30 vol% YSZ suspension

8.1 Direct Coagulation Casting of Ceramic Suspension

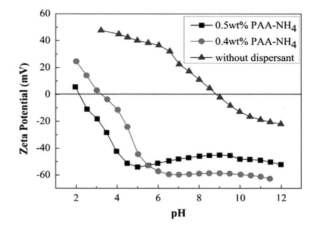

Fig. 8.67 Zeta potential of YSZ suspensions with and without ammonium polyacrylate at different pHs

8.8. The IEP moves to lower pH with addition of ammonium polyacrylate due to the absorption of ions on the YSZ particles. The IEPs of YSZ suspensions with 0.4 and 0.5 wt% PAA–NH$_4$ are 3.1 and 2.2, respectively. Stable suspension can be prepared in the pH range of 9–10. The suspension with 0.4 wt% PAA–NH$_4$ shows higher zeta potential value than that with 0.5 wt%. Therefore, dispersant concentration of 0.4 wt% is used for preparation of YSZ suspensions in subsequent experiments.

Figure 8.68 shows the viscosity of YSZ suspension with different solid loadings, (a) at different shear rates and (b) at the shear rate of 100 s^{-1}. The suspensions exhibit a shear-thinning behavior at a wide range of shear rate. The viscosity increases with the increase of the solid loading. The viscosity of 50 vol% YSZ suspension is 0.62 Pa s at the shear rate of 100 s^{-1} which is still low enough for colloidal processing. Usually, the viscosity of the suspension used for colloidal forming should be below 1 Pa s at

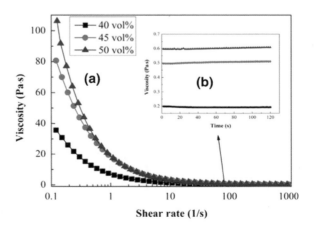

Fig. 8.68 Viscosity of YSZ suspension with different solid loadings, **a** at different shear rates and **b** at the shear rate of 100 s^{-1}

Fig. 8.69 Relative viscosity of YSZ suspension as a function of volume fraction at 100 s^{-1}

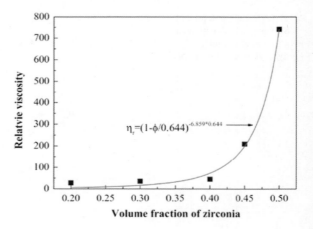

the shear rate of 100 s^{-1} which is low enough for degassing and casting (Si et al. 2010).

The Dougherty–Krieger (D–K) equation is commonly used to express the relationship between the relative viscosity (η_r) and volume fraction (φ) for hard sphere system (Krieger 1972):

$$\eta_r = \left(1 - \frac{\varphi}{\varphi_m}\right)^{-[\eta]\varphi_m} \quad (8.17)$$

where φ_m is the maximum packing fraction, which has the order of 0.64 for random packing, and $[\eta]$ is the asymptotic slope of the curve zero, and is normally 2.5. Usually, φ_m and $[\eta]$ are chosen as variables at the same time to fit the experimental results and to analyze the suspension structures. The relative viscosity of YSZ suspensions as a function of volume fraction at the shear rate of 100 s^{-1} is presented in Fig. 8.69. Relative viscosities were simulated as a function of volume fraction of the particles using Krieger–Dougherty's equation. Using this model, relative viscosity at the shear rate of 100 s^{-1} can be expressed as

$$\eta_r = \left(1 - \frac{\varphi}{0.644}\right)^{-6.859*0.644} \quad (8.18)$$

It is indicated in Fig. 8.69 that the relative viscosity of the suspension increases rapidly above 50 vol% and the preparation and handling of the suspension will be difficult at higher volume fractions of the powders. Therefore, YSZ suspension of 50 vol% was chosen in order to avoid high viscosity of the suspension which may impact on the properties of sintered ceramics.

Coagulation of concentrated YSZ suspension: Fig. 8.70 shows the effect of magnesium concentration on the viscosity of 50 vol% YSZ suspension. The viscosity of the suspension is very low for magnesium concentrations below 14 mmol/l. However, rapid increase in viscosity is observed for magnesium concentrations higher than

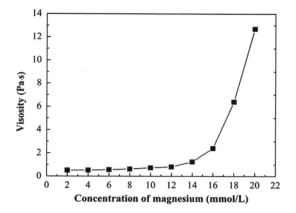

Fig. 8.70 Effect of magnesium concentration on the viscosity of 50 vol% YSZ suspension

14 mmol/l. Therefore, the critical coagulation concentration (CCC) of magnesium for 50 vol% YSZ suspensions is 14 mmol/l. When the concentration of magnesium reaches 20 mmol/l, high viscosities exceeding 13 Pa s are observed transforming the concentrated suspension from liquid to solid. In this sense, 50 vol% YSZ suspension can be coagulated by adding magnesium with concentration of 20 mmol/l.

Magnesium citrate shows solubility sensitive to pH in the alkaline region with minor dissolution in aqueous solution at room temperature at pH above 10, but the solubility increases gradually with the decrease of pH (Yang and Sigmund 2003; Xu et al. 2014b). Figure 8.71 shows the effect of magnesium citrate concentrations on the viscosities of 50 vol% YSZ suspensions at pH of 9.0 and 10.2. The viscosity measurement lasts for 30 min after addition of magnesium citrate at 25 °C with a shear rate of 100 s^{-1}. At pH 10.2, the viscosity of such a suspension changes rarely with increase of magnesium citrate up to 1.0 wt%. At pH 9.0, the behavior is rather different. The viscosity of the suspension remains low up to magnesium

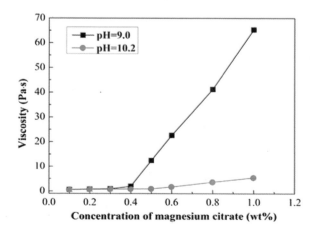

Fig. 8.71 Influence of magnesium citrate concentration on the viscosity of 50 vol% YSZ suspension at different pHs with a holding time of 30 min

citrate concentrations of 0.4 wt%. However, the viscosity increases rapidly with concentrations of magnesium citrate above 0.4 wt% and the viscosity of suspension with magnesium concentration above 0.6 wt% exceeds 25 Pa s. The increase in viscosity of the suspension with magnesium citrate is attributed to the coagulation of the suspension due to the decomposition of magnesium citrate at pH below 10. The results suggest that 50 vol% YSZ suspensions can be coagulated by adding sufficient amount of magnesium citrate at pH below 10.

Glycerol diacetate can hydrolyze and produce two acetic acids and a glycerol in the alkaline region. The pH of the solution shifts to lower pH due to the generation of acetic acid (Gauckler et al. 1999a; Flory 1953). In this study, concentration of 2 vol% glycerol diacetate is chosen to adjust the pH of YSZ suspension at the pH of 10.2. Figure 8.72 shows the effect of glycerol diacetate on pH and viscosity of 50 vol% YSZ suspension. The initial pH of suspension before glycerol diacetate addition is set at 10.2 adjusted by ammonia. It is observed that the pH of the suspension decreases gradually after the addition of glycerol diacetate and the pH is at pH¼ 8.9 after 90 min. The viscosity has a minor change with the addition of glycerol diacetate which indicates the viscosity of the suspension is not influenced by pH value in the range of 9–10.

Figure 8.73 shows the effect of magnesium citrate and glycerol diacetate (2 vol%) on the viscosity of 50 vol% YSZ suspension. The initial pH of the suspension before magnesium citrate and glycerol diacetate addition is set at 10.2 adjusted by ammonia. It is observed that the viscosity increases slowly within 10 min as the pH of the suspension is still above 10. Rapid increase in viscosity is observed in subsequent treatment, and the viscosity increases to ca. 30 and 40 Pa s for the suspensions with 0.8 and 1.0 wt% magnesium citrate, respectively. Then the viscosity increases slowly again after 50 min. The results suggest that controlled coagulation of 50 vol% YSZ

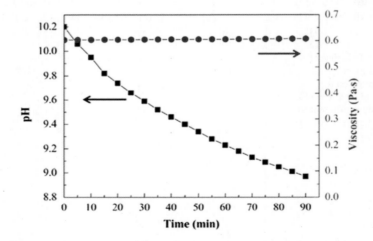

Fig. 8.72 Effect of glycerol diacetate on pH and viscosity of 50 vol% YSZ suspension

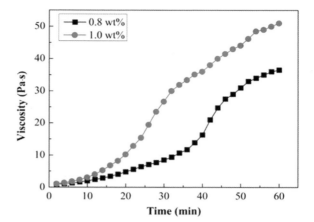

Fig. 8.73 Effect of magnesium citrate and glycerol diacetate on the viscosity of 50 vol% YSZ suspension

suspension can be realized by adding enough amount of magnesium citrate using glycerol diacetate as a pH regulator at room temperature.

Properties of green body and ceramics: Fig. 8.74 shows the compressive strength of the green bodies with different amounts of magnesium citrate. The concentration of glycerol diacetate is 2 vol%. The compressive strength of the sample with 0.5 wt% magnesium citrate is 1.5 MPa. The compressive strength increases to 2.5 MPa when the concentration of magnesium citrate is 1.0 wt%. The increase of magnesium citrate content provides more magnesium ions during the coagulation process and enhances the strength of the green bodies. However, further addition of magnesium citrate decreases the strength which may be due to the inhomogeneous distribution of magnesium citrate in the samples which does not dissolve during the coagulation process.

High green strength of DCC–HVCI ensures the fabrication of near-net-shape green bodies without deformation for handling. Complex shape green bodies and

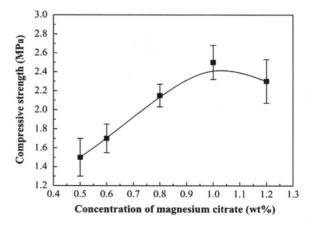

Fig. 8.74 Effect of magnesium citrate concentration on the compressive strength of the wet-coagulated bodies

sintering ceramics have been prepared by DCC–HVCI using magnesium citrate as coagulating agent with glycerol diacetate as a pH regulator (shown in Fig. 8.75). Smooth surface and no cracking are observed both in green and sintering samples. Figure 8.76 shows the microstructures of the fractured surface of the green body and sintered ceramics prepared by DCC–HVCI using 1.0 wt% magnesium citrate and 2 vol% glycerol diacetate. Homogenous microstructures have been obtained by this coagulation method to guarantee the good mechanical properties and high reliability of the ceramics. The shrinkage and relative density of dried samples are 1.6–2.9 and 61.1–62.3%, respectively. The relative density of ceramics sintered at 1450 °C is 98.9–99.2% of theoretical density of YSZ. Flexural strength of YSZ ceramics prepared by DCC–HVCI using 1.0 wt% magnesium citrate and 2 vol% glycerol diacetate is 869 ± 84 MPa.

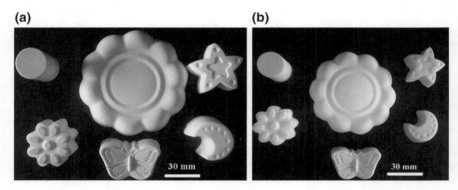

Fig. 8.75 Photographs of **a** green bodies and **b** sintering YSZ prepared by DCC–HVCI using 1.0 wt% magnesium citrate and 2 vol% glycerol diacetate

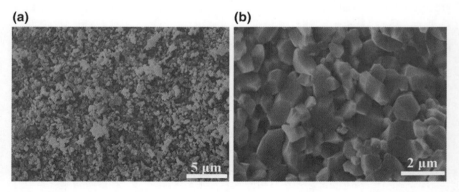

Fig. 8.76 Microstructures of fractured surface of **a** green and **b** sintered samples prepared by DCC–HVCI using 1.0 wt% magnesium citrate and 2 vol% glycerol diacetate

8.1 Direct Coagulation Casting of Ceramic Suspension

(4) Enhanced piezoelectric properties of PZT ceramics prepared by direct coagulation casting via high valence counter-ions (DCC–HVCI)

Lead zirconate titanate (PZT) ceramics are known to be a ferroelectric as well as a ferroelastic material which are widely used in piezoelectric actuators, transducers, and ferroelectric memory devices (Surowiak et al. 2001; Chu et al. 2004; Wang et al. 2001). One of the most common methods for preparing PZT ceramics is direct powder pressing and subsequent sintering. However, there may exist some heterogeneity in the microstructures which deteriorates the properties of the ceramics. With the growing awareness of the detrimental effect of different types of heterogeneities on the properties of ceramics, a new concept called colloidal processing has been successfully proposed to improve the product reliability (Alford et al. 1987; Lewis 2000; Cho and Dogan 2001).

During the last two decades, several colloidal forming methods for PZT ceramics have been developed. Tape casting has been widely applied to prepare PZT thin layers and graded ceramics (Smay and Lewis 2001; Navarro et al. 2004a, b). To prepare near-net-shape ceramics with complex geometry, gel casting and direct coagulation casting (DCC) have been used. Guo et al. reported a series of works on gel casting of PZT using an acrylamide gelling system (Guo et al. 2003a, b, c, d, 2004). High mechanical strength green bodies had been obtained (18 MPa) which provided a way for machining of the green bodies without cracking and deformation (Guo et al. 2004). However, the main component of the monomer acrylamide system is a neurotoxin, which limits the application of acrylamide system (Yang et al. 2011a). Low toxic systems, for example, epoxy resin, have been developed in recent years (Olhero et al. 2012; Xie et al. 2014, 2015). PZT pillar arrays with good properties have been prepared. Natural gelling systems, such as egg white (Roy et al. 2007a, b) and agarose (Ewais and Safari 2010), have also been extensively investigated to prepare PZT ceramics. Compared with gel casting, direct coagulation casting merits in low or no organic binder and a special binder burnout process is avoided (Gauckler et al. 1999a). Direct coagulation casting has also been applied to prepare PZT ceramics. Prabhakaran et al. investigated direct coagulation casting of PZT suspensions using MgO as coagulating agent (Prabhakaran et al. 2010). However, the wet strength of the green bodies is not high enough for handling. Recently, we proposed a novel ceramic forming method called direct coagulation casting via controlled release of high valence counter-ions (DCC–HVCI) using magnesium citrate as coagulating agent and glycerol diacetate as a pH regulator. It has been applied to prepare structural ceramics, such as alumina and zirconia ceramics (Xu et al. 2014b, 2015a, b). The concentrated suspensions coagulate at room temperature and the ceramics show homogeneous microstructures with good properties. This inspires us to apply the new method to prepare functional materials, such as PZT.

In this section, direct coagulation casting of PZT suspension via DCC–HVCI using magnesium citrate as coagulating agent and glycerol diacetate as pH controller was reported. The effect of dispersant on the dispensability of PZT suspensions was investigated by measuring the viscosity and the zeta potential of the suspensions. The solid loading of the suspension has been optimized. Influence of concentrations of

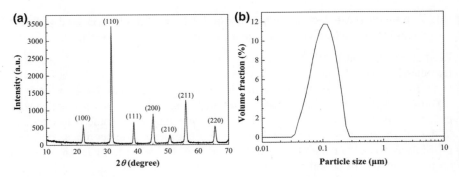

Fig. 8.77 **a** XRD pattern and **b** particle distribution of as-received PZT powder

glycerol diacetate and magnesium citrate on the rheological behaviors and pH value of PZT suspensions was evaluated. PZT green bodies and sintered ceramics have been prepared by DCC–HVCI, and the properties of the samples were measured.

PZT-5H powder (PbZr$_{0.52}$Ti$_{0.48}$O$_3$, Baoding HongSheng Acoustic Electron Apparatus Co., Ltd., China) with a mean particle size of 0.18 μm and theoretical density of 7.45 g/cm^3 was used as ceramic powder. The XRD pattern and particle distribution of as-received PZT powder is shown in Fig. 8.77. Solution with 30 wt% ammonium polyacrylate (PAA–NH$_4$) purchased from Zibo Jinghe Chemical Dyestuff Co., Ltd., China was used as dispersant. Chemical purity magnesium citrate [Mg$_3$(C$_6$H$_5$O$_7$)$_2$·10H$_2$O] with 99.0% purity was used as the coagulating agent. Glycerol diacetate (GDA) was used to tailor the pH value of the PZT suspensions for coagulation. The molecule weights of magnesium citrate and glycerol diacetate are 703.4 and 176.17, respectively. Both magnesium citrate and glycerol diacetate were purchased from Beijing Hengye Zhongyuan Chemical Co., Ltd.

Suspensions with different solid loadings were prepared by mixing the PZT powder, water, and ammonium polyacrylate in polyethylene containers for 24 h. Zirconia balls with diameter of 5–10 mm were used as grinding media. The mass ratio between grinding media and PZT powder is 1:2. Ammonia solution was utilized to adjust the pH value of the suspensions. The initial pH value of the PZT suspension is 10.6. To coagulate the suspension, the suspensions were degassed in vacuum condition for 15 min at room temperature. Different amounts of magnesium citrate were added to the suspensions and mixed thoroughly by continuing the tumbling process for another 20 min to improve the homogeneity of the suspensions. Then glycerol diacetate was added into the suspensions. The prepared suspensions were cast into a plastic mold. To avoid the evaporation of water during the coagulation process, a rubber seal was used. The coagulated wet samples were demolded after being treated at different temperatures for 1–2 h. The green bodies were dried at 80 °C for 24 h and then were sintered at 1250–1280 °C for 2 h at a heating rate of 5 °C/min, followed by natural cooling to room temperature. To avoid Pb loss during sintering, the green bodies were put into corundum crucibles containing PbZrO$_3$ powder to produce an excess PbO atmosphere.

Fig. 8.78 Zeta potential of PZT suspensions with and without ammonium polyacrylate at different pHs

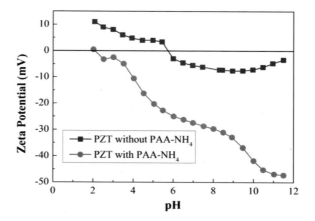

Preparation of high solid loading and low viscosity PZT suspensions: In colloidal processing, the colloidal behavior of the suspension is one of the most essential factors. A stable suspension containing well-dispersed powder particles will ensure a dense and homogeneous powder compact which would be beneficial to the final properties of the ceramics. Figure 8.78 shows the zeta potential of PZT suspensions with and without ammonium polyacrylate at different pHs. It is observed that PZT suspension without PAA–NH$_4$ shows low zeta potential value with an isoelectric point (IEP) at pH around 6. The low zeta potential value indicates the poor dispersion of the PZT powder without a dispersant. Previous reports showed that the leaching of Pb, Zr, and Ti ions would also impact the properties of PZT suspension (Bakarič et al. 2015; Noshchenko et al. 2014). The zeta potential increases remarkably after the addition of PAA–NH$_4$. The IEP moves to lower pH region (pH = 2) as the adsorption of dispersant on PZT particles. Higher zeta potential value (−40 mV) is obtained at pH range 10–12. It is indicated that the addition of PAA–NH$_4$ is beneficial to the stable dispersion of PZT powder in the alkaline region.

In addition to pH value, the dispersant amount also has great effect on the colloidal behavior of the suspension (Traiphol et al. 2010; Suntako et al. 2009). Figure 8.79 shows the effect of concentration of ammonium polyacrylate on viscosity and zeta potential of 30 vol% PZT suspension. The pH value of the suspension is set at 10.6. The viscosity decreases gradually with the increase of PAA–NH$_4$ content. The minimum value appears when the PAA–NH4 content is 0.5 wt% (based PZT powder). Further addition of ammonium polyacrylate increases the viscosity of the suspension as the existence of excess amount of PAA–NH$_4$ which does not adsorb on the particle surface. The zeta potential curve shows similar change with the viscosity variation with a minimum value of −38 mV when the PAA–NH$_4$ content is 0.5 wt%. Based on the results above, stable PZT suspensions can be prepared by the addition of 0.5 wt% PAA–NH$_4$ at pH of 10.6. Therefore, dispersant concentration of 0.5 wt% is used for the preparation of PZT suspensions in subsequent experiments.

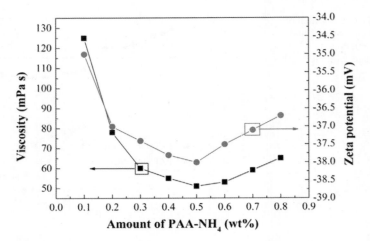

Fig. 8.79 Effect of concentration of ammonium polyacrylate on the viscosity and zeta potential of 30 vol% PZT suspension

In colloidal processing, suspension with high solid loading is desirable to ensure low shrinkage during the drying and sintering process which can decrease the possibility of cracking or deformation, and hence improves the properties of the ceramics. The viscosities of PZT suspensions with different solid loadings at the shear rate of 100 s^{-1} are plotted in Fig. 8.80. The viscosity increases with the increase of the solid loading. The viscosities of 54 and 56 vol% PZT suspensions are 0.41 and 0.48 Pa s, respectively, which are still low enough for colloidal processing. However, when the solid loading is 57 vol%, the viscosity increases to 1.2 Pa s. It has been difficult to prepare PZT suspension with solid loading of 58 vol% because of extremely high viscosity. Usually, the viscosity used for colloidal forming should be below 1 Pa s at

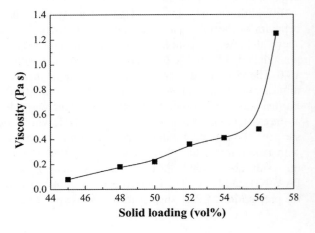

Fig. 8.80 Viscosities of PZT suspensions with different solid loadings at the shear rate of 100 s^{-1}

8.1 Direct Coagulation Casting of Ceramic Suspension

the shear rate of $100\ s^{-1}$ which is low enough for degassing and casting. Therefore, PZT suspension with solid loading of 56 vol% is chosen in subsequent experiments.

Coagulation of concentrated PZT suspension: High valence counter-ions (HVCIs) are high valence ions with opposite charge to that of the ceramic particles in the suspensions (Xu et al. 2012a). The present PZT suspension is negatively charged, so high valence cations are the HVCIs. In order to investigate the influence of HVCIs on the viscosity of PZT suspensions, magnesium chloride and lead nitrate solutions with concentration of 1 M were added into the suspensions. The viscosity of 56 vol% PZT suspensions with different concentrations of magnesium and lead is presented in Fig. 8.81. It is observed that the viscosity of the suspensions maintains at a low value when the concentration of HVCIs is lower than the critical coagulation concentration. The critical coagulation concentrations of magnesium and lead are 15 and 35 mM, respectively. The viscosity reaches 16 Pa s when the concentration of magnesium is 30 mM. It is also found that the critical coagulation concentration of lead is higher than that of magnesium. It is mainly attributed to the hydrolysis of lead at high pH value (pH = 10.6) which consumes part of the lead. The main compounds are lead oxide (Baes et al. 1986).

Magnesium citrate shows solubility sensitive to pH in the alkaline region and the solubility increases gradually with the decrease of pH value below pH 10 (Yang and Sigmund 2003). It has been applied to coagulate alumina and yttria-stabilized zirconia suspensions assisted by the hydrolysis of glycerol diacetate (Xu et al. 2014b, 2015a, b). Figure 8.82 shows the influence of magnesium citrate concentration on the viscosity of 56 vol% PZT suspension at different temperatures. The viscosities of the suspensions are in the range 1–3 Pa s at 25 °C with concentrations of magnesium citrate below 1.0 wt% (based on PZT powder). The viscosity increases to 9–15 Pa s at 60 °C with the same magnesium citrate concentrations. The suspensions show a semisolid state and the samples are not rigid enough for handling. Figure 8.83 shows the pH value of PZT suspension treated at 60 °C. The initial pH value of the PZT suspension is 9.8. The pH value decreases to 9.5 after 30 min. The decrease of pH value of PZT suspension at 60 °C is attributed to the evaporation of ammonia.

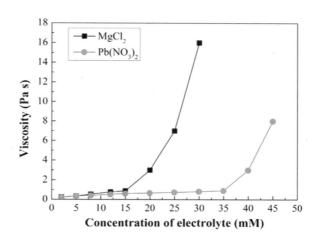

Fig. 8.81 Effect of magnesium and lead concentrations on the viscosities of 56 vol% PZT suspension

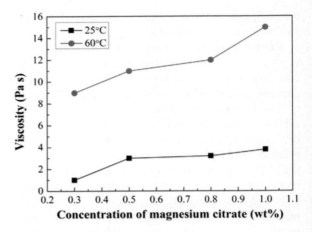

Fig. 8.82 Influence of magnesium citrate concentration on the viscosities of 56 vol% PZT suspension at different temperatures

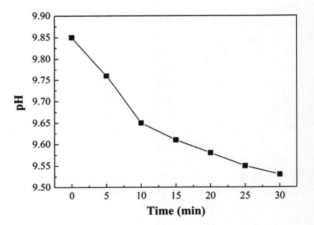

Fig. 8.83 The pH value of PZT suspension treated at 60 °C

The increase in viscosity of the suspension is mainly attributed to two factors. Firstly, the pH value approaching to the IEP of PZT contributes to part of the increase of the viscosity. The main factor is the decomposition of magnesium citrate with the decrease of pH value. Previous report showed that the concentration of free magnesium ions decomposed from magnesium citrate was around 20 mM at pH of 9.5 (Xu et al. 2014b). The concentration is higher than the critical coagulation concentration of magnesium ions in the present PZT suspension. Magnesium ions releasing from magnesium citrate compress the electric double layer of PZT particles, leading to the increase of the viscosity of the suspension.

Treatment at 60 °C can lower the pH value of the PZT suspension which promotes the decomposition of magnesium citrate and increases the viscosity of the suspension. However, the viscosity is not sufficiently high to coagulate the suspension. To further lower the pH value, 2 vol% glycerol diacetate (based on suspension volume) was added. Figure 8.84 shows the effect of glycerol diacetate on the viscosity and pH value of 56 vol% PZT suspension. It is observed that the viscosity of the suspension changes rarely at 25 °C. At 60 °C, the viscosity increases gradually and reaches 3 Pa s after 30 min. The pH values of the suspensions decrease to 10.1 and 8.9 at 25 and 60 °C, respectively. The decrease of the pH value is due to the generation of acetic acid from glycerol diacetate. The increase of the viscosity at 60 °C is attributed to the shift of pH to lower value where the zeta potential of the suspension is lower, as shown in Fig. 8.78. However, the sole addition of glycerol diacetate cannot lead to the coagulation of the suspension.

In order to coagulate the PZT suspension, magnesium citrate and glycerol diacetate were added into the suspension simultaneously. Figure 8.85 shows the effect of

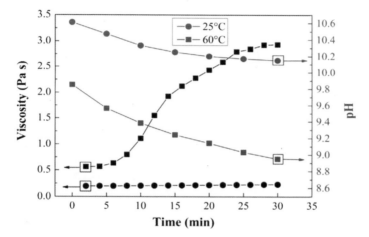

Fig. 8.84 Effect of glycerol diacetate on the viscosity and pH value of 56 vol% PZT suspension at different temperatures

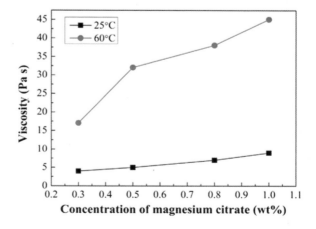

Fig. 8.85 Effect of magnesium citrate and glycerol diacetate on the viscosity of 56 vol% PZT suspension

magnesium citrate and glycerol diacetate on viscosity of 56 vol% PZT suspension. The concentration of glycerol diacetate is 2 vol% based on the suspension volume. The viscosity was measured 30 min after the addition of glycerol diacetate and magnesium citrate. The viscosity of the suspension treated at 25 °C is similar to that with the addition of magnesium citrate alone and the value is around 4–7 Pa s. Viscosities of suspensions with 0.5–1.0 wt% magnesium citrate and 2 vol% glycerol diacetate increase to 32–45 Pa s when the treated temperature increases to 60 °C. The high viscosity transfers the suspensions into rigid coagulated bodies. The coagulating mechanism is that the increase in temperature promotes the hydrolysis of glycerol diacetate which lowers the pH value of the suspensions. The shift of pH value helps to decompose magnesium citrate and release magnesium ions. The free magnesium ions compress the electric double layer of PZT particles which decreases the repulsion potential energy of the system and, hence, increases the viscosity of the suspensions.

Properties of green bodies and PZT ceramics: It is indicated from the results above that 56 vol% PZT suspensions can be coagulated with the addition of 0.5–1.0 wt% magnesium citrate and 2 vol% glycerol diacetate at 60 °C. The coagulating time is in the range 60–90 min at 50–70 °C. Figure 8.86 shows the stress–strain behaviors of the wet-coagulated bodies, (a) with 1.0 wt% magnesium citrate at different coagulating temperatures and (b) with different magnesium citrate concentrations at 60 °C. The strength decreases with the increase of coagulating temperature but increases with the increase of magnesium citrate content. The strength is in the range 1.5–2.8 MPa which is sufficiently high for demolding and handling without deformation. Complex shape green bodies have been prepared by DCC–HVCI using magnesium citrate as coagulating agent with glycerol diacetate as a pH regulator (shown in Fig. 8.87). Smooth surface and no cracking are observed in the green samples. Figure 8.88 shows the microstructure of the fractured surface of PZT ceramics prepared by DCC–HVCI using 0.8 wt% magnesium citrate and 2 vol% glycerol diacetate. Homogenous microstructures have been obtained by this coagulation method

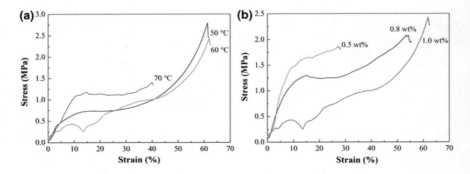

Fig. 8.86 Stress–strain behaviors of the wet-coagulated bodies, **a** with 1.0 wt% magnesium citrate at different coagulating temperatures and **b** with different magnesium citrate concentrations at 60 °C

8.1 Direct Coagulation Casting of Ceramic Suspension

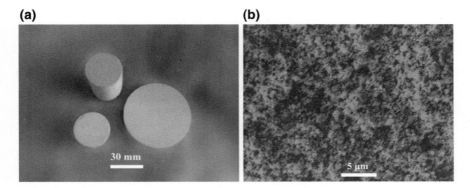

Fig. 8.87 **a** Photograph and **b** microstructure of green bodies prepared by DCC–HVCI using 0.8 wt% magnesium citrate and 2 vol% glycerol diacetate

Fig. 8.88 Microstructure of fractured surface of PZT ceramics prepared by DCC–HVCI using 0.8 wt% magnesium citrate and 2 vol% glycerol diacetate

which guarantees the good piezoelectric properties of the ceramics. Table 8.1 compares the properties of PZT ceramics prepared by different forming methods. The relative densities of ceramics prepared by DCC–HVCI and sintered at 1250–1280 °C are 99.2–99.7% of theoretical density of PZT which is higher than samples prepared by die pressing. PZT ceramics prepared by DCC–HVCI also show better piezoelectric properties compared with die pressing method, with relative permittivity of 2954–2699, dielectric loss of 2.3–2.7%, and d_{33} of 589–625 pC/N. The good properties of PZT suggest that the present DCC–HVCI method is a good colloidal method for preparing both structural and functional ceramics with high performance.

8.1.5 Direct Coagulation Casting via High Valence Counter-Ions from Chelation Reaction

(1) Direct coagulation casting of alumina via controlled release of calcium from ammonium polyphosphate chelate complex

The controlled release of high valence counter-ions to coagulate the concentrated suspensions in a controllable manner is the key technology for DCC–HVCI. To date, we have developed several methods for controlling the release of HVCIs. Calcium ions were controlled by the release from calcium iodate or reactions between calcium phosphate and hydrochloric acid by increasing the temperature which increased the viscosity of the concentrated alumina suspension (Xu et al. 2012a, b, c). However, the calcium is not suitable for some ceramic systems which limit the applications of the method. Thermo-sensitive liposomes were used to encapsulate HVCIs, and controlled release of HVCIs was achieved by a phase transition of the liposomes to coagulate the suspensions (Yang et al. 2013). Liposomes can encapsulate different salts for different ceramic systems and provide a wider application range of the method. However, the size of liposomes is ten times larger than that of ceramic particles which leads to defects in the microstructure that may deteriorate the mechanical properties of the sintered samples. More recently, a series work has been reported on DCC–HVCI of ceramics using citrate salts as coagulating agents assisted by glycerol diacetate (Xu et al. 2014a, b, 2015a, b). Different ceramic systems can be coagulated by changing citrate salt which contains different HVCIs. The method is one of the promising DCC–HVCI methods that can be widely applied. However, citrate salts used for coagulation are insoluble powders and the particle sizes are difficult to control. Usually, the particle sizes of ceramic powders used for suspension preparation in colloidal forming are in the submicrometer range. Therefore, coagulating agents with submicrometer size are preferred to avoid fewer aggregations and defects which could enhance the properties of the ceramics. Ammonium polyphosphate (APP) is a soluble phosphate salt with good chelating properties in the base region (van Wazer and Callis 1958). As an inorganic chelate complex, it is likely to decompose with the decrease of pH to acidic region and the increase of temperature during which part of the chelated cations is released.

In this section, a new colloidal forming method via controlled release of high valence counter-ions using APP chelate complex as a source of HVCI was proposed. The preparation procedure is shown in Fig. 8.89. Suspensions with 55 vol% alumina powder and TAC as dispersant of suspension with 0.3 wt% based on alumina powder was mixed in the deionized water and milled for 12 h in polyethylene containers. Zirconia balls with diameters of 5–10 mm were used as grinding media. The mass ratio between grinding media and alumina powder is 1:2. The chelate complex was prepared by mixing 20 wt% APP and 10–15 wt% $CaCl_2$ into the water to obtain chelate complex suspensions. The chelate complex suspensions were ball milled for different times. The alumina suspension was degassed under vacuum condition for 15 min. The APP chelate complex suspensions were added into the alumina suspensions and mixed thoroughly by continuing the tumbling process for another

8.1 Direct Coagulation Casting of Ceramic Suspension

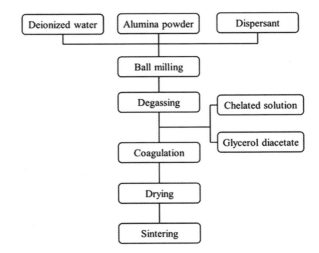

Fig. 8.89 Preparation procedure of ceramics via DCC–HVCI using APP chelate complex as the coagulating agent

15 min to obtain a homogeneous suspension. Then, glycerol diacetate was added into the suspension. The suspension was cast into a plastic mold. The samples were placed in a water bath at 40–70 °C for a period of time at a heating rate of 5 °C/min and then demolded. The green body was dried at 80 °C for 24 h. The dried samples were sintered at 1600 °C for 2 h at a heating rate of 5 °C/min.

Properties of APP chelate complex suspension: The dissolution characteristics of APP are mainly dependent on the length of the chain of APP. The molecular structure of APP is $(NH_4)_{n+2}P_nO_{3n+1}$, and it is soluble when the "n" value is in the range 10–20. It is observed that the solution of 20 wt% APP can dissolve completely without any particles suspended in the solution (Fig. 8.90). The complete dissolution of APP ensures the uniform distribution of chelate complex in the solution.

To investigate the chelating ability of APP, solutions of 20 wt% APP with 10 and 15 wt% calcium chloride were prepared. When calcium chloride was added into the solution of APP, a chelate complex forms rapidly through the reaction of chelation between APP and calcium. The chelation reaction can be expressed as follows:

$$2(NH_4)_{n+2}P_nO_{3n+1} + (n+2)Ca^{2+} \rightarrow Ca_{n+2}(P_nO_{3n+1})_2 \\ + (2n+4)NH_4^+ (10 < n < 20) \quad (8.19)$$

To investigate the chelating properties of APP, the free calcium was measured (Table 8.6). As calculated from Eq. (8.19), the concentration of calcium ions is 901 mM in the suspension with 20 wt% APP and 10 wt% calcium chloride. The concentration of free calcium ions is 2 mM in the chelated suspension after 12 h which is much lower than the critical coagulation concentration of calcium in alumina suspension (10 mM) (Xu et al. 2012a, c). It is indicated that most of the calcium ions were chelated in the complex. However, for a suspension with 20 wt% APP and 15 wt% calcium chloride, it was difficult to disperse homogeneously. Also, the concentration

Fig. 8.90 The solubility and chelating behavior of APP

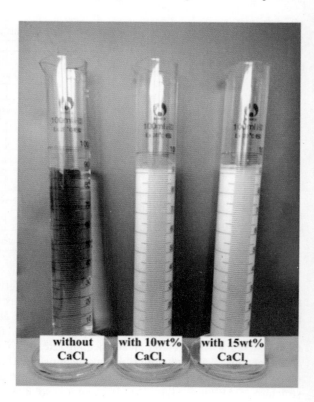

Table 8.6 Calcium ions in the chelate complex suspension

Concentration of calcium chloride	APP (wt%)	Theoretical calcium (mM)	Free calcium (mM)
10	20	901	2
15	20	1351	150

of free calcium ions is 150 mM in the chelated solution which is much higher than the critical coagulation concentration of calcium in alumina suspension, though most of the calcium ions were chelated. A chelate complex suspension prepared from 20 wt% APP and 10 wt% $CaCl_2$ was used in subsequent experiments.

The particle size and the concentration of free calcium ions in the chelated solution as the controlled coagulating agent are crucial factors in the DCC–HVCI process. However, it is difficult to find out a good method to solve the problem. Coagulation agents with particle sizes that are in the same range of ceramic particles which can be dispersed uniformly in the suspension are also acceptable. Figure 8.91a shows that the size of the chelate complex decreases gradually with the increase of the milling time. No evident decrease in the size of the chelate complex is observed after 48 h,

8.1 Direct Coagulation Casting of Ceramic Suspension

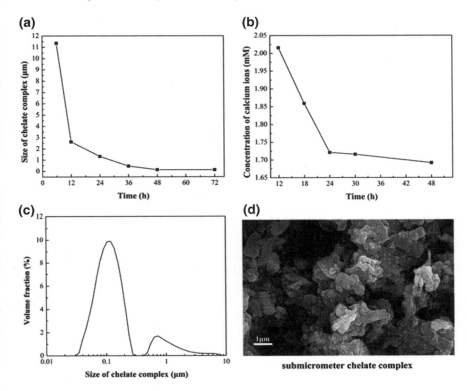

Fig. 8.91 **a** Size of the chelate complex and **b** concentration of calcium as milling time (time is the stirring time after mixing APP and $CaCl_2$); **c** size distribution and **d** microstructure of the chelate complex (milling time = 48 h)

with an average particle size of 0.13 μm. In addition, Fig. 8.91b demonstrates that the chelate complex suspension has a relatively low concentration of calcium ions after milling for 24 h and the concentration of free calcium ions was approximately 1.7 mM in the chelated solution. Therefore, a chelate complex milled for 48 h was used in subsequent experiments. Figure 8.91c, d, respectively, shows the size distribution as well as the microstructure of the chelate complex prepared with 20 wt% APP and 10 wt% $CaCl_2$ for 48 h. It is found that a submicrometer chelate complex has been obtained. The average particle size is 0.13 μm with a narrow particle size distribution which is at the same level of alumina particles (0.33 μm).

Controlled release of calcium and coagulation of alumina suspensions: It is indicated that 50–55 vol% alumina suspension can be coagulated by using 10–20 mM calcium ions (Xu et al. 2012a, c). For the chelate complex suspension prepared from 20 wt% APP and 10 wt% $CaCl_2$, theoretical calcium is 901 mM when all of the calcium ions are released from the chelate complex. In this sense, the concentration of calcium ions is 27 mM in the suspension with an addition of 3 vol% APP chelate complex which is enough for the coagulation of the suspensions. Therefore, the suspension with the addition of 3 vol% chelate complex suspension is suitable for the process.

To investigate the influence of the APP chelate complex suspension on the viscosity of 55 vol% alumina suspension, 3 vol% solutions were added into the alumina suspension at different temperatures (Fig. 8.92). The initial pH value of the suspension after adding APP chelate complex suspension is 10.5. The suspensions with chelate complex are stable at 25 °C, and the viscosity is about 1.7 Pa s. At 60 °C, the viscosity changes very slightly upon the addition of APP chelate complex suspension and increases from 1.7 to 3.2 Pa s after 2 h. The increase in viscosity is mainly attributed to the release of calcium ions from the chelate complex and then calcium ions compress the electric double layer of alumina particles. However, the suspension is still in a fluid state despite increasing the temperature. From the analysis above, the chelate complex is well dispersed in the alumina suspension even at a temperature of 60 °C. To our knowledge, glycerol diacetate is weak acid ester which can hydrolyze and produce acetic acid and glycerol in the alkaline region. The generation of acetic acid lowers the pH value of the solution. The hydrolysis reaction of glycerol diacetate is shown as follows (Flory 1953; Haas 1989):

$$C_7H_{12}O_5 + 2H_2O \rightarrow 2C_2H_4O_2 + C_3H_8O_3 \tag{8.20}$$

Figure 8.93 shows the pH values of 55 vol% alumina suspensions with different amounts of glycerol diacetate treated at different temperatures. The initial pH value is set at 10.6. The pH decreases to approximately 8.0–8.5 at 60 °C for 2 h, while the value is 9.6 at 25 °C. A higher concentration of glycerol diacetate is beneficial to decrease the pH value. It is also indicated that the hydrolysis rate can be promoted by increasing the temperature. To get a lower pH value, higher temperature is preferred.

Figure 8.94 shows the viscosities of 55 vol% alumina suspensions with 3–5 vol% glycerol diacetate treated at 25 and 60 °C. It is observed that the viscosities of

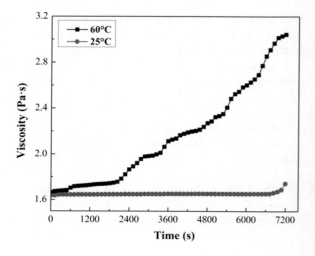

Fig. 8.92 Viscosity (shear rate = 10 s^{-1}) of 55 vol% alumina suspension with 3 vol% APP chelate complexes treated at 25 and 60 °C

Fig. 8.93 pH values of 55 vol% alumina suspensions with different amounts of glycerol diacetate treated at different temperatures

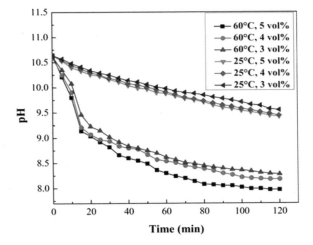

Fig. 8.94 Viscosities (shear rate = 10 s^{-1}) of 55 vol% alumina suspensions with different amounts of glycerol diacetate at different temperatures

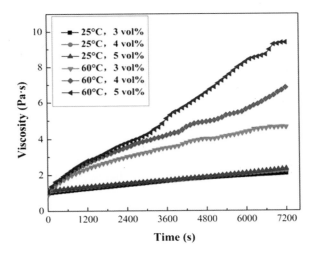

suspensions at 25 °C increase to approximately 2.0 Pa s after holding for 2 h. The viscosities of suspensions increase to 4.1, 6.2, and 9.6 Pa s with 3–5 vol% glycerol diacetate at 60 °C which is higher than the viscosity at 25 °C. The increase in viscosity is mainly attributed to the hydrolysis of glycerol diacetate which lowers the pH value to approach the IEP of alumina with absorbed TAC (approximately pH = 4). However, suspensions with the addition of glycerol diacetate are still in a fluid state.

To control further release of calcium ions in the chelated solution to increase the viscosity of the alumina suspension, APP chelate complex suspension and glycerol diacetate are added simultaneously into the suspensions. Figure 8.95a shows the viscosity of 55 vol% alumina suspensions with the addition of 3 vol% APP chelate

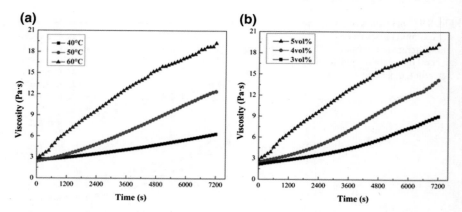

Fig. 8.95 **a** Viscosities (shear rate = 10 s^{-1}) of 55 vol% alumina suspensions with 3 vol% APP chelate complex suspension and 5 vol% glycerol diacetate at different temperatures. **b** Viscosities (shear rate = 10 s^{-1}) of 55 vol% alumina suspensions with 3 vol% APP chelate complex suspension and different concentrations of glycerol diacetate (GDA) at 60 °C

complex suspension and 5 vol% glycerol diacetate at different temperatures. It was found that the growth rate of suspension viscosity was accelerated by the rise of temperature. The suspension is still in a fluid state at 40 °C when the viscosity of the suspension increases to 6 Pa s after holding for 2 h. The viscosities at 50 and 60 °C increase remarkably and reach 12 and 19 Pa s, respectively; within 2 h, the suspension can transfer to a semisolid state. However, it is difficult to coagulate the suspension at 50 °C at a rapidly heating rate. The samples tend to show inhomogeneous and cracking phenomenon. Therefore, there is a great relationship between the heating temperature and the coagulation time. With the increase of coagulation temperature, the coagulation time is getting shorter. Glycerol diacetate can rapidly hydrolyze at higher temperature which promotes the release of calcium ions from the chelate complex. When the pH value shifts to a lower value, the chelate complex decomposes as follows:

$$Ca_{n+2}(P_nO_{3n+1})_2 + (2n+4)H^+ \rightarrow 2H_{n+2}P_nO_{3n+1} + (n+2)Ca^{2+} \quad (8.21)$$

Figure 8.95b shows the viscosities of 55 vol% alumina suspensions with 3 wt% APP chelate complex suspension and different proportions of glycerol diacetate treated at 60 °C. For a suspension with 3 vol% APP chelate complex suspension and 3 vol% glycerol diacetate, the viscosity increases to 9 Pa s for 2 h which is not high enough to coagulate the suspension. The increase in viscosity is mainly attributed to the release of calcium ions from the chelate complex. The increase of glycerol diacetate can effectively increase the viscosity of the suspension. It is attributed to the hydrolysis of glycerol diacetate which promote the decomposition of the chelate complex at lower pH value. The suspension by the addition of 3 vol% APP chelate

8.1 Direct Coagulation Casting of Ceramic Suspension

complex and 5 vol% glycerol diacetate at 60 °C is most suitable for coagulation and wet-coagulated samples can be more easily demolded.

Properties of green bodies and ceramics: Fig. 8.96 illustrates the de-moldable time and the compressive strength of wet green bodies which are prepared from DCC–HVCI using 3 vol% APP chelate complex suspensions and 5 vol% glycerol diacetate at different temperatures. The samples of diameter 30 mm and height 37 mm are used in the following experiment. The demoldable time almost linearly decreases with the increase of temperature, whereas accelerated reaction rates were observed above 70 °C, leading to expedited demoldable time. At a temperature of 60 °C, the coagulation time reduces to 2 h, in which the production could achieve a good performance. The compressive strength of the wet green bodies decreases slightly with increasing the temperature. The strength is in the range between 1.8 and 3.7 MPa. High strength of wet-coagulated samples prepared by the present method has enabled us to fabricate near-net-shape ceramics.

Complex-shaped and large size wet-coagulated samples have been prepared by this new method which is shown in Fig. 8.97a, b. The coagulated samples are dried at 80 °C for 24 h and a burnout process can be avoided. Figure 8.97c shows the microstructure of the green body prepared from 55 vol% suspensions with 3 vol% APP chelate complex suspension and 5 vol% glycerol diacetate coagulated at 60 °C. A homogeneous microstructure is observed in the sample without large pores. After being entirely dried in an oven, the green body was sintered at 1600 °C for 3 h to obtain a component. Table 8.7 compares the density of ceramics prepared from different particle sizes of chelate complex. It is observed that the reduction of particle size increases the density of the ceramics which may be attributed to the fewer large

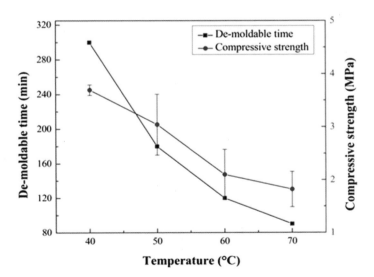

Fig. 8.96 The demoldable time and compressive strength of wet green bodies from 55 vol% alumina suspensions coagulated at different temperatures

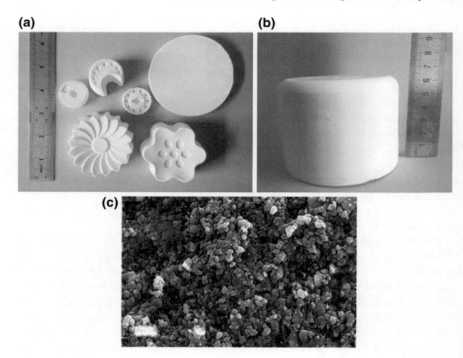

Fig. 8.97 **a, b** Photographs and **c** microstructure of green bodies prepared from 55 vol% alumina suspension with 3 vol% APP chelate complex suspension and 5 vol% GDA coagulated at 60 °C

Table 8.7 Relationship between the particle size of the chelate complex and density of ceramics

Milling time (h)	Particle size (μm)	Relative density (%)
24	1.32	92.74
36	0.46	95.48
48	0.13	97.86

defects in the ceramics caused by the coagulating agent. The ceramics prepared from 55 vol% suspensions with 3 vol% APP chelate complex suspension milled for 48 h and 5 vol% glycerol diacetate show a relative density of 97.86%, and the flexural strength was compacted at 388 ± 23 MPa and Weibull modulus of 14. The flexural strength prepared by this method is lower than that of other DCC methods because of lower relative density, yet higher than that by gel casting (Gauckler et al. 1999a; Binner et al. 2006; Graule et al. 1995; He et al. 2011). The SEM photograph of sintered samples is displayed in Fig. 8.98. Alumina ceramics show relatively uniform microstructure without large pores.

Fig. 8.98 Microstructure of alumina ceramics sintered at 1600 °C

(2) **Reliable high strength alumina fabricated by DCC–HVCI using submicron calcium citrate complex**

In this section, a modified DCC–HVCI method for alumina suspensions via controlled release of calcium from calcium citrate complex assisted by pH shift in the presence of glycerol diacetate is investigated.

The preparation procedure is shown in Fig. 8.99. First, different amounts of tri-ammonium citrate and calcium chloride were added into the water to obtain chelated suspensions. Then alumina powder was added to the chelated suspensions to prepare 50 vol% alumina suspensions after milling for 12–60 h in polyethylene containers. The compositions of alumina suspensions are listed in Table 8.8. Tri-ammonium citrate was used as chelating agent and dispersant. Part of tri-ammonium citrate reacted with calcium chloride to form a complex. An excess of 0.3 wt% tri-ammonium citrate based on alumina powder was designed to act as the dispersant of the suspensions. The pH value of the suspensions was adjusted to 10.5 using ammonia. The alumina

Fig. 8.99 Preparation procedure of alumina via DCC–HVCI using submicron calcium citrate complex

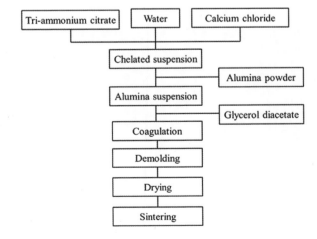

Table 8.8 Compositions of alumina suspensions with different amounts of tri-ammonium citrate and calcium chloride

No.	Al$_2$O$_3$ (g)	H$_2$O (g)	CaCl$_2$ (g)	TAC (g)	Solid loading (vol%)	Calcium (mM)
1	800	200	1.33	4.36	50	30
2	800	200	1.55	4.68	50	35
3	800	200	1.78	5.02	50	40

suspensions were degassed under vacuum condition for 20 min. Then, glycerol diacetate was added into the suspensions and mixed thoroughly for another 15 min to improve the homogeneity of the suspension. The suspensions were cast into a plastic mold. The samples were placed into a water bath at 40–70 °C at a heating rate of 5 °C/min and held for a period of time. The wet-coagulated samples were demolded and dried at 80 °C for 24 h. The green bodies were sintered at 1550 °C for 2 h at a heating rate of 5 °C/min.

Properties of chelated complex suspensions: In order to investigate the chelating behavior of tri-ammonium citrate and calcium chloride, 4.36 g tri-ammonium citrate and 1.33 g calcium chloride were added into 200 ml water (Suspension no. 1 in Table 8.8). The experiment was carried out at 25 °C. Upon mixing, following reactions between tri-ammonium citrate and calcium chloride occur (Wang and Wu 1960):

$$2(NH_4)_3C_6H_5O_7 + 3CaCl_2 \rightleftharpoons Ca_3(C_6H_5O_7)_2 + 6NH_4Cl \quad (8.22)$$

$$Ca_3(C_6H_5O_7)_2 + 2H_2O \rightleftharpoons 2CaHC_6H_5O_7 + Ca(OH)_2 \quad (8.23)$$

It is indicated that calcium is chelated by tri-ammonium citrate and forms calcium citrate and calcium hydrogen citrate complexes. It is observed that the mixture is still clear after 1 h, suggesting the slow reaction rates between tri-ammonium citrate and

Fig. 8.100 Complex suspensions prepared by tri-ammonium citrate and calcium chloride with different reaction times, **a** 1 h, **b** 12 h, and **c** 48 h

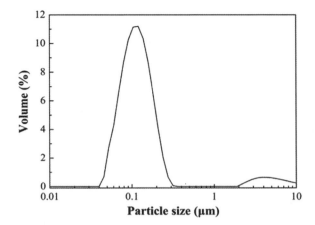

Fig. 8.101 Particle size distribution of calcium citrate complex with a reaction time of 48 h

calcium chloride at room temperature, as shown in Fig. 8.100a. After 12 h as shown in Fig. 8.100b, the mixture becomes slightly turbid, indicating the formation of a complex. When the treatment time is 48 h, the mixture is highly turbid with white color (Fig. 8.100c). The particle size distribution of calcium citrate complex has been measured. Figure 8.101 shows the particle size distribution of calcium citrate complex with a reaction time of 48 h. The complex size shows a sharp peak distribution with an average particle size of 129 nm, although there are some agglomerations with particle sizes in the range of 2–10 μm. It is indicated that submicron calcium citrate complex can be prepared with a reaction time of 48 h. The small particle size of the complex ensures the homogeneity of the suspension which is beneficial to the ceramics.

Effect of milling time on the viscosity of alumina suspensions: In order to investigate the effect of milling time on the viscosity of alumina suspensions, the viscosity measurement was taken every 12 h from 12 to 60 h. As shown in Fig. 8.102a, the viscosity of the suspensions decreases with milling time and tends to stable after 48 h. It is also noted that the viscosity increases with the increase of calcium concentration. For suspensions No. 1 in Table 8.8, the initial concentration of calcium in the suspension is 30 mM. The concentrations of free calcium in the suspensions after treating 1 h, 12 h, and 48 h are 25 mM, 14 mM, and 6 mM, respectively. Most of the calcium has been chelated by tri-ammonium citrate after 48 h, and the concentration of free calcium is lower than the critical coagulation concentration of calcium in 50 vol% alumina suspension (Xu et al. 2012a). The viscosities of suspensions with 30, 35, and 40 mM calcium are 0.43, 0.54, and 0.86 Pa s, respectively. The viscosities are all below 1 Pa s, which is suitable for the colloidal forming process. As shown in Fig. 8.102b, the viscosities of the suspensions show a shear-thinning behavior and then exhibit a slight shear-thickening behavior. The critical shear rate of the suspension decreases with the increase of calcium concentration. The critical shear rates of suspensions with 30, 35, and 40 mM calcium are 774, 494, and 315 s^{-1}, respectively. The phenomenon can be attributed to an order–disorder transition of the alumina particle microstructure (Bergström 1998). Shear-thinning behavior indicates that the

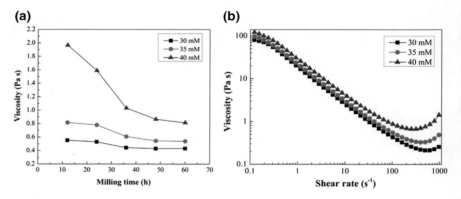

Fig. 8.102 Viscosities of 50 vol% alumina suspensions with different concentrations of calcium, **a** effect of milling time on the viscosity at shear rate of $100\ s^{-1}$, **b** the viscosity at different shear rates with a milling time of 48 h

flow brings about a more favorable two-dimensional structure arrangement of the particles. Due to the broken of the unstable two-dimensional arrangement above the critical shear rate, the viscosity increases again and the suspension exhibits a shear-thickening behavior (Cai et al. 2005). It is suggested that the increase of calcium leads to the instability of the suspension. Also, the increase in viscosity at a higher shear rate indicates that alumina particles in these suspensions are easy to agglomerate. Hence, the suspensions should be handled at a lower shear rate to ensure the homogeneity of the suspensions, hence to improve the properties of the final ceramics. A shear rate of $10\ s^{-1}$ will be selected in the subsequent experiments.

Coagulation of alumina suspensions: To examine the stability of the alumina suspensions, the suspensions were treated at 25 and 60 °C, respectively. As shown in Fig. 8.103a, the viscosity of the alumina suspensions is very stable at 25 °C. The

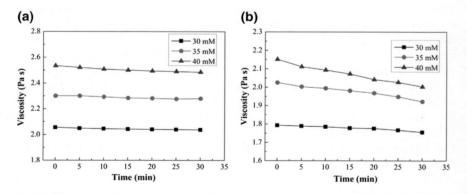

Fig. 8.103 Viscosities of alumina suspensions with different concentrations of calcium at different temperatures, **a** 25 °C and **b** 60 °C

8.1 Direct Coagulation Casting of Ceramic Suspension

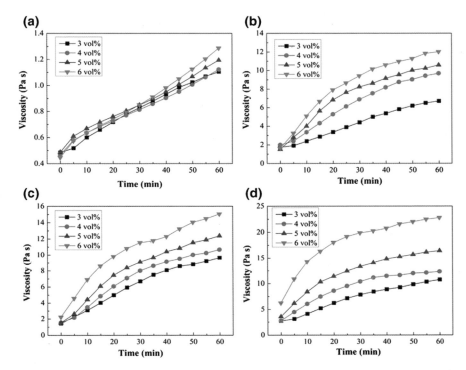

Fig. 8.104 Viscosities of alumina suspensions with different concentrations of calcium and glycerol diacetate at 60 °C, **a** without calcium, **b** 30 mM calcium, **c** 35 mM calcium, and **d** 40 mM calcium

viscosities are 2.05, 2.31, and 2.53 Pa s for suspensions with 30, 35, and 40 mM calcium, respectively. The viscosity of the suspensions decreases slightly when the temperature increases to 60 °C (Fig. 8.103b). The increase of temperature leads to the evaporation of ammonia in the suspension which decreases the pH value of the suspension. The initial pH value of alumina suspensions is 10.5. The pH value of the suspensions is after treating at 60 °C for 30 min. It is indicated in our previous research that the viscosity of alumina suspension tends to be lower at pH between 9 and 10 than that above 10 (Xu et al. 2014b). However, the change is not significant compared with that at room temperature. In this sense, heat treatment alone cannot lead to the coagulation of alumina suspension, indicating the stable state of calcium citrate complex in the suspension at 25 and 60 °C.

It is noted that calcium citrate and magnesium citrate have been applied to coagulate alumina and zirconia suspensions assisted by glycerol diacetate (Xu et al. 2014a, b, 2015a, b). Calcium citrate and magnesium citrate are stable with minor solubility at pH above 10. The hydrolysis of glycerol diacetate leads to the shift of pH to a lower value which helps the decomposition of calcium citrate and magnesium citrate. Free calcium or magnesium compresses the electrical double layer of the particles, leading to the in-situ coagulation of the suspension. Figure 8.104 shows the viscosities of alumina suspensions with different concentrations of calcium and glycerol diacetate at 60 °C. Figure 8.104a shows the influence of glycerol diacetate on the viscosity of suspensions without calcium. It is observed that the viscosity increases slightly from 0.45–0.53 Pa s to 1.13–1.35 Pa s, indicating that the addition of glycerol diacetate alone cannot result in the coagulation of the suspensions. For suspension with 30 mM calcium (Fig. 8.104b), the initial viscosity is 1.84 Pa s. The viscosity increases with the increase of glycerol diacetate amount and the treating time. The viscosity reaches 6 Pa s when the concentration of glycerol diacetate is 3 vol% after 60 min. The viscosity of suspension with 6 vol% increases to 12 Pa s for 1 h. The increase of viscosity with the increase of glycerol diacetate can be attributed to two factors. On one hand, the hydrolysis of glycerol diacetate leads to the decomposition of calcium citrate complex, hence compressing the electric double layer which is the dominating factor for the coagulation of the suspension. On the other hand, the shift of pH to lower value approaches the IEP of alumina absorbed with citrate (IEP = 4–5). The surface potential of the particles decreases which leads to the coagulation of the suspension. However, the latter is much weaker as the pH value is not low enough (pH = 7–8). The case of suspensions with 35 and 40 mM calcium is similar to that with 30 mM, as shown in Fig. 8.104c, d. Higher concentration of glycerol diacetate promotes the coagulation of the suspension. It is noted the initial viscosity of suspension with 40 mM calcium is 3–5 Pa s which is much higher than the other two. Also, with the increase of calcium concentration, the viscosity of the suspensions with equal glycerol diacetate content increases. The alumina suspensions can be completely coagulated within 2–3 h at 60 °C, with more time at lower temperatures (40–50 °C).

Properties of green body and alumina ceramics: In order to investigate the coagulating temperature on the properties of wet-coagulated samples, 50 vol% alumina suspension with 35 calcium was chosen. As shown in Fig. 8.105, the compressive strength and linear shrinkage of wet-coagulated body decrease with the increase of coagulating temperature. The wet strength and linear shrinkage are in the range 1.1–2.4 MPa and 0.9–2.0%, respectively. Higher strength can be obtained at a lower coagulating temperature as larger shrinkage occurs to enhance the green density. However, longer coagulating time is necessary to accomplish, which is a drawback for the process. The suspensions tend to be unstable for a long coagulation time. Also, large shrinkage may increase the possibility of cracking during the coagulation process. Yet, high evaporation of the water from the suspension is observed at a coagulating temperature above 70 °C. It is found that the suspension is most suitable to coagulate at 60 °C, with coagulating time, compressive strength, and shrinkage of 2 h, 2.2 MPa, and 1.1%, respectively. Figure 8.106 shows the complex shape green bodies and alumina ceramics prepared by 50 vol% alumina suspension with 35 mM calcium and 5 vol% glycerol diacetate. It is observed that the complex shape parts with a smooth surface are obtained. No cracking or deformation is found in both green bodies and ceramics which can be attributed to the homogeneous coagulation of the suspension and high wet strength of green bodies.

8.1 Direct Coagulation Casting of Ceramic Suspension

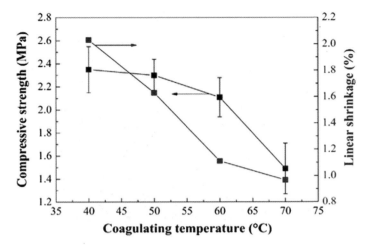

Fig. 8.105 Compressive strength and linear shrinkage of wet-coagulated bodies at different coagulating temperatures

Figure 8.107 shows the microstructures of the green bodies prepared by 50 vol% alumina suspensions with (a) 30, (b) 35, (c) 40 mM calcium, and 5 vol% glycerol diacetate. No agglomeration or large pore has been observed in the green body. There is no great difference in the microstructures of green bodies coagulated by suspensions with different concentrations of calcium. However, there are some differences in the microstructures of the ceramics. Figure 8.108 shows the microstructures of alumina ceramics prepared by 50 vol% alumina suspensions with different concentrations of calcium and 5 vol% glycerol diacetate. Although no large pores are observed in all ceramics, alumina prepared by 50 vol% alumina suspensions with 35 mM calcium and 5 vol% glycerol diacetate shows a denser microstructure. More pores are observed when the concentration of calcium increases to 40 mM which is attributed to the higher initial viscosity of the suspension. Table 8.9 compares the properties of alumina prepared by 50 vol% alumina suspensions with different concentrations of calcium and 5 vol% glycerol diacetate. The shrinkage of the ceramics is in the range 18.4–18.93% which is typical shrinkage for ceramics prepared by 50 vol% alumina suspension. The density of alumina increases with the increase of calcium and then decreases. Alumina prepared by 50 vol% alumina suspensions with 35 mM calcium and 5 vol% glycerol diacetate shows higher density and flexural strength, 3.92 g/m^3 and 455 ± 17 MPa, respectively. Figure 8.109 shows the Weibull analysis for alumina ceramics prepared by DCC–HVCI with different concentrations of calcium. High Weibull modulus of 30 has been obtained by 50 vol% alumina suspensions with 35 mM calcium and 5 vol% glycerol diacetate, indicating that reliable ceramics can be fabricated by this new method. However, the increase of calcium (40 mM) leads to lower flexural strength and poor reliability because of the high initial viscosity of the suspension, with flexural strength and Weibull modulus of 421 ± 58 MPa and 8, respectively. Table 8.10 summarizes the mechanical properties of alumina ceramics

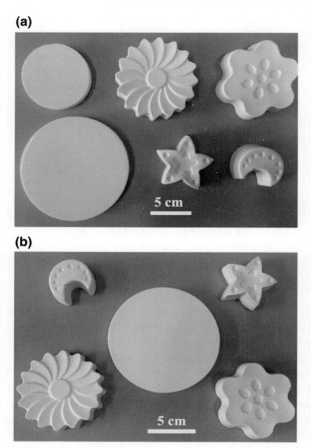

Fig. 8.106 a Complex shape green bodies and **b** alumina ceramics prepared by 50 vol% alumina suspension with 35 mM calcium and 5 vol% glycerol diacetate

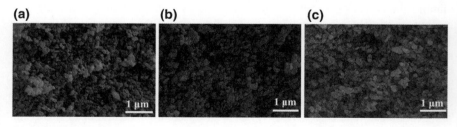

Fig. 8.107 Microstructures of green body prepared by 50 vol% alumina suspension with **a** 30, **b** 35, **c** 40 mM calcium and 5 vol% glycerol diacetate

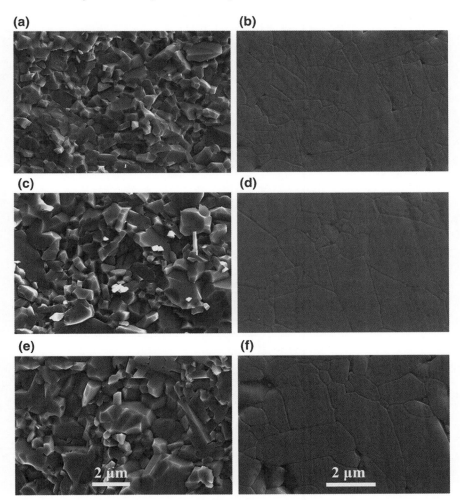

Fig. 8.108 Microstructures of alumina ceramics prepared by 50 vol% alumina suspensions with different concentrations of calcium and 5 vol% glycerol diacetate, **a** and **b** 30 mM, **c** and **d** 35 mM, **e** and **f** 40 mM

Table 8.9 Properties of alumina prepared by DCC–HVCI using calcium citrate complex

Calcium (mM)	Shrinkage (%)	Density (g/cm^3)	Strength (MPa)	Weibull modulus
30	18.42	3.91	444 ± 34	14
35	18.93	3.92	455 ± 17	30
40	18.78	3.88	421 ± 58	8

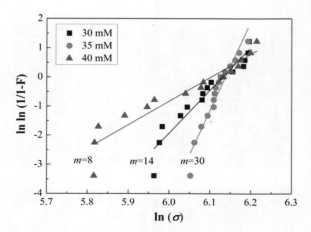

Fig. 8.109 Weibull analysis for alumina ceramics prepared by 50 vol% alumina suspensions with different concentrations of calcium and 5 vol% glycerol diacetate

Table 8.10 Comparison of the mechanical properties of alumina ceramics fabricated by different colloidal forming methods

Process	Density (g/cm³)	Strength (MPa)	Weibull modulus	References
Aqueous injection molding, agar	3.81	240	6.2	Rak and van Tilborg (1991)
Slip casting	3.92	422	–	Davies and Binner (2000)
Centrifugal slip casting	3.97	540	24	Huisman et al. (1995)
Gel casting, egg white	–	314	13.4	He et al. (2011)
Gel casting, curdlan	3.92	347	–	Xu et al. (2015a, b)
In-situ coagulation molding	3.92	451	10	Binner et al. (2006)
Direct coagulation casting via pH shift	3.97	485	8.5	Graule et al. (1995)
Direct coagulation casting and hot isostatic pressing	3.98	683	47	Gauckler et al. (1999a)
Direct coagulation casting via HVCI	3.92	455	30	Current work

8.1 Direct Coagulation Casting of Ceramic Suspension

fabricated by different colloidal forming methods (Gauckler et al. 1999a; Binner et al. 2006; Rak and van Tilborg (1991); Davies and Binner 2000; Huisman et al. 1995; He et al. 2011; Xu et al. 2015a, b; Graule et al. 1995). The strength of alumina is the range 240–540 MPa using pressureless sintering method. High strength of 683 MPa was obtained by direct coagulation casting and hot isostatic pressing (Gauckler et al. 1999a). It is observed that the Weibull modulus of alumina is below 10 prepared by most of the colloidal forming methods. However, high Weibull modulus ($m = 30$) is achieved by our present method besides direct coagulation casting and hot isostatic pressing ($m = 47$), indicating that DCC–HVCI can prepare high reliable ceramics with moderate strength. The facile coagulation method can be applied to prepare various ceramic systems by changing the high valence cations.

(3) **Highly reliable yttria-stabilized zirconia prepared by DCC–HVCI using calcium polyphosphate**

Yttria-stabilized zirconia (YSZ), also known as tetragonal zirconia polycrystals (TZP), possesses excellent biocompatibility, wear resistance, high chemical and corrosion resistance, and mechanical properties which make it a strong candidate for biomedical and structural applications (Evans and Cannon 1986; Laberty-Robert et al. 2003; Ozkurt and Kazazoğlu 2010). However, high-performance and reliable complex-shaped ceramic products are highly demanded that traditional shaping methods could not meet. It is widely recognized that colloidal processing could solve the problems by preparing ceramics from homogeneous suspensions (Alford et al. 1987; Lewis 2000). In the past two decades, several colloidal processing methods (gel casting, direct coagulation casting, temperature-induced forming, etc.) for advanced ceramics have been developed (Young et al. 1991; Gauckler et al. 1999a; Ewais et al. 2002; Franks and Lange 1996). Among them, direct coagulation casting (DCC) is considered to be an excellent method for reliable ceramics due to binder free in the green bodies which could enhance the mechanical properties of the ceramics (Baader et al. 1996a, b). Alumina ceramics with high Weibull modulus ($m = 47$) has been prepared by pH shift method (Graule et al. 1995). However, there are some drawbacks such as low wet strength, long coagulation time, and easy cracking which have prevented DCC from industrial application (Yang et al. 2002).

Prabhakaran et al. (2008a, b) reported a novel near-net-shape direct coagulation casting method using MgO. Mg^{2+} ion reacted with ammonium poly (acrylate) which lead to insufficient dispersant surface coverage and the coagulation of suspension. The coagulation time was less than 30 min at 60 °C. The compressive strength of the coagulated samples increased up to 0.2 MPa. In recent years, a modified DCC method via controlled release of high valence counter-ions with the combination of the Derjaguin–Landau–Verwey–Overbeek (DLVO) theory and the Schulze–Hardy rule has been developed (Xu et al. 2012c). High valence counter-ions were controlled release via increasing temperature or pH shift which can destabilize high solid loading ceramic suspensions and near-net-shape ceramic components have been prepared (Xu et al. 2014a b). To increase the homogeneity of the suspension during the coagulation process, submicron chelate complexes have been used as coagulating agents. With ammonium polyphosphate (APP) chelate complex as the coagulating agent, alumina ceramics with Weibull modulus of 14 has been obtained (Yang et al. 2016). More excitingly, alumina ceramics with high Weibull modulus (m = 30) has been prepared using calcium citrate complex assisted by pH shift in the presence of glycerol diacetate (Xu et al. 2016). The above results indicate that DCC–HVCI using submicron chelate complex as coagulating agent is a promising method for preparing highly reliable near-net-shape ceramics.

In this section, a new colloidal forming method via controlled release of high valence counter-ions using calcium polyphosphate chelate complex as a source of HVCI was proposed. Submicrometer coagulating agent was prepared by sodium tripolyphosphate (STPP) and calcium chloride to ensure the good dispersion of the coagulating agent in the suspension. The DCC–HVCI process for suspension with the addition of coagulating agent is controlled by shifting the pH value and increasing temperature. The process is based on the concept that the hydrolysis of glycerol diacetate assists the release of calcium ions from chelate complex which compress the electric double layer of ceramic particles, leading to the destabilization of the suspension. The effects of chelate complex and glycerol diacetate on the rheological behaviors of YSZ suspensions were studied. The properties of the coagulated samples and ceramics have been characterized.

YSZ suspensions were prepared using YSZ powder, tri-ammonium citrate, and deionized water milling for 12 h in polyethylene containers. Zirconia balls with diameters of 5–10 mm were used as grinding media. The mass ratio between grinding media and YSZ powder is 1:2. The chelate complex suspension was prepared by mixing 10 wt% STPP and 5 wt% $CaCl_2$ in water for 24 h to obtain a homogeneous state. YSZ suspension was degassed under vacuum for 20 min to remove the bubbles. STPP chelate complex suspensions were then added to YSZ suspensions and mixed thoroughly by continuing the tumbling process for another 15 min to obtain a homogenous suspension. Then, glycerol diacetate was added into the suspension. The suspension was cast in a plastic mold. The samples were placed in a water bath at 40–70 °C for a period of time with a heating rate of 5 °C/min and then demolded. The green body was dried at 80 °C for 24 h. The dried samples were sintered at 1450 °C for 3 h at a heating rate of 5 °C/min.

It is reported that there are several dispersants for YSZ suspensions (Greenwood and Kendall 1999; Wang et al. 1999; Rao et al. 2007; Briscoe et al. 1998). Among them, tri-ammonium citrate stabilizes YSZ suspensions via electrostatic stabilization mechanism. Citrate adsorbs onto the particle surface and forms a unimolecular layer at the optimum concentration. Figure 8.110 shows the effect of tri-ammonium citrate addition on the viscosity of the 30 vol% aqueous YSZ suspensions. The suspension showed a minimum viscosity of 0.05 Pa s with 0.2 wt% tri-ammonium citrate. The viscosity of the suspension increased marginally with further addition of tri-ammonium citrate and reached 0.4 Pa s with 0.8 wt% tri-ammonium citrate. With tri-ammonium citrate addition of 0.1 wt%, citrate cannot completely cover the surface of YSZ particles and causes agglomeration and flocculation. With tri-ammonium citrate addition of 0.2 wt%, citrate sufficiently covers the surface of the particles resulting in lower viscosity. However, excessive tri-ammonium citrate will lead to high ionic strength in suspension which compresses the electric double layer

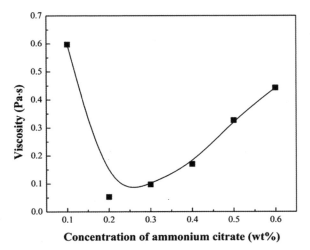

Fig. 8.110 Influence of tri-ammonium citrate on the viscosity of yttria-stabilized zirconia suspension

between ceramic particles, leading to the increase in viscosity of YSZ suspension. Based on the analysis above, the optimum concentration of tri-ammonium citrate is selected as 0.2 wt%. 50 vol% YSZ suspension with viscosity of 0.32 Pa s has been prepared by adding 0.2 wt% tri-ammonium citrate at pH = 9.5 which is suitable for DCC process.

To examine the chelating ability of sodium tripolyphosphate, solutions of 10 wt% sodium tripolyphosphate and 5 wt% calcium chloride were prepared. When calcium chloride is added into the solution of sodium tripolyphosphate, calcium polyphosphate forms through the reaction between sodium tripolyphosphate and calcium chloride which is expressed as follows:

$$2Na_5P_3O_{10} + 5CaCl_2 \rightarrow Ca_5(P_3O_{10})_2 + 10NaCl \qquad (8.24)$$

To get low free calcium concentration in the suspension, the suspension is milled for different times. Figure 8.111 shows the influence of milling time on the concentration of calcium and particle size of calcium polyphosphate. It is observed that the free calcium concentration decreases with the increase in milling time. The concentration of free calcium ions is 6 mmol/l in the suspension after 12 h which is lower than the critical coagulation concentration of calcium in YSZ suspension. It is also found that submicron calcium polyphosphate has been obtained after milling for 10 h. The average particle size is 0.18 μm which is at the same level of YSZ particles (0.13 μm).

It has been proved that ceramic suspensions could be coagulated with 20 mmol/l calcium ions (Xu et al. 2014a). Theoretically, for calcium polyphosphate suspension prepared from 10 wt% STPP and 5 wt% calcium chloride, the concentration of calcium is 450 mmol/l when all of the calcium ions are released from the chelate complex. In this sense, 2–4 vol% calcium polyphosphate suspension based on the

Fig. 8.111 Influence of milling time on the concentration of calcium and particle size of calcium polyphosphate

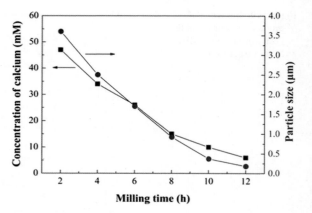

volume of YSZ suspension is enough for in-situ coagulation casting. Figure 8.112 shows the viscosities of YSZ suspensions with 3 vol% calcium polyphosphate suspension at different temperatures. The suspension is stable at 25 °C, and the viscosity is in the range 2.2–2.5 Pa s. At 60 °C, the viscosity increases to 7.5 Pa s after 14 min. The increase in viscosity is mainly attributed to the release of calcium ions from the chelate complex, and then calcium ions compress the electric double layer which decreases the zeta potential of the suspension. However, the suspension is still in a fluid state despite increasing the temperature to 60 °C.

Glycerol diacetate is weak acid ester which can hydrolyze and produce acetic acid and glycerol in the alkaline region (Flory 1953). The generation of acetic acid leads to lower pH value of the solution. Figure 8.113 shows the viscosity of YSZ suspension with 1 vol% glycerol diacetate treated at 25 and 60 °C. It is observed

Fig. 8.112 Viscosities of yttria-stabilized zirconia suspensions with 3 vol% sodium tripolyphosphate chelate complex suspension at different temperatures

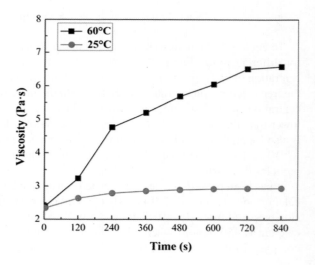

8.1 Direct Coagulation Casting of Ceramic Suspension

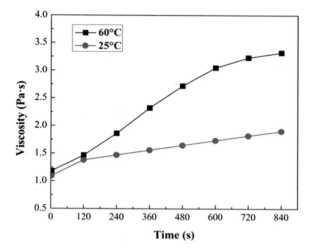

Fig. 8.113 Viscosities of yttria-stabilized zirconia suspensions with 1 vol% glycerol diacetate at different temperatures

that the viscosity of suspension at 25 °C increases to ca. 1.5 Pa s after 14 min. The viscosity of suspension increases to 3.3 Pa s at 60 °C which is higher than the viscosity at 25 °C. The increase in viscosity is mainly attributed to the hydrolysis of glycerol diacetate which lowers the pH value to approach the IEP of YSZ with absorbed TAC (*ca.* pH = 4.5). However, the solely addition of glycerol diacetate could not lead to the coagulation of YSZ suspension.

To coagulate the YSZ suspension, calcium polyphosphate suspension and glycerol diacetate are added together. Figure 8.114 shows the viscosity of suspension with the addition of 3 vol% calcium polyphosphate suspension and 3 vol% glycerol diacetate at different temperatures. It is observed that the viscosities of suspensions at 40 and

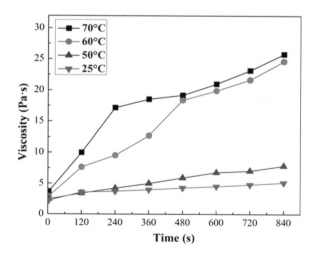

Fig. 8.114 Viscosities of yttria-stabilized zirconia suspensions with 3 vol% calcium polyphosphate suspension and 1 vol% glycerol diacetate at different temperatures

Fig. 8.115 Viscosities of yttria-stabilized zirconia suspensions with different concentrations of calcium polyphosphate at 60 °C

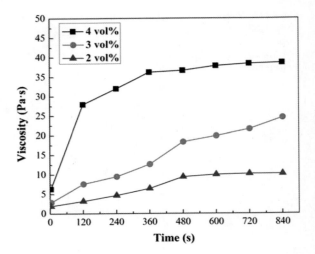

50 °C increase to 4–7 Pa s after holding for 14 min. The suspension is still in a fluid state. The cases at 60 and 70 °C are rather different. The viscosity increases remarkably and reaches 20.0 Pa s within 14 min which makes the suspension transfer to a semisolid state. Glycerol diacetate can rapidly hydrolyze at higher temperature which promotes the release of calcium ions from calcium polyphosphate. When the pH value shifts to a lower value, calcium polyphosphate decomposes as follows:

$$Ca_5(P_3O_{10})_2 + 10H^+ \rightarrow 2H_5P_3O_{10} + 5Ca^{2+} \quad (8.25)$$

Figure 8.115 shows the viscosities of YSZ suspensions with different concentrations of calcium polyphosphate suspension at 60 °C. The concentration of glycerol diacetate is 1 vol%. For YSZ suspension with 2 vol% calcium polyphosphate suspension, the viscosity increases to 6 Pa s for 14 min which is not high enough to coagulate the suspension. The increase in calcium polyphosphate suspension can effectively increase the viscosity of the suspension. When the concentration of calcium polyphosphate suspension is 3 vol%, the viscosity increases to above 20 Pa s for 14 min. It is assumed that YSZ suspension with 2 vol% calcium polyphosphate suspension has higher viscosity (35 Pa s). However, the initial viscosity of the suspension is also high (6 Pa s) which is not beneficial to the homogeneity of the suspension, leading to lower reliability.

Figure 8.116 shows the viscosities of YSZ suspensions with different concentrations of glycerol diacetate at 60 °C. It is expected that the hydrolysis of glycerol diacetate promotes the decomposition of the chelate complex at lower pH value. However, the suspension with 2–3 vol% glycerol diacetate shows high viscosity at the initial stage which may impact the coagulation process.

Based on the analysis above, YSZ suspension with the addition of 3 vol% calcium polyphosphate suspension and 1 vol% glycerol diacetate is most suitable for

Fig. 8.116 Viscosities of yttria-stabilized zirconia suspensions with different concentrations of glycerol diacetate at 60 °C

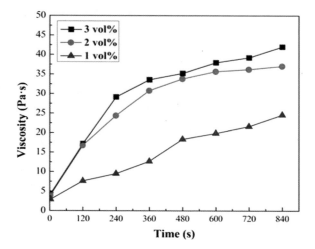

Table 8.11 Effect of coagulation temperature on coagulation time and properties of yttria-stabilized zirconia green bodies

Coagulation temperature (°C)	50	60	70
Coagulation time (min)	60	20	15
Wet compressive strength (MPa)	2.12 ± 0.19	2.64 ± 0.35	2.48 ± 0.23
Coagulation shrinkage (%)	2.26	2.75	2.83

coagulation. Table 8.11 shows the effect of coagulation temperature on coagulation time and properties of the green bodies which were prepared by DCC–HVCI using calcium polyphosphate. It is observed that the coagulation time shortens obviously from 60 min at 50 °C to 20 min at 60 °C with the increase in the coagulation temperature. This is due to the slow decomposition rate of calcium polyphosphate at 50 °C. The coagulation time is reduced to 15 min at 70 °C. The compressive strength of wet green bodies is in the range of 2.12–2.64 MPa. The linear shrinkage of coagulated bodies increases at elevated temperature which indicates high calcium concentration to compress the electric double layer of the particles. Figure 8.116 shows the microstructures of the green bodies prepared by DCC–HVCI. No agglomeration or large pore has been observed in the green body which demonstrates the homogeneity of the suspension during the coagulation process and is beneficial to the properties of YSZ ceramics (Fig. 8.117).

Figure 8.118 shows the photo and microstructure YSZ ceramics by DCC–HVCI. It can be observed that no crack is in the sample and the microstructure is uniform which further proves the homogeneous coagulation of the suspension. Table 8.12 compares the properties of YSZ prepared by DCC–HVCI with different concentrations of calcium polyphosphate. The shrinkage of the ceramics is in the range 17.82–18.47%. The relative density of YSZ increases with the increase in calcium polyphosphate and

Fig. 8.117 Microstructure of fracture surface of yttria-stabilized zirconia green body prepared by DCC–HVCI

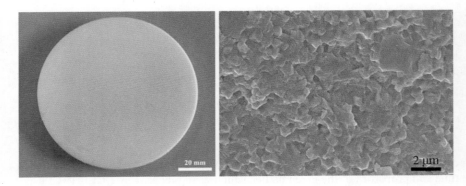

Fig. 8.118 Photo and microstructure yttria-stabilized zirconia ceramics prepared by DCC–HVCI

Table 8.12 Properties of yttria-stabilized zirconia prepared by DCC–HVCI using calcium polyphosphate

Calcium polyphosphate (vol%)	Sintering shrinkage (%)	Relative density (%)	Flexural strength (MPa)
2	17.82	98.2	785 ± 71
3	18.24	98.9	826 ± 48
4	18.47	98.4	793 ± 62

then decreases. YSZ prepared by DCC–HVCI using 3 vol% calcium polyphosphate suspension and 1 vol% glycerol diacetate shows higher relative density and flexural strength, 98.9% and 826 ± 48 MPa, respectively. Figure 8.119 shows the Weibull modulus of YSZ prepared by DCC–HVCI using 3 vol% calcium polyphosphate suspension and 1 vol% glycerol diacetate. High Weibull modulus of 21 has been obtained, indicating that reliable ceramics can be fabricated by this new method.

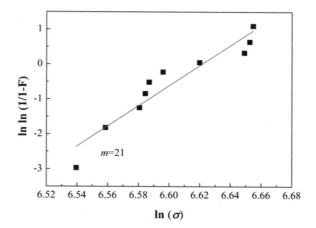

Fig. 8.119 Weibull modulus of yttria-stabilized zirconia prepared by DCC–HVCI using 3 vol% calcium polyphosphate suspension and 1 vol% glycerol diacetate

8.2 Dispersion Removal Coagulation Casting

Direct coagulation casting (DCC) has been systematically studied for two decades. The coagulation of suspension in DCC process relies on the decrease of electrostatic repulsion force between particles to reduce the interparticle total potential energy. In the direct coagulation casting process, the zeta potential was usually decreased via adjusting pH value and increasing ionic strength. Two traditional approaches to achieve the DCC process include I: adjusting the pH of suspension (ΔpH) to isoelectric point (IEP) and II: increasing the ionic strength (ΔI) to compress the electric double layer. However, the route of zeta potential decrease has the third method. The electrostatic stabilized suspension can be directly coagulated via shifting the IEP of the suspension (DCC–ΔIEP). Nowadays, we achieve the third approach. The novel forming method was named dispersion removal coagulation casting (DRCC). Previously, we had reported some colloidal forming methods utilizing the properties of dispersant. The suspensions can be in-situ coagulated by dispersant reaction, hydrolysis, removal, and so on. Some of dispersion removal methods can be classified as DCC–ΔIEP but others cannot. This section aims to induce the novel DRCC method and the amendment of direct coagulation casting via shifting IEP method using the route of dispersion removal.

8.2.1 Dispersant Reaction Method

(1) Raw materials, preparation, and characterization

Commercially available silicon nitride powder (Vesta Ceramics AB, Sweden) with an average particle diameter of 0.8 μm and specific surface area of 9.39 m²/g and α-SiC powder (Qinhuangdao Eno High-Tech Material Development Co., Ltd. China) with average particle diameter of 0.5 μm and specific surface area of 9.62 m^2/g were used. For liquid phase sintering, CT3000SG alumina powder (average particle diameter: 0.33 μm, surface area: 8.08 m^2/g, Almatis, Ludwigshafen, Germany) and yttrium oxide powder (99.95%, average particle diameter: 0.39 μm, specific surface area: 0.3 m^2/g, Shanghai Junyu Ceramic-molded Product Co., Ltd., China) were employed as sintering additives. Ammonium peroxydisulfate ((NH_4)$_2S_2O_8$) and ammonium carbonate ((NH_4)$_2CO_3$) were used to leach the silicon carbide powder which were produced from Sinopharm Chemical Reagent Co., Ltd., China. Tetramethylammonium hydroxide (TMAOH) aqueous solution with concentration of 10 wt% was used to disperse the suspension. Glycerol diacetate (GDA) was used as a coagulation agent. Both were purchased from Hengye Zhongyuan Chemical Co., Ltd., Beijing, China.

In DCC process, suspension with high solid loading and low viscosity was desirable. Silicon nitride and silicon carbide suspensions with different solid loadings were prepared by tumbling the powder, sintering additives, water, dispersant, ammonia, and grinding media in polyethylene container for 24 h. Ammonium peroxydisulfate and ammonium carbonate were used to soak and leach the as-received silicon carbide powder by a vacuum machine till pH = 2.5–3 and pH = 10–10.5, respectively. Then, wet powder was washed and leached by deionized water till pH = 7.5–8. The treated powder was dried in a drying closet at 80 °C for 24 h. The as-received silicon carbide powder is abbreviated as AR, and the acid and base leached silicon carbide powders are abbreviated as AL and BL in all figures and tables, respectively. TMAOH was used as dispersant with 0.1–1.0 wt% based on powder. Sintering additives were 2 wt% Al_2O_3 and 5 wt% Y_2O_3 based on silicon nitride powder and 4 wt% Al_2O_3 and 6 wt% Y_2O_3 based on silicon nitride powder. Agate balls with diameter of 5–10 mm were used as grinding media. The mass ratio between grinding media and ceramic powder was 1:2. Ammonia was used to adjust the pH value of the suspension to ca. 10.5–11.

The suspension was maintained at 25 °C for 5 h and degassed under vacuum condition for 15–20 min. Various amounts of glycerol diacetate were added to the suspension and mixed thoroughly by continuing the tumbling process for another 20 min to increase the homogeneity of the suspension. Then, the suspension was cast into a plastic mold. To prevent the evaporation of water during the coagulation process, a seal was used. The samples were placed in a water bath with a heating rate of 5 °C/min and treated at different temperatures for a period of time and then demolded. The green bodies were dried at 80 °C for 24 h. For the pressureless sintering process, the dried samples were sintered at 1850 °C for 2 h at a heating rate of 5 °C/min, under nitrogen atmosphere (0.1 MPa). For the hot pressing sintering process, the samples were put into a graphite mold with a diameter of 50 mm and sintered at 1800 °C for 2 h under uniaxial pressure of 20 MPa at a heating rate of 5 °C/min, under nitrogen atmosphere (0.1 MPa).

The phases of silicon carbide powder and ceramics were analyzed by X-ray diffraction (XRD) method using Cu Kα radiation (D8ADVANCE, Bruker, Karlsruhe, Germany). Wavelength-dispersive sequential X-ray fluorescence spectrometer (XRF-1800, Shimadzu, Tokyo, Japan) with a Rhodium target was used to analyze the chemical composition of silicon carbide powder. X-ray photoelectron spectroscopy

(XPS) microprobe using monochromated micro-focused Al K-Alpha radiographic source with spot size of 200–900 μm (ESCALAB 250Xi, Thermo Fisher Scientific, USA) was used to analyze the degree of surface oxidation of silicon carbide powder. The pH values of suspension were measured by pH meter (LE438, Mettler, Toledo, Switzerland) in the temperature range of 25–70 °C. The rheological properties of the suspension were measured using a rheometer (KINEXUS PRO, Malvern Instruments, Worcestershire, UK) attached with C25 R0634 SS spindle and PC25 C0138 AL cylinder. In viscosity measurement, the shear rates were in the range of 0.1–1000 s^{-1}. Zeta potential was measured by a Zeta Potential Analyzer (CD-7020, Colloidal Dynamics Co., Ltd., Ponte Vedra Beach, FL, USA) via the electroacoustic measurement technique with a stirring speed of 300 r/min. In zeta potential measurement, 10 vol% silicon nitride suspension was prepared. 1 mol/l HCl and NaOH were used to adjust the pH value. The density of the green bodies was determined on the cut and fine ground rectangular bars by measuring the dimensions and weight. The density of the sintered samples was measured using the water displacement technique. To investigate the degree of oxidation of the silicon nitride powder after being in the suspension for different periods of time, a wavelength-dispersive sequential X-ray spectrometer (XRF-1800, Shimadzu, Tokyo, Japan) with a Rhodium target was used to analyze the chemical composition of silicon nitride powder. Suspension with dispersant and coagulated wet green body by glycerol diacetate was dried in stove at 80 °C and milled. For wet compressive strength measurements, cylindrical bodies with 25.5 mm in diameter and a height between 25 and 30 mm were cast. After coagulating at different temperatures for different times, the samples were demolded, and their wet green strength was measured immediately in a mechanical testing machine (AG-IC20KN, Shimadzu, Tokyo, Japan) with a crosshead speed of 0.5 mm/min. Using the same mechanical testing machine, the flexural strength of the specimens with dimensions of 3 mm × 4 mm × 36 mm was measured via the three-point bending test. The samples were polished before measurement. The microstructures of the green body and plasma-etched (by a CF_4 gas) and fracture surfaces of sintered samples were observed by a field emission scanning electron microscope (FESEM) (MERLIN VP Compact; Carl Zeiss, Jena, Germany).

(2) **Effect of dispersant on rheological of silicon nitride suspension**

As we know, the surface of silicon nitride powder is easy to form Si–O bonds, which affect the rheology of suspension. Several studies had investigated the effect of TMAOH on the dispersion of silicon nitride suspension. Millan et al. showed that concentrated aqueous silicon nitride slips could be stabilized at pH 11–11.5 using TMAOH as dispersant (Millan et al. 2000). Moreno et al. showed that TMAOH was useful because it formed a partially oxidize layer with Si–O–N species which acted as a protective screen onto the silicon nitride particles, thus difficulty a further oxidation (Moreno et al. 1998). In this section, TMAOH is selected as dispersant for the preparation of silicon nitride suspension. The zeta potential of silicon nitride suspensions at different pHs is shown in Fig. 8.120. It can be seen that the IEP of silicon nitride suspension without dispersant is 4. There is a high zeta potential of −30 mV in pH range 10–12. The IEP moves to five after adding dispersant (TMAOH)

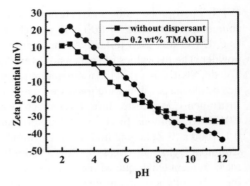

Fig. 8.120 Zeta potential of 10 vol% silicon nitride suspension at different pHs

which is slightly different from the IEP without dispersant. Silicon nitride suspension with dispersant has higher zeta potential of −40 mV in pH range 10–12 than the suspension without dispersant. The results indicate that the absorption of TMAOH on silicon nitride particles enhances the repulsive forces and therefore increases the zeta potential value.

Figure 8.121 shows the effect of different concentrations of TMAOH on zeta potential of 10 vol% silicon nitride suspension and viscosity of 30 vol% silicon nitride suspension at pH = 11. Absolute potential of silicon nitride suspension increases with the increase of dispersant content and then decreases with further addition. The maximum value appears with the concentration of dispersant at 0.2 wt%. The viscosity of silicon nitride suspension decreases first and then increases with the increase of dispersant content. Due to the addition of dispersant, the electrostatic repulsive force between particles increases gradually. The silicon nitride suspension achieves stability when the concentration of TMAOH reaches a certain amount. Excess additives will result in the increase of ionic strength in the system which reduces the repulsion force among particles. The silicon nitride particles are easy to reunite and aggregation which lead to the increase of the viscosity. The minimum

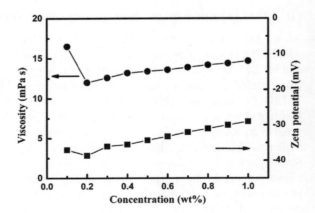

Fig. 8.121 Effect of TMAOH concentration on zeta potential of 10 vol% silicon nitride suspension and viscosity of 30 vol% silicon nitride suspension at pH = 11

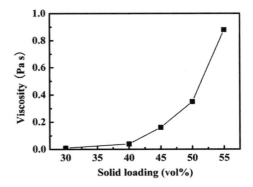

Fig. 8.122 Effect of solid loading on the viscosity of silicon nitride suspension with 0.2 wt% TMAOH at the shear rate of 100 s^{-1}

value of the viscosity appears when the concentration of TMAOH is at 0.2 wt%. Based on the analysis above, the concentration of TMAOH is selected as 0.2 wt%.

Figure 8.122 shows the effect of different solid loadings on the viscosity of silicon nitride suspension with 0.2 wt% TMAOH. The viscosity of the suspension increases gradually with the increase of solid loading. The viscosities of the suspension with solid loadings of 45 vol% and 50 vol% are 0.16 Pa s and 0.35 Pa s, respectively. The low viscosity of silicon nitride suspension indicates the well dispersion of silicon nitride by addition of TMAOH. The viscosity increases to 0.92 Pa s when the solid loading of the suspension is 55 vol%. Generally, the viscosity of the suspension for colloidal forming should be below 1 Pa s at the shear rate of 100 s^{-1}. In order to prevent the high viscosity, silicon nitride suspension with solid loading of 50 vol% is chosen in subsequent experiments.

(3) **Influence of GDA on viscosity of silicon nitride suspension**

Glycerol diacetate has been commonly used in practical applications for tailoring the pH internally in systems with sodium silicate binders (Flory 1953; Haas 1989). Also, glycerol diacetate has been used as pH regulator to assist the decomposition of citrate salts by lowering the pH value of alumina suspensions in our previous studies (Xu et al. 2014a; Moreno et al. 1998). In order to investigate the influence of glycerol diacetate on viscosity of silicon nitride suspension, 2 vol% glycerol diacetate is added into 50 vol% silicon nitride suspension (shown in Fig. 8.123). The viscosity of the suspension at 25 °C increases slightly after the addition of glycerol diacetate. The viscosity increases to 9 Pa s when the treatment time is 60 min. However, the case at 60 °C is rather different. The viscosity of the suspension increases rapidly upon the addition of glycerol diacetate. The viscosity increases to 15 Pa s when the treatment time is 20 min. High viscosity of 50 Pa s is obtained 40 min upon the addition of glycerol diacetate. The results indicate that the suspension with 2 vol% glycerol diacetate can be coagulated at 60 °C for 30–40 min.

In order to explore the coagulation mechanism of the silicon nitride suspension, the effect of treating temperature on the pH value of silicon nitride suspension with 2 vol% glycerol diacetate was investigated (shown in Fig. 8.124). It can be seen that the pH value of suspension decreases gradually with increase of the treatment time.

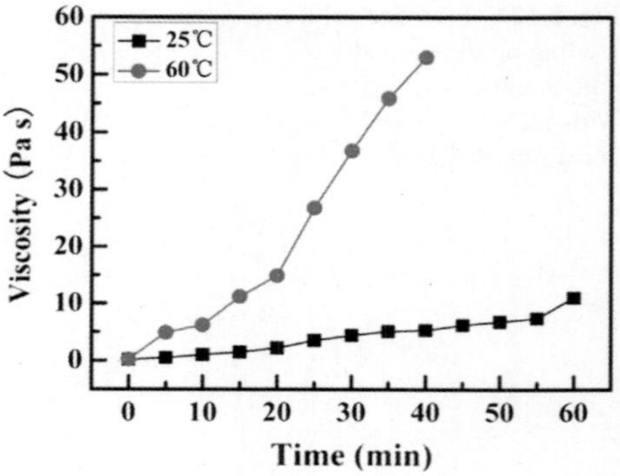

Fig. 8.123 Effect of treating temperature on the viscosity of 50 vol% silicon nitride suspension with 2 vol% glycerol diacetate

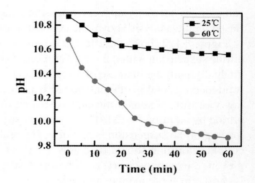

Fig. 8.124 Effect of treating temperature on the pH value of 50 vol% silicon nitride suspension with 2 vol% glycerol diacetate

The pH values of the suspensions at 25 and 60 °C are 10.6 and 9.8, respectively, when treatment time is 60 min.

The effect of temperature on viscosity and pH of silicon nitride suspension without glycerol diacetate is shown in Fig. 8.125. It shows that the pH value decreases to 9.6 after 60 min at 60 °C. The pH value decreases gradually with treatment time, which

Fig. 8.125 Viscosity and pH value of 50 vol% silicon nitride suspension without glycerol diacetate at 60 °C

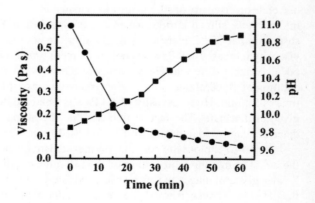

results in the increase of the viscosity of the suspension, as the pH value moves to the IEP of silicon nitride. But the viscosity of the suspension increases from 0.15 to 0.55 Pa s which is not high enough to coagulate the suspension. Therefore, it should be other factors that contribute to the increase in viscosity of the silicon nitride suspension.

As we know, glycerol diacetate is a weak acid ester which can hydrolyze and produce acetic acid in alkaline region. The hydrolysis equation is listed as follows (Flory 1953):

$$(CH_3COOCH_2)_2CHOH + 2H_2O \leftrightarrow (CH_2OH)_2CHOH + 2CH_3COOH \quad (8.26)$$

In the present research, tetramethylammonium hydroxide is a strong base which is added in the silicon nitride suspension as dispersant. The acid–base reaction between acetic acid and tetramethylammonium hydroxide will occur after the hydrolysis of glycerol diacetate. The reaction equation is listed as follows:

$$(CH_3)_4NOH + CH_3COOH \rightarrow CH_3COON(CH_3)_4 + H_2O \quad (8.27)$$

Tetramethylammonium hydroxide is consumed gradually by this reaction which causes the decrease in concentration of dispersant. The silicon nitride particles' surface charge decreases which results in the coagulation of silicon nitride suspension. Therefore, a creative direct coagulation casting method for silicon nitride suspension is proposed through the reaction between tetramethylammonium hydroxide and acetic acid from hydrolysis of glycerol diacetate.

To further investigate the new coagulation method, the effects of temperature and the concentration of glycerol diacetate on the viscosity of silicon nitride suspension are shown in Fig. 8.126. The treatment time is set at 30 min. The viscosity of the suspension increases slightly with 2.0 vol% glycerol diacetate to 6 Pa s at 25 °C. However, there is a great viscosity variation of suspension with the increase of glycerol diacetate concentration at 60 °C. The viscosity reaches 25 Pa s with the amount of glycerol diacetate higher than 1.0 vol%. The increase in viscosity is due

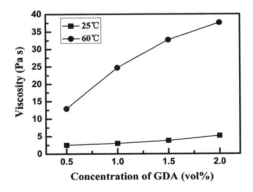

Fig. 8.126 Effect of treating temperature on the viscosity of 50 vol% silicon nitride suspension with different concentrations of glycerol diacetate for 30 min

Table 8.13 Effect of different glycerol diacetate concentrations of silicon nitride suspensions and green bodies

Concentration of GDA (vol%)	0.5	1.0	1.5	2.0	2.5
Coagulation	No	Yes	Yes	Yes	Yes
Coagulation time (min)		40	30	20	20
Green samples		Cracking	Flawless	Flawless	Flawless

to the reaction between tetramethylammonium hydroxide and acetic acid. Glycerol diacetate is a hydrolyzable ester, and the rate of hydrolysis reaction is influenced by temperature. Higher temperature promotes the hydrolysis rate and produces more acetic acid. In this sense, we can adjust the hydrolysis rate of glycerol diacetate by controlling temperature.

Moreover, as shown in Table 8.13, the silicon nitride suspension can be coagulated by adding 1.0 vol% glycerol diacetate after 40 min. But the green bodies are crackle. Though the flawless green bodies can be coagulated by adding 1.5 vol% after 30 min, the same result by adding 2 vol% can be achieved after 20 min which is the shortest time. In addition, the viscosity of silicon nitride suspension is less than 1 Pa s in the period of ten minutes after addition of glycerol diacetate at 25 °C. Because of low viscosity of suspension in the initial stage, there is enough time to degas and mold at room temperature. In conclusion, direct coagulation casting of silicon nitride suspension can be achieved with sufficient amount of glycerol diacetate at elevated temperature.

(4) **Properties of silicon nitride ceramics**

Table 8.14 shows the chemical composition of silicon nitride powder. The oxygen content of as-received silicon nitride powder is 1.54 wt%. The oxygen content of dried dispersive suspension with TMAOH increases to 1.87 wt% which is higher than as-received powder. Then the oxygen content of coagulated green body by glycerol diacetate decreases slightly to 1.84% which may due to the evaporation of water produced from the reaction between tetramethylammonium hydroxide and acetic acid. In short, the degree of oxidation of the silicon nitride powder after being in the suspension for different periods of time is microvariation.

Table 8.15 shows the coagulation time and properties of the green bodies prepared by direct coagulation casting via dispersant reaction method at different coagulation temperatures. It is observed that the coagulation time shortens observably with the

Table 8.14 The XRF of the silicon nitride powder after being in the suspension for different periods of time (wt%)

Element	Si (%)	N (%)	O (%)	Al (%)	Y (%)	Fe (%)
As-received	56.22	42.04	1.54	0.027	0.002	0.079
Suspension	56.02	41.90	1.87	0.036	0.002	0.097
Green body	55.54	42.45	1.84	0.039	0.003	0.086

8.2 Dispersion Removal Coagulation Casting

Table 8.15 Coagulation time and properties of green bodies prepared by glycerol diacetate at different coagulation temperatures

Coagulation temperature (°C)	Coagulation time (min)	Green samples	
		Shrinkage (%)	Compressive strength (MPa)
40	30	2.34	0.42
50	25	1.92	0.45
60	20	1.83	0.56
70	15	1.67	0.68

increase of the coagulation temperature. The coagulation time shortens from 45 min at 30 °C to 15 min at 60 °C, which is shorter than the method of direct coagulation casting by high valence counter-ions (Xu et al. 2012a, b). The linear shrinkage of coagulated bodies also decreases with increase of the coagulation temperature. Thus, the crackability will reduce with lower linear shrinkage of green body though the density will decrease due to the linear shrinkage decreasing. The compressive strength of green bodies increases from 0.42 MPa to 0.65 MPa, which is one-third of the method of direct coagulation casting by high valence counter-ions (DCC–HVCI), though higher than coagulation by monovalent ions and pH shift to IEP (Xu et al. 2012c; Balzer et al. 2001). The compressive strength increases with coagulation temperature, as the hydrolysis rate of glycerol diacetate is higher at higher temperature which promotes the reaction between tetramethylammonium hydroxide and acetic acid.

Figure 8.127a shows the photograph of green complex components which are coagulated by dispersant reaction method. There are smooth surfaces of bodies without defects such as micropore and bubble. Figure 8.127b shows the microstructure of green body which observed that homogeneous microstructure has been obtained. Figure 8.128 shows the microstructures of (a) polished and etched surface and (b) fracture surface of sintered ceramics which are prepared from silicon nitride suspension with 2 vol% glycerol diacetate. It is observed that the well-grown columnar

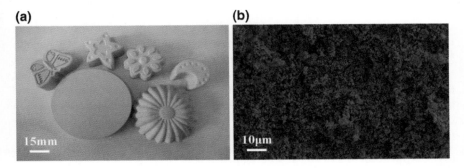

Fig. 8.127 a Photographs of complex components and b microstructure of green body coagulated by glycerol diacetate

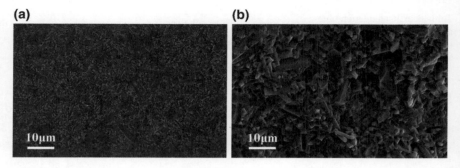

Fig. 8.128 Microstructures of **a** polished and etched (by a CF$_4$ gas) surface and **b** fracture surface of sintered ceramics coagulated by glycerol diacetate

crystalline grains are presented in the sample. There is complex interleaving between crystalline grains and high ratio of length to diameter with itself. It suggests that the new processing method is a promising method to prepare near-net-shape ceramics with good mechanical properties.

Furthermore, the relative density of sintered samples is 98.78% which are measured by water displacement technique. The flexural strength of sintered samples is 817 ± 75 MPa which are measured by three-point bending test. There is a comparison on flexure strength of silicon nitride ceramics prepared by different methods which is shown in Table 8.16 (Yang and Huang 2010; Torre and Bigay 1986; Sano et al. 1995; Agarwala et al. 1996; Wang et al. 2011; Liu et al. 2001). Though it can be seen that the flexural strength of silicon nitride ceramics prepared by DCC via dispersant reaction is the highest in which prepared by colloidal forming processes with pressureless sintering, the properties of silicon nitride ceramics prepared by DCC via dispersant reaction should be increased with powder treatment and much more effective sintering processes in expectation.

Direct coagulation casting of silicon nitride suspension via dispersant reaction has been reported. The suspension is destabilized by the reaction between tetramethylammonium hydroxide and acetic acid which is produced from the hydrolysis of glycerol diacetate. It is indicated that the suspension can be coagulated by glycerol diacetate after treating at 60 °C for 30 min. Green bodies with compressive strength of 0.42–0.68 MPa are obtained by heating the silicon nitride suspension with 2 vol% glycerol diacetate at 40–70 °C for 15–30 min. The sintered silicon nitride ceramics have homogeneous microstructures and well-grown crystalline grain with relative density of 98.78% and flexural strength of 817 ± 75 MPa. The results suggest the new coagulation method with low coagulation temperature and short coagulation time is a promising colloidal processing method for preparing high-performance near-net-shaped ceramics.

(5) **Preparation of silicon carbide suspension**

As we know that it is easy to form Si–O bond on the surface of silicon carbide particle which impacts the dispersion of silicon carbide powder in disperse medium, several

8.2 Dispersion Removal Coagulation Casting

Table 8.16 Flexure strength comparison of silicon nitride ceramics prepared by different methods

Forming method	Sintered process	Powders treatment	Flexure strength (MPa)	References
Dry press	Gas pressure sintering (GPS)		582.7	Yang and Huang (2010)
Isostatic pressing			771 ± 66	Torre and Bigay (1986)
Slip casting	GPS		974 ± 57	Torre and Bigay (1986)
Pressure casting	GPS		612	Sano et al. (1995)
Fused deposition of ceramics (FDC)	GPS		824 ± 110	Agarwala et al. (1996)
Gel casting	Pressureless sintering		730	Wang et al. (2011)
	GPS	Oxidized powder	785.16	Yang and Huang (2010)
	GPS	Coated additive	814.5 ± 68.5	Yang and Huang (2010)
Gel casting and cool isostatic pressing (CIP)	GPS	Coated additive	868.9 ± 59.8	Yang and Huang 2010
	Two-stage GPS		944.7 ± 29.5	Yang and Huang (2010)
DCC via increase ions strength	Pressureless sintering		758.4	Liu et al. (2001)
DCC via dispersant reached	Pressureless sintering		817 ± 75	Present work

researches investigated the oxidation of silicon carbide particle surface (Fukada and Nicholson 2004; Adar et al. 1988). Ammonium peroxydisulfate and ammonium carbonate could be used to leach the silicon carbide powder. The XRD pattern of the silicon carbide powders before and after treating is shown in Fig. 8.129. It can be seen from Fig. 8.129 that there is no phase transformation among the three kinds of powder.

The XRF analysis of the silicon carbide powders before and after treating is listed in Table 8.17. It can be observed that the oxygen content of as-received silicon carbide powder is 1.39%. The oxygen contents of silicon carbide powder leached by acid and base solutions decrease to 0.86% and 0.87%, respectively. The other elements (including Na^+, Mg^{2+}, Ca^{2+}, Al^{3+}) contents of silicon carbide powder leached by acid and base solutions also decrease. Due to the silicon carbide suspension with dispersant is negatively charged, the decrease of high valence counter-ions can be conducive to the preparation of the suspension with high solid loading. It indicates

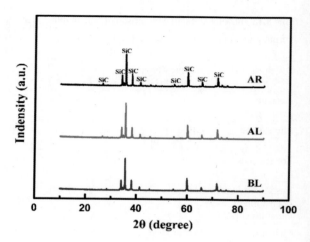

Fig. 8.129 XRD of silicon carbide powders before and after treating

Table 8.17 XRF analysis of silicon carbide powders before and after treating

Element	Si%	C%	O%	Others%
AR	59.3	38.8	1.39	0.51
AL	59.6	39.1	0.87	0.43
BL	58.4	40.3	0.86	0.44

that silicon carbide powder chemical treatment by both acid and base solutions can reduce the ion contents and oxidation degree of silicon carbide powder without phase transformation.

X-ray photoelectron spectroscopy analyzer (XPS) can be used for the analysis of chemical bond on the solid surface with the depth of analysis in about 10 nm. According to relevant reports (Kaplan and Parrill 1986; Harris et al. 2013), the degree of silicon carbide powder surface oxidation can be analyzed by XPS. Figure 8.130 shows the XPS analysis of (a) three kinds of silicon carbide powder and (b) Si_2p core level spectrum after curve fitting. It can be seen from Fig. 8.130b that there are two binding energy peaks on 100.4 and 102.7 mV. Through consulting NIST X-ray Photoelectron Spectroscopy Database and relevant reports (Parrill and Chung 1991; Delplancke et al. 1991; Yu and Hantsche 1993), the binding energy vicinity 100.4 mV is oriented Si–C and the binding energy vicinity 102.7 mV is oriented Si–O.

The assignments of XPS peaks of silicon carbide powder are shown in Table 8.18. It is clear that all of oxygen elements are used to form silicon–oxygen bond. The percentage of AR on $SiC:SiO_xC_y$ is 88.7:11.3. The percentages of AL and BL on $SiC:SiO_xC_y$ are 89.3:10.7 and 90.3:9.7, respectively. The bind of Si–C in AL and BL silicon carbide powders increases while the bind of Si–O in AL and BL silicon carbide powders decreases. It is indicated that powder chemical treatment is an effective method to reduce the degree of silicon carbide particle surface oxidation.

8.2 Dispersion Removal Coagulation Casting

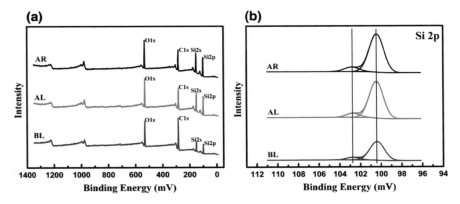

Fig. 8.130 XPS analysis of **a** silicon carbide powders before and after treating and **b** Si2p core level spectrum (silicon carbide and oxycarbide) after curve fitting

Table 8.18 Assignment of XPS peaks of silicon carbide powders

Peak	Binding energy (mV)	Percentage (%)			Species
		AR	AL	BL	
Si2p	100.4	88.7	89.3	90.3	SiC
Si2p	102.7	11.3	10.7	9.7	SiO_xC_y/SiO_2
O1s	532.4				SiO_2

Figure 8.131 shows the effect of powder chemical treatment with acid and base solutions on zeta potential of 10 vol% silicon carbide suspension. It can be seen that the IEP of as-received silicon carbide suspension is pH = 2 which is near the IEP of silica (Sun et al. 2000). The IEP of suspensions prepared using leached silicon

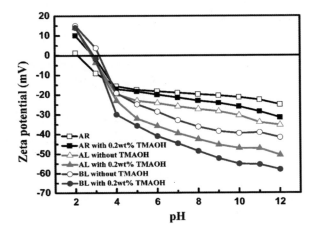

Fig. 8.131 Effect of dispersant and powder treatment with acid and base solutions on zeta potential of 10 vol% silicon carbide suspension at different pHs

carbide powder moves to pH = 3. It is deemed to the real IEP of silicon carbide particle actually. The AL silicon carbide suspension shows higher zeta potential of −30 mV than as-received silicon carbide suspension in pH range 10–12. However, the BL silicon carbide suspension shows the highest zeta potential of −40 mV in this pH range. It is indicated that the surface charging of the silicon carbide particle can be increased by chemical treatment, and the effect of BL on zeta potential of silicon carbide suspension has an advantage over AL. The effect of 0.2 wt% TMAOH on zeta potential of silicon carbide suspensions is also shown in Fig. 8.131. It can be seen that the zeta potential of silicon carbide suspensions with TMAOH is higher than the suspensions without dispersant; there is higher zeta potential of −60 mV of BL than zeta potential of −50 mV of AL in pH range of 10–11. Therefore, TMAOH was used to disperse the silicon carbide suspension in subsequent experimental procedures.

Figure 8.132 shows the effect of dispersant contents on zeta potential of 10 vol% silicon carbide suspension and viscosity of 35 vol% silicon carbide suspension at pH = 10.5. It can be seen from Fig. 8.132 that the absolute potential of both AL and BL silicon carbide suspension increases with the increase of dispersant content and then decreases with further addition. The maximum value appears at the concentration of 0.2 wt%. The viscosity of both AL and BL silicon carbide suspension decreases first and then increases with the increase of dispersant content. The minimum value of the viscosity appears when the concentration of TMAOH is 0.2 wt% which is selected in subsequent experiment.

Figure 8.133 shows the relative viscosity of AL and BL silicon carbide suspensions as a function of volume fraction at the shear rate of 100 s^{-1}. The Dougherty–Krieger (D–K) equation is commonly used to express the relationship between the relative viscosity (η_r) and volume fraction (φ) for hard sphere system (Krieger 1972):

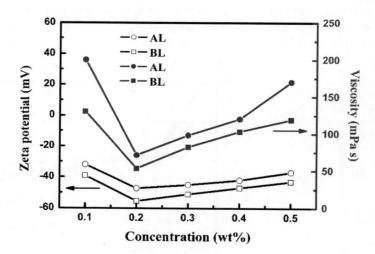

Fig. 8.132 Effect of dispersant concentrations on zeta potential of 10 vol% silicon carbide suspension and viscosity of 35 vol% silicon carbide suspension at pH = 10.5

Fig. 8.133 Relative viscosity of silicon carbide suspensions as a function of volume fraction at the shear rate 100 s^{-1}

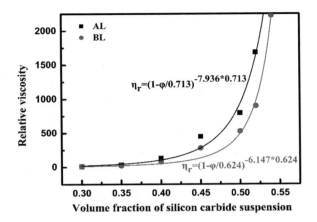

$$\eta_r = \left(1 - \frac{\varphi}{\varphi_m}\right)^{-[\eta]\varphi_m} \quad (8.28)$$

In this equation, φ_m is the maximum packing fraction, which has the order of 0.64 for random packing, and $[\eta]$ is the asymptotic slope of the curve zero which is normally 2.5. Usually, φ_m and $[\eta]$ are chosen as variables at the same time to fit the experimental results and to analyze the suspension structures (Xu et al. 2015a, b). The relative viscosities of AL and BL silicon carbide suspensions are simulated as a function of volume fraction of the particles using the D–K equation. Equation (8.29) is for suspension treated by AL and Eq. (8.30) is for suspension treated by BL.

$$\eta_r = \left(1 - \frac{\varphi}{0.713}\right)^{-7.936*0.713} \quad (8.29)$$

$$\eta_r = \left(1 - \frac{\varphi}{0.624}\right)^{-6.147*0.624} \quad (8.30)$$

It is indicated that the relative viscosities of suspensions increase rapidly above AL 50 vol% and BL 52 vol%. The viscosities of AL 50 vol% and BL 52 vol% silicon carbide suspensions are 0.71 Pa s and 0.80 Pa s, respectively. The preparation and handling of the suspensions will be difficult at higher volume fractions of the powders. Therefore, the silicon carbide suspensions of AL 50 vol% and BL 52 vol% were chosen in order to avoid high viscosity of the suspension which may impact on the properties of sintered ceramics.

In order to compare the influence of acid and base treatment with the same solid loading on properties of silicon carbide ceramics, the BL 50 vol% suspension is also prepared and studied. Combined with the above description, different solid loadings of silicon carbide suspensions with high zeta potential and low viscosity can be prepared using both AL and BL powders and 0.2 wt% TMAOH addition.

(6) Fabrication of silicon carbide ceramics

Figure 8.134 shows the effect of 2 vol% GDA on the viscosity of silicon carbide suspensions at different temperatures. Though the viscosities of three kinds of suspension increase slowly from 0.5 Pa s to 2 Pa s at 25 °C, they increase fleetly in 30 min to about 20 Pa s at 60 °C and followed a slow rise to their maximum values. The maximum values of AL and BL 50 vol% are 19.03 Pa s and 21.68 Pa s, respectively. The maximum value of BL 52 vol% is 23.93 Pa s. The increase of viscosity is due to the tetramethylammonium hydroxide reacted with acetic acid which hydrolyzed from glycerol diacetate at elevated temperature, and then desorbed from silicon carbide particles (Gan et al. 2016). The viscosities are high enough to coagulate the silicon carbide suspensions at 60 °C. Figure 8.135 shows the silicon carbide green bodies which prepared using chemical treatment powder by DCC via dispersant reaction. Complex shape ceramic parts can be produced by this method.

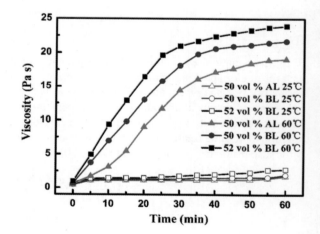

Fig. 8.134 Effect of 2 vol% GDA on viscosity of silicon carbide suspensions at different temperatures

Fig. 8.135 Silicon carbide green bodies prepared using chemical treatment powder by DCC dispersant reaction

8.2 Dispersion Removal Coagulation Casting

In order to study the phase transformation of silicon carbide ceramics prepared by liquid phase sintering, the X-ray diffraction was used to analyze the sintered sample. The XRD pattern of the silicon carbide ceramics is shown in Fig. 8.136. It can be seen that there are two phases in sintered sample with SiC as the main phase and YAG ($Y_3Al_5O_{12}$) as the second phase which has a melting point at 1760 °C. The YAG is uniformly distributed in silicon carbide sintered ceramics. It is indicated that enough content of YAG could be formed at 1950 °C which can promote the densification of silicon carbide ceramics (Sciti and Bellosi 2000).

(7) **Properties of silicon carbide ceramics**

Table 8.19 shows the effect of powder treatment and coagulation temperature on the coagulation time. The coagulation time shorten gradually with the increase of temperature. There is a same coagulation time of 50 vol% AL and BL suspensions, and a shorter coagulation time of 52 vol% BL suspension. The shortest coagulation time of suspension is 52 vol% BL at 70 °C about 15 min. The shrinkage and compressive strength of green bodies are also listed in Table 8.19. It can be seen that the shrinkage of green bodies reduces and compressive strength increases gradually with the increase of temperature. There are lower shrinkage and higher compressive strength of 50 vol% BL than AL. Due to higher solid loading, there are lowest shrinkage of 1.27% and highest compressive strength of 1.13 MPa of 52 vol% BL at 70 °C. Though compressive strength of 50 vol% AL is lowest, it is high enough to handle in subsequent experiments.

Figure 8.137(a) AL 50 vol% and (b) BL 52 vol% shows the microstructures of silicon carbide green bodies. Both the green bodies without agglomeration are prepared with homogeneous microstructure. Figure 8.137(c) AL 50 vol% and (d) BL 52 vol% shows the microstructures of silicon carbide ceramic fracture surfaces. It can be seen that dense silicon carbide ceramics can be prepared using AL and BL

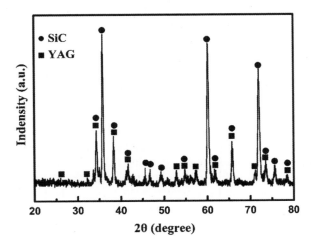

Fig. 8.136 XRD of silicon carbide ceramics sintered at 1950 °C

Table 8.19 Effect of powder treatment and coagulation temperature on coagulation time and properties of silicon carbide green bodies

Powder treatment	Coagulation temperature (°C)	Coagulation time (min)	Shrinkage (%)	Compressive strength (MPa)
50 vol% AL	50	40	1.93	0.41
	60	30	1.87	0.55
	70	20	1.61	0.68
50 vol% BL	50	40	1.86	0.54
	60	30	1.72	0.57
	70	20	1.57	0.73
52 vol% BL	50	30	1.39	0.89
	60	20	1.30	1.07
	70	15	1.27	1.13

silicon carbide powders sintered at 1950 °C. Due to the adoption of Al_2O_3 and Y_2O_3 as sintering additives and YAG was formed, the silicon carbide ceramics showed intergranular fracture. The local regions of the ceramic fracture surface are shown in Fig. 8.137(e) AL 50 vol% and (f) BL 52 vol%. The grain size of BL 52 vol% silicon carbide ceramics should be slightly larger. The polished and etched surfaces of ceramics are shown in Fig. 8.137 (g) AL 50 vol% and (h) BL 52 vol%. It can be seen that the microstructure of ceramics from BL 52 vol% suspension is more compact and smooth. It is also proved that the preparation of silicon carbide ceramics using BL powder is better than AL.

Figure 8.138 shows the properties of silicon carbide ceramics prepared using these three kinds of powder at different temperatures. With the same solid loading, there is a weak effect on the properties of ceramics between AL and BL. The highest flexural strengths of AL and BL 50 vol% silicon carbide ceramics are 669 ± 37 MPa and 672 ± 45 MPa, respectively. Due to the higher zeta potential and lower viscosity of BL suspension, there is a slightly higher flexural strength and relative density of BL 52 vol% silicon carbide ceramics coagulated at 60 °C about 697 ± 30 MPa and 99.3%. Therefore, in contrast to acid and base treatment of silicon carbide ceramics, a higher solid loading of the suspension can be prepared using base treatment which is better than acid treatment.

Table 8.20 lists the properties of silicon carbide ceramics which are prepared by different methods. It can be seen from the table that dense silicon carbide ceramics can be prepared by all methods; however, the flexural strength of silicon carbide ceramics prepared by liquid phase pressureless sintering (LPPS) is higher than solid phase pressureless sintering (SPPS). By colloidal forming, the silicon carbide ceramics with the highest flexural strength of 697 ± 30 MPa can be prepared by powder chemical treatment and LPPS in our experiment. Moreover, though there is a higher strength about 710 MPa of silicon carbide ceramics which prepared by liquid phase hot pressure sintering (LPHS), the strength with error of ±105 MPa is also higher

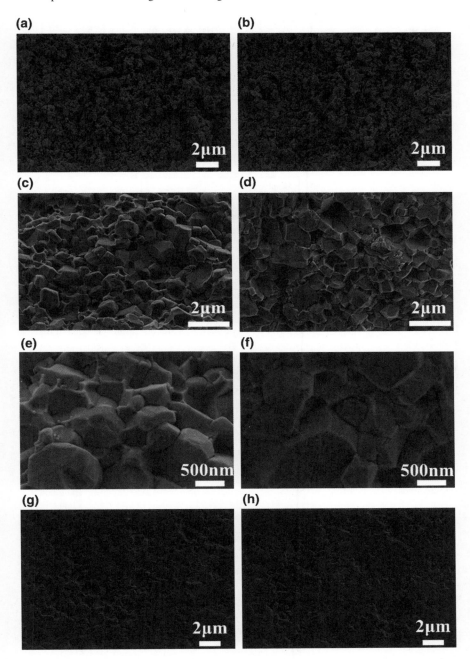

Fig. 8.137 Microstructures of silicon carbide green bodies **a** 50 vol% AL and **b** 52 vol% BL, microstructures of silicon carbide sintered ceramic fracture surfaces **c**, **e** 50 vol% AL and **d**, **f** 52 vol% BL, polished and etched (by a CF_4 gas) surfaces **g** 50 vol% AL and **h** 52 vol% BL

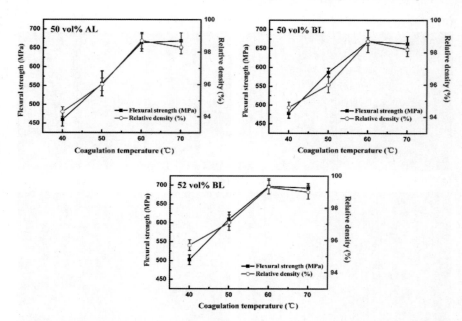

Fig. 8.138 Effect of coagulation temperature on properties of silicon carbide ceramics sintered at 1950 °C for 2 h

Table 8.20 Properties comparison of silicon carbide ceramics prepared by different methods

Powder treatment	Forming method	Sintered process	Flexural strength (MPa)	Relative density (%)	References
No treatment	DCC	SPPS		98.5	Si et al. (2010)
No treatment	Slip casting	LPPS		99.6	Gubernat et al. (2015)
No treatment	Gel casting	SPPS	520 ± 65	Up to 98	Jiang and Zhang (2011)
No treatment	Gel casting	SPPS	531 ± 38	Up to 98	Zhang et al. (2013)
Additives chemical treatment	Gel casting	LPPS	694 ± 48	Up to 99	Zhang et al. (2015)
No treatment		LPHS	712 ± 105	99.4	Sciti and Bellosi (2000)
SiC powder chemical treatment	DCC dispersant reaction	LPPS	697 ± 30	99.3	Present work

8.2 Dispersion Removal Coagulation Casting

than the samples prepared by LPPS. It is indicated that the silicon carbide ceramics prepared by LPPS is more stable.

In this section, dense silicon carbide ceramics was prepared using chemical treatment powder by DCC via dispersant reaction method and liquid phase sintering. In order to prepare well-dispersed suspension, the silicon carbide powder was surfaced chemical treated using ammonium peroxydisulfate and ammonium carbonate in advance. It was found that the treatment leads to a shift of IEP and a good dispersion behavior of the silicon carbide suspension. After DCC process and drying, green samples with homogeneous microstructure can be obtained. The sample with relative density of 99.3% can be prepared using 52 vol% base treatment powder sintered at 1950 °C for 2 h. The effect of acid and base treatment on the properties of sintered ceramics was not significant. It is impact on the viscosity of suspension and the properties of green body, and the solid loading was more affected by the acid and base treatment. The results showed that chemical treatment of silicon carbide powder was effective for the dispersion of suspension preparation and the DCC process.

8.3 Dispersant Hydrolysis Method

This section aims to introduce a novel in-situ coagulation method without coagulation agent and adjusting pH value for yttria-stabilized zirconia (YSZ) suspension via dispersant hydrolysis. Sodium tripolyphosphate (STPP) is used as dispersant to prepare electrostatic stabilized YSZ suspension. Influences of STPP contents on the dispersion and pH value of YSZ suspension were investigated. It indicated that there was a well-dispersed YSZ suspension with the addition of 0.3 wt% STPP at pH = 10. Influence of coagulation temperature on coagulation process and properties of green body was investigated. The sufficiently high viscosity suspension to coagulate was achieved at 60–80 °C. The coagulation mechanism was different from traditional direct coagulation casting. The suspension was coagulated by directly shifting the isoelectric point to the original state without increasing the ionic strength and adjusting the pH value. It was proposed that the YSZ suspension could be destabilized via decrease of zeta potential by sodium tripolyphosphate hydrolyzing at elevated temperature. Coagulated samples with wet compressive strength of 3.60 MPa could be demolded without deformation by treating 50 vol% YSZ suspension with 0.3 wt% STPP at 60 °C for 30 min. Dense YSZ ceramics with flexural strength of 887 ± 110 MPa and relative density of 98.9% had been prepared by this method sintered at 1450 °C for 3 h.

(1) **Raw materials, preparation, and characterization**

Figure 8.139 shows the particle distribution and microstructure of YSZ powder. Tetragonal zirconia powder (OZ-3Y, stabilized with 3 mol% Y_2O_3) with average particle size of 0.13 μm and specific surface area of 8.53 m^2/g was produced from Guangdong Orient Zirconic Ind. Sci. & Tech. Co., Ltd., China. Sodium tripolyphosphate was used as dispersant with different concentrations. $Na_4P_2O_7$, NaH_2PO_4, and

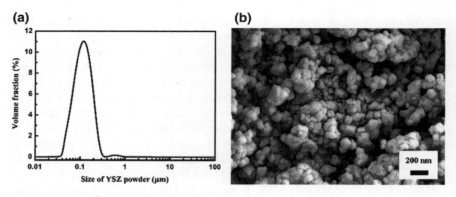

Fig. 8.139 **a** Particle distribution and **b** Microstructure of YSZ powder

Na_2HPO_4 were used in proportion as the hydrolysates of sodium tripolyphosphate to investigate the effect of hydrolysates on viscosity of alumina suspension.

Figure 8.140 shows the flowchart of the in-situ coagulation process via dispersant hydrolysis. Different solid loadings of YSZ suspensions were prepared by tumbling the YSZ powder, dispersant, water, and grinding media for 24 h. Grinding media was zirconia balls with diameter of 5–10 mm. The mass ratio between YSZ powder and grinding media was 2:1. The suspensions were degassed for 15–20 min under vacuum condition. Then, the YSZ suspensions were cast into plastic molds. The molds were placed in a water bath which was set a heating rate of 5 °C/min. The heating was lasted a period of time at different temperatures and then demolded. The samples were dried at 80 °C for 24 h. Finally, the green bodies were sintered at 1450 °C for 3 h.

Fig. 8.140 Flowchart of the in-situ coagulation process via dispersant hydrolysis

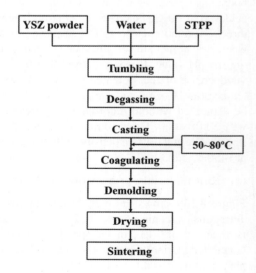

8.3 Dispersant Hydrolysis Method

LE438 pH meter (Mettler, Toledo, Switzerland) was employed to measure the pH value of YSZ suspension. Zeta Potential Analyzer (CD-7020, Colloidal Dynamics Co., Ltd., Ponte Vedra Beach, FL, USA) was employed to measure the zeta potential of YSZ suspensions with a stirring speed setting at 300 r/min. 10 vol% YSZ suspension was used in the test. HCl (1 mol/l) and NaOH (3 mol/l) aqueous solutions were prepared and used to adjust the pH value of the suspensions. In order to analyze the influence of pH value on zeta potential of alumina suspension, the ionic strength and conductivity of the suspension were maintained a constant via adding NaCl (analytic reagent). Rheometer (KINEXUS PRO, Malvern Instruments, Worcestershire, UK) attached with PC25 C0138 AL cylinder and C25 R0634 SS spindle was employed to measure the rheological properties of the YSZ suspensions. FTIR spectra were collected on Vertex 70v Fourier Spectrophotometer (Bruker, Germany) using pressed KBr pellets as reference substance. AG-IC20KN (Shimadzu, Tokyo, Japan) testing machine with a crosshead speed of 0.5 mm/min was employed in the wet compressive strength test. The samples were made into a cylindrical with diameter of 25.5 mm and a height between 25 and 30 mm. The measurement was immediate after demolding. For the flexural strength, the sintered specimens were prepared with dimension of 3 mm × 4 mm × 36 mm and measured via three-point bending test method using the same testing machine. The pressure side of the samples was polished before the measurement. The density of the sintered ceramics was measured using the water displacement method. The microstructure of YSZ powder, green bodies, and ceramics were observed by FESEM (MERLIN VP Compact; Carl Zeiss, Jena, Germany). For observing the microstructure of YSZ ceramics after sintering, the specimens were thermally etched at 1400 °C for 30 min.

(2) **Effect of dispersant on rheological of yttria-stabilized zirconia suspension**

Sodium tripolyphosphate (STPP) was selected as dispersant to prepare YSZ suspension. Figure 8.141 shows the effect of dispersant on zeta potential of 10 vol% YSZ suspension. The zeta potential was measured at constant ionic strength of 50 mmol/l

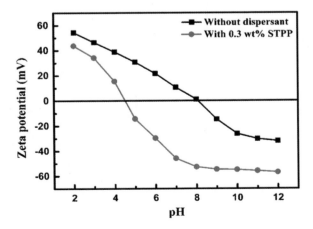

Fig. 8.141 Effect of dispersant on zeta potential of 10 vol% YSZ suspension

and conductivity of 3.5 mS/cm. It indicates that the isoelectric point (IEP) of YSZ suspension without dispersant is pH = 8. A high zeta potential of 45 mV appeared at pH = 3. The IEP moves to pH = 4.5 after the addition of sodium tripolyphosphate. It is obviously different from the suspension without dispersant. With dispersant, YSZ suspension shows a higher absolute zeta potential of −60 mV in pH range of 8–12. It indicates that the absorption of STPP enhances the surface negative charge on YSZ particles since the sodium tripolyphosphate is an anionic surfactant. Therefore, it increases the electrostatic repulsion among the YSZ particles and the absolute potential value of the suspension in alkaline environment.

Figure 8.142 shows the effect of the STPP content on zeta potential of 10 vol% YSZ suspensions and viscosity of 45 vol% YSZ suspensions. The zeta potential was measured at a constant pH value of 11. It can be seen the absolute zeta potential firstly increases and then decreases with the increase of dispersant. Correspondingly, the viscosity of YSZ suspension decreases with the increase of dispersant content and then increases with further addition. The extremum appears with the content of STPP at 0.3 wt%. Ionic dispersant can make the increase of the surface charge, thereby increasing the electrostatic repulsion between particles. The YSZ suspension achieves stability when the content of STPP reaches a certain amount. But an excessive addition will lead to increase in ionic strength and viscosity of suspension. Therefore, the content of STPP is selected as 0.3 wt%.

Figure 8.143 shows the relative viscosity of YSZ suspensions as a function of volume fraction at the shear rate of 100 s^{-1}. The Dougherty–Krieger (D–K) equation is commonly used to express the relationship between the relative viscosity (η_r) and volume fraction (φ) for hard sphere system:

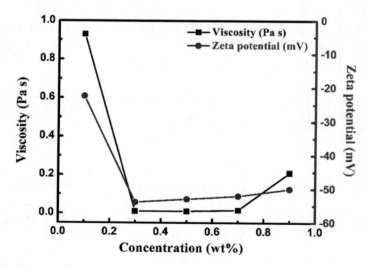

Fig. 8.142 Effect of STPP content on zeta potential of 10 vol% YSZ suspension and viscosity of 45 vol% YSZ suspension

8.3 Dispersant Hydrolysis Method

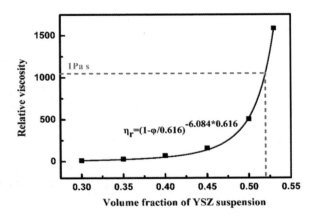

Fig. 8.143 Relative viscosity of YSZ suspensions as a function of volume fraction at the shear rate 100 s^{-1}

$$\eta_r = \left(1 - \frac{\varphi}{\varphi_m}\right)^{-[\eta]\varphi_m} \quad (8.31)$$

In this equation, φ_m is the maximum packing fraction, which has the order of 0.64 for random packing, and [η] is known as intrinsic viscosity and calculated as 2.5 for the perfect hard spheres. Usually, φ_m and [η] are chosen as variables at the same time to fit the experimental results and analyze the suspension structures (Xu et al. 2015a, b). The relative viscosities of YSZ suspensions are simulated as a function of volume fraction of the particles using the D–K equation. It can be expressed as

$$\eta_r = \left(1 - \frac{\varphi}{0.616}\right)^{-6.048*0.616} \quad (8.32)$$

According to the fitted data, it is found that the maximum solid loading of the suspension can be prepared is 61.6 vol%. When the solid loading reaches a certain amount, the rheological properties of suspension will get worse with further increasing due to the fact that the particles in the solution are too crowded. An excessive solid loading will result in unable to prepare the suspension and unable to measure the viscosity. It can be seen that the relative viscosity of YSZ suspensions increased slowly when the solid loading is below 45 vol% and increases rapidly above 50 vol%. It indicates that the particles begin to become crowded when the solid loading is higher than 50 vol%. Generally, the viscosity of suspension less than 1 Pa s is considered to meet the requirements of casting processes (Si W.J., et al., 2010). The viscosity of water is calculated by 0.8904 × 10^{-3} Pa s, and the relative viscosity of 1 Pa s is 1123. The relative viscosities of 50 vol% and 52.5 vol% YSZ suspensions are 505 and 1572, respectively. In order to avoid high viscosity of the suspension which may hinder the preparation process and impact on the properties of sintered ceramics, 50 vol% YSZ suspension was chosen even if the maximum solid loading is 61.6 vol%.

Fig. 8.144 Effect of coagulation temperature and hydrolysates (HS) of dispersant on the viscosity of YSZ suspensions

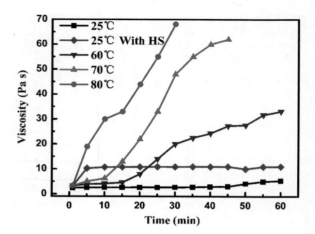

(3) **Fabrication of yttria-stabilized zirconia ceramics**

To investigate the influence of temperature on viscosity of YSZ suspensions, the viscosity of 50 vol% YSZ suspension without coagulation agent at different temperatures has been shown in Fig. 8.144. The viscosity of the suspension at 25 °C increases slightly to 5 Pa s when the treatment time is 60 min. However, the viscosity of the suspension increases rapidly at 60 °C to 20 Pa s for 30 min. A higher viscosity of 50 Pa s is obtained within 20 min upon the temperature increase to 80 °C. The results indicate that the suspension without coagulation agent can be coagulated at elevated temperature for a short time. Moreover, the effect of hydrolysate (HS) of dispersant on the viscosity of YSZ suspensions is shown in Fig. 8.144 which is discussed in the coagulation mechanism part.

Figure 8.145 investigates the pH value and zeta potential of 10 vol% YSZ suspension at 60 °C. It indicates that the pH value of suspension decreases slightly with treatment time from 9.9 to 9.0 after 60 min at 60 °C. The absolute potential also decreases from −60 to −15 mV. There are no complex chemical reactions from addition of any coagulation agents and pH regulators. Therefore, the coagulation mechanism can be attributed to that sodium tripolyphosphate is an alkaline phosphate which can hydrolyze in heating environment above 50 °C. The hydrolysis equations are listed as follows:

$$Na_5P_3O_{10} + H_2O \rightarrow Na_4P_2O_7 + NaH_2PO_4 \quad (8.33)$$

$$Na_5P_3O_{10} + H_2O \rightarrow Na_3HP_2O_7 + Na_2HPO_4 \quad (8.34)$$

$$Na_3HP_2O_7 + H_2O \rightarrow NaH_2PO_4 + Na_2HPO_4 \quad (8.35)$$

Over the hydrolysis processes above, the hydrolysis equations of sodium tripolyphosphate can be combined and listed as follows:

8.3 Dispersant Hydrolysis Method

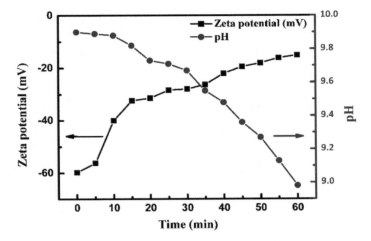

Fig. 8.145 Zeta potential and pH value of 10 vol% YSZ suspension at 60 °C

$$2Na_5P_3O_{10} + 3H_2O \rightarrow Na_4P_2O_7 + 2NaH_2PO_4 + 2Na_2HPO_4 \quad (8.36)$$

It is apparent that the hydrolysates from STPP include $Na_4P_2O_7$, NaH_2PO_4, and Na_2HPO_4. There is no well dispersion of these hydrolysates for YSZ particles which lead to the decrease of the zeta potential of YSZ suspension from −60 to −15 mV.

Figure 8.146 shows the IR spectra of STPP and YSZ powders, suspension, and green body. It can be seen that the IR spectrum of all samples in the vOH region exhibits three OH bands at 3419 and 1634 cm^{-1} (bound water) (Mekhemer 1998; Silva and Lameiras 2000). The Zr–11O bands exhibit at 500–650 cm^{-1} which is similar to the research by reference (Tonsuaadu et al. 2012). The molecular structural of

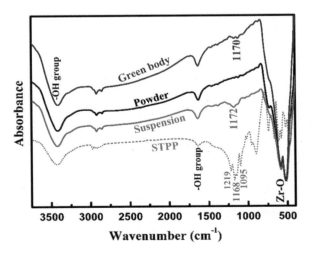

Fig. 8.146 IR spectra of STPP, YSZ powder, and YSZ suspension with STPP and green body

STPP and the absorption peak types of STPP is listed in Table 8.21. It can be seen that there is a very strong νP=O peak at 1168 cm^{-1} and two strong peaks at 1095 and 1219 cm^{-1} of STPP, respectively. In STPP, νP=O is usually symmetrical stretching vibration with a band, if asymmetrical can appear two bands (Klahn et al. 2004). It is clear that the νP=O peak at 1168 cm^{-1} is not symmetrical. YSZ suspension with STPP shows the obviously same νP=O peak at 1172 cm^{-1}, but not at 1095 cm^{-1} or 1219 cm^{-1}. It indicates that the absorption cannot be in the form of these two functional groups' absorption peaks at 1095 and 1219 cm^{-1}. Therefore, the absorption of STPP on the YSZ powder is caused by the functional group's absorption peaks at 1168 cm^{-1} due to the P–O. Moreover, there are two weak peaks of YSZ green body exhibited at 1170 cm^{-1}. It is due to the hydrolysis of sodium tripolyphosphate which results in the fracture of P–O and the weakening of the spectral band. Due to the fracture of the P–O, the adsorption of the sodium tripolyphosphate on the surface of YSZ particle will disappear and the dispersion of suspension will lose gradually. It can result in the increase of viscosity in Fig. 8.144 and the decrease of zeta potential of the suspension in Fig. 8.145 at elevated temperature. Eventually, the hydrolysis of sodium tripolyphosphate and the adsorption on the YSZ particle lead to the coagulation of suspension.

To better exhibit the effect of sodium tripolyphosphate on YSZ particle surface after hydrolyzing, Fig. 8.147 shows the schematic illustration of the effect of STPP on YSZ particles and coagulation process. The YSZ suspension with positive charged particles could be prepared at pH = 5 without STPP and pH regulator. Suspension with stronger surface negative electricity particles was achieved after adding STPP without other additives while the pH value increased to ca. 10 due to the alkaline of STPP. After heating at 60 °C, the surface charge of YSZ decreased obviously due to the sodium tripolyphosphate hydrolysis, and the pH value decreased slightly to ca. 9. The significant decrease of the surface charge would lead to the electrostatic

Table 8.21 The molecular structural and absorption peak types of STPP

Functional group	Wavenumber (cm^{-1})	Peak intensity	Type (remarks)
[STPP molecular structure diagram]			Molecular structural of STPP
P–O–P	1095	Strong	
[P–O–P diagram with =O groups]	1168	Very strong	νP=O
(RO)$_2$P–O–P(RO)$_2$ with =O	1219	Strong	νP=O

8.3 Dispersant Hydrolysis Method

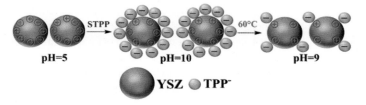

Fig. 8.147 Schematic illustration of the effect of STPP on YSZ particles and coagulation process

repulsion of the YSZ particles reduced and the dominant position of the van der Waals force. Thus, the YSZ particles lost stability and then the suspension coagulated in-situ rapidly.

Figure 8.148 shows the change route of zeta potential and pH by in-situ coagulation via dispersant hydrolysis (In-situ-DH) and DCC via pH shift to IEP (DCC–ΔpH) which are made with arrow. As we know, there are two methods to achieve the direct coagulation casting of ceramic suspensions which include either shift the pH value to IEP of ceramic powder system (DCC–ΔpH) or increase the ionic strength of the suspension (Baader et al. 1996a, b). The alumina suspension with sodium tripolyphosphate as dispersant also can be coagulated by the DCC–ΔpH method, and the change route of zeta potential is shown in Fig. 8.148 made with arrows. In this paper, the suspension is coagulated in-situ via dispersant hydrolysis which has a phenomenon similar to DCC–ΔpH that the zeta potential of suspension after coagulating is close to IEP. Figure 8.148 shows the change route of zeta potential and pH value of YSZ suspension by in-situ coagulation via dispersant hydrolysis (In-situ-DH) which is also made with arrows. For DCC–ΔpH method, the change route of zeta potential is followed by the change of pH value, and the effect of dispersant is persistent with the coagulation process. For in-situ-DH method, there is subtle change on pH value from 10 to 9 and a distinct change on zeta potential from

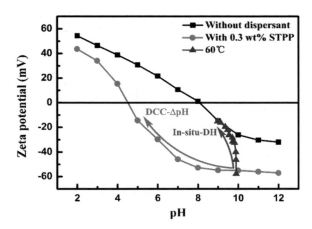

Fig. 8.148 The change routes of zeta potential and pH by in-situ coagulation via dispersant hydrolysis (In-situ-DH) and DCC via pH shift to IEP (DCC–ΔpH) which are made with arrows

−60 to −15 mV after heating at 60 °C for 60 min. The zeta potential regains the original state of YSZ suspension without dispersant at pH = 9. It indicates that a different and shorter change route of zeta potential with pH of in-situ-DH than DCC–ΔpH due to the dispersant (STPP) hydrolysis to achieve the same objective which is close to the IEP. However, the IEP is the IEP of the suspension without dispersant rather than the suspension with dispersant, so that the fundamental reason of the coagulation process can be explained as the fact that the IEP is directly shifted by the dispersant hydrolysis. Furthermore, the hydrolysates of sodium tripolyphosphate are ions and cannot adsorb on the alumina particle surface which have been proved from the IR spectra analysis. Although these hydrolysates are only small quantities, they are still remained in the suspension which may increase the ionic strength and influence the coagulation process of suspension. Thus, to analyze the effect of the hydrolysates on coagulation process, Fig. 8.144 shows the viscosity of suspension with these hydrolysates (HS) at 25 °C which are added in proportion according to the content of sodium tripolyphosphate in suspension. It can be seen that there is an obvious increase on the viscosity of suspension after adding the hydrolysates for 5 min. Then, the viscosity maintains the fixed value of 10 Pa s in 1 h. The viscosity is not high enough to achieve the coagulation of suspension. It indicates that although the hydrolysates can really affect the viscosity of the suspension, the suspension cannot be coagulated by increasing ionic strength. The main reason for the coagulation of suspensions is still caused by the hydrolysis of the dispersant and directly shifting the isoelectric point of the suspension to the original state without dispersant.

Figure 8.149 shows the photograph of YSZ green bodies and sintered samples. It can be seen from Fig. 8.149 that the YSZ green bodies without defects such as micropore and bubble could be prepared by in-situ-DH method. The smooth surface sintered sample without cracks was prepared by this method and sintered at 1450 °C for 3 h.

(4) **Properties of yttria-stabilized zirconia ceramics**

Table 8.22 shows the effect of coagulation temperature on coagulation time and properties of the green bodies which were prepared by in-situ coagulation via dispersant hydrolysis method. It is observed that the coagulation time shortens observably from

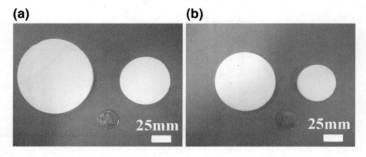

Fig. 8.149 Photographs of YSZ **a** green bodies and **b** sintered samples

8.3 Dispersant Hydrolysis Method

Table 8.22 Effect of coagulation temperature on coagulation time and properties of YSZ green bodies

Coagulation temperature (°C)	50	60	70	80
Coagulation time (min)	60	30	20	15
Wet compressive strength (MPa)	3.01	3.25	3.60	3.34
Dry compressive strength (MPa)	1.24	1.41	1.87	1.60
Coagulation shrinkage (%)	2.41	2.33	2.17	2.09
Drying shrinkage (%)	2.50	2.43	2.25	2.18

60 min at 50 °C to 30 min at 60 °C with the increase of the coagulation temperature. This is due to the slow hydrolysis rate of sodium tripolyphosphate at 50 °C. The coagulation time is reduced to 15 min at 80 °C with further increasing coagulation temperature. The compressive strength of wet green bodies increases from 3.01 MPa to 3.60 MPa with the increase of coagulation temperature from 50 to 70 °C due to the hydrolysis rate of sodium tripolyphosphate accelerated. It decreases to 3.34 MPa with further increase to 80 °C which is an excessive temperature causing the moisture evaporation and non-uniform coagulation. On account of the hydrolysis rate of dispersant accelerated with the increase of coagulation temperature, the coagulating rate is quickened, and the linear shrinkage of coagulated bodies shows a downward trend.

Figure 8.150a shows the microstructure of YSZ green body. It can be seen that homogenous microstructure without pores has been obtained by this method. Figure 8.150b shows the polished surface of YSZ ceramics after thermal etching at 1400 °C for 30 min. Dense YSZ ceramics with 0.5–2 μm zirconia crystal grains was obtained. Figure 8.150c, d shows the microstructure of YSZ fracture surface of sintered ceramics before and after thermal etching, respectively. It is observed that the well-grown tetragonal zirconia crystal grains are presented in the samples which is shown a transcrystalline rupture. Moreover, the relative density of sintered samples is 98.9% which are measured by water displacement technique. The flexural strength of sintered samples is 877 ± 110 MPa which are measured by three-point bending test.

In-situ coagulation of YSZ suspension via dispersant hydrolysis without coagulation agent has been reported. YSZ suspension with electrostatic stabilization is prepared by adding 0.3 wt% sodium tripolyphosphate (STPP) as dispersant. The suspension is destabilized by the decrease of zeta potential closed to IEP from dispersant hydrolysis without coagulation agent and adjusting pH value. It indicated that the viscosity of the suspension could increase to around 30 Pa s after treating at 60 °C for 30 min. YSZ green bodies with compressive strength of 3.01–3.60 MPa are obtained by heating the YSZ suspension at 60–80 °C for 15–30 min. Dense YSZ ceramics with relative density of 98.9% and flexural strength of 887 ± 110 MPa was obtained. The results suggest that the novel in-situ coagulation method with a short coagulation time and without any coagulation agent and pH regulator is a simple and promising colloidal forming method.

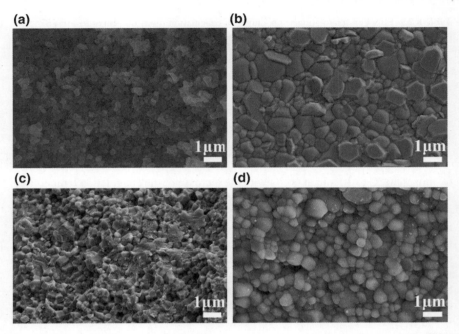

Fig. 8.150 Microstructures of YSZ **a** green body, **b** polished surface of sintered ceramics after thermal etching, fracture surface of sintered ceramics **c** before and **d** after thermal etching at 1400 °C for 30 min

8.4 Dispersant Separation Method

This part introduces a dispersion removal coagulation casting method for nonaqueous alumina suspension via dispersant separation. High solid loading and low viscosity alumina suspensions were prepared using normal octane as solvent. Influence of oleic acid on the dispersion of suspension was investigated. There was a well-dispersed alumina suspension with 1.3 wt% oleic acid. Influence of gelation temperature on the coagulation process and properties of green body was investigated. The sufficiently high viscosity to coagulate the suspension was achieved at −20 °C. The gelation temperature was controlled between the melting point of dispersant and solvent. The gelation mechanism is proposed that alumina suspension is destabilized by dispersant separating out from the solvent and removing from the alumina particles surface. The alumina green body with wet compressive strength of 1.07 MPa can be demolded without deformation by treating 53 vol% alumina suspension at −20 °C for 12 h. After being sintered at 1550 °C for 3 h, dense alumina ceramics with relative density of 98.62% and flexural strength of 371 ± 25 MPa have been obtained by this method.

8.4 Dispersant Separation Method

Table 8.23 Molecular structure and selected physical data of dispersant and solvent

Name	Oleic acid	Octane
Molecular formula	$C_{18}H_{34}O_2$	C_8H_{18}
Structure	(CH₂)₆–COOH / (CH₂)₆–CH₃	
Molecular weight (g/mol)	282.46	114.23
Density (g/cm³)	0.891	0.7
Melting point (°C)	13.4	−56.8
Boiling point (°C)	350	125.6
Vapor pressure (kPa)		1.33
Viscosity (mPa s)		0.288

(1) Raw materials, preparation, and characterization

CT3000SG alumina powder (Almatis, Ludwigshafen, Germany) with average particle size of 0.33 μm and specific surface area of 8.08 m²/g was used. Oleic acid and octane were used as dispersant and solvent to prepare the alumina suspension, respectively. The molecular structure and selected physical data of oleic acid and octane are listed in Table 8.23.

Figure 8.151 shows the flowchart of the low-temperature gelation process. Low viscosity alumina suspensions with different solid loadings were prepared by tumbling the alumina powder, dispersant, and solvent for 24 h. Zirconia balls with diameter of 5–15 mm were used as grinding media. The mass ratio between grinding media and alumina powder was 1:2. Due to the low surface tension of solvent, the suspension can be directly cast into a plastic mold without degassing process. The samples were placed in low-temperature environment at different temperatures for

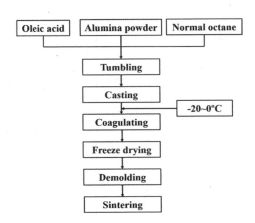

Fig. 8.151 Flowchart of the gelation process

a period of time. Then, the coagulated bodies were dried in vacuum for 15–20 min and then demolded. The dried samples were sintered at 1550 °C for 3 h at a heating rate of 5 °C/min.

Zeta potential of the nonaqueous alumina suspension was measured by a Zeta Potential Analyzer (CD-7020, Colloidal Dynamics Co., Ltd., Ponte Vedra Beach, FL, USA) with a stirring speed of 300 r/min. In zeta potential measurement, 10 vol% alumina suspension was employed. Acetic acid and triethylenediamine were used to adjust the pH value of the suspension. Both of them are analytical reagent. Rheometer (KINEXUS PRO, Malvern Instruments, Worcestershire, UK) attached with C25 R0634 SS spindle and PC25 C0138 AL cylinder was employed to measure the rheological properties of the suspension. After gelation and drying process, the samples were demolded, and their compressive strength was measured in an AG-IC20KN (Shimadzu, Tokyo, Japan) testing machine with a crosshead speed of 0.5 mm/min. Cylindrical bodies with 25.5 mm in diameter and a height between 25 and 30 mm were prepared for compressive strength measurement. Using the same mechanical testing machine, the flexural strength of the ceramic specimens with dimension of 3 mm × 4 mm × 36 mm were measured via the three-point bending test. The pressure side of the samples was polished before measurement. The density of the sintered ceramics was measured using the water displacement technique. The microstructure of green bodies and sintered ceramics was observed by a field emission scanning electron microscope (MERLIN VP Compact; Carl Zeiss, Jena, Germany). For observing the microstructure of alumina ceramics after sintering, the samples were thermal etching at 1500 °C for 30 min.

(2) **Effect of dispersant on rheological of alumina suspension**

Generally, stability mechanism of ceramic suspension includes electrostatic stabilization, steric stabilization, and electrosteric stabilization (Lewis 2000). In addition, there is another unusual stability mechanism of ceramic suspension which is called semisteric stabilization (Horn 2010). The semisterically stability ceramic suspensions use organic solvent as the dispersion medium. The dispersion mechanism can be considered as the facts that the ceramic particles change to lipophilicity in nonpolar solvent (wetting) and the modification of Hamaker constant decreases the van der Waals attractive potential energy due to the effect of dispersant (Zhou et al. 2001).

As we know, oleic acid is one of the fatty acids which are commonly used as a dispersant for ceramic powder in tape casting technology (Moreno and Cordoba 1997; Moloney et al. 1995). It is not a long-chain molecule (see Table 8.23) dispersant and cannot achieve steric stabilization of suspension. In previous studies, there are many reviews which have shown that oleic acid dispersed ceramic suspension is due to the semisteric stabilization (Moreno and Cordoba 1997; Moloney et al. 1995; Ruckenstein 1993; Liu 1999; Zurcher and Graule 2005).

To investigate the effect of oleic acid on dispersion of alumina powder, Fig. 8.152 shows the effect of oleic acid concentration on viscosity of 53 vol% alumina suspension. Due to the high viscosity of the suspension, the data of the oleic acid content lower than 0.5 wt% cannot be measured. It can be seen in Fig. 8.152 that the viscosity of alumina suspension decreases gradually with the addition of oleic acid. The lowest

8.4 Dispersant Separation Method

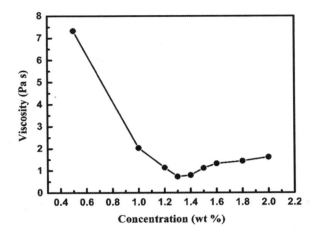

Fig. 8.152 Effect of oleic acid content on the viscosity of 53 vol% alumina suspensions

viscosity of 0.73 Pa s appears when the content of oleic acid reaches 1.3 wt%. It indicates that the alumina suspension can achieve sufficient stability with 1.3 wt% oleic acid in octane. Furthermore, it can be also seen in Fig. 8.152 that excess additives of oleic acid result in the increase of viscosity. It can be explained by dispersant–dispersant and/or dispersant–solvent interactions due to the free-floating dispersant molecules (Zurcher and Graule 2005). The alumina particles are easy to reunite and aggregate which leads to the increase of the viscosity. The minimum value of the viscosity appears when the concentration of oleic acid is at 1.3 wt%. Based on the analysis above, the content of dispersant is selected as 1.3 wt%.

The Dougherty–Krieger (D–K) equation is commonly used to express the relationship between the relative viscosity (η_r) and volume fraction (φ) for hard sphere system (Krieger 1972).

$$\eta_r = \left(1 - \frac{\varphi}{\varphi_m}\right)^{-[\eta]*\varphi_m} \tag{8.37}$$

Figure 8.153 shows the relative viscosity of alumina suspensions as a function of volume fraction at the shear rate of 100 s^{-1}. The experimental data points are fitted to an empirical expression relating the relative viscosity to the volume fraction of solids loading. The best fit is obtained with $\varphi_m = 0.6058$ and $\eta = -7.1483$. The viscosity of the solvent (octane) is calculated as $0.288*10^{-3}$ Pa s, and the relative viscosity of 1 Pa s is 3472. It indicates that the relative viscosity of suspension increases rapidly above 45 vol%. The relative viscosities of 53 vol% and 55 vol% alumina suspensions are 2550 and 4909, respectively. Generally, the viscosity of suspension less than 1 Pa s is considered to meet demand for casting process (Si et al. 2010). The handling of the suspensions will be difficult at higher volume fractions. Therefore, the alumina suspension of 53 vol% was chosen to avoid the high viscosity of the suspension which may impact on the properties of sintered ceramics.

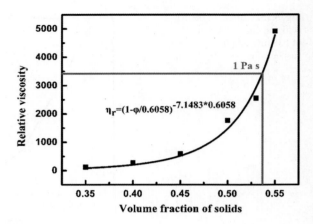

Fig. 8.153 Relative viscosity of alumina suspensions as a function of volume fraction at the shear rate 100 s^{-1}

$$\eta_r=(1-\varphi/0.6058)^{-7.1483*0.6058}$$

(3) **Fabrication of alumina ceramics**

To investigate the influence of temperature on viscosity of alumina suspensions, the viscosity of 53 vol% alumina suspension without coagulation agent at different temperatures is shown in Fig. 8.154. The viscosity of the suspension at 25 °C increases slightly to 4 Pa s after 24 h. However, the results at low-temperature environment are rather different. The viscosity of the suspension increases gradually to 8 Pa s at 0 °C for 24 h. Sufficiently higher viscosity of 20 Pa s for coagulation is obtained under the temperature decrease to −20 °C. The results indicate that the suspension without gelation agent can form gel and be coagulated finally in a low-temperature environment.

In order to analyze the mechanism of gelation, the physical condition of oleic acid at different temperatures is investigated as shown in Fig. 8.155. It can be seen in Fig. 8.155a that oleic acid is liquid at room temperature. However, the oleic acid

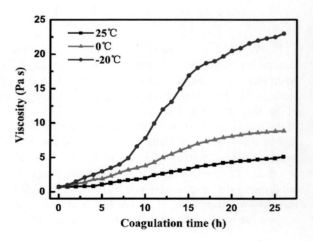

Fig. 8.154 Effect of temperature on the viscosity of alumina suspensions

8.4 Dispersant Separation Method

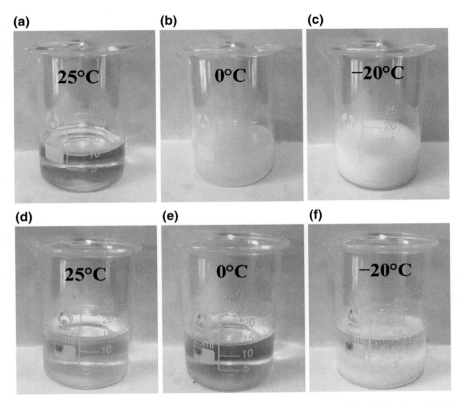

Fig. 8.155 Photographs of the physical condition of oleic acid at **a** 25 °C, **b** 0 °C, and **c** −20 °C after 12 h; the dissolution of 5 ml oleic acid in 10 ml octane at **d** 25 °C, **e** 0 °C, and **f** −20 °C after 12 h

converts into solid state at 0 °C in Fig. 8.155b due to the fact that the ambient temperature is lower than its melting point (13.4 °C), but it still has few flowabilities. When the ambient temperature decreases to −20 °C in Fig. 8.155c, oleic acid is completely transformed into a solid without mobility. The dissolution of oleic acid in octane is also investigated in Fig. 8.155. It is clear that oleic acid can be completely dissolved in octane at room temperature in Fig. 8.155d and slightly separated out at 0 °C in Fig. 8.155e. However, it is almost completely precipitated from octane at −20 °C in Fig. 8.155f. It indicates that the dissolution of oleic acid in octane decreases with the decrease of ambient temperature. As we know, the melting point of oleic acid is 13.4 °C which is higher than the gelation temperature. Therefore, the gelation can be summarized that the dispersant gradually changes into the solid state and the solubility of dispersant decreases in solvent at low temperature, so that the gelation mechanism is proposed that the semisterically stable alumina suspension is destabilized by dispersant removal via dispersant separated out from solvent at low temperature. Moreover, due to the removal of dispersant, the wetting effect of

Table 8.24 Effect of gelation temperature on coagulation time and properties of green bodies

Gelation temperature (°C)	Coagulation time (h)	Green body Linear shrinkage (%)	Compressive strength (MPa)
0	Not gelling		
−5	48	0.82	0.48
−10	36	0.6	0.52
−15	30	0.52	0.89
−20	24	0.49	1.07

oleic acid on alumina particles is lost. The Hamaker constant of the system returns to the original high. It is the fundamental reason of the suspension solidification. Furthermore, though the ambient temperature is lower than melting point of dispersant, liquid–solid transition and the decrease of solubility are gradual processes which may affect the gelation time and properties of green bodies.

(4) **Properties of alumina ceramics**

Through the above experiments, the gelation mechanism of alumina suspension is analyzed. However, the purpose of this experiment is to prepare high-performance alumina ceramics. So, the effect of ambient temperature on the coagulation process and properties of alumina green body are worth researching. Table 8.24 lists the effect of gelation temperature on coagulation time and the property data of green bodies. The alumina suspension cannot be coagulated at 0 °C. This is because the oleic acid can still mostly dissolve in octane and disperse the alumina particle. The suspension began to form a gel gradually; when the temperature decreased to −5 °C. Yet, the gelation process is not only the lengthiest for 48 h to demold, but also the compressive strength of the green body is the lowest to 0.48 MPa at −5 °C. With the gelation temperature decreasing to −20 °C, the gelation time shortens to 12 h, and the compressive strength also increases to 1.07 MPa. It indicates that gelation of the suspension occurs at less than −5 °C; gelation time shortens and the compressive strength increases with the decrease of temperature which is due to the increase of dispersant separating rate out from the solvent and the precipitated amount of dispersant. In addition, the shrinkages of the green bodies are 0.49–0.82% which are lower than that prepared by most in-situ coagulation method at elevated temperature (Xu et al. 2012c, 2015a, b; Gan et al. 2016, 2017). It may be ascribed to the low vapor pressure of octane which prevents evaporation of solvent around the whole gelation processes.

Figure 8.156 shows the photographs of alumina (a) green bodies and (b) sintered samples. It can be seen from Fig. 8.156 that the complex-shaped alumina green bodies could be prepared by this method without defects such as micropores and bubbles. Surface smooth sintered alumina ceramics without cracks was prepared by this method and sintered at 1550 °C for 3 h. Also, some big size green bodies have been prepared by this method. There is a slightly sunken phenomenon like typical

8.4 Dispersant Separation Method

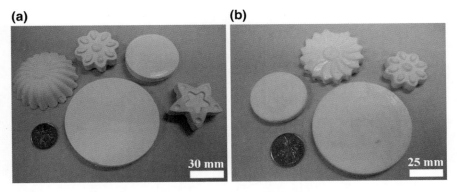

Fig. 8.156 Photographs of alumina **a** green bodies and **b** sintered ceramics at 1550 °C

structure changes in freeze casting in the middle position of the big size samples. The sunken phenomenon of green body may be due to the vacuum environment of the freeze-drying process.

Figure 8.157 shows the effect of gelation temperature on flexural strength and relative density of alumina ceramics. Relative density of sintered samples increases from 97.3 to 98.62% with the decrease of gelation temperature. The flexural strength of sintered samples also increases with the decrease of gelation temperature from 310 ± 22 to 371 ± 25 MPa. It indicates that the effect of temperature on properties of ceramics is distinct. Also, the trend of the influence on the ceramics is similar to that for green bodies which indicates the uniformity of suspension coagulation can

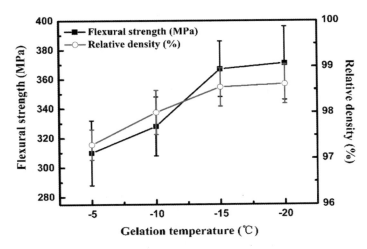

Fig. 8.157 Effect of the gelation temperatures on flexural strength and relative density of alumina ceramics

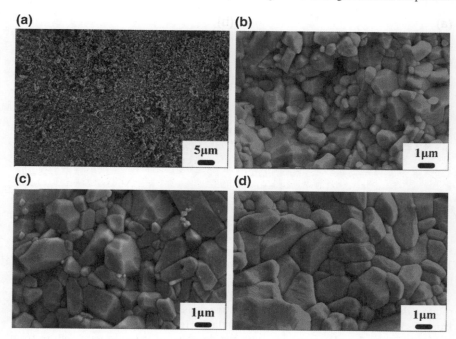

Fig. 8.158 Microstructures of alumina **a** green body and fracture surface of sintered ceramics with gelation temperature **b** −5 °C, **c** −15 °C, and **d** −20 °C thermal etched at 1500 °C for 30 min

be inherited from green bodies to the sintered ceramics and amplified due to the role of internal stress.

Figure 8.158a shows the microstructure of alumina green body. It can be seen that homogenous microstructure of green body without pores has been obtained by this method. Figure 8.158b–d shows the fracture surface of sintered ceramics gelled at different temperatures after thermal etching at 1500 °C for 30 min. It can be seen obviously that the ceramics gelled at −5 and −15 °C is not as dense as that gelled at −20 °C. The source of the holes is derived from green body which owing to the longer gelation time. The above results will affect the mechanical properties of ceramics. As a result, the dense alumina ceramics gelled at −20 °C with well-grown grains of 0.5–1 μm was obtained. These results suggest that the new coagulation process is a promising method to prepare near-net-shape ceramics with homogenous microstructures and good mechanical properties.

This method is a novel temperature-induced gelation method for nonaqueous alumina suspension. The semisterically stability 53 vol% alumina suspension can be prepared using 1.3 wt% oleic acid as dispersant in octane. The viscosity of the alumina suspension increases to around 20 Pa s after treating at −20 °C for 24 h. The alumina suspension is destabilized by the dispersant removing from the alumina particles surface without gelation agent at low temperature. The loss of the wetting effect of dispersant on the alumina particles and the recurrence of the Hamaker

8.4 Dispersant Separation Method

constant of the system are the fundamental reasons. The gelation process cannot occur above −5 °C due to the high solubility of the dispersant in the solvent. Green body with compressive strength of 1.07 MPa can be obtained by freezing the alumina suspension for 12 h. There are homogeneous microstructure and well-grown crystalline grains in sintered ceramics with relative density of 98.62% and flexural strength of 371 ± 25 MPa. The results suggest that the novel temperature-induced gelation method without any gelation agent for nonaqueous ceramic suspension is a simple and promising colloidal forming method for preparing high-performance near-net-shape ceramics.

References

Adar, J. H., Mutsuddy, B. C., & Drauglis, E. J. (1988). Stabilization of silicon carbide whisker suspension: i influence of surface oxidation in aqueous suspension. *Advanced Ceramic Materials, 3*, 231–234.

Adolfsson, E. (2008). Gelcasting of zirconia using agarose. *Journal of the American Ceramic Society, 89*, 1897–1902.

Agarwala, M. K., Bandyopadhyay, A., Weeren, R., et al. (1996). Rapid fabrication of structural components. *American Ceramic Society Bulletin, 75*, 60–66.

Alford, N. M., Birchall, J. D., & Kendall, K. (1987). High strength ceramics through colloidal control to remove defects. *Nature, 330*, 51–53.

Baader, F. H., Graule, T. J., & Gauckler, L. J. (1996a). Direct coagulation casting–a new green shaping technique. Part I. Processing Principles. *Indian Ceramic, 16*(1), 31–35.

Baader, F. H., Graule, T. J., & Gauckler, L. J. (1996b). Direct coagulation casting–a new green shaping technique. Part II. Application to Alumina. *Indian Ceramic, 16*(1), 36–40.

Baes, C. F., & Mesmer, R. E. (1986). *The hydrolysis of cations*. Malabar: Krieger Publishing Co.

Bakarič, T., Budič, B., Malič, B., et al. (2015). The influence of pH dependent ion leaching on the processing of lead–zirconate–titanate ceramics. *Journal of the European Ceramic Society, 35*, 2295–2302.

Balzer, B., Hruschka, M. K. M., & Gauckler, L. J. (1999). Coagulation kinetics and mechanical behavior of wet alumina green bodies produced via DCC. *Journal of Colloid and Interface Science, 216*(2), 379–386.

Balzer, B., Hruschka, M. K. M., & Gauckler, L. J. (2001). In situ rheological investigation of the coagulation in aqueous alumina suspensions. *Journal of the American Ceramic Society, 84*(8), 1733–1739.

Barabási, A.-L., Albert, R., & Schiffer, P. (1999). The physics of sand castles: maximum angle of stability in wet and dry granular Media. *Physics A, 266*(1–4), 366–371.

Bergström, L. (1998). Shear thinning and shear thickening of concentrated ceramic suspensions. *Colloid Surface A, 133*, 151–155.

Binner, J. G., McDermott, A. M., Yin, Y., et al. (2006). In situ coagulation moulding: A new route for high quality, net–shape ceramics. *Ceramics International, 32*(1), 29–35.

Briscoe, B. J., Khan, A. U., & Luckham, P. F. (1998). Stabilising zirconia aqueous suspensions using commercial polyvalent electrolyte solutions. *Journal of the European Ceramic Society, 18*(14), 2169–2173.

Cai, K., Huang, Y., & Yang, J. (2005). Alumina gelcasting by using HEMA system. *Journal of the European Ceramic Society, 25*(7), 1089–1093.

Chelvi T. P. Ralhan R., 1995. Designing of thermosensitive liposomes from natural lipids for multimodality cancer therapy. *International Journal of Hyperthermia, 11*(5), 685–695.

Cho, J. M., & Dogan, F. (2001). Colloidal processing of lead lanthanum zirconate titanate ceramics. *Journal of Materials Science, 36*, 2397–2403.

Chu, S. Y., Chen, T. Y., Tsai, I. T., et al. (2004). Doping effects of Nb additives on the piezoelectric and dielectric properties of PZT ceramics and its application on SAW device. *Sensor and Actuators A: Physical, 113*, 198–203.

Davies, J., & Binner, J. G. P. (2000). Plastic forming of alumina from coagulated suspensions. *Journal of the European Ceramic Society, 20*, 1569–1577.

Delplancke, M. P., Powers, J. M., Vandentop, G. J., et al. (1991). Preparation and characterization of amorphous SiC: H thin films. *Journal of Vacuum Science and Technology, 9*, 450–455.

Evans, A. G., & Cannon, R. M. (1986). Overview no. 48: toughening of brittle solids by martensitic transformations. *Acta Metallurgica, 34*, 761–800.

Ewais, E., & Safari, A. (2010). Gelation of water–based PZT slurries in the presence of ammonium polyacrylate using agarose. *Journal of the European Ceramic Society, 30*, 3425–3434.

Ewais, E., Zaman, A. A., & Sigmund, W. (2002). Temperature induced forming of zirconia from aqueous slurries: mechanism and rheology. *Journal of the European Ceramic Society, 22*(16), 2805–2812.

Fanelli, A. J., Silvers, R. D., Frei, W. S., et al. (1989). New aqueous injection molding process for ceramic powders. *Journal of the American Ceramic Society, 72*(10), 1833–1836.

Flory, P. J. (1953). *Principles of polymer chemistry*. London: Cornell University Press.

Frankel, D. A., Lamparski, H., Liman, U., et al. (1989). Photoinduced destabilization of bilayer vesicles. *Journal of the American Ceramic Society, 111*(26), 9262–9263.

Franks, G. V., & Meagher, L. (2003). The isoelectric points of sapphire crystals and alpha–alumina powder. *Colloids and Surfaces A: Physicochemical and Engineering Aspects, 214*(1–3), 99–110.

Franks, G. V., Johnson, S. B., Scales, P. J., et al. (1999). Ion–specific strength of attractive particle networks. *Langmuir, 15*(13), 4411–4420.

Franks, G. V., & Lange, F. F. (1996). Plastic–to–brittle transition of saturated, alumina powder compacts. *Journal of the American Ceramic Society, 79*(12), 3161–3168.

Fukada, Y., & Nicholson, P. S. (2004). The role of Si–O species in the colloidal stability of silicon–containing ceramic powders. *Journal of the European Ceramic Society, 24*, 17–23.

Gaber, M. H., Hong, K., Huang, S. K., et al. (1995). Thermo–sensitive sterically stabilized liposomes: formulation microscopy, and invitro studies on mechanism of doxorubicin release by bovine serum and human plasma. *Pharmacological Research, 12*(10), 14016.

Gan, K., Xu, J., Lu, Y. J., et al. (2017). Preparation of silicon carbide ceramics using chemical treated powder by DCC via dispersant reaction and liquid phase sintering. *Journal of the European Ceramic Society, 37*(3), 891–897.

Gan, K., Xu, J., Zhang, X. Y., et al. (2016). Direct coagulation casting of silicon nitride suspension via a dispersant reaction method. *Ceramics International, 42*(3), 4347–4353.

Garvie, R. C., Hannink, R. H., & Pascoe, R. T. (1975). Ceramic steel? *Nature, 258*, 703–704.

Gaucker, L. J., Graule, T. J., & Baader, F. H. (1999b). Enzyme catalysis of alumina forming. *Key Engineering Materials, 159–160*, 135–150.

Gauckler, L. J., Graule, T. J., & Baader, F. H. (1999a). Ceramic forming using enzyme catalyzed reactions. *Materials Chemistry and Physics, 61*(1), 78–102.

Graule, T. J., Baader, F. H., & Gaucker, L. J. (1995). Casting uniform ceramics with direct coagulation. *ChemTech, 25*(6), 31–37.

Greenwood, R., & Kendall, K. (1999). Selection of suitable dispersants for aqueous suspensions of zirconia and titania powders using acoustophoresis. *Journal of the European Ceramic Society, 19*, 479–488.

Gubernat, A., Zych, L., & Wierzba, W. (2015). SiC products formed by slip casting method. *The International Journal of Applied Ceramic Technology, 12*(5), 957–966.

Guo, D., Cai, K., Huang, Y., et al. (2003c). Water based gelcasting of lead zirconate titanate. *Materials Research Bulletin, 38*, 807–816.

Guo, D., Cai, K., Li, L. T., et al. (2003a). Application of gelcasting to the fabrication of piezoelectric ceramic parts. *Journal of the European Ceramic Society, 23*, 1131–1137.

References

Guo, D., Cai, K., Li, L. T., et al. (2003b). Gelcasting of PZT. *Ceramics International, 29*, 403–406.
Guo, D., Li, L. T., Cai, K., et al. (2004). Rapid prototyping of piezoelectric ceramics via selective laser sintering and gelcasting. *Journal of the American Ceramic Society, 87*, 17–22.
Guo, D., Li, L. T., Gui, Z. L., et al. (2003d). Anti–crack machining of PZT ceramics for fabricating piezocomposites by using gelcasting technique. *Materials Science and Engineering B, 99*, 25–28.
Haas, P. A. (1989). Gel processes for preparing ceramics and glasses. *Chemical Engineering Progress, 854*, 44–52.
Halter, W. E. (1999). Surface acidity constants of α–Al_2O_3 between 25 and 70 °C. *Geochimica et Cosmochimica Acta, 63*(19–20), 3077–3085.
Halter, W. E. (2003). Surface acidity constants of α–Al_2O_3 Between 25 and organic dispersants in aqueous alumina suspensions. *Journal of the European Ceramic Society, 23*(6), 913–919.
Hannink, R. H. J., Kelly, P. M., & Muddle, B. C. (2000). Transformation toughening in zirconia–containing ceramics. *Journal of the American Ceramic Society, 83*, 461–487.
Harris, A. J., Vaughan, B., Yeomans, J. A., et al. (2013). Surface preparation of silicon carbide for improved adhesive bond strength in armour applications. *Journal of the European Ceramic Society, 33*, 2925–2934.
He, X., Su, B., Zhou, X., et al. (2011). Gelcasting of alumina ceramic using an egg white protein binder system. *Ceramics–Silikaty, 55*(1), 1–7.
Hesselbarth, D., Tervoort, E., Urban, C., et al. (2001). Mechanical properties of coagulated wet particle networks with alkali swellable thickners. *Journal of the American Ceramic Society, 84*(8), 1689–1695.
Hidber, P. C., Graule, T. J., & Gauckler, L. J. (1996). Citric acid–A dispersant for aqueous alumina suspensions. *Journal of the American Ceramic Society, 79*(7), 1857–1867.
Horn, R. G. (2010). Surface forces and their action in ceramic materials. *Journal of the American Ceramic Society, 73*(5), 1117–1135.
Hornbaker, D. J., Albert, R., Albert, I., et al. (1997). What keeps sandcastles standing? *Nature, 387*, 765.
Hotza, D., & Greil, P. (1995). Aqueous tape casting of ceramic powders. *Materials Science and Engineering A, 202*(1–2), 206–217.
Hsu, J. P., & Kuo, Y. C. (1997). The critical coagulation concentration of counterions: spherical particles in asymmetric electrolyte solutions. *Journal of Colloid and Interface Science, 185*(2), 530–537.
Huisman, W., Graule, T., & Gauckler, L. J. (1995). Conditions to obtain reliable high strength alumina via centrifugal casting, ceramic processing, science and technology. *Ceramic Transactions, 51*, 451–455.
Hunter, R. J. (1987). *Foundations of colloid science*. Oxford: Clarendon Press.
Isrealachvili, J. N. (1991). *Intermolecular and surface forces* (2nd ed.). London: Academic Press.
Jaeger, H. M., & Nagel, S. R. (1992). Physics of the granular state. *Science, 255*, 1523–1531.
Jia, Y., Kanno, Y., & Xie, Z. P. (2002). New gel–casting process for alumina ceramics based on gelation of alginate. *Journal of the European Ceramic Society, 22*(12), 1911–1916.
Jiang, D. L., & Zhang, J. X. (2011). Properties of carbide ceramics from gelcasting and pressure–Less sintering. In *Advance engineerig ceramics and composites: IOP conference series materials science and engineering* (Vol. 18, p. 202001).
Johnson, S. B., Franks, G. V., Scales, P. J., et al. (2000). Surface chemistry–Rheology relationships in concentrated mineral suspensions. *The International Journal of Mineral Processing, 58*, 267–304.
Kaplan, R., & Parrill, T. M. (1986). Reduction of SiC surface oxides by a Ga molecular beam: LEED and electron spectroscopy studies. *Surface Science, 165*, 45–52.
Karaman, M. E., Pashley, R. M., Waite, T. D., et al. (1997). A comparison of the interaction forces between model alumina surfaces and their colloidal properties. *Colloids and Surfaces A: Physicochemical and Engineering Aspects, 129–130*, 239–255.
Klahn, M., Mathias, G., Kotting, C., et al. (2004). IR spectra of phosphate ions in aqueous solution: predictions of a DFT/MM approach compared with observations. *Journal of Physical Chemistry A, 108*(108), 6186–6194.

Kono, K., Hayashi, H., & Takagishi, T. (1994). Temperature–sensitive liposomes: Liposomes bearing poly(N–isopropylacrylamide). *Journal of Controlled Release, 30*(1), 69–75.

Kosmac, T., Novak, N., & Sajko, M. (1997). Hydrolysis–assisted solidification (HAS): A new setting concept for ceramic net–shaping. *Journal of the European Ceramic Society, 17*(2–3), 427–432.

Krall, A. H., & Weitz, D. A. (1998). Internal dynamics and elasticity of fractal colloidal gels. *Physical Review Letters, 80*(4), 778–781.

Krieger, I. M. (1972). Rheology of monodisperse lattices. *Advances in Colloid and Interface Science, 3*(2), 111–136.

Laberty-Robert, C., Ansart, F., Deloget, C., et al. (2003). Dense yttria stabilized zirconia: Sintering and microstructure. *Ceramics International, 29*(2), 151–158.

Lange, F. F. (1982). Transformation toughening. *Journal of Materials Science, 17,* 225–234.

Lewis, J. A. (2000). Colloidal processing of ceramics. *Journal of the American Ceramic Society, 83*(10), 2341–2359.

Li, M. (2003). *Concise handbook of chemical data.* Beijing: Chemical Industry Press.

Lindner, L. H., Eichhorn, M. E., Eibl, H., et al. (2004). Novel temperature–sensitive liposomes with prolonged circulation time. *Clinical Cancer Research, 10,* 2168–2178.

Liu, D. M. (1999). Effect of dispersants on the rheological behavior of zirconia–wax suspensions. *Journal of the American Ceramic Society, 82,* 1162–1168.

Liu, X. J., Huang, L. P., Gu, H. C., et al. (2001). Direct coagulation casting of silicon nitride ceramics. *Journal of Inorganic Materials, 16,* 877–882.

Luther, E. E., Yanez, J. A., Franks, G. V., et al. (1995). Effect of ammonium citrate on the rheology and particle packing of alumina slurries. *Journal of the American Ceramic Society, 78*(6), 500–1495.

Mekhemer, G. A. H. (1998). Characterization of phosphate zirconia by XRD, Raman and IR spectroscopy. *Colloids and Surfaces A, 141*(2), 227–235.

Metcalfe, I. M., & Healy, T. W. (1990). Charge–regulation modelling of the schulze–hardy rule and related coagulation effects, Faraday Discuss. *Chemical Society, 90,* 335–344.

Millan, A. J., Nieto, M. I., & Moreno, R. (2000). Aqueous injection moulding of silicon nitride. *Journal of the European Ceramic Society, 20*(14), 2661–2666.

Moloney, V. M. B., Parris, D., & Edirisinghe, M. J. (1995). Rheology of zirconia suspensions in a nonpolar organic medium. *Journal of the American Ceramic Society, 78*(12), 3225–3232.

Moreno, R., & Cordoba, G. (1997). Oil–related deflocculants for tape casting slips. *Journal of the European Ceramic Society, 17*(2–3), 351–357.

Moreno, R., Salomoni, A., & Castanho, S. M. (1998). Colloidal filtration of silicon nitride aqueous slips. Part I: Optimization of the slip parameters. *Journal of the European Ceramic Society, 18,* 405–416.

Mustafa, S., Dilara, B., Neelofer, Z., et al. (1998). Temperature effect on the surface charge properties of $\gamma-Al_2O_3$. *Journal of Colloid and Interface Science, 204*(2), 93–284.

Navarro, A., Alcock, J. R., & Whatmore, R. W. (2004a). Aqueous colloidal processing and green sheet properties of lead zirconate titanate (PZT) ceramics made by tape casting. *Journal of the European Ceramic Society, 24,* 1073–1076.

Navarro, A., Whatmore, R. W., & Alcock, J. R. (2004b). Preparation of functionally graded PZT ceramics using tape casting. *The Journal of Electroceramics, 13,* 413–415.

Needham, D., & Dewhirst, M. W. (2001). The development and testing of a new temperature–sensitive drug delivery system for the treatment of solid tumors. *Advanced Drug Delivery Reviews, 53*(3), 285–305.

Noshchenko, O., Kuscer, D., Mocioiu, O. C., et al. (2014). Effect of milling time and pH on the dispersibility of lead zirconate titanate in aqueous media for inkjet printing. *Journal of the European Ceramic Society, 34,* 297–305.

Olhero, S. M., Garcia-Gancedo, L., Button, T. W., et al. (2012). Innovative fabrication of PZT pillar arrays by a colloidal approach. *Journal of the European Ceramic Society, 32,* 1067–1075.

Ozkurt, Z., & Kazazoğlu, E. (2010). Clinical success of zirconia in dental applications. *Journal of Prosthodontic, 19*(1), 64–68.

References

Parrill, T. M., & Chung, Y. W. (1991). Surface analysis of cubic silicon carbide (001). *Surface Science, 243*, 96–112.

Prabhakaran, K., Joseph, K., Sooraj, R., et al. (2010). Magnesia induced coagulation of aqueous PZT powder suspensions for direct coagulation casting. *Ceramics International, 36*, 2095–2101.

Prabhakaran, K., Kumbhar, C. S., & Raghunath, S. (2008a). Effect of concentration of ammonium poly(acrylate) dispersant and MgO on coagulation characteristics of aqueous alumina direct coagulation casting slurries. *Journal of the American Ceramic Society, 91*(6), 1933–1938.

Prabhakaran, K., Melkeri, A., Gokhale, N. M., et al. (2009). Direct coagulation casting of YSZ powder suspensions using MgO as coagulating agent. *Ceramics International, 35*, 1487–1492.

Prabhakaran, K., Raghunath, S., & Melkeri, A. (2008b). Mechanical properties of wet–coagulated alumina bodies prepared by direct coagulation casting using a MgO coagulating agent. *Journal of the American Ceramic Society, 91*(11), 3608–3612.

Prabhakaran, K., Raghunath, S., & Melkeri, A. (2008c). Novel coagulation method for direct coagulation casting of aqueous alumina slurries prepared using a poly(acrylate) dispersant. *Journal of the American Ceramic Society, 91*(2), 615–619.

Rak, Z., & Van, Tilborg P. (1991). Aqueous injection moulding process. *Euro-Ceramics II, 1*, 409–441.

Rao, S. P., Tripathy, S. S., & Raichur, A. M. (2007). Dispersion studies of sub–micron zirconia using Dolapix CE 64. *Colloids and Surfaces A, 302*(1), 553–558.

Roy, S., Rao, B. C., & Subrahmanyam, J. (2007a). Water–based gelcasting of lead zirconate titanate and evaluation of mechanical properties of the gelcast samples. *Scripta Materialia, 57*, 817–820.

Roy, S., Rao, B. C., & Subrahmanyam, J. (2007b). Evaluation of mechanical properties of gelcast lead zirconate titanate disks sintered at different temperatures. *Scripta Materialia, 57*, 1024–1027.

Ruckenstein, E. (1993). On the stability of concentrates non–aqueous dispersions. *Colloids and Surfaces, 69*(4), 271–275.

Sabin, J., Prieto, G., Messina, P. V., et al. (2005). On the Effect of Ca^{2+} and La^{3+} on the Colloidal Stability of Liposomes. *Langmuir, 21*(24), 10968–10975.

Sano, S., Oda, K., Ohshima, K., et al. (1995). Slip casting of silicon nitride and mechanical properties of sintered body V vacuum–pressure–assisted slip casting of silicon nitride powder produced by imide decomposition. *Journal of the Ceramic Society of Japan, 103*, 939–943.

Schenker, I., Filser, F. T., Aste, T., et al. (2008). Microstructures and mechanical properties of dense particle gels: Microstructural characterization. *Journal of the European Ceramic Society, 28*(7), 1443–1449.

Sciti, D., & Bellosi, A. (2000). Effects of additives on densification, microstructure and properties of liquid–phase sintered silicon carbide. *Journal of Materials Science, 35*(15), 3849–3855.

Sharma, D., Chelvi, T. P., Kaur, J., et al. (1998). Thermosensitive lipo–somal taxol formulation: heat–mediated targeted drug delivery in murine melanoma. *Melanoma Research, 8*(3), 240–244.

Shen, Z. G., Chen, J. F., Zou, H. K., et al. (2004). Dispersion of nanosized aqueous suspensions of barium titanate with ammonium polyacrylate. *Journal of Colloid and Interface Science, 275*, 158–164.

Si, W. J., Graule, T. J., Baader, F. H., et al. (2010). Direct coagulation casting of silicon carbide components. *Journal of the American Ceramic Society, 82*(5), 1129–1136.

Silva, V., & Lameiras, F. S. (2000). Synthesis and characterization of composite powders of partially stabilized zirconia and hydroxyapatite. *Materials Characterization, 45*(1), 51–59.

Smay, J. E., & Lewis, J. A. (2001). Structural and property evolution of aqueous– based lead zirconate titanate tape–cast layers. *Journal of the American Ceramic Society, 84*, 2495–2500.

Sprycha, R. (1989). Electrical double layer at alumina/electrolyte interface i. Surface charge and zeta potential. *Journal of Colloid and Interface Science, 127*(1), 1–11.

Sun, J., & Gao, L. (2003). Influence of forming methods on the microstructure of 3Y–TZP specimens. *Ceramics International, 29*, 971–974.

Sun, J., Gao, L., & Guo, J. K. (2000). Surface properties of silicon carbide powders and rheological properties of their slurries. *Journal of Inorganic Materials, 15*, 426–430.

Suntako, R., Laoratanakul, P., & Traiphol, N. (2009). Effects of dispersant concentration and pH on properties of lead zirconate titanate aqueous suspension. *Ceramics International, 35*, 1227–1233.

Surowiak, Z., Kupriyanov, M. F., & Czekaj, D. (2001). Properties of nanocrystalline ferroelectric PZT ceramics. *Journal of the European Ceramic Society, 21*, 1377–1381.

Tadros, T. F. (1996). Correlation of viscoelastic properties of stable and flocculated suspension with their interparticle interactions. *Advances in Colloid and Interface Science, 68*, 97–200.

Tarì, G., Olhero, S. M., & Ferreira, J. M. F. (2000). Influence of temperature on stability of electrostatically stabilized alumina suspensions. *Journal of Colloid and Interface Science, 231*(2), 221–227.

Teot, S., & Daniels, S. L. (1969). Flocculation of negatively charged colloids by inorganic cations and anionic polyelectrolytes. *Science and Technology, 3*(9), 825–829.

Tiller, F. M., & Tsai, C.-D. (1986). Theory of filtration of ceramics: I, Slip casting. *Journal of the American Ceramic Society, 69*(12), 882–887.

Tonsuaadu, K., Zalga, A., Beganskiene, A., et al. (2012). Thermoanalytical study of the YSZ precursors prepared by aqueous sol–gel synthesis route. *The Journal of Thermal Analysis and Calorimetry, 110*(1), 77–83.

Torre, J. P., & Bigay, Y. (1986). Fabrication of silicon nitride parts by slip casting. *The Ceramic Engineering and Science, 7*, 893–899.

Traiphol, N., Suntako, R., & Chanthornthip, K. (2010). Roles of polymeric dispersant charge density on lead zirconate titanate aqueous processing. *Ceramics International, 36*, 2147–2153.

Valdivieso, A. L., Bahena, J. L. R., Song, S., et al. (2006). Temperature effect on the zeta potential and fluoride adsorption at the $\alpha-Al_2O_3$ aqueous solution interface. *Journal of Colloid and Interface Science, 298*(1), 1–5.

Wang, F., Xie, Z. P., Jia, C., et al. (2011). Study on the process of Si_3N_4 ceramic with high properties by gel–casting. *Journal of Synthetic Crystals, 40*, 743–747.

Wang, J., & Gao, L. (1999). Deflocculation control of polyelectrolyte–adsorbed ZrO_2 suspensions. *Journal of Materials Science Letters, 18*(23), 1891–1893.

Wang, J., Gao, L., Sun, J., et al. (1999). Surface characterization of NH_4PAA-stabilized zirconia suspensions. *Journal of Colloid and Interface Science, 213*, 552–556.

Wang, K., & Wu, B. (1960). *Instability constant of complex* (pp. 22–168). Beijing: Science Press.

Wang, X. X., Murakami, K., Sugiyama, O., et al. (2001). Piezoelectric properties, densification behavior and microstructural evolution of low temperature sintered PZT ceramics with sintering aids. *Journal of the European Ceramic Society, 21*, 1367–1370.

van Wazer, J. R., & Callis, C. F. (1958). Metal complexing by phosphates. *Chemical Reviews, 58*(6), 1011–1046.

Wen, N., Cai, K., Xu, J., et al. (2011). Influence of high valence counter–ions on the rheology of alumina suspension. *Journal of the Chinese Ceramic Society, 39*(11), 1768–1772.

Wyss, H. M., Hütter, M., Müller, M., et al. (2002). Quantification of microstructures in stable and gelated suspensions from cryo–SEM. *Journal of Colloid and Interface Science, 248*(2), 340–346.

Wyss, H. M., Tervoort, E. V., & Gauckler, L. J. (2005). Mechanics and microstructure of concentrated particle gels. *Journal of the American Ceramic Society, 88*(9), 48–2337.

Wyss, H. M., Tervoort, E., Meier, L. P., et al. (2004). Relation between microstructure and mechanical behavior of concentrated silica gels. *Journal of Colloid and Interface Science, 273*(2), 455–462.

Xie, R., Liu, C., Zhao, Y., et al. (2015). Gelation behavior and mechanical properties of gelcast lead zirconate titanate ceramics. *Journal of the European Ceramic Society, 35*, 2051–2056.

Xie, R., Zhao, Y., Zhou, K., et al. (2014). Fabrication of fine–scale 1–3 piezoelectric arrays by aqueous gelcasting. *Journal of the American Ceramic Society, 97*, 2590–2595.

Xie, Z. P., Ma, J. T., Xu, Q., et al. (2004). Effects of dispersants and soluble counter–ions on aqueous dispersibility of nano–sized zirconia powder. *Ceramics International, 30*, 219–224.

Xu, J., Gan, K., Yang, M., et al. (2015a). Direct coagulation casting of yttria–stabilized zirconia using magnesium citrate and glycerol diacetate. *Ceramics International, 41*(4), 5772–5778.

Xu, J., Qu, Y., Qi, F., et al. (2014b). Direct coagulation casting of alumina using magnesium citrate as coagulating agent with glycerol diacetate as pH regulator. *Journal of Materials Science, 49*, 5564–5570.

Xu, J., Qu, Y., Xi, X., et al. (2012c). Properties of alumina coagulated bodies prepared by direct coagulation casting via high valence counter ions (DCC–HVCI). *Journal of the American Ceramic Society, 95*(11), 3415–3420.

Xu, J., Wen, N., Li, H., et al. (2012a). Direct coagulation casting of alumina suspension by high valence counter ions using $Ca(IO_3)_2$ as coagulating agent. *Journal of the American Ceramic Society, 95*(8), 2525–2530

Xu, J., Wen, N., Qi, F., et al. (2012b). Direct coagulation casting of positively–charged alumina suspension by controlled release of high valence counter ions from calcium phosphate. *Journal of the American Ceramic Society, 95*(7), 2155–2160.

Xu, J., Yang, M., Gan, K., et al. (2016). Reliable high strength alumina fabricated by DCC–HVCI using submicron calcium citrate complex. *Ceramics International, 42*, 8030–8037.

Xu, J., Zhang, Y., Gan, K., et al. (2015b). A novel gelcasting of alumina suspension using curdlan gelation. *Ceramics International, 41*(9), 10520–11025.

Xu, J., Zhang, Y., Qu, Y., et al. (2014a). Direct coagulation casting of alumina suspension from calcium citrate assisted by pH shift. *Journal of the American Ceramic Society, 97*(4), 1048–1053.

Yang, J., & Huang, Y. (2010). *Novel colloidal forming of ceramics.* Beijing: Springer.

Yang, J., Huang, Y., Meier, L. P., et al. (2002). Direct coagulation casting via increasing ionic strength. *Key Engineering Materials, 224–226*, 631–636.

Yang, J., Huang, Y., Si, W., et al. (1997b). Study on direct coagulation casting process of a–Al_2O_3. *Journal of the Chinese Ceramic Society, 25*(4), 514–519.

Yang, J., Xie, Z., & Huang, Y. (1997a). Effect of the soluble ions in the Si_3N_4 powder on the solids volume loading of suspension. *Journal of the Chinese Ceramic Society, 25*(6), 679–686.

Yang, J., Xu, J., Wen, N. et al. (2011b). A method for coagulating ceramics suspension via controlled release of high valence counter ions. CN Patent 2011102914263.

Yang, J., Xu, J., Wen, N., et al. (2013). Direct coagulation casting of alumina suspension via controlled release of high valence counter ions from thermo-sensitive liposomes. *Journal of the American Ceramic Society, 96*(1), 62–67.

Yang, J., Yu, J., & Huang, Y. (2011a). Recent developments in gelcasting of ceramics. *Journal of the European Ceramic Society, 31*(14), 2569–2591.

Yang, M., Xu, J., Gan, K., et al. (2016). Direct coagulation casting of alumina via controlled release of calcium from ammonium polyphosphate chelate complex. *Journal of Materials Research and Technology, 31*(1), 154–162.

Yang, Y., & Sigmund, W. M. (2003). A new approach to prepare highly loaded aqueous alumina suspensions with temperature sensitive rheological properties. *Journal of the European Ceramic Society, 23*(2), 253–261.

Yatvin, M. B., Kretz, W., Horwitz, B. A., et al. (1980). pH–sensitive liposomes: Possible clinical implications. *Science, 210*, 1253–1255.

Yatvin, M. B., Weinstein, J. N., Dennis, W. H., et al. (1978). Design of liposomes for enhanced local release of drugs by hyperthermia. *Science, 202*, 1290–1293.

Young, A. C., Omatete, O. O., Janney, M. A., et al. (1991). Gel–casting of alumina. *Journal of the American Ceramic Society, 74*(3), 612–618.

Yu, X. R., & Hantsche, H. (1993). Vertical differential charging in monochromatized small spot X–ray photoelectron spectroscopy. *Surface and Interface Analysis, 20*, 555–558.

Zhang, J. X., Jiang, D. L., Lin, Q. L., et al. (2013). Gelcasting and pressureless sintering of silicon carbide ceramics using Al_2O_3–Y_2O_3 as the sintering additives. *Journal of the European Ceramic Society, 33*(10), 1695–1699.

Zhang, J. X., Jiang, D. L., Lin, Q. L., et al. (2015). Properties of silicon carbide ceramics from gelcasting and pressureless sintering. *Materials and Design, 65*(20), 12–16.

Zhang, J. X., Ye, F., Sun, J., et al. (2005). Aqueous processing of fine ZrO_2 particles. *Colloids and Surfaces A: Physicochemical and Engineering Aspects, 254*, 199–205.

Zhou, Z. W., Scales, P. J., & Boger, D. V. (2001). Chemical and physical control of the rheology of concentrated metal oxide suspensions. *Chemical Engineering Science, 56*(9), 2901–2920.

Zurcher, S., & Graule, T. (2005). Influence of dispersant structure on the rheological properties of highly concentrated zirconia dispersions. *Journal of the European Ceramic Society, 25,* 863–873.

Appendix A
The Testing, Analyzing and Sintering Methods Used in Authors' Research

1. The flexural strength was measured by standard three-point bending method for ceramics on samples of size 4 mm × 3 mm × 40 mm with a loading speed of 0.5 mm/min. Weibull modulus was derived from 20 such tests on each material.

   Fracture toughness K_{IC} was measured by four-point bending test with a single edge notch beam (SENB), on an universal test machine (Model AG-2000G, Shimazu, Japan) with a sample size of 4 mm × 6 mm × 30 mm and a loading speed of 0.1 mm/min. Each set of data involves 4–5 samples.
2. Zeta potentials of ceramic powders were determined by a Zetaplus Analyzer (Brookhaven Instrument Corp., USA) using an electrophoretic light-scattering method at 25 °C. 1 mmol/L NaCl was used as the solvent and the powder concentration was maintained at 0.05 vol.%. The pH adjustment was carried out by using HCl and NaOH solutions.
3. The particle size distributions were measured by an X-ray centrifugal sedimentation technique (Instrument type: BI-XDC, made in Brookhaven Instrument Co., New York, USA).
4. Rheological behaviors of the slurry were tested using a rheometer (Model MCR300, Paar Physica USA, Glen Allen, Pa.).
5. The surface tension and wetting angel of the slurry were determined (Mode Processor Tensionmeter K12, Kruss USA, Charlotte, N.C).
6. SEM (Model S-450, Hitachi Corp., Tokyo) was used to characterize the microstructures.
7. Thermogravimetric analysis was carried out with the thermogravimetric analyser (TGA92-18, SETARAM Corp., France).
8. Density of specimens was determined by the Hg immersion method or by the Archimedes principle.

Notes: The above testing and analyzing methods are not specially indicated in the book. In order to avoid the iterative introductions in the every part, they are list altogether here. The other testing and analyzing methods used were marked in every chapter.

Appendix B
The Raw Materials Used in Authors' Research

1. Alumina, Henan Xinyuan Alumina Industry Co. Ltd., >99.7 wt% purity, a mean particle size 2.9 μm, and specific surface area 0.434 m^2/g.
2. Monomer: acrylamide (AM), Mitsui Toatsu Chemical Inc., Japan, chemical purity.
3. Crosslinker: N, N'-methylenebisacrylamide (MBAM), Beijing Hongxing Chemical Factory, chemical purity.
4. Catalyst: N, N, N', N'-tetramethylethylenediamine (TEMED), Xingfu Fine Chemical Institute, chemical purity.
5. Initiator: ammonium persulfate, $(NH_4)_2S_2O_8$, Beijing Chemical Reagent Company, chemical purity.
6. Dispersant: ammonium citrate (AC), the Beijing Chemical Reagents Company, chemical purity.
7. Dispersant: tri-ammonium citrate (TAC), the Beijing Chemical Reagents Company, chemical purity.

Notes: The above raw materials (alumina, monomer, crosslinker, catalyst and initiator) are not specially indicated in the book. In order to avoid the iterative introductions about raw materials in the every part, they are list altogether here. The other raw materials used were marked in every chapter.

Index of Scholars

A

Abbas E. N., 186, 221
Aboaf J. A., 309
Agarwals M. K., 360, 402
Aksay I. A., 135, 224
Alford N. M., 79, 134
Allen S. D., 403
Amott S., 80, 81, 134
Ando M., 403
Anifantis N., 403
Araki K., 371, 402, 404
Armstrong B. L., 309
Asad U. K., 200, 221
Askay I. A., 222
Athena T., 222

B

Baader F. H., 222, 224, 225, 306, 307
Baes C. F., 176, 221
Baets P. J. M., 134
Baldovino D., 307
Balzer B., 166, 173, 221
Baratoon M. I., 307
Barnes H. A., 120, 121, 127, 128, 134, 351–353, 357, 364, 403
Barsoum M., 303, 306
Baudin C., 136
Bauer W., 201, 219, 222
Bear A. D., 222
Beck E., 136
Bell N. S., 223
Bender J., 66, 75, 158, 221
Bengisu M., 357
Bergstrom L., 142, 148, 158, 174, 221, 307, 308
Bernd B., 68, 75

Biundo G. Lo, 75
Blackburn S., 135, 137, 403
Bleier A., 223
Boersma W. H., 134
Boger D. V., 223
Boiteux Y., 309
Bossel C., 140, 187, 221, 329, 357
Bowen H. K., 223, 357
Boyer S. M., 303, 307
Braccini I., 102, 134
Brandon D. G., 75, 307
Brandt J., 223
Breckenridge R. G., 76
Brian J. B., 221
Brichalli J. D., 134
Bridgwater J., 137
Brinker C. J., 173, 221
Briscoe B. J., 22, 40, 63, 75, 223
Brow K., 270, 307
Brownlow J. M., 76
Bruneau A., 75
Buckley R., 222
Burnfield K. E., 20, 75
Busca G., 241, 250, 307, 309

C

Cai K., 124, 126, 134, 135, 357
Calvert P. D., 76
Cao K., 128, 134
Carisey T., 19, 75, 307
Carlstrom E., 76
Carniglia S. C., 176, 222
Cartwright J. A., 222
Castanho M., 269, 307
Cesarano III J., 135
Cesio A. M., 309

Chang A. S., 365, 403
Chartier T., 23, 40, 75, 403
Chen Ruifeng, 161, 222, 382, 403
Chen X. B., 335, 357
Chen Y. L., 77, 95, 122, 130, 135, 136, 403
Chen Yirui, 222
Cheng H. M., 357
Chong J. S., 222
Christiansen E. B., 222
Christos A., 197, 222
Chryssolouris G., 394, 403
Chu Jinyu, 338, 357
Clarke D. R., 307
Clegg D. W., 222
Coimbra M. A., 136
Coleman M. M., 136
Collyer A. A., 222
Cook R. F., 285, 307
Corbel S., 135, 403
Corfe A. G., 403
Costa R. O. R., 132, 135
Courts A., 136
Crimp M. J., 289, 307
Cui X. M., 40, 66, 68, 75

D
Dai J. Q., 250–252, 273, 307
Dai Chunlei
Dakskobler A., 16
Dalgleish B. J., 308
Dalton P. D., 135
Danforth S. C., 309, 360, 403
Daufenbach J., 403
De Jonghe L. C., 307
De Kruif C. G., 198, 213, 222
Decker C., 136
Decourcelle S., 403
Dewhurst D. N, 173, 222
DiChiara R. A., 77
Diz H. M. M., 403
Doreau F., 40, 41, 43, 76, 77
Draskovich B. S., 221
Du H. Q., 332, 357
Dufaud O., 132, 135
Dutta J., 221, 357

E
Edirisinghe M. J., 136, 137
Egashira M., 20, 75
Epstien J., 308
Erauw J. P., 16, 222

Evans J. R. G., 16, 136, 137, 224
Ezis A., 264, 292, 307

F
Fagerholm H., 255, 281, 307
Fanelli A. J., 88, 135
Fang Y., 307
Feng Hanbao, 76
Ferreira J. M., 136, 224
Ferreira J. M. F., 76, 223, 363, 403
Fierro J. L. G., 307
Fiori C., 19, 75
Flitsch R., 309
Flynn L., 135
Fonseca A. T., 224
Ford R. W., 22, 75
Frei W. S., 135
Friedrich H., 308
Fukasawa T., 371, 403
Fukuda J., 76
Fulmer A., 134
Fu Wenqing, 404

G
Gabriele S., 309
Gao L., 309, 358
Garcia A. B., 309
Garg A. K., 256, 307
Gaucker L. J., 165, 181, 222
Gault R. S., 403
Gedde U. W., 135
George W. Scherer, 22, 77
Gilissen R., 2, 16
Giridhar R.V., 302, 307
Giuliano T., 158, 222
Goh S. H., 132, 135
Gonzenbach U. T., 404
Grader G., 75
Graeffe M., 307
Graff G. L., 77
Granja M. F. L., 66, 76
Grauckler L. J., 76
Graule T., 31, 76
Graule T. J., 174, 222, 223, 225, 306, 307
Greil P., 20, 76, 77, 226, 230, 241, 247, 255, 307–309
Griffin C., 403
Griffin M. L., 360, 403
Guedes M., 75
Guha S., 372, 403
Guo D., 127, 135
Guo J. K., 127, 358

Index of Scholars

Guo Ruisong, 185, 222
Gutierrez C. A., 40, 41, 43, 76
Gu T. R., 137
Gyurk W. J., 77

H

Hackley V. A., 260, 264, 308
Haggerty S., 309
Halloran J. W., 307, 402–404
Harame D. L., 247, 308
Harn Y. P., 221
Hausselt J., 222
Heartling C., 77
Hedberg E. L., 135
Hedenqvist M. S., 135
Heegh H., 358
Heinrich J., 310
Helbig M. H., 165, 167, 222
Hench L. L., 76, 307
He Panfa, 77
Herschel H., 198, 222
He Wenjin, 404
Hill D. J., 135
Hinczewski C., 360, 403
Hoffman R. L., 80, 135, 222
Hong Xiaoyin, 76
Horn R. G., 127, 128, 135
Ho Sah-Jai, 16
Hotza D., 19, 20, 22, 23, 40, 41, 76
Houiet R., 221, 357
Howatt G. N., 018, 76
Hruschka M. K. M., 221
Huang D., 136
Huang Yong (Huang Y.), 16, 17, 19, 70, 75, 77, 134, 135–137, 139, 140, 158, 187, 188, 222, 223, 281, 298, 307–310, 329, 348, 357, 358
Hutter M., 222
Hutton J. F., 134, 357, 403
Huzzard R. J., 130, 135, 403
Hyatt E. P., 19, 76

I

Ioanna L., 222
Israelachvili J., 350, 357

J

James R. O., 308
Jams S. Reed, 136
Janarthanan V., 136

Janney M. A., 16, 195, 222–224, 308–310, 329, 357, 358, 403, 404
Jennings H. M., 303, 308
Jiang D. L., 77
Jin J., 63, 76, 89, 135
Johansson L., 307
Johansson L. S., 307
Johnson R. E., 307
Joshi R. N., 256, 308
Jurge G. H., 71

K

Kaiser G., 309
Kangutkar P., 306
Kannan T. S., 136
Karagiannis S., 403
Karlsson G. E., 133, 135
Kedall K., 134
Khan A. U., 219, 223
Khoury I. A., 77
Kiggans J. O., 356, 357
Kim B. H, 308
Kim J. S., 256, 308
Kingery W. D., 152, 223, 329, 357
Kiratzis N. E., 128, 135
Kirby G. H., 223, 358
Kisailus D., 309
Kitahara A., 234, 270, 308
Kita K., 20, 76
Koczak M. J., 306
Koh Y. H., 371, 403
Koumoto K., 358
Kristoffer Krnel, 224
Kristoffersson A., 22, 40, 41, 76
Krstic V. D., 358
Krug S., 2, 16
Kulicke W. M., 222
Kulig M., 76, 247, 308
Kumar V., 218, 222
Kuo C. K., 105, 135

L

Laarz E., 246, 308
Lai M. J., 309
Landham R. R., 19, 76, 195, 223
Lange F. F., 79, 135, 335, 357
Lang F. F., 158, 223
Lang L., 218, 223
Laugier W. A., 307
Laugier-Wert A. H., 75
Laven J., 134
Lee E. J., 403

Lee H. D., 19, 76
Lee S. Y., 135
Le H. R. (Le Huirong), 222
Lennart Bergstrom, 142, 148, 158, 174, 221, 307, 308
Lenninger G., 308
Leong Y. K., 141, 223
Lesca S., 223
Leung D. K., 76, 223
Levine S., 234, 308
Lewis J. A., 223
Li B. G., 134
Li B. R., 329, 357
Li L., 403
Lim L. C., 158, 223
Lin M., 135
Lin M. C, 309
Lin M. J., 76
Liu D. M. (Liu Den-Mo), 2, 16, 158, 223
Liu K. S., 333, 357
Liu Q., 133, 135
Liu X. L. (Liu Xiaolin), 223
Liu X. Q., 357
Liu Y. F., 348, 357
Liu Z. W., 135
Longergan L., 222
Lorenzelli V., 307, 309
Low D. K. Y., 395, 403
Luckham P. F., 135, 223
Lumpp J. K., 395, 403
Luo Guanghua, 404
Lyckfeldt O., 136, 153, 205, 206, 208, 223

M
Ma Jan, 223
Ma J. C., 358
Ma J. M., 137, 357
Ma J. T., 200, 223
Ma K., 136
Malak K. M., 133, 135
Malghan S. G., 308
Ma L. (Ma L. G., Ma Liguo), 10, 16, 140, 187, 222–224, 307, 308, 329, 357, 404
Mandal B. M., 403
Manupin G. D., 77
Marchant J. Q., 358
Masayoshi Fuji, 404
Ma Tian, 224
Mcmillim S., 403
Mcnulty T. F., 364, 403
Medowski G. O., 20, 76

Meireles M., 358
Meng G. Y., 371, 403
Menon M., 371, 372, 403
Mesmer R. E., 176
Messing G. L., 77
Metev S. M., 397, 399, 400, 403
Mezzasalma S., 226, 234, 255, 308
Miao H. Z., 223
Michael Scheffler, 372, 386, 403
Mikulasek P., 332, 358
Millan A. J., 80, 135, 136
Mister R. E., 19, 22, 40, 76
Montgomery F. C., 357
Moreno R., 19, 25, 31, 45, 76, 77, 136, 307
Morissette S. L., 153, 223
Morris J. Jr., 21, 77
Moskala E. J., 132, 136
Moulson A. J., 301, 307, 308
Mueller M. E, 404
Murray A. J., 395, 403
Murthy J., 404
Myers D., 34, 77

N
Nagata K., 20, 77
Nagel A., 389, 403
Nahass P., 19, 22, 76, 77, 223
Nakanishi K., 136
Natansohn S., 309
Natarajan K. A., 224
Neale G., 281
Newnham R. E., 77
Nieto M. I., 136
Niihara K, 310
Nilsen K., 241, 250, 309
Nitzsche R., 308
Norman J., 75
Nottelmann H., 222
Novak S., 2, 16
Novich B. E., 77
Nunn S. D., 140, 187, 223, 309, 329, 358

O
Odian G. G., 195, 223
Ohji T., 403
Ohmura H., 76
Olhero S. M., 130, 136
Oliveira F. J., 310
Omatate O. O., 298, 309
Onoda Jr., 76
Orita N., 395, 403

Index of Scholars

Ouyang S. X., 75
Ozkan N., 75

P

Pagnoux C., 40, 41, 43, 75, 77
Paik U., 308
Pantano C. G., 307
Paolo Colombo, 403
Parks G. A., 308
Pasto A. E., 309
Paugh R. J., 307
Paul F. L., 221
Pavithran C., 309, 358
Penarroya R., 75
Perera D. I., 133, 136
Peterson B. C., 75
Petzow G., 308, 309
Peuckert M., 246, 309
Pham D. T., 403
Philip Molyneux, 215, 223
Pierre A. C., 113, 115, 136
Pober R. L., 76, 77
Popper P., 264, 292, 309
Porcile G., 307
Portu G., 75
Prabhakaran K., 298, 309, 348, 358
Pratt P. L., 308
Pugh R. J., 358

Q

Qian Xingnan, 77
Qin Zhengqi, 17, 21, 37, 77
Quinn G. D., 223

R

Rahaman M. N., 137, 158, 223, 246, 307
Raider S. I., 263, 309
Ramis G., 241, 250, 309
Ramousse S., 403
Rao R. R., 111, 115, 136
Ray B., 403
Rechard E., 18
Reed J. S., 79, 136
Rhine W. E., 77
Ribitsch V., 16
Ritzhaupt-Kleissl H. J., 222
Rolf W., 309
Ronald R. A., 308
Roopa H. N., 136
Rose K., 307
Rosenholm J. B., 307

Rourke W. J., 309
Ruddlesden S. N., 309
Runk R. B., 76
Russel W. B., 223

S

Sahling O., 76
Sanchez R. T., 226, 255, 309
Santacruz I., 136
Santhiya D., 224
Sasa Novak, 181, 224
Scherer G. W., 173, 221, 224
Schoichet M. S., 135
Schonholzer U. P., 222
Schubert H., 308
Schuetz J. E., 77
Schulz W., 309
Schwartz B., 18, 77
Schwelm M., 309
Scott W., 134
Sekino T., 310
Shanefield D. J., 76, 403
Shanks R. A., 136
Shanti N., 379, 404
Sharp C. M., 404
Sheldon B.W., 309, 302
Shih T. S., 403
Shih W. H., 224, 256, 309
Shih W. Y., 224, 309
Shimizu Y., 75
Sigmund W. M., 148, 158, 223
Silva R. F., 310
Silvers R. D., 135
Si W. J., 165, 223
Smith A. L., 308
Smith D. J., 19, 40, 77
Smolders A., 16, 222
Song J. H., 214, 224, 403
Sreienitz I., 358
Stadelmann H., 250, 309
Starov V., 352, 358
Steinborn G., 224, 358
Sterklow R. A., 358
Stetson H. N., 18, 77
Stevens R., 224
Strauss H., 335, 358
Strehlow P. A., 16
Studart A. R., 371, 404
Studer K., 132, 136
Subramanian S., 224
Su L. (Su Liang), 16, 224, 404
Sun J., 309

Sun L., 329, 358
Suresh B., 63, 77
Sutch R. D., 76

T
Tadros T. F., 77
Takata M., 358
Takatsuki S., 75
Takeaki Kato, 404
Tamai Y., 133, 136
Tanaka H., 136
Tanaka T., 224
Tang Q., 136, 137, 222, 281, 308, 309, 357
Tang S. Q., 357
Tan Mengkwang, 44, 76
Tari G., 75, 136, 224, 404, 158, 371
Teng W. D., 113, 136
ter Maat J. H. H., 16
Tervoort E., 404
Theodore P., 20, 76
Thyagarajan G., 132, 136
Tieggs T. N., 223
Tiegs T. N., 357
Tomandl G., 136
Tomaz Kosmac, 224
Tuan W. H., 285, 309

U
Uhlmann D. R., 223, 357

V
Varnell D. F., 136
Vasconcelos W. L., 135
Veiko V. P., 403
Venkataswamy K., 77
Vlajic M. D., 348, 358

W
Waack R., 19, 77
Waesche B., 140, 141, 187, 224, 329, 358
Wagner N. J., 221
Wakeman R. J., 358
Walers K., 134, 357, 403
Wang Aiguo, 390, 404
Wang Chang'an, 403
Wang H. T., 403
Wang J. (Wang Jian), 150, 224, 338, 358
Wang P. S., 308
Wang X. Z., 357
Ward A. G., 88, 89, 136

Watanab A., 308
Wautier H., 309
Wayne N. J., 23, 77
Wber K., 116, 136
Weeren R., 402
Wei H., 357
Wei Wen-Cheng J., 2, 16
Westmoreland C. G., 223
Whistler R. L., 80, 81, 136
White K. W., 307
Whittaker A. K., 135
Wilcox D. L., 77
Wolfgang M. S., 224
Wong P. M., 223
Wotting G., 310
Wu Chundu, 357
Wu J. G., 136
Wu Rong-Yuan, 16

X
Xiang Junhui, 70, 76, 77
Xiao Huashan, 390, 404
Xie Z. P. (Xie Zhipeng), 63, 77, 79, 88, 95,
    103, 135–137, 222, 223, 307–310,
    357, 358, 403, 404

Y
Yamashita K., 336, 358
Yang J. F., 287, 310
Yang J. (Yang J. L., Yang Jinlong), 3, 83, 86,
    134–137, 153, 222–224, 256, 308–
    310, 329, 357, 358, 403, 404
Yang T., 222
Yang X., 128, 137
Yang Y. (Yang Yunpeng), 142, 222, 224
Yang Zhengfang, 222
Yilmaz E., 357
Yi Z. Z., 111, 113, 137, 348, 358
Yoshikawa S., 19, 77
Young A. C., 140, 141, 181, 187, 194, 195,
    223–225, 258, 278, 310, 358, 360,
    404
Yu F., 307
Yu J., 134
Yu Z. Y., 75

Z
Zeng Y. P., 40, 41, 77
Zhang Fazhi, 371, 404
Zhang J. G., 79, 137
Zhang Liming, 158, 222, 224

Index of Scholars

Zhang T., 88, 90, 137
Zhang X. L., 357
Zhao Baolu, 404
Zhao L., 182, 223, 224
Zhdanov V., 358

Zheng W. J., 403
Zhou L. J., 111, 113, 137, 223, 283, 307, 336
Zhou Ninghuai, 263, 292, 310
Zhou Z. K., 89, 137
Ziegler G., 303, 310

Index of Terms

A

Acid cleaning, 238, 244, 245
Acrylamide (AM) C$_2$H$_3$CONH$_2$, 153, 190, 263, 330
Acrylic acid, 10, 11, 15, 125, 169, 174
Activation energies, 129
Agarose, 79–85, 87, 88, 90, 95
Agglomerations, 139, 158, 159, 161–163, 485
Alumina powder, 38, 41, 42, 46, 56, 64, 80–82, 84, 86, 89, 91, 94–96, 103, 110, 119, 124, 125, 130, 142, 145, 153, 159, 166, 178, 180, 182, 196, 204, 210–213, 216, 221, 287, 363, 364, 367, 368
Ammonium citrate, 31, 32, 37, 45, 46, 64, 111–113, 119, 124, 142, 153, 159, 166, 167, 169, 188, 342, 367, 368
Ammonium persulphate (AP) (NH$_4$)$_2$S$_2$O$_8$, 182, 551
Aqueous ceramic suspensions, 83, 84, 142
Aqueous colloidal injection molding, 1, 13, 15
Atomic force microscope (AFM), 323

B

Ball milling, 10, 12, 21, 37, 38, 142, 147, 148, 159, 166, 171, 180, 206, 227, 232, 239, 408
Binder agent, 18
Binders, 8, 20, 40, 41, 43, 44, 55, 60, 62–64, 68, 74, 218, 348, 371, 400
Breakdown strength, 311, 329, 335, 336
Bulk density, 58, 61, 201, 209, 383, 385, 386

C

Calcinations, 234, 238, 245, 252, 278
Calcined rutile mixture, 330, 331, 334, 335
Calcining, 81, 333
Calcium phosphate, 64, 103, 111, 116
Ceramic sheet, 17, 18, 38, 40, 71, 75, 360
Ceramic slurry, 1, 10–12, 15, 20, 22, 27–30, 40, 41, 79, 89, 91, 116, 312
Ceramic substrates, 17, 40
Ceramic suspension, 4, 7, 15, 63, 67, 80–84, 88, 95, 125, 127, 132, 139, 141, 153, 158, 159, 163, 167, 174, 175, 177, 178, 181, 188, 189, 192, 255, 314, 350, 363, 367, 387, 437, 440, 493, 495, 529, 534, 541
Ceramics with special porous structures, 372
Coagulation forming, 100
Cold isostatic pressing (CIP), 2
Colloidal behaviors, 112
Colloidal chemistry, 120, 329, 348, 350
Colloidal forming, 1, 3, 15, 101, 139–141, 145, 153, 155, 158, 159, 163, 165, 181, 182, 187, 225, 311
Colloidal injection molding (CIM), 1–4, 7, 13, 15, 338
Colloidal injection molding of ceramics (CIMC), 1
Computer-aided design (CAD), 360
Concentrated Si$_3$N$_4$ suspensions, 359, 387, 388, 393
Concentrated suspensions, 127, 179, 198, 201, 206, 213, 225, 232, 243, 246, 353, 388
Contact angle, 51, 52, 321–323
Counter-ion, 139, 178, 179, 181, 234, 235, 245, 246, 248

Cross-linker, 20, 23, 38, 119, 124, 128, 133, 187

D

Defoamer, 21, 34, 37, 40, 54, 56, 74
Deionization, 233–238
Deionized water, 20, 52, 56, 59, 64, 82, 91, 96, 103, 111, 117, 119, 125, 145, 159, 166, 187, 196, 212, 213, 226, 230, 232–235, 237, 250, 256, 257, 270, 276, 278, 288, 289, 293, 330, 331, 342, 348, 363, 367, 388, 391
Deposition, 5, 360, 395, 397
Differential thermal analysis (DTA), 34, 35, 38, 153–155, 278, 279
Diffuse reflectance infrared Fourier transform (DRIFT), 241
Direct coagulation casting (DCC), 167, 405–407, 411, 423, 424, 426, 432, 442, 451, 456–458, 465, 474, 492, 493, 501, 507–510, 521, 529
Dispersants, 23, 44–48, 111–113, 119, 121, 125–127, 141, 153, 174, 226, 314, 342, 343, 345, 346, 350, 352, 363
Dispersion removal, 405, 407, 501, 532

E

Elastic modulus, 8, 9, 80, 140, 184, 187, 188, 191
Electrophoretic light scattering (ELS), 81
Environmental scanning electron microscope (ESEM), 159
Enzyme catalysis, 79, 95–97
Exfoliation elimination effect, 131
Exothermic reaction, 10, 67, 153, 264, 292

F

Flexural strength, 8, 13, 130, 145, 146, 150, 152, 188, 195, 197, 201, 203, 204, 209, 210, 212, 219, 295, 304, 305, 347, 360
Flow behavior, 331
Forming process, 1, 3, 10, 13, 18, 22, 30, 51, 69, 75, 79, 84, 85, 91, 96, 97, 104, 111, 112, 139–141, 149, 158, 165, 174, 187, 190, 226, 228, 311, 315, 337, 359, 369
Free radical, 11, 12, 20, 25, 26, 30, 131, 153, 157, 182–184, 199, 390, 391, 393, 394

Freeze-gel-casting, 359, 372, 373, 375, 377, 379–382, 386
Fused deposition modeling (FDM)
Fused deposition of ceramics (FDC), 360

G

Gas-pressure sintering (GPS), 259
Gelation forming, 79, 80, 83, 86–88
Gelation time, 1, 99
Gel-bead forming, 315
Gel-casting, 1–3, 7, 10–12, 14, 15, 18, 19, 23, 54, 63, 70, 73, 79, 80, 87, 88, 90, 102, 103, 106, 110, 111, 115, 119, 120, 122–125, 131, 134, 139–141, 149–154, 157, 174, 181, 185–187, 190, 192, 194–196, 200, 203, 208–211, 214, 219, 221, 225–228, 230, 232, 255, 256, 258, 259, 262–267, 270, 275, 276, 278, 281–284, 287, 288, 291, 294–296, 298, 300, 304–306, 311, 329–331, 333, 335, 337, 338, 342, 347, 348, 350, 353–357, 359, 360, 362, 363, 366–368, 371–373, 375, 377, 379–383, 385–388, 393–396, 401, 402, 457, 458, 465, 482, 492, 493, 511, 520
Gel network, 7, 8, 81, 88–90, 101, 124, 166, 173, 174, 187, 191, 194
Gel-pen-ball, 311
Gel point, 1, 81, 85, 91
Gel-tape-casting, 17, 19, 20, 23, 28, 33, 37–40, 47, 54, 63, 64, 68, 70, 71, 73–75
Generation mechanisms, 158, 159, 163
Glass transition temperature, 44, 56, 60, 64, 74
Glycerol, 23, 49, 64, 66
Green body, 3, 6, 7, 9, 13–15, 20, 22, 49, 54, 70, 73, 79, 81, 85–91, 93–96, 100–103, 106–108, 110–112, 116–125, 129–131, 139–141, 146, 149, 152–155, 157–159, 165–174, 181–195, 200, 201, 203, 204, 208–210, 212, 214, 216–221, 225–228, 230, 232, 258, 259, 262–265, 267, 268, 273, 274, 276, 278, 279, 283–285, 287, 288, 291–295, 297–300, 304–306, 329, 330, 335, 336, 339, 342, 346–349, 354, 356, 359–362, 368, 371, 372, 374–378, 380, 383, 385, 387, 392–402, 405–408, 414–419, 421–423, 426, 428–431, 439, 440, 442, 449–451, 456–458, 463–466,

472, 473, 475, 481, 482, 484, 489, 490, 493, 494, 499, 500, 502, 503, 508–510, 516–519, 521–523, 527, 530–532, 534, 538–541
Green body strength, 22, 81, 380, 400
Green tape, 17, 20, 22, 23, 25, 27–29, 32–38, 40, 41, 44, 45, 49, 50, 52–56, 58, 60–63, 65, 66, 68, 69, 71, 73–75

H

High dielectric constant, 311, 329, 338
High performance ceramics, 359
High valence counter-ions, 405–408, 414–418, 421, 423, 428, 430–433, 437, 439–442, 450, 451, 456, 463–466, 469, 472–476, 481, 483, 489, 491–494, 499–501, 509, 511
Hydrostatic pressure, 6, 14
Hydroxyethyl acrylate (HEA), 140
2-hydroxyethyl methacrylate (HEMA), 124, 129, 131–134

I

Idle time, 139, 153, 166, 169–171, 173, 182, 390
Industrialization, 1, 2, 10, 15, 203, 361
Initiator, 4, 6, 10, 12, 15, 20, 23, 25, 28, 30, 31, 37, 38, 70–73, 119, 124, 125, 128, 129, 131, 133, 139, 141, 145, 146, 153, 159, 181, 182, 186–189, 195, 196, 203, 204, 208, 211, 215, 226–229, 258, 262, 264, 267, 276, 278, 288, 291, 292, 295–297, 299, 329, 330, 338, 346, 348, 360, 363, 367, 369, 372, 374, 377, 378, 387, 390, 391, 393–396
Injection molding, 1–5, 7, 8, 13–15, 80, 87, 88, 140, 187, 338, 348, 406
Inner stress, 6–9, 14, 15, 70, 139, 140, 181–188, 190, 191, 194
In-situ coagulation, 405–407, 488, 492, 496, 521, 522, 529–531, 538
Ion conductivity constants, 139, 176, 178–181
Ionic conductance, 174
Isoelectric point (IEP), 203, 241

L

Laminated Object Manufacturing (LOM), 360
Laser machining technology, 394, 395

Lead Zirconate Titanate (PZT), 311
Lead Zirconate Titanate (PZT) $Pb(Zr_{0.52}Ti_{0.48})O_3$, 311
Liquid medium, 115, 158, 159, 240, 241, 249, 313, 377
Low toxicity system, 79

M

Mechanical properties, 13, 117, 140–142, 149–152, 163, 187, 259, 265, 267, 269, 273–275, 320, 321, 370, 372, 377, 378, 382, 383, 401
Mechanical strength, 40, 111, 121, 145, 152, 396
Microbeads, 311–314, 316–319, 323, 328
Microcracks, 7, 342, 395, 397, 400–402
Monomer, 1, 4, 6, 7, 10, 15, 20, 23, 25, 26, 28, 30, 31, 34, 37, 38, 40, 45, 47, 50, 49, 51, 54, 63, 73, 79, 119, 120, 124, 125, 128, 129, 131, 139–141, 145, 146, 153, 154, 157, 159, 183, 187, 188, 190, 191, 195, 199, 201, 203, 205, 206, 208–212, 216, 217, 221, 226, 227, 258, 267, 276, 288, 290, 295–301, 304–306, 311, 329, 330, 335, 338, 348, 350, 352, 360, 362, 364, 367, 372, 387, 389, 390, 395, 396, 406, 465
Monomer polymerization, 1, 20, 25, 389
Monomer system, 195, 201, 203, 210, 221
Morphotropic Phase Boundary (MPB), 348

N

New colloidal forming processes
N, N'-methylenebisacrylamide (MBAM) $(C_2H_3CONH)_2CH_2$, 124
N, N, N', N'-tetramethylethylenediamine (TEMED), 124
Non-aqueous solvents, 19, 63
Non-oxide ceramics, 225

O

Organic monomer, 1, 20, 23, 40, 45, 49, 54, 195, 206, 295, 299–301, 338
Ozone generator, 311, 338–341

P

Plasticizer, 18, 20, 21, 23, 25, 32, 33, 37, 38, 46, 49, 50, 56, 60, 64, 66, 68, 73, 74, 187
Polyacrylamide, 25, 154, 195, 221, 377, 393

Polyacrylic acid-NH$_4$ (PAA–NH$_4$), 42
Polyelectrolyte, 42, 113, 114, 124–128, 145, 174
Polyethylene Glycol (PEG), 23, 140, 203
Polymerization rate, 28–31, 70, 73
Polymerization temperature, 28, 29
Polymerization times, 31
Poly (methacrylic acid) (PMAA), 81, 82, 84, 334
Polyvinyl Alcohol (PVA), 24, 41, 133, 355, 400
Polyvinyl Butyral (PVB), 400
Poly (vinylpyrrolidone) (PVP), 140
Porous alumina ceramics, 381, 382
Powder surface modification, 226
Pressure induced forming, 6, 14
Pressure-induced solidification, 187, 387

R

Refractory nozzle, 311, 342, 346, 347
Rheological behavior, 19, 20, 32, 38, 41, 43, 44, 46, 56, 60, 66, 68, 74, 80–82, 92, 100, 139, 141–143, 145, 147, 148, 153, 167, 198, 206, 214, 245, 252, 255, 289–291, 334, 335, 342, 343, 345, 346, 348, 350, 363, 368, 389, 394, 451, 466, 494
Rheological properties, 12, 22, 47, 50, 60, 65, 80, 81, 89, 92, 96, 99, 103, 106, 107, 111, 113, 114, 119, 120, 125, 142, 145, 146, 196, 198, 204, 210, 212–214, 216, 221, 238, 243–246, 248, 250, 252, 254, 262, 281, 290, 294, 363, 367, 368, 407, 408, 411, 442, 451, 453, 458, 503, 523, 525, 534
Rheology, 45, 82, 83, 90, 91, 99, 141, 246, 249, 503
Rutile capacitor, 311, 329

S

Scanning electron microscopy (SEM), 13
Sedimentation behavior, 115
Sediment stability, 343, 345
Selective laser sintering (SLS), 360
Shear rate, 33, 42, 43, 46, 49, 51, 56, 57, 60, 61, 65, 66, 81, 83, 99, 103, 111, 113, 115, 120, 121, 127, 128, 142–144, 146–149, 175, 180, 181, 196, 198, 199, 201, 204–206, 212–215, 243–245, 248, 272, 290, 291, 330, 335, 343, 345, 346, 351, 352, 364, 368, 407, 408, 442, 459–461, 468, 478–480, 485, 486, 503, 505, 514, 515, 524, 525, 535, 536
Shear viscosity, 42
SiC suspension, 112–116, 289–291, 293
Silicon carbide (SiC), 79, 112, 165, 348
Silicon nitride bonded silicon carbide (SNBSC, Si$_3$N$_4$ bonded SiC ceramics), 298
Silicon nitride (Si$_3$N$_4$), 187, 226, 227, 250, 251, 276, 298
Sintered body, 2, 107, 109, 146, 156, 164, 201, 295, 337, 378, 397
Sintering additives, 187, 331, 332
Slip casting, 1, 2, 63, 209, 305, 350
Sodium alginate, 63–68, 74, 79, 103, 104, 107, 111–113, 116, 118, 119
Solid freeform fabrication (SFF), 359
Solidification properties, 10
Solid loading, 8, 10, 12, 22, 40, 41, 49, 51, 55, 79, 83, 85, 86, 88, 127, 130, 139, 140, 142–145, 147–150, 152, 163, 164, 181, 184, 187, 188, 192, 196, 198, 207, 212–214, 405–407, 418, 420, 425, 433, 451, 458, 459, 465, 467, 468, 484, 493, 502, 505, 511, 515, 517, 518, 521, 522, 525, 532, 533
Solvents, 18–22, 37, 40, 63
Stability, 10, 11, 20, 31, 37, 41, 43–45, 47, 51, 63, 74, 111, 113, 115, 127, 147, 161, 203, 204, 214, 256, 267, 299, 321, 322, 334, 342, 343, 345, 346, 350, 353, 407, 434, 486, 504, 524, 529, 534, 535, 540
Stabilization mechanism, 405, 494
Standard deviation (SD or S. D.), 7, 130, 356, 399
Stereo lithographic (SLA), 360
Styrene-acrylic latex, 41, 55, 58, 60, 62, 64, 66, 68, 74
Surface exfoliation, 124, 131, 134
Surface group, 234, 247, 249–253, 388
Surface modification, 225, 226, 238, 264, 270, 276, 279, 321, 322
Surface tension, 22, 33, 34, 51, 54, 74, 312
Surfactant, 23, 33, 34, 37, 40, 51–54, 74, 125, 350

T

Tape-casting, 17, 19, 20, 23, 25, 28, 33, 37–40, 47, 54, 55, 63, 64, 68, 70, 71, 73–75

Temperature gradient, 7, 14, 140, 183–185, 187, 194, 371, 374, 375, 380–382, 386, 387, 392
Temperature-induced solidification, 6
Tert-butyl alcohol (TBA), 372
Tetragonal zirconia polycrystals (TZP), 319
Tetramethylammonium hydroxide (TMAH) ($CH_3)_4NOH$), 47
Thermogravimetric analysis (TGA), 87
Thin-wall rutile tube, 311, 338, 339
Transmission electron microscopy (TEM), 259–261, 279–281
Tri-ammonium citrate (TAC), 124, 153
Tungsten carbide (WC), 320

U

Ultra low density, 382, 383, 385, 386
Ultrasonic accelerated solidification, 389
Ultrasonic effects, 359, 387

W

Water-based binder, 41
Water-based ceramic slurry, 20
Water-based system, 18, 20–22
Water-soluble polymer, 140, 195, 203

Weibull modulus, 13, 139, 145, 150, 152, 265, 266, 275, 283–285

X

X-ray diffraction (XRD), 260, 278–280, 301, 303, 305, 306, 333, 334, 346–348
X-ray photoelectron spectroscopy (XPS), 269

Y

Yttrium aluminum garnet (YAG), 395, 396, 401

Z

Zeta potential (zeta-potential or ς potential), 31, 32, 41, 42, 45–49, 51, 81–83, 96, 99, 100, 111–113, 120, 121, 125, 126, 146, 147, 175–177, 197, 199, 203–205, 212, 213, 226, 230–237, 241, 246–248, 257, 260, 261, 281, 288–290, 330, 331, 334, 335, 349–351, 353, 363, 364, 388
Zirconia-toughened alumina (ZTA), 150

Postscript

When this book will be published we want to express the appreciation to our graduate students. This book "Novel Colloidal Forming of Ceramics" is alsoderived from their diligent work in our group. Their names and degrees are listed as below.

(1) Ph D students and their research directions

No.	Ph D student	Periods	Research direction
1	Junhui Xiang	1997–2002	Aqueous gel tape casting
2	Longjie Zhou	1997–2002	Ceramic colloidal forming
3	Xiaolin Liu	1997–2002	Ceramic colloidal forming
4	Jianqing Dai	1997–2002	Ceramic colloidal forming
5	Liguo Ma	1998–2002	Ceramic colloidal forming
6	Zhiyong Yu	1999–2003	Ceramic freeform fabrication and computer-assisted forming technique
7	Jingtao Ma	1999–2003	Ceramic colloidal forming
8	Kai Cai	1999–2003	Colloidal forming technique for ceramics microbeads
9	Ma Tian	2000–2004	Ceramic colloidal forming
10	Yang Song	2001–2005	Forming technique for superconducting wires and tapes
11	Liming Zhang	2001–2005	Ceramic colloidal forming
12	Xuemin Cui	2002–2004	Ceramic freeform fabrication process
13	Ruifeng Chen	2002–2007	New processing for porous super-light and high-strength ceramics

(2) Master students and their research directions

No.	Master student	Periods	Research direction
14	Yali Chen	1996–1998	Application of natural macro-molecule gel in ceramicin-situ solidification
15	Yu Jia	1997–2000	Application of sodium alginate in-situ solidification ofceramic suspension
16	Chunlei Ma	1999–2002	Aqueous gel tape casting of alumina
17	Lei Zhao	2000–2003	Ceramic colloidal forming
18	Liang Su	2000–2003	Pressure-induced colloidal injection molding for ceramics
19	Chunlei Dai	2001–2004	Key techniques in gelcasting process
20	Tiechao Wang	2002–2005	New RTV rutile composites for ozone generator
21	Jiesheng Luo	2002–2005	Injection molding of super-fine zirconia
22	Yan Gao	2003–2005	Injection molding of zirconia and SiC and debindering process through water extraction
23	Xiaojun Liu	2003–2006	Injection molding of alumina
24	Xinyue Zhang	2004–2006	Properties and characterization of ceramic microbeads
25	Ming Yue	2004–2006	Wear behavior of the ceramic microbeads prepared bygel-beads forming
26	Xiaoqing Xi	2004–2007	Key techniques and equipments for industrializedproduction of high performance ceramic microbeads
27	Huang Lin	2005–2007	Particle-stabilized porous ceramics by gelcasting
28	Yu Ding	2006–2008	Influences of forming techniques on ceramic reliability
29	KeZeng	2006–2008	New process of laser machining of ceramic green bodies
30	Yuanyuan Cui	2007–2010	New process of laser machining of ceramic green bodies